# NATO ASI Series

## Advanced Science Institutes Series

*A series presenting the results of activities sponsored by the NATO Science Committee, which aims at the dissemination of advanced scientific and technological knowledge, with a view to strengthening links between scientific communities.*

The Series is published by an international board of publishers in conjunction with the NATO Scientific Affairs Division

| | |
|---|---|
| A **Life Sciences**<br>B **Physics** | Plenum Publishing Corporation<br>London and New York |
| C **Mathematical and Physical Sciences**<br>D **Behavioural and Social Sciences**<br>E **Applied Sciences** | Kluwer Academic Publishers<br>Dordrecht, Boston and London |
| F **Computer and Systems Sciences**<br>G **Ecological Sciences**<br>H **Cell Biology**<br>I **Global Environmental Change** | Springer-Verlag<br>Berlin Heidelberg New York<br>London Paris Tokyo Hong Kong<br>Barcelona Budapest |

PARTNERSHIP SUB-SERIES

| | |
|---|---|
| 1. **Disarmament Technologies** | Kluwer Academic Publishers |
| 2. **Environment** | Springer-Verlag/Kluwer Acad. Publishers |
| 3. **High Technology** | Kluwer Academic Publishers |
| 4. **Science and Technology Policy** | Kluwer Academic Publishers |
| 5. **Computer Networking** | Kluwer Academic Publishers |

*The Partnership Sub-Series incorporates activities undertaken in collaboration with NATO's Cooperation Partners, the countries of the CIS and Central and Eastern Europe, in Priority Areas of concern to those countries.*

## NATO-PCO DATABASE

The electronic index to the NATO ASI Series provides full bibliographical references (with keywords and/or abstracts) to about 50000 contributions from international scientists published in all sections of the NATO ASI Series. Access to the NATO-PCO DATABASE compiled by the NATO Publication Coordination Office is possible in two ways:

- via online FILE 128 (NATO-PCO DATABASE) hosted by ESRIN, Via Galileo Galilei, I-00044 Frascati, Italy.

- via CD-ROM "NATO Science & Technology Disk" with user-friendly retrieval software in English, French and German (© WTV GmbH and DATAWARE Technologies Inc. 1992).

The CD-ROM can be ordered through any member of the Board of Publishers or through NATO-PCO, Overijse, Belgium.

Series I: Global Environmental Change, Vol. 50

**Springer**
*Berlin*
*Heidelberg*
*New York*
*Barcelona*
*Budapest*
*Hong Kong*
*London*
*Milan*
*Paris*
*Santa Clara*
*Singapore*
*Tokyo*

# Gravity Wave Processes

## Their Parameterization in Global Climate Models

Edited by

Kevin Hamilton

Princeton University
Geophysical Fluid Dynamics Laboratory
Princeton, NJ 08542, USA

With 189 Figures (2 Colour Plates) and 15 Tables

Springer

Published in cooperation with NATO Scientific Affairs Division

Proceedings of the NATO Advanced Research Workshop "Gravity Wave Processes and Their Parameterization in Global Climate Models", held in Santa Fe, USA, April 1–5, 1996

**Library of Congress Cataloging-in-Publication Data**

Gravity wave processes : their parameterization in global climate models / edited by Kevin Hamilton.
p. cm. -- (NATO ASI series. Series I, Global environment change ; vol. 50)
"Published in cooperation with NATO Scientific Affairs Division."
"Proceedings of the NATO Advanced Research Workshop "Gravity Wave Processes and Their Parameterization in Global Climate Models", held in Santa Fe, USA, April 1-5, 1996"--T.p. verso.
Includes bibliographical references and index.
ISBN-13: 978-3-642-64495-5 e-ISBN-13: 978-3-642-60654-0
DOI: 10.1007/978-3-642-60654-0

1. Climatology--Mathematical models--Congresses. 2. Gravity waves--Congresses. 3. Atmosphere, Upper--Mathematical models--Congresses. I. Hamilton, Kevin, 1956- . II. North Atlantic Treaty Organization. Scientific Affairs Division. III. NATO Advanced Research Workshop "Gravity Wave Processes and Their Parameterization in Global Climate Models" (1996 : Santa Fe, [New Mexico]) IV. Series.
QC851.G64 1997
551.51'5--dc21 96-37587
CIP



Typesetting: Camera ready by authors/editor
Printed on acid-free paper
SPIN: 10499104 31/3137 - 5 4 3 2 1 0

# PREFACE

Gravity waves can act to transfer mean horizontal momentum from the ground to levels aloft in the atmosphere or from one layer of the atmosphere to another. Flow over topography can generate stationary gravity waves that break nonlinearly in the troposphere and lower stratosphere. Such waves transfer momentum from the breaking region to the earth's surface, and this process is thought to act as a significant drag on the eastward mean winds in the midlatitude troposphere. It is known that numerical simulation models of the global atmosphere run without any attempt to impose such a drag tend to produce results characterized by unrealistically intense eastward surface winds in midlatitudes. Other processes (such as convection, jet stream instabilities etc.) can produce gravity waves with nonzero horizontal phase speeds and which act to transfer mean momentum between the troposphere and the middle atmosphere. Comprehensive numerical models of the atmosphere generally produce unrealistic simulations of the extratropical stratospheric/mesospheric circulation unless some account is taken of the effects of these gravity waves. If credible simulations of climate (and predictions of climate response to anthropogenic forcing) are to be obtained, some physically justifiable parameterization of the momentum transport due to unresolvable gravity waves needs to be formulated. This issue is now recognized as one of the most important challenges in dynamical meteorology.

In order to begin addressing this problem a NATO Advanced Research Workshop was held in Santa Fe, U.S.A., from April 1-5, 1996. The meeting was attended by a total of 36 scientists from 10 countries. The workshop was notable in bringing together experts in observations and theory of gravity waves along with representatives of several climate modelling groups. This volume includes 26 papers representing contributions from 30 of the meeting participants and serves as an excellent summary of the scientific issues discussed.

I would like to thank the Scientific Affairs Division of NATO for providing the bulk of the financial support for the meeting. Additional support was provided by the World Climate Research Programme as part of its initiative on Stratospheric Processes and their Role in Climate (SPARC) and by the Solar-Terrestrial Energy Program. Finally I wish to thank the attendees for their enthusiastic participation at the workshop (even through some of the very long sessions that were scheduled!) and for submitting the excellent papers that make up this volume.

Kevin Hamilton

**CONTENTS**

# Some Problems Relating to the Observed Characteristics of Gravity Waves in the Middle Atmosphere

**Isamu Hirota**
*Department of Geophysics*
*Kyoto University*
*Sakyoku, Kyoto 606*
*Japan*

**ABSTRACT.** A critical review is presented of the understanding of gravity wave characteristics based on earlier observational studies since the 1980's. Following comments on the basic idea of the gravity wave dynamics and the capability of observational techniques, results of rocketsonde, balloonsonde and MU radar data analyses are briefly recalled. It is stressed that the "climatology of gravity wave sources" is still open to question, even though we have had some case studies of various processes for the generation of gravity waves in the lower atmosphere. Another important question is the vertical distribution of gravity wave activities relating to wave breaking (and/or dissipation) processes in the stratosphere and mesosphere, in connection with seasonal variations of the background field. Finally critical comments are added on the current attempt of the gravity wave parameterization in GCMs by emphasizing the need for many more detailed observations in the future.

## 1. Introduction

Through the progress made in observing the middle atmosphere in the last three decades, the climatology of the global circulation has been established (e.g., CIRA model), and our understanding of the mechanism for the generation and maintenance of the mean zonal wind and temperature field has been substantially developed in terms of large-scale dynamics as well as radiative and photo-chemical processes.

In addition to this, around 1980, it became widely recognized that vertically propagating, internal gravity waves play an important role in determining the large-scale wind field in the middle atmosphere through their momentum transport and deposition. In their pioneering works Lindzen (1981), Matsuno (1982) and Holton (1982) showed the effect of the breaking of gravity waves on the mean zonal flow, and in the last decade there have been many attempts to parameterize the gravity wave effect in the numerical modelling of the middle atmosphere on a global basis.

It should be noted, however, that these numerical models assume *a priori* physical parameters of gravity waves such as wave amplitudes and frequencies without having substantial evidence for these in the actual atmosphere. In the present paper, therefore, I should like to review our observational studies to clarify what properties of gravity waves are known and what is needed for future studies.

## 2. Difficulties in observations of gravity waves

Since internal gravity waves have larger amplitudes in association with their vertical propagation in such a way as $|v| \sim \rho_0^{-1/2} \sim \exp(z/2H)$, it is easier to separate them from the

NATO ASI Series, Vol. I 50
Gravity Wave Processes
Their Parameterization in Global Climate Models
Edited by Kevin Hamilton

background flow and large-scale waves in the middle atmosphere than in the troposphere. MST radars, lidars, rockets and balloons have been used as observational tools for this purpose.

However, these techniques are essentially based on single-station observations to obtain the physical quantity of gravity waves as a function of height and time at a fixed location (the only exception is the recent satellite observation under development). Because of this limitation, observable parameters at each station are vertical wavelength and Doppler-shifted frequency, in addition to the mean field. Horizontal wavelength and intrinsic frequency, for example, have to be estimated indirectly with the aid of theoretical considerations (e.g., dispersion relation) by assuming the dominance of monochromatic waves.

Moreover, in view of the inhomogeneous distribution of wave sources and intermittent occurrence of generation processes, we have to be careful in detecting representative characteristics of gravity waves in both case studies and long-term statistics.

## 3. Early statistics for morphology and climatology

In order to have an overview of the gravity wave activity in the middle atmosphere on a global basis, though in a rather crude manner, Hirota (1984) first presented the climatology of gravity waves with the aid of meteorological rocket observations gathered at many stations over several years. By applying a high-pass filter to daily rocket data in the height range of 20 - 65km, wind and temperature fluctuations with vertical scale $\leq$ 10km are extracted, and their variances (r.m.s.) were used as a rough measure of the gravity wave activity. A large seasonal variation with a maximum in winter at high latitudes as well as evidence for a semiannual cycle at low latitudes were found from these statistics. This sort of climatology has been recently revised by Eckermann, Hirota and Hocking (1994) by expanding the data period to 1977 -1987.

Hirota and Niki (1985) made a statistical study based on the same rocket data to show that the short vertical scale wind fluctuations are due mainly to upward propagating inertia-gravity waves, as judged from the rotation of horizontal wind vectors in hodographs. Note that the degree of elliptic polarization of hodographs gives us information about the intrinsic frequency $\omega$ as well as the direction of horizontal propagation of gravity waves. Results of the statistics indicate that the dominant values of $f/\omega$ ($f$ : Coriolis parameter) fall into the range of $0.2 \sim 0.4$ in spite of the seasonal change of the mean flow. This suggests a wide variety of horizontal phase velocities.

It should be reminded, of course, that rocket soundings are too infrequent to resolve the behaviour of gravity waves in time and are too sparse in geographical distribution to resolve the continent-ocean contrast and the difference between the northern and southern hemispheres.

## 4. Wave sources in the lower atmosphere

Our goal in the climatological study of gravity waves is to give full information about their effects on the global circulation of the middle atmosphere in terms of gravity wave forcing of the mean flow through momentum deposition. The wave variance shown in early climatology is not necessarily the forcing itself but merely a rough measure of the dominant structure and behaviour of gravity waves propagating from below. Hence our next step should be to observe wave generation processes in the lower atmosphere.

It has been assumed that gravity waves should be generated in the troposphere by the effect of topography, synoptic disturbances, convective activity, geostrophic adjustment around the jet stream and shear instabilities, but in these cases details are poorly known and rely on a few case studies.

As regards the topographic effect on gravity wave generation, Sato (1990) made a detailed analysis of vertical wind disturbances in the troposphere and lower stratosphere as observed by the MU radar at Shigaraki, Japan in wintertime when the tropospheric westerlies are dominant. In a statistical approach, she showed that the vertical wind disturbances are due mainly to gravity waves generated by the effect of mountain trains located to the west of the MU radar site. It is quite interesting to note that, among 8 observation periods within 1985-1987, the gravity wave intensity is not uniform but rather intermittent, i.e., active periods and quiet periods exist, suggesting that the generation of gravity waves by topography is not simply a function of surface winds.

Concerning wave generation from the mid-latitude jet stream, Hirota and Niki (1986) made a case study of inertia-gravity waves around the intense westerlies over Japan in winter using the MU radar. From their hodograph analysis, they found that the direction of horizontal wind vector rotation, namely the direction of vertical propagation, is opposite above and below the height of the jet core ($\sim$ 11km). This fact strongly supports the theoretical idea of the jet stream as a wave source. This analysis was extended by Kitamura and Hirota (1989) over the Japanese islands (27° - 45°N), with the aid of operational rawinsounde observations at 18 stations for the year of 1986. Results of the statistical analysis revealed that the gravity waves with vertical wavelengths of 2 $\sim$ 5km are dominant in the lower stratosphere and their distribution in a meridional plane along 140°E is closely related to the subtropical jet over Japan. It is also noteworthy that this work proves for the first time the usefulness of routine balloon observations for the study of gravity waves.

However, recent study by Sato (1994) based on 3-year data of the MU radar claimed that the geostrophic adjustment at the jet axis is not necessarily the main generation mechanism of gravity waves, since most of these in the lower stratosphere propagate meridionally toward the jet. Therefore, we have to say that there still remains a problem with the generation process of geostrophic adjustment around the jet stream.

Another important candidate gravity wave source is convection associated with synoptic-scale disturbances. By using the MU radar, Sato (1993) made a detailed analysis of small-scale disturbances during the passage of a typhoon (tropical cyclone) over Japan and showed clear evidence of vertically propagating gravity waves in the lower stratosphere with a characteristic period of several hours which appeared in association with the intense convective motions (rain bands) in the typhoon. It is also noted that her result represents quite typical behaviour but the intense convection, observed in the typhoon, is a rather fragmentary phenomenon in the extratropics. (Gravity wave generation from convections in the tropics will be discussed by Sato in this volume.)

A statistical study on tropospheric sources of gravity wave excitation was also made by Fritts and Nastrom (1992) based on data obtained aboard commercial aircraft flying over the western part of USA and eastern Pacific. Their study showed the relative contribution of various sources such as topography, jet stream, front and convection.

As was mentioned earlier, the global distribution of wave sources is, however, quite inhomogeneous in space and time, and hence our knowledge of the "climatology of gravity wave sources" is still very poor. Although recent progress in satellite observations (cf. Wu and Waters, 1996) would be promising for this purpose, it is emphasized that many more detailed observations, using various techniques and covering wide regions of latitudes and longitudes and across long timespan is of great importance. In this regards, the use of a global network of routine balloon observations is recommended.

## 5. Wave sources and sinks in the middle atmosphere

In considering the gravity wave-mean flow interaction, the most important quantity is the forcing term that acts to accelerate or decelerate the background flow, namely the divergence of momentum fluxes associated with the breaking and/or dissipation of gravity waves. Before making a quantitative estimate of the momentum flux divergence, it is therefore interesting to observe the vertical distribution of gravity wave intensity (variance) in the middle atmosphere, in connection with the seasonal variation of the mean flow profile.

To estimate the levels at which gravity waves propagating from the troposphere will break (or dissipate) in the stratosphere and mesosphere, Hirota and Niki (1985) made the statistical estimation of wave amplitudes at several rocket stations for each season in the height region of 30 - 60km. In the course of a year it was found that the amplitude decays gradually with increasing height compared with the energy-conserved profile $\sim \exp(z/2H)$. No specific level is observed at which the amplitude decreases suddenly, suggesting either uniform decay or a random distribution of breaking levels.

On the other hand, the result of a statistical study of seasonal variations based on the combination of the MU radar and French lidar data (Tusda et al., 1994) indicates that the energy density of gravity waves shows an annual cycle in the stratosphere with a maximum in winter, as in Hirota (1984), whereas in the upper mesosphere the annual cycle with a maximum in summer dominates. This fact strongly suggests that the decay process around and above the stratopause level is different between winter and summer. In view of the comparable speeds of the wintertime westerlies and summertime easterlies, it is hard to attribute the difference between the two seasons simply to the effect of Doppler-shift in the vertical propagation of stationary gravity waves. Rather there is a possibility of additional sources in the stratosphere and mesosphere for the generation of gravity waves which are sensitive to the seasonally varying mean zonal wind profiles. To prove this, careful inspection of individual observations (snap shots) for the vertical profile of gravity waves would be required rather than the long-term statistics.

## 6. Comments on the gravity wave parameterization in GCMs

Throughout this brief review of the observational studies, it has been stressed that our knowledge of the characteristic features of gravity waves in the middle atmosphere is not yet well established, mainly because of their inhomogeneity, anisotropy, intermittency, and wide variety of wave parameters.

Nevertheless, there have been various attempts to parameterize the effect of gravity waves on the mean flow: In numerical models gravity wave parameters are assumed without substantial evidence for these in the actual atmosphere. Therefore, before evaluating the results of this, we should consider what the scientific significance of the parameterization in numerical modelling of the atmosphere is, under the present situation of inadequate observational evidence.

Generally speaking, our goal is to understand the nature of atmospheric motions in terms of the dynamics of various waves and large-scale flows on the rotating globe. To this purpose, observations and theories are essentially indispensable, and the numerical modelling is simply a supporting tool: If idealized (or unrealistic) assumptions of gravity waves, such as uniform sources and isotropic propagation etc., were used, it does not make sense to show a quantitative comparison of numerical results with observations of the real atmosphere. Instead an inference should be drawn from the inevitable difference between the idealized numerical result and the real observation, so as to deepen our understanding of the nature of the atmosphere. On the other hand, if the aim was to have a good simulation of the real atmosphere, it might be better to use practical and empirical methods, such as "tuning", without speaking of the physics. The operational numerical weather prediction is, in fact, a typical example of this approach. It is, however, not science but technology.

In conclusion, I should like to recall an old phrase of Shakespeare (1603);

*There are more things in heaven and earth, Horatio,*
*Than are dreamt of in your philosophy.*

Replace the words "dreamt of" to "parameterized", and "philosophy" to "numerical modelling". This is, I believe, the heart of science.

**Acknowledgments**

The author would like to thank Prof. J. P. Mathews for his comments on the draft of this paper. Thanks are also due to Miss Eriko Edo for typing the manuscript.

**References**

Eckermann, S. D., I. Hirota and W. K. Hocking, 1994: Gravity wave and equatorial morphology of the stratosphere derived from long-term rocket soundings. Q. J. R. Met. Soc., **112**, 149-186.

Fritts, D. C. and G. D. Nastrom, 1992: Sources of meaoscale variability of gravity waves. Part II, Frontal. convective, and jet stream excitation. J. Atm. Sci., **49**, 111-127.

Hirota, I., 1984: Climatology of gravity waves in the middle atmosphere. J. Atm. Terr. Phys., **46**, 767-773.

Hirota, I. and T. Niki, 1985: A statistical study of inertia-gravity waves in the middle atmosphere. J. Met. Soc. Japan, **63**, 1055-1066.

Hirota, I. and T. Niki, 1986: Inertia-gravity waves in the troposphere and stratosphere observed by the MU radar. J. Met. Soc. Japan, **64**, 995-999.

Holton, J. R., 1982: The role of gravity wave-induced drag and diffusion in the momentum budget of the mesosphere. J. Atm. Sci., **39**, 791-799.

Kitamura, Y. and I. Hirota, 1989: Small-scale disturbances in the lower stratosphere revealed by daily rawinsonde observations. J. Met. Soc. Japan, **67**, 817-831.

Lindzen, R. S., 1981: Turbulence and stress owing to gravity wave and tidal breakdown. J. Geophys. Res., **86C**, 9707-9714.

Matusno, T., 1982: A quasi one-dimensional model of the middle atmosphere circulation interacting with internal gravity waves. J. Met. Soc. Japan, **60**, 215-226.

Sato, K., 1990: Vertical wind disturbances in the troposphere and lower stratosphere observed by the MU radar. J. Atm. Sci., **47**, 2803-2817.

Sato, K., 1993: Small-scale wind disturbances observed by the MU radar during the passage of typhoon Kelly. J. Atm. Sci., **50**, 518-537.

Sato, K., 1994: A statistical study of the structure, saturation and sources of inertio-gravity waves in the lower stratosphere observed with the MU radar. J. Atm. Terr. Phys., **56**, 755-774.

Shakespeare, W., 1603: The tragicall historie of Hamlet, Prince of Denmarke.

Tsuda, T., Y. Maekawa, T. Nakamura, R. A. Vincent, A. H. Manson, C. E. Meak and R. L. Wilson, 1994: Variations of the gravity wave characteristics with height, season and latitude revealed by comparative observations. J. Atm. Terr. Phys., **56**, 555-568.

Wu, D. L. and J. W. Waters, 1996: Satellite observations of atmospheric gravity waves. To be appeared in Geophys. Res. Lett.

# Gravity-Wave Parameters in the Lower Stratosphere

R. A. Vincent, and S. J. Allen
Department of Physics and Mathematical Physics
University of Adelaide
Adelaide 5005
Australia

S. D. Eckermann
Computational Physics, Inc.
Fairfax, VA 22031
USA

## Abstract

Observations made in the lower stratosphere with high-resolution radiosondes routinely launched from Macquarie Island (55°S, 159°E) are used to study parameters important in gravity-wave parameterization schemes. The combination of horizontal wind and temperature data allow wave amplitudes and directions of horizontal propagation to be determined. Significant enhancements in wave energy are found to occur near the inertial frequency and there is appreciable directional anisotropy, especially in winter. Estimates of vertical fluxes of horizontal momentum are made from the covariances between the wind and temperature perturbations.

## 1. Introduction

The use of radiosondes has proved very fruitful for the study of gravity waves in the troposphere and lower stratosphere. The regular nature of radiosonde launchings by national meteorological agencies means that they have the potential to provide much needed information on gravity wave activity in the lower atmosphere on a global scale. Hamilton and Vincent (1995) highlighted the potential benefits of the high-vertical resolution (~50 m) data that are becoming available, although not always archived. Allen and Vincent (1995) (hereinafter AV95) used high-resolution temperature data taken with the Australian Bureau of Meteorology's (ABM) network of radiosondes to develop a climatology of gravity wave energy in the troposphere and lower stratosphere over continental Australia and the Southern Ocean. They found a significant latitudinal gradient in wave activity, with the highest levels of activity in the tropics. Ogino *et al.* (1995) using data from radiosondes launched from ships reported similar latitudinal behavior in the northern hemisphere.

An important application for such data is to help initialize and constrain numerical circulation models which incorporate gravity-wave parameterization schemes. Typically, such models are initialized at heights near 20 km in the lower stratosphere. The requirements for each

NATO ASI Series, Vol. I 50
Gravity Wave Processes
Their Parameterization in Global Climate Models
Edited by Kevin Hamilton

scheme vary. Information which might be used include wave amplitudes, phase speeds, directions of propagation or degrees of anisotropy, and spectral characteristics. The Lindzen (1981) scheme uses a discrete set of waves with prescribed amplitudes and ground-based phase velocities. More broad-band spectral models have recently been introduced. The "Doppler-spread" scheme of Hines (1996) utilizes an incident spectrum in vertical wavenumber, *m*, and wave-induced variances as a function of azimuthal direction in the horizontal. Non-linearities produce a Doppler spreading in *m* of the initial spectrum, the spreading being induced by the irregular winds produced by the waves themselves. The Fritts and Lu (1993) scheme uses a model three-dimensional spectral form to describe the gravity wave field so that the total energy spectrum is assumed to be separable in intrinsic frequency, $\hat{\omega}$, vertical wavenumber, *m*, and direction, $\phi$,, viz:

$$E(\mu,\hat{\omega},\phi) = E_o A(\mu) B(\hat{\omega}) \Phi(\phi) \tag{1}$$

$$A(\mu) = A_0 \frac{\mu^s}{1+\mu^{s+t}} \tag{2}$$

$$B(\hat{\omega}) = B_o \hat{\omega}^{-p} \tag{3}$$

where $\mu = m/m_*$, $m_*$ is the characteristic wavenumber, $f^2 < \hat{\omega}^2 < N^2$, and $f$ and $N$ are the inertial and Brunt-Vaisala frequencies, respectively.

Using radiosonde temperature data AV95 estimated the total energy $E_o$, or equivalently the rms velocity amplitude, as a function of season and latitude. A limitation of their work was the absence of accompanying wind measurements so that it was not possible to determine wave parameters other than temperature perturbations. However, some sites in the ABM network do take and archive wind measurements made with the same nominal resolution as the temperature measurements. One such site is Macquarie Island (55°S, 159°E) located in the storm tracks of the Southern Ocean. Using routine radiosonde observations made in 1993 and 1994 we discuss some of the important properties of gravity waves in the lower stratosphere which can be derived from the wind and temperatures, including wave energies, periods, directions of propagation and phase speeds.

## 2. Observations

Standard Vaisala RS80-15 radiosondes were used in this study, with flights being made twice daily. Measurements of pressure, temperature and relative humidity were made at 2-s intervals, but the measurements pass through quality control and smoothing algorithms such that data are finally recorded at 10 s intervals. Given that the balloons rise at about 5 $ms^{-1}$ this corresponds approximately to a 50 m height resolution. Here all the data are interpolated using a third-order cubic spline to 50 m intervals. The finite response time of the temperature sensor means that the amplitude of the highest wavenumber temperature fluctuations are reduced relative to lower frequency components, an effect that becomes more acute at higher altitudes. Details on how the effect can be compensated for are given by AV95.

Horizontal winds were determined with the Digicora system which locates the position of the balloon using radio transmitters of the global VLF NAVAID network. Wind values are reported at 10 s intervals, but because smoothing routines are used, the 10 s values are effectively oversampled. In the data analyzed here, cubic spline smoothing algorithms are used within running windows of 130 s duration which corresponds to vertical intervals of 650 m. Fluctuations with vertical scales smaller than twice these intervals will be attenuated in some manner. The rms random errors are approximately 0.5 $ms^{-1}$ in each wind component. Both the smoothing and relatively large rms errors limit the study of high-frequency gravity waves with Digicora winds, but valuable information is obtained about dominant scale gravity waves.

A final point is that it is assumed throughout that the rate of ascent of the balloons is sufficiently fast that the gravity wave field is sampled without significant distortion due to the finite time it takes the balloons to cover the region of interest and the accompanying horizontal drift. This should be the situation if the mean winds are less than about 50 $ms^{-1}$ (Gardner and Gardner, 1993), which is generally the case here although mean winds may on occasion exceed 60 $ms^{-1}$ at heights above 20-25 km above Macquarie island in winter.

## 3. Vertical Wavenumber Spectra

It is desirable to characterize the spectral indices, $s$ and $t$, in (2), the so-called modified-Desaubies form (VanZandt and Fritts, 1989) of the vertical wavenumber spectrum. AV95 showed that $t \sim 3$ in the troposphere and lower stratosphere, in good agreement with measurements made with a wide variety of techniques in the lower and middle atmosphere. However, estimating $s$, the low-$m$ spectral index, is more difficult.

One constraint on the resolution of low-wavenumber waves is imposed by the difficulty of distinguishing the background wind and temperature profiles from large wavelength waves. The background parameters are usually estimated by fitting and removing a low-order polynomial to the individual profiles. To illustrate the problem, Figure 1 shows a vertical profile of the zonal wind component obtained on May from a typical mid-winter sounding in the 12-25 km height range at Macquarie Island (hereinafter referred to as MI). Wavelike features are seen superimposed on the background profile. Second and third-order polynomial fits are also shown. It is not clear which fit is to be preferred as best representing the background wind. In this study we have chosen to use a third-order fit, but recognize that in doing so the amplitude of longer vertical wavelength, higher intrinsic phase speed, waves will be reduced relative to shorter-scale waves and the resulting m-spectrum will be distorted at the smallest resolved values of $m$.

The limited height range acts as a further constraint on estimating $s$. Typically, in the lower stratosphere $\lambda_*$, the vertical wavelength corresponding to $m_*$ is about 2-3 km (Allen and Vincent, 1995) which means that a height coverage of ~20 km is required to resolve adequately the low-$m$ region. Given that most routine radiosondes have a ceiling of 30 km at best, and the lowest tropopause altitudes are about 10 km it is clearly difficult to obtain the required height coverage in the stratosphere. At MI the upper limit of radiosonde flights is about 25 km which gives a height coverage of about 13 km above the tropopause which is usually located at 10-12 km altitude. This means that the first independent spectral estimate occurs at a wavenumber of $2\pi/13$ $km^{-1}$ and this spectral estimate will affected by the polynomial fitting process, which acts as a high-pass filter, and spectral windowing effects. Even the second spectral estimate at $2\pi/6.5\ km^{-1}$ may be affected.

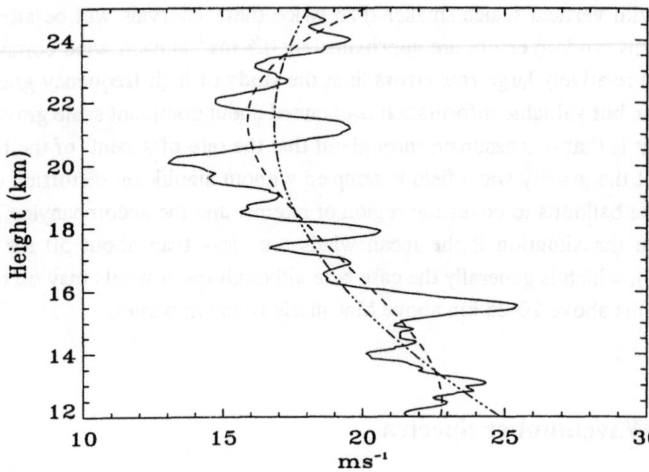


Figure 1. Zonal wind component (solid line) measured at Macquarie Island on 2 May 1993 at 10 UT. Second (chain) and third-order (dashed) polynomial fits are also shown.

Figure 2 illustrates these effects for the mean power spectra of the temperature perturbation normalized to the background temperature, $\hat{T}'$, for all soundings that reached 25 km in winter (May-August) and summer (November-January) at MI. In all, about 145 profiles were available for analysis in each season. For comparison the dashed line indicates the saturation limit due to Smith *et al.* (1987) and given by the expression

$$S_{\hat{T}'}(m) = \frac{N^4}{10g^2m^3} \tag{4}$$

The spectra agree reasonably well with the saturation limit, and show an approximately $m^{-3}$ variation at high wavenumbers, although it should be noted that the correction factor required to compensate for the response time of the temperature sensor has not been applied (see AV95 for details). Applying the correction tends to reduce the spectral slope somewhat. The main point is that the positive slope at wavelengths greater than 6.5 km in Figure 2a is an artifact. The first spectral ordinate is not well determined because of the use of a polynomial to remove the background temperature profile and should not be plotted. Spectra plotted in area-preserving form (without the first spectral ordinate) are displayed in Figure 2b. The diagram emphasizes the fact that the dominant fluctuations occur at wavelengths ~2-4 km and it clearly shows that wave energies in winter are approximately twice as large as those in summer.

Finally, it is noted that a limited height coverage itself imposes a significant constraint on the gravity waves that can be studied, restricting the study of waves with long vertical wavelengths or, equivalently, high intrinsic phase speeds. At MI, measurements over a ~15 km height range means that only waves with phase speeds less than about 50 $ms^{-1}$ can be resolved. In tropical regions the high tropopause near 18 km allows a "window" of only 7-10 km, which means that only waves with phase speeds less than ~22-30 $ms^{-1}$ can be studied. Eckermann *et al.* (1995) have discussed this issue in more detail.

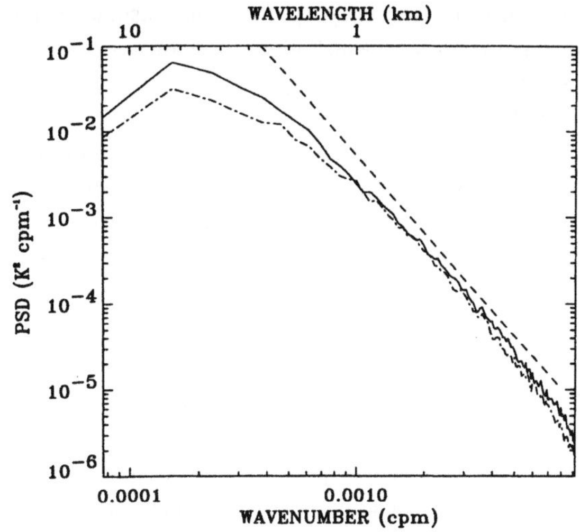


Figure 2a. Mean vertical-wavenumber power spectra of normalized temperature fluctuations at Macquarie island in winter (solid) and summer (chain). The saturation limit due to Smith *et al.*, (1987) is shown for comparison (dashed).

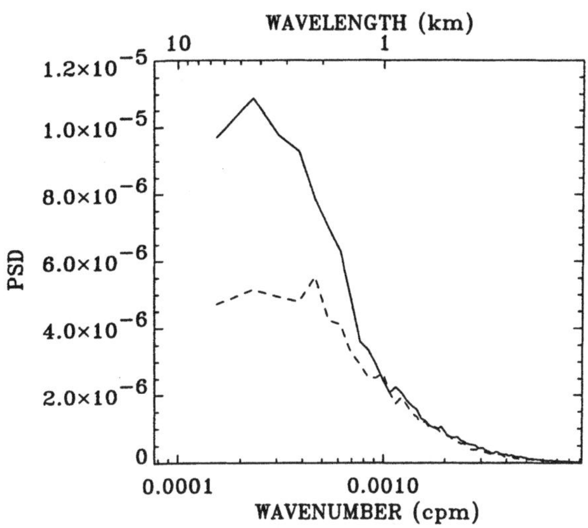


Figure 2b. Mean vertical-wavenumber spectra for winter and summer plotted in area-preserving form.

## 4. Intrinsic-Frequency Spectral index

The spectral index for the intrinsic frequency spectrum has usually been estimated from ground-based radar or lidar measurements, which give values of $p \sim 5/3$ in the middle atmosphere. It is probable that the actual value of $p$ is larger than this because Doppler-shifting due to mean-wind effects will tend to move energy to higher observed frequencies and hence reduce the slope of the spectrum (Fritts and VanZandt, 1987; Vincent and Eckermann, 1990).

The low sounding rate of radiosonde observations of 2 per day precludes these observations being used to find $p$ directly, but it is possible to obtain an indirect estimate from the wave energy. Assuming that (3) is correct, then the kinetic energy (KE) of the wave motions is related to the potential energy (PE) by

$$\left(\overline{u'^2} + \overline{v'^2}\right) = p\frac{g^2}{N^2}\overline{\hat{T}'^2}\,. \tag{5}$$

Height-profiles of KE, PE, and their ratio are shown in Figure 3 for winter and summer at MI. It is clear that the ratio in winter is on average about 2.25 and is closer to a value of 2.5 in summer. These results could indicate that the spectral index is greater than 2. An alternative explanation is that there is an excess of energy near $f$, as argued by Nastrom *et al.* (1996). High-frequency gravity waves have their energy equally partitioned between kinetic and potential, but near-inertial frequency waves oscillate almost horizontally and wave energy is essentially kinetic.

## 5. Gravity-wave propagation directions

Information about the vertical direction of wave propagation can be obtained from the wind data alone. The Coriolis effect causes the perturbation wind vector associated with a gravity waves propagating energy upward (downward) to rotate anticyclonically (cyclonically) with increasing height. An anticyclonic rotation is anticlockwise in the southern hemisphere. The rotary-spectral technique is a convenient way of decomposing the wind field into anticlockwise (AW) and clockwise (CW) components (e.g. Vincent, 1984). With the MI data it is found that in the lower stratosphere (~20 km altitude) typically 60-80% of the energy is contained in the AW component, with similar percentages inferred for upward wave energy. These values are likely to be underestimates because gravity wave motions are in general elliptically polarized with the motions becoming more linearly polarized as the intrinsic frequency increases. An elliptically polarized wave field will partly decompose into both AW and CW components, even if all the wave energy is upward propagating (Vincent, 1984; Eckermann and Vincent, 1989).

Both wind and temperature data are required to determine the direction of horizontal propagation. The major axis of the perturbation velocity ellipse is aligned along the horizontal direction of propagation and with wind data alone it is possible to determine only the mean orientation of the ellipse which can done objectively using the Stokes-parameters technique

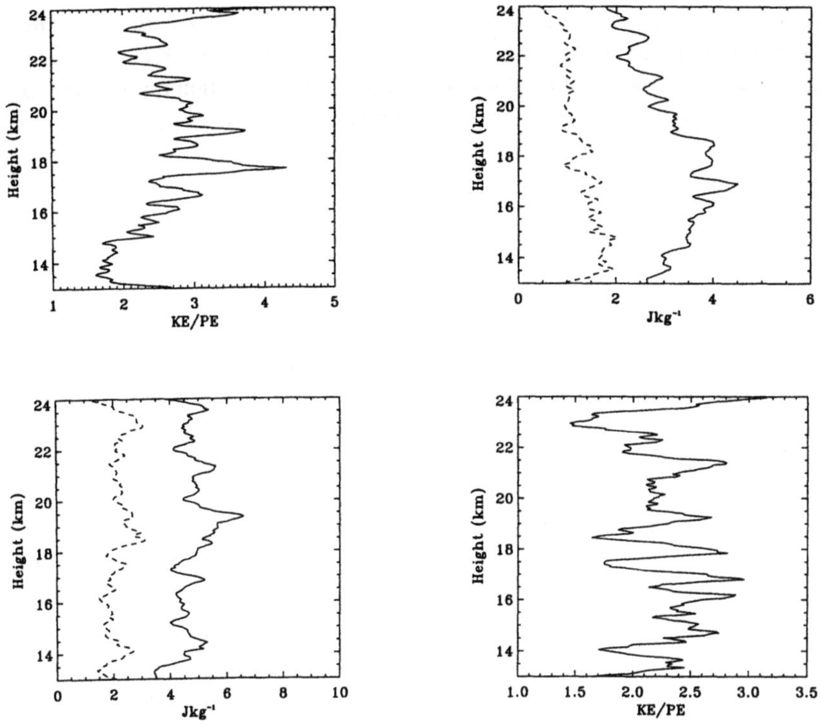


Figure 3. The left-hand panels show height profiles of the mean kinetic energy per unit mass (solid lines) and potential energy (dashed) for summer (top) and winter (bottom). Height profiles of the ratios of the kinetic to potential energy are shown in the right-hand panels.

(Vincent and Fritts, 1987; Eckermann and Vincent, 1989). This method uses the analogy between a partially-polarized gravity wave and a partially-polarized electromagnetic wave to statistically describe the degree of polarization and the mean parameters of the polarized part of the wave. This method gives the mean ellipse orientation and hence the direction of propagation with an ambiguity of 180°. This can be resolved by comparing the sign of the phase between $U'$ and $T'$, where $U'$ is the perturbation velocity parallel to the major axis of the ellipse. For a monochromatic wave these components are 90° out-of-phase. Hamilton (1991) used the sign of the quantity $\eta = \overline{U'T'_z}$ to find the dominant direction of propagation, where the subscript denotes the height derivative and the overbar an average over all heights. Alternatively, the quadrature spectrum between $U'$ and $T'$ could be used. A particularly elegant method is to bring the $T'$ field into phase with $U'$ by Hilbert transforming $T'$ in the appropriate sense and then taking the reverse transform to produce $T'_{+90}$ (Eckermann, *private communication*). In this way all spectral components of $T'$ are shifted by 90° without their amplitude being changed. The sign of $\overline{U'T'_{+90}}$ gives the direction of propagation.

Using the above method the mean directions of propagation of gravity waves were determined for each radiosonde observation in the height range 12-25 km. Applying the Stokes-parameters method determined the mean orientation of the ellipse, with directions measured counterclockwise from east. Then $\overline{U'T'_{+90}}$ was computed to resolve the directional ambiguity.

In order to obtain a measure of the dominant direction of propagation for a particular group of flights, the directions were binned into 30° segments and then the energy-weighted average was formed, so that for the *i*th segment

$$\Phi_i = \frac{\Sigma E_j}{E_T} \tag{6}$$

Here

$$E_j = \frac{1}{2}(\overline{u'^2} + \overline{v'^2} + \frac{g^2}{N^2}\overline{\hat{T}'^2}) \tag{7}$$

is the wave energy for a single sounding averaged over the 12-25 km height interval. $E_T$ is the total energy computed for all soundings in the sequence. Histograms of $\Phi$ plotted in polar form for winter and summertime soundings at MI are shown in Figure 4. These plots can be taken to represent the "angular spectra" of wave energy in these seasons averaged over all heights in the 12-25 km range and over all wavenumbers. It might be argued that in the high-wavenumber tail of the *m*-spectrum that the linear relationship between $U'$ and $T'$ no longer holds. However, an analysis of the cross-spectrum between $U'$ and $T'$ shows that over 95% of the $\overline{U'T'_{+90}}$ covariance occurs at wavelengths longer than 1 km. This is not really surprising given the 650 m smoothing that is implicit in the winds.

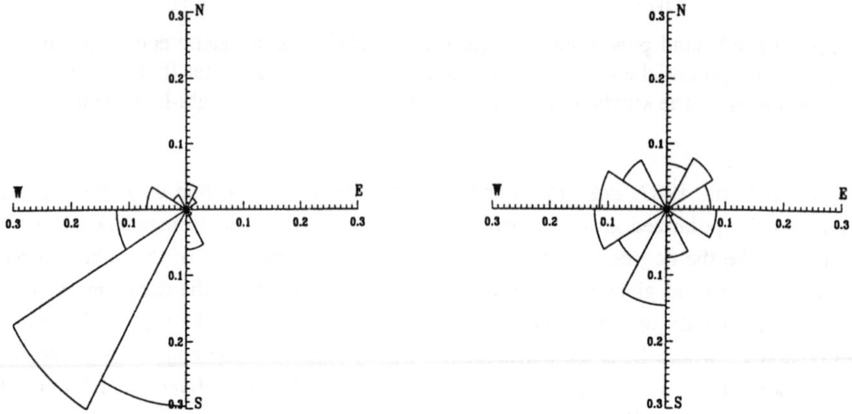


Figure 4. Angular distribution of dominant directions of horizontal propagation for winter (left) and summer (right).

In winter it is clear that the wave energy is directed essentially south-westward. In summer wave energy is directed more uniformly as function of azimuth, although about 60% is westward directed and there is a bias toward the southwest. These results suggest that a significant degree of directional anisotropy can exist in the lower stratosphere, a result which if confirmed for other stations has implications for the initialization of the various parameterization schemes.

## 6. Wave Parameters

The limited range of propagation directions in winter may occur because of a dominant source/and or the effects of wave filtering by the background wind as the waves propagate up from the lower atmosphere. However, in order to determine sources by, for example ray-tracing techniques (e.g. Marks and Eckermann, 1995), it is necessary to know the intrinsic phase velocity of the waves, which in turn requires knowledge of the intrinsic frequency.

In principle, $\hat{\omega}$ may be obtained from the axial-ratio of the wind hodograph which for a monochromatic wave is just $|\hat{\omega}/f|$ (e.g. Gossard and Hooke, 1975). The wind and temperature fluctuations often look quasi-monochromatic (see Figure 1), and the values of *d*, the degree of polarization determined from the Stokes-parameter analysis for each radiosonde flight are often significantly greater than 0.5; a value of $d$=1 indicates a monochromatic wave while $d$=0 means a random wavefield. For the MI soundings in the lower stratosphere the mean values of *d* are 0.54 and 0.67 for summer and winter, respectively (Allen, 1996). Only 5% of soundings in summer and 20% in winter had values of *d* less than 0.4. This suggests a significant degree of polarization, especially in summer, and imply that $\hat{\omega}$ can be derived from the axial-ratios.

Eckermann and Hocking (1989), however, cautioned against a simple interpretation of hodograph analyses. They showed that a broad spectrum of waves can give rise to seemingly "monochromatic" waves, supporting their arguments with both analytic and numerical analyses. They argued that using the axial ratio can give more information on horizontal anisotropy in wave directions than on $|\hat{\omega}/f|$ and applied their results to the interpretation of rocket observations in the middle and upper stratosphere.

While the cautions of Eckermann and Hocking (1989) should be heeded, there are indications that, at least in the lower stratosphere, the wave field may be a mixture of quasi-monochromatic waves and a broad spectrum. As noted in section 4, the kinetic to potential energy ratios suggest that a preponderance of wave energy near the inertial frequency. If quasi-monochromatic waves are present then the use of the Stokes parameters technique may give useful information on wave parameters such as frequency and wavelength. In order to test this hypothesis the numerical analysis of Eckermann and Hocking (1989) was repeated, but with some differences. In their gravity-wave computer experiment a grid was set up in $(\hat{\omega}, m)$ space, each grid point defining a single wave whose amplitude was set by integrating the two-dimensional spectrum defined by (2) and (3) over a rectangle centered on the grid point and with dimensions equal to the grid spacing. Each wave was assigned a random starting phase and was allowed to propagate in time and space, with wave propagation upward ($m < 0$). Wave azimuths were assigned randomly within prescribed limits, so that the wave field could be made as anisotropic as desired. With this model a wave field with known attributes was obtained and a simulated sounding obtained from which the $\hat{\omega}/f$ ratio could be inferred. Repeating the process many times gave a distribution of $\hat{\omega}/f$ which could be compared with values inferred from actual rocket soundings.

A similar model was used here, but with the following difference. In the Eckermann and Hocking (1989) analysis they were particularly concerned with the superposition of linearly polarized waves and so their $\hat{\omega}$ values were chosen linearly in the range $f < \hat{\omega} < N$. This meant that the smallest value of $\hat{\omega}$ was about $10f$. Here a logarithmic distribution in $(\hat{\omega}, m)$ space was used so that the smallest $\hat{\omega}$ was $2f$, giving more prominence to inertial scale waves. To model the radiosonde soundings as closely as possible a height range of 12 km was used, with the $m$ values selected from the range $2\pi/6$ km$^{-1}$ to $2\pi/0.12$ km$^{-1}$; $m_*$ was set at $2\pi/3$ km$^{-1}$ consistent with the stratospheric values from AV95. Amplitudes were calculated using equations (2) and (3) with $s$=1, $t$=3, and $p$=5/3. In all, 36 waves were used, with each wave assigned a random phase and direction of horizontal propagation, $\phi$, which was selected randomly within an azimuthal range extending from 0° to a limiting value, $\phi_{lim}$. For isotropic propagation $\phi_{lim}$=360°. In this way a typical sounding was simulated and the resulting profile analyzed with the standard Stokes parameters program to give a value for the dominant axial ratio or equivalently, $|\hat{\omega}/f|$. By carrying out 1000 such statistical realizations for a given $\phi_{lim}$ a probability density function (pdf) of $|\hat{\omega}/f|$ was built up, and then compared with the distributions of $|\hat{\omega}/f|$ derived from the radiosonde observations.

Before the ellipse axial ratios were converted to $|\hat{\omega}/f|$, the transverse-shear effect was taken into account. Hines (1989) pointed out that a shear in the wind component transverse to the direction of propagation will change the axial ratio, AXR. The polarization ratio becomes

$$AXR = \left| \frac{f}{\hat{\omega}} - \frac{1}{N}\frac{dV_T}{dz} \right| \tag{8}$$

where $V_T$ is the mean wind velocity component transverse to the direction of propagation. The axial ratios were corrected by using the mean shear averaged over the full 12-25 km height interval. In practice, the effect was not very significant, ranging between 6% in summer to 14% in winter.

Histograms or pdfs of $|\hat{\omega}/f|$ inferred from both winter and summertime soundings are compared with results of the numerical experiments in Figure 5. The histograms are normalized to unit area. The difference between the pdfs for the simulated data for the winter and summertime cases is due to the different range of azimuths used in each simulation. To compare with the more isotropic summertime observations a value of 360° was used for $\phi_{lim}$. To simulate winter conditions, with their restricted range of azimuths (see Figure 4), a value of $\phi_{lim}$=90° was used. The restricted range of azimuths causes the oscillations to be more aligned, and there is a greater chance of obtaining a larger axial ratio and hence a larger $|\hat{\omega}/f|$ ratios than in the summer simulations.

It is clear that in the observations there is an increased chance of finding values of $\hat{\omega}$ in the range $f < \hat{\omega} < 2f$ compared with that found for the simulated data. Conversely, there is reduced probability of observing intrinsic frequencies with $\hat{\omega} > 2f$ relative to the frequencies deduced from the simulations. The differences are especially distinct in summer. Given the large number of observations (~145) for each season the statistical differences between the observed and simulated distributions are highly significant. The bias to inertial frequencies becomes even more pronounced if the comparisons are restricted to those soundings for which the degree of polarization, $d$, is greater than 0.5. These results support the suggestion based on the KE/PE ratios that, as well as a wide-band spectrum of waves, there are often quasi-monochromatic inertia gravity waves present in the lower stratosphere over MI.

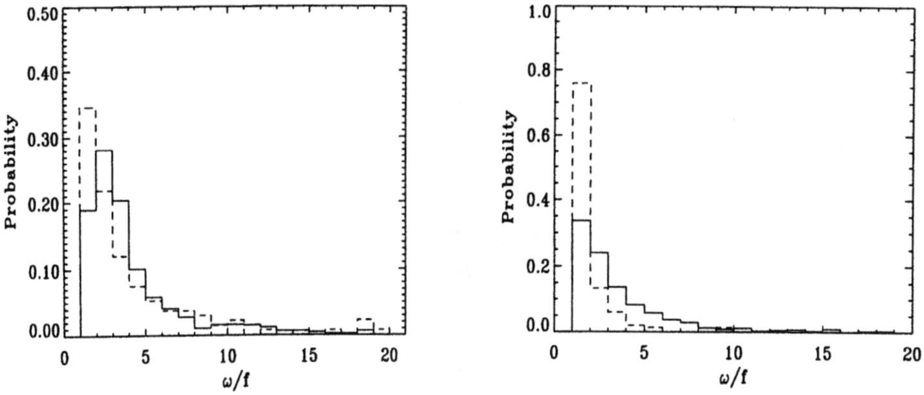


Figure 5. Probability density functions of $\hat{\omega}/f$ inferred from radiosonde soundings (dashed) and numerical simulations (solid) for winter (left) and summer (right).

The presence of these dominant motions provides the opportunity to derive important wave parameters such as horizontal wavelengths and phase speeds. The horizontal wavenumber $k_h$ is determined from each sounding using the dispersion relation

$$m^2 = \frac{N^2 k_h^2}{(\hat{\omega}^2 - f^2)} \tag{9}$$

using the inferred values of $\hat{\omega}$ and $m$. Using the known direction of propagation derived from the wind and temperature covariances the horizontal intrinsic phase velocity is obtained from

$$\underline{\hat{c}} = \frac{\hat{\omega}}{k_h}(\sin\phi, \cos\phi), \tag{10}$$

where $\phi$ is the azimuth of propagation measured clockwise from north. Other important quantities are the ground-based horizontal phase velocity, $\underline{c}$, and the ground-based horizontal group velocity, $\underline{c}_{gh}$, defined as

$$\underline{c} = (\hat{\omega} + \underline{k}_h \cdot \underline{U}) \frac{\underline{k}_h}{|\underline{k}_h|^2} \tag{11}$$

$$\underline{c}_{gh} \approx \underline{U} + \underline{\hat{c}} \tag{12}$$

Figure 6 shows plots of $\underline{\hat{c}}$, $\underline{c}$, and $\underline{c}_{gh}$ in vector form for soundings in winter. The intrinsic phase speeds are typically 10 $ms^{-1}$ and waves are propagating toward the south west. The ground-based phase speeds are about 15 $ms^{-1}$ and directed toward the north-east, while the mean horizontal group velocity is about 35 $ms^{-1}$. It is apparent that $\underline{c}_{gh}$ is dominated by the strong zonal winds and are directed south of east. Since it is the group velocity that is important for ray-tracing this suggests that the wave source in winter is situated upstream of MI.

Table 1 summarizes the main parameters of the observations at MI for winter and summer. Because of the asymmetric nature of the frequency distributions the median values of the intrinsic frequency distributions were used, rather than the mean values. The figures in brackets give the values for those soundings when the degree of polarization, $d$, was greater than 0.5, and the wavefield appeared to be strongly polarized.

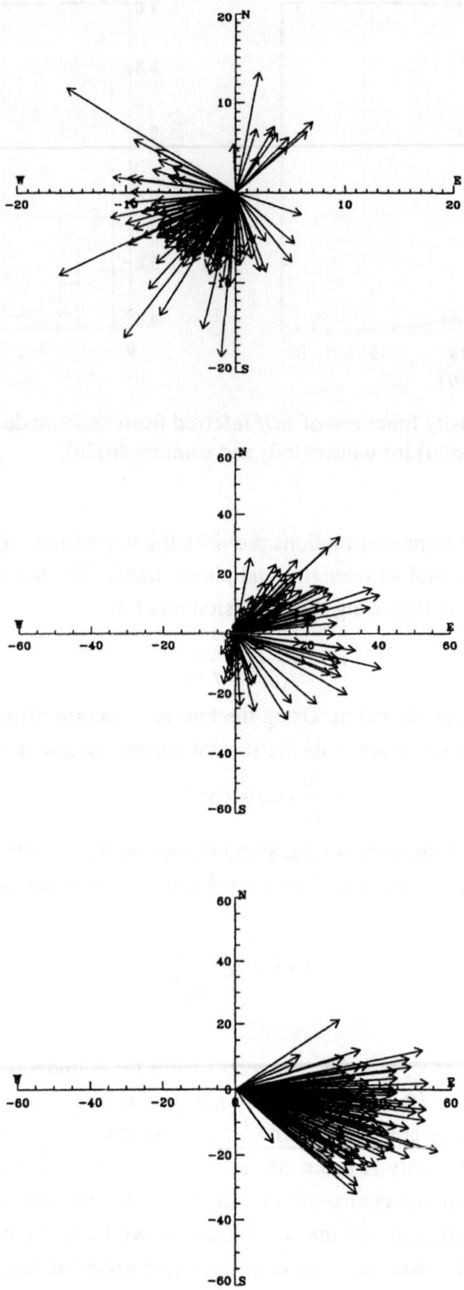


**Figure 6. Intrinsic phase velocity vector (top), ground-based phase velocity (middle) and horizontal group velocity (bottom) for soundings in winter. Units are in $ms^{-1}$.**

Table 1. Parameters of dominant waves in the 12-25 km height range. Values in brackets correspond to soundings for which $d > 0.5$ (see text).

| | $\lvert\hat{\omega}/f\rvert$ | $2\pi/\overline{m}$ (km) | $2\pi/k_h$ (km) | $\lvert\hat{c}\rvert$ (ms$^{-1}$) |
|---|---|---|---|---|
| Winter | 3.0 (2.2) | 2.1 (2.2) | 181 (236) | 8 (9) |
| Summer | 1.5 (1.5) | 2.1 (2.9) | 373 (391) | 10 (10) |

## 7. Discussion

Observations made with high-resolution radiosonde observations launched routinely from Macquarie Island located in the Southern Ocean are used to study gravity waves in the lower stratosphere (12-25 km). The combination of wind and temperature data means that climatologies of wave parameters of particular relevance to gravity wave parameterization schemes, such as wave amplitudes and directions of propagation, can be achieved. Seasonal variations in wave activity and directions are found, as is evidence for enhanced energy near the inertial frequency. It is suggested that the enhancements may be due to quasi-monochromatic inertia-gravity waves (IGW) information is derived on frequencies, horizontal wavelengths and phase velocities.

These results are complementary to those given in AV95 since they fill a latitudinal gap between Hobart (43°S, 147°E) and the Antarctic stations located near 69°S that they used to study latitudinal and seasonal variations in wave activity. AV95 used the quantity $E_o$, energy density per unit mass, as a measure of wave activity, but since only temperature data were available they converted these to energy density by using the gravity wave polarization relationships and the model spectrum described by (1)-(3). The values for MI in the lower stratosphere are given in Table 2, where $E_{oT}$ denotes the energy density computed from temperatures alone, with $p = 5/3$, while $E_o$ denotes energies derived from both wind and temperature data. The quantity

$$\sigma_T = (\overline{u'^2} + \overline{v'^2})^{1/2} \tag{13}$$

is the total rms perturbation velocity, which is one of the input parameters in the Hines (1996) parameterization scheme.

Table 2. Seasonal mean energy densities and amplitudes in the 12-25 km height range.

| | $E_{oT}$ (J kg$^{-1}$) | $E_o$ (Jkg$^{-1}$) | $\sigma_T$ (ms$^{-1}$) |
|---|---|---|---|
| Winter | 5.5 | 6.7 | 3.0 |
| Summer | 3.2 | 4.2 | 2.4 |

The values of $E_{oT}$ fit well with the latitudinal and seasonal variations determined by AV95, They are somewhat smaller than the corresponding seasonal values determined for Hobart, but larger than the corresponding Antarctic values. The fact that $E_o$ is larger than $E_{oT}$ is an indication of the apparent dominance of inertia-gravity waves (IGW) as discussed above and possibly indicate that the energy densities estimated by AV95 for other latitudes are somewhat too small.

Evidence for large-scale IGW in the lower stratosphere comes from a range of sources. Using radiosonde data Thompson (1978) found strong evidence for waves with $\omega \sim f$ in the summer lower stratosphere over southern Australia. Nastrom *et al.* (1996) using high-resolution radiosonde observations taken at the Flatland site in central Illinois (40°N) concluded that there was an enhancement in gravity-wave energy near $f$ in all seasons at this location. MST radar observations also show strong evidence for IGW. Sato (1994) carried out a climatological study using the MU radar (35°N) in Japan while Cho (1995) used the Arecibo radar to study large-scale IGW over Puerto Rico (19°N). The wave properties determined in these studies are in broad agreement with the values determined here. Wave motions in the lower stratosphere are dominated by vertical wavelengths ~ 2- 3 km, horizontal wavelengths are typically in the range 300 to 1000 km and horizontal intrinsic phase speeds are about 10 ms$^{-1}$.

The source of the waves is unknown. The ground-based horizontal group velocities have magnitudes of ~40 ms$^{-1}$ and are directed toward south of east. Taken together with the small vertical groups speeds of about 0.1 ms$^{-1}$ this suggests that the waves originate far to the northwest of MI. O'Sullivan and Dunkerton (1995) studied the excitation and propagation of IGW in a high-resolution 3D nonlinear numerical model. The waves were generated spontaneously in the upper tropospheric jetstreams distorted by baroclinic instability. O'Sullivan and Dunkerton (1995) noted the potential importance of these waves in constituent mixing through the formation of dynamical instabilities during wave breakdown. Dynamical instability occurs when the rms wave amplitude exceeds a critical value given by

$$V_c' = \frac{2\delta_-^{1/2}}{(1+\delta_-^{1/2})}\hat{c} \tag{14}$$

where $\delta_- = (1 - f^2/\omega^2)$ (Fritts and Rastogi, 1985; Fritts, 1989). From Table 1 $|\omega/f| \sim 1.5$-3, which gives break down when $V_c' \sim 0.8$-0.9 $\hat{c}$, that is about 7-9 ms$^{-1}$. The rms wave amplitudes shown in Table 2 are 2-3 times too small satisfy this criterion, but it should be recognized that these values are averages over a 13 km altitude range and the saturation criterion could well be exceeded locally. Routine radiosonde measurements provide the scope to explore further this possibility and the contribution that IGW might make to cross-isentropic mixing in the lower stratosphere.

The significant degree of wave anisotropy evident in Figure 4, especially in winter, has implications for the implementation of gravity-wave parameterization schemes in models. Directional anisotropy enters the Fritts and Lu (1993) scheme through the factor $\Phi(\phi)$ in equation (1) and propagation directions are a key element in the Hines (1996) scheme. Because of the paucity of information on wave directions it is common in implementations of these schemes to assume that the wave spectrum is identical in all directions at some initial level (e.g., Mengel *et al.*, 1995). Further information is required to determine if the degree of anisotropy found here is common at other locations and how it varies with season.

The anisotropy strongly suggests that there is a significant flux of momentum associated with the wave field. The rotary spectral analyses show that the waves are predominantly upward propagating in the lower stratosphere and the bias toward southwestward propagation evident in Figure 4 implies negative values for the vertical fluxes per unit mass of zonal ($\overline{u'w'}$) and meridional ($\overline{v'w'}$) momentum. Although direct measurements of $w'$ are not possible from these balloon data, estimates of the pseudo-momentum fluxes can be made if we use the gravity wave polarization relations between $u'$, $w'$ and $T'$. For a monochromatic wave it can be shown that the upward flux of zonal momentum is

$$\overline{u'w'} = -\frac{\omega\, g}{N^2}\,\overline{u'\hat{T}'_{+90}}\,\delta_{-}(\omega)\,. \tag{15}$$

In practice, we are dealing with either a spectrum of waves or a mixture of quasi-monochromic waves and a spectrum. If a spectrum of the form given by (3) exists then $\omega$ may be replaced by the spectral average value, $\bar{\omega}$, (Eckermann *et al.*, 1996), where

$$\bar{\omega} = N\frac{(p-1)}{(2-p)}\hat{f}^{\,p-1}(1-\hat{f}^{\,p-2}) \tag{16}$$

for $p \neq 2$

or

$$\bar{\omega} = -f\ln(\hat{f}) \tag{17}$$

if $p = 2$, and where $\hat{f} = f/N$. Momentum fluxes evaluated using (15) with $\bar{\omega}$ evaluated with values of $p$ given by 5/3 and 2 are shown in Figure 7 for the wintertime stratosphere over MI. The horizontal bar indicates the standard error of the mean of the 145 observations which went into the average. Two values of the fluxes are shown in Figure 7 as the precise value of $p$ is unknown, but clearly the flux is a strong function of $p$, with the flux magnitude decreasing as the spectral slope increases. This is expected, since the steeper the slope of the frequency spectrum means that waves with near-inertial frequencies contribute relatively more to the energy, but contribute relatively less to the momentum flux than high-frequency waves. Table 3 summarizes the flux estimates for both winter and summer for a mean altitude of 18 km.

Table 3. Mean momentum fluxes in summer and winter assuming a spectrum of waves with $p$=5/3.

| | $\overline{u'w'}$ ($m^2s^{-2}$) | $\overline{v'w'}$ ($m^2s^{-2}$) | $\overline{\rho_o u'w'}$ ($Nm^{-3}$) | $\overline{\rho_o v'w'}$ ($Nm^{-3}$) |
|---|---|---|---|---|
| Winter | -0.07 | -0.02 | 9.3 $10^{-3}$ | 1.1 $10^{-2}$ |
| Summer | -0.08 | -0.01 | 2.6 $10^{-3}$ | 1.3$10^{-3}$ |

Given the qualifications, how realistic are these fluxes? The values are similar in magnitude to more direct estimates made by Fritts *et al.* (1990), Murayama *et al.* (1994), and Sato (1994) who used the dual-beam technique (Vincent and Reid, 1983) with the MU radar. From data taken over several years Murayama *et al.* (1994) found a clear annual cycle in the $\overline{u'w'}$ fluxes which were negative at all times of the year. Averaged over the 5 min-21 hr period and 15.5-17 km height ranges the values were about -0.2 $m^2s^{-2}$in winter and about -0.03 $m^2s^{-2}$in summer. This suggests that the values shown in Figure 7 and Table 3 are not unreasonable.

The flux profiles shown in Figure 7 are almost constant with height which suggests that there may be significant momentum deposition. However, the expressions for the fluxes have been derived using linear wave theory and if wave breaking or dissipation occurs the phase relationships between the temperature and wind fluctuations will change. This aspect needs to be considered further, as does the effects of vertical shears in the mean winds. It is noted that Sato and Dunkerton (1996) used a similar technique to that described here to study momentum

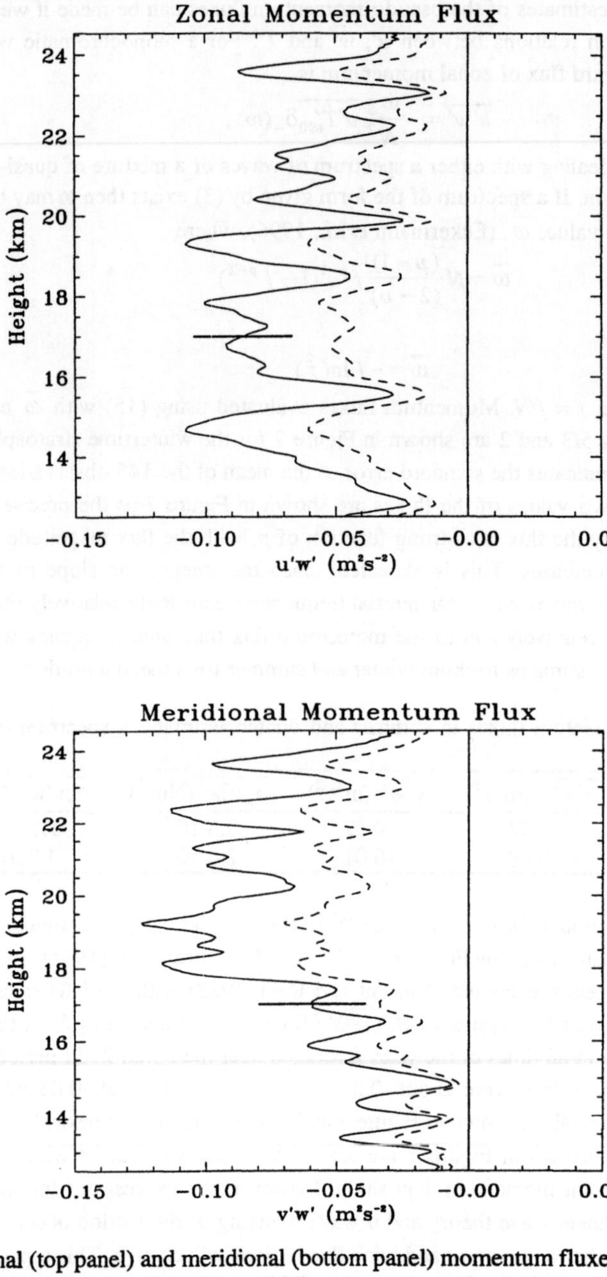


**Figure 7. Zonal (top panel) and meridional (bottom panel) momentum fluxes in winter inferred from wind and temperature data. The solid line shows the estimate for *p=5/3* while the dashed line indicates the value for *p*=2.**

fluxes associated with equatorially trapped gravity waves with radiosonde data taken at Singapore (1°N). They also discuss a more indirect method of obtaining fluxes based on theoretical and numerical work by Dunkerton (1995) which examines the covariance between wind and temperature perturbations of gravity waves propagating in background having vertical shear. The shears change the phase relationship between $u'$ and $T'$ leading to non-zero covariances. For waves in the 1-3 day period band Sato and Dunkerton (1996) found very significant $\overline{u'T'}$ covariances in strong regions of shear associated with the QBO, but only relatively small values of the quadrature component. In contrast, here we find that the $\overline{u'T'_{+90}}$ covariances are at least a factor of 4 larger than the $\overline{u'T'}$ covariances, which suggests that shear effects were not very important.

While it may be difficult to estimate the body forces associated with wave breaking or dissipation the results do have implications for wave effects in the upper stratosphere and mesosphere. In winter, the mean zonal wind shear is positive whereas the waves are propagating against the wind, which means that their intrinsic frequencies will increase as they propagate upward. The negative $\overline{u'w'}$ fluxes mean that when the waves eventually break in the upper middle atmosphere they will exert a westward drag. In summer, in contrast, the zonal winds change from westward to eastward at heights above 25 km and the waves are likely to encounter critical layers in the 25-30 km height range. The small negative $\overline{u'w'}$ fluxes imply that the body forces associated with wave breaking will lead to an acceleration of the mean flow in the summer middle stratosphere, as recent modeling suggests (Alexander and Rosenlof, 1996).

Further work is clearly required to place the momentum flux estimates based on combined wind and temperature data on a firmer footing. Nevertheless, the results reported here hold the promise that climatologies of momentum fluxes on a global scale wherever high-resolution radiosonde data are available.

## Acknowledgments

We acknowledge the provision of data by the Australian Bureau of Meteorology, and especially thank the staff at Macquarie Island who provided the data on a weekly basis. This work was supported by the Australian Research Council. SDE's research was performed in-part under contract number N00014-95-C-2152 of the Naval Research Laboratory.

## References

Alexander M. J., and K. H. Rosenlof, 1996: Non-stationary gravity wave forcing of the stratospheric zonal mean wind, *J. Geophys. Res.*, (Submitted).

Allen, S. J., 1996: Gravity wave motions in the troposphere and lower stratosphere, *Ph. D. Thesis*, University of Adelaide, pp253.

Allen S. J., and R. A. Vincent, 1995: Gravity wave activity in the lower atmosphere: Seasonal and latitudinal variations, *J. Geophys. Res., 100,* 1327-1350.

Cho, J. Y. N., 1995: Inertia-gravity wave parameter estimation from cross-spectral analysis, *J. Geophys. Res., 100,* 18,727-18,737.

Dunkerton, T. J., 1995: Horizontal buoyancy flux of internal gravity waves in vertical shear, *J. Meteor. Soc. Japan, 73,* 747-755.

Eckermann, S. D., and R. A. Vincent, 1989: Falling sphere observations of anisotropic gravity wave motions in the upper stratosphere over southern Australia, *Pure Appl. Geophys., 130,* 509-532.

Eckermann, S. D., and W. K. Hocking, 1989: Effect of superposition on measurements of atmospheric gravity waves: A cautionary note and some reinterpretations, *J. Geophys. Res., 94,* 6333-6339.

Eckermann S. D., I. Hirota, and W. K. Hocking, 1995: Gravity wave and equatorial wave morphology of the stratosphere derived from long-term rocket soundings, *Quart. J. Roy. Met. Soc.,121,* 149-186.

Eckermann, S. D., S. J. Allen, and R. A. Vincent, 1996: Computation of gravity-wave Eliassen-Palm fluxes from contemporaneous horizontal-velocity and temperature data: A feasibility study, (Paper in preparation).

Fritts, D. C., 1989: Gravity wave saturation processes, *Pure Appl. Geophys., 130,* 343-371.

Fritts, D. C., and P. K. Rastogi, 1985: Convective and dynamical instabilities due to gravity wave motions in the lower and middle atmosphere: Theory and observations, *Radio Sci., 20,*1247-1277.

Fritts, D. C., and T. E. VanZandt, 1987: Effects of Doppler shifting on the frequency spectra of atmospheric gravity waves, *J. Geophys. Res., 92,* 9723-9732.

Fritts, D. C., and W. Lu, 1993: Spectral estimates of gravity wave energy and momentum fluxes, II Parameterization of wave forcing and variability, *J. Atmos. Sci., 50,* 3695-3713.

Fritts, D. C., T. Tsuda, T. E. VanZandt, S. A. Smith, T. Sato, S. Fukao, and S. Kato, 1990: Studies of velocity fluctuations in the lower atmosphere using the MU radar, II Momentum fluxes and energy densities, *J. Atmos. Sci., 47,* 51-66.

Gardner, C. S. and N. F. Gardner, 1993: Measurement distortion in aircraft, space shuttle, and balloon observations of atmospheric density and temperature perturbation spectra, *J. Geophys, Res., 98,* 1023-1033.

Gossard, E E., and W. H. Hooke, 1975: *Waves in the Atmosphere,* Elsevier Science, New York.

Hamilton, K., 1991: Climatological statistics of stratospheric inertia-gravity waves deduced from historical rocketsonde wind and temperature data, *J. Geophys. Res., 96,* 20,831-20,839.

Hamilton, K., and R. A. Vincent, 1995: High-resolution radiosonde data offer new prospects for research, *EOS, 76,* 497.

Hines, C. O., 1989: Tropopausal mountain waves over Arecibo: A case study, *J. Atmos. Sci., 46,* 476-488.

Hines, C. O., 1996: Doppler-spread parameterization of gravity wave momentum deposition in the middle atmosphere, part I: Basic Formulation, *J. Atmos. Terr. Phys.,* submitted, 1996.

Lindzen, R. S., 1981: Turbulence and stress wing to gravity wave and tidal breakdown, *J. Geophys. Res, 86,* 9707-9714.

Marks, C. J. and S. D. Eckermann, 1995: A three-dimensional nonhydrostatic ray-tracing model for gravity waves: formulation and preliminary results for the middle atmosphere, *J. Atmos. Sci., 52,* 1959-1984.

Mengel, J. G., H. G. Mayr, K. L. Chan, C. O. Hines, C. A. Reddy, N. F. Arnold, and H. S. Potter, 1995: Equatorial Oscillations in the middle atmosphere generated by small scale gravity waves, *Geophys. Res. Lett., 22,* 3027-3030.

Murayama, Y., T. Tsuda, and S. Fukao, 1994: Seasonal variation of gravity wave activity in the lower atmosphere observed with the MU radar, *J. Geophys. Res., 99,* 23,057-23,069.

Nastrom, G. D., T. E. VanZandt and J. M. Warnock, 1996: Vertical wavenumber spectra of wind and temperature from high-resolution balloon soundings in the lower atmosphere over Illinois, *J. Geophys. Res.,* (Submitted).

Ogino, S., M. D. Yamanaka, and S. Fukao, 1995: Meridional variation in of lower-stratospheric gravity wave activity: A quick look at Hakuho-Maur J-COARE cruise rawinsonde data,' *J. Meteorol. Soc. Japan,73,*407-413.

O'Sullivan, D. and T. J. Dunkerton, 1995: Generation of inertia-gravity waves in a simulated life cycle of baroclinic instability, *J. Atmos. Sci., 52,* 3695-3716.

Sato, K., 1994: A statistical study of the structure, saturation and sources of inertio-gravity waves in the lower stratosphere observed with the MU radar, *J. Atmos. Terr. Phys., 56,* 755-774.

Sato, K. And T. J. Dunkerton, 1996: Estimates of momentum flux associated with equatorial Kelvin and gravity waves, *J. Geophys. Res.,* (Submitted).

Smith, S. A., D. C. Fritts, and T. E. VanZandt, 1987: Evidence for a saturated spectrum of atmospheric gravity waves, *J. Atmos. Sci., 44,* 1404-1410.

Thompson, R. O. R. Y., 1978:Observations of inertial waves in the stratosphere, *Quart. J. Roy. Met. Soc., 101,* 691-698.

VanZandt T. E., and D. C. Fritts, 1989: A theory of enhanced saturation of the gravity wave spectrum due to increases in atmospheric stability, *Pure Appl., Geophys., 130,* 399-420.

Vincent, R. A., 1984: Gravity-wave motions in the mesosphere, *J. Atmos. Terr. Phys., 46,* 119-128.

Vincent, R. A. And I. M. Reid, 1983: HF Doppler measurements of mesospheric gravity wave momentum fluxes, *J. Atmos. Sci., 40,* 1321-1333.

Vincent, R. A., and D. C. Fritts, 1987: A climatology of gravity wave motions in the mesopause region at Adelaide, Australia, *J. Atmos. Sci., 44,* 748-760.

Vincent, R. A., and S. D. Eckermann, 1990:VHF radar observations of mesoscale motions in the troposphere: Evidence for gravity wave Doppler shifting, *Radio Sci., 25,* 1019-1037.

# Measurements of Intermittency in the Upper Stratosphere and Mesosphere

R. J. Sica and M. D. Thorsley
Department of Physics
The University of Western Ontario
London, Ontario N6A 3K7
Canada

## 1. Introduction

Measurements of the gravity wave spectrum in the upper stratosphere and mesosphere are scarce compared to those in the troposphere and lower thermosphere, which are easier to probe via in situ and remote sensing techniques. This paucity of measurements impacts the parameterization of gravity waves in general circulation models of the middle atmosphere, since the effects of gravity waves are not adequately constrained by measurements. The development of a larger power-aperture product lidar at The University of Western Ontario has allowed the measurement of the gravity wave spectrum at high temporal-spatial resolution. These measurements have been used to impact three key areas relevant to parameterizations: the thermodynamic perturbations caused by gravity waves, the variability of the spatial and temporal spectra and the estimation of the eddy diffusion coefficient. After a brief introduction to the instrumentation (Section 2), Section 3 describes an extension of the initial measurements of superadiabatic lapse rates by Sica and Thorsley (1996a). A taste of the spatial and temporal spectral retrievals possible with the lidar measurements is given in Section 4. In

NATO ASI Series, Vol. I 50
Gravity Wave Processes
Their Parameterization in Global Climate Models
Edited by Kevin Hamilton

Section 5, the spectra are used to calculate energy dissipation and the eddy diffusion coefficient.

## 2. Instrumentation

The Purple Crow Lidar is a monostatic lidar system which can simultaneously measure both Rayleigh and sodium resonance fluorescence backscatter. The lidar is located just outside of London, Ontario, Canada at the Delaware Observatory (42°52' N; 81°23' W). The Purple Crow Lidar enjoys a large power-aperture product by the use of a high power transmitter and large aperture receiver. The transmitter is a Nd:YAG laser operating at the second harmonic, with a output energy of nominally 600 mJ/pulse at 20 Hz. The receiver is a 2.65-m diameter liquid mercury mirror, which is coupled to a photon counting system by a optical fiber. Sica et al. (1995) show that the liquid mirror has a performance similar to a traditional glass telescope of the same area. The measurements are obtained every 1 min in 24 m range bins. Temperatures are found from the density profiles as described by Chanin and Hauchecorne (1984). The density fluctuations are determined by the methodology of Gardner et al. (1989b).

## 3. Measurements of Superadiabatic Lapse Rates

The methodology for determining layers of stability and potential instability is, by necessity, tedious (Sica and Thorsley 1996a). In principle one could simply take a temperature profile, differentiate it and be done. In practice, despite the large return photocount signal, the temperature changes sought are extremely small. Furthermore, the expectation is, that higher in the atmosphere more unstable regions will exist, as the amplitude of the waves increases with height. Unfortunately, the lidar return signal decreases for the same reason, and in fact, faster than the gravity waves grow due to the exponential decrease of density with height. Hence, extreme care must be taken in the determination of stable and potentially unstable regions.

The photon noise in the temperature profiles, if sufficiently large, can cause the appearance of superadiabatic layers in an otherwise stable temperature profile. To help determine if the lapse rates measured are due solely to noise, a synthetic data set for each night is made, or "cloned", using the average density profile from the measurements for the shape of the photocount profile. Gaussian noise is added and lapse rates are calculated after which each individual measurement and cloned data point (and their associated error) is then treated as the mean and variance of a Gaussian distribution. Distributions in which 80% of the probability lies below zero are classified as negative and vice versa for positive. Distributions which fall in neither category are indeterminate. With the differences in lapse rates and their signs available for both the measurements and the cloned data, the two distributions can be statistically tested to determine whether or not they are likely to arise from the same parent population. Kupier's variant of the Kolmogorov-Smirnov test is used to test whether or not the measurements and the cloned data are from the same distribution, while the $F$-test is used to test for significantly different variances between the data sets (Press et al. 1992).

On some nights the variance of the lapse rate measurements is significantly different from the photocount noise. The difference between these variances is a measure of atmospheric variability. To illustrate this variability, a representative grouping of nights from the Purple Crow Lidar's database is sampled (Table 1). The photon noise on all these

**Table 1. Four groups used to study the occurrence of positive and negative layers. The group name is in the first row, while subsequent cells contain the nights in the group (year-month-day format).**

| Oct94 | Jun95 | Oct95 | May96 |
|---|---|---|---|
| 94/10/7 | 95/6/1 | 95/10/12 | 96/5/13 |
| 94/10/12 | 95/6/13 | 95/10/13 | 96/5/31 |
| 94/10/23 | 94/6/15 | 95/10/17 | 96/6/1 |
| 94/10/24 | | | |

nights is similar. The atmospheric variability calculated for each group is shown in Figure 1. The Oct94 group has essentially no variability above 40 km, as the photon noise dominates. However, the other groups

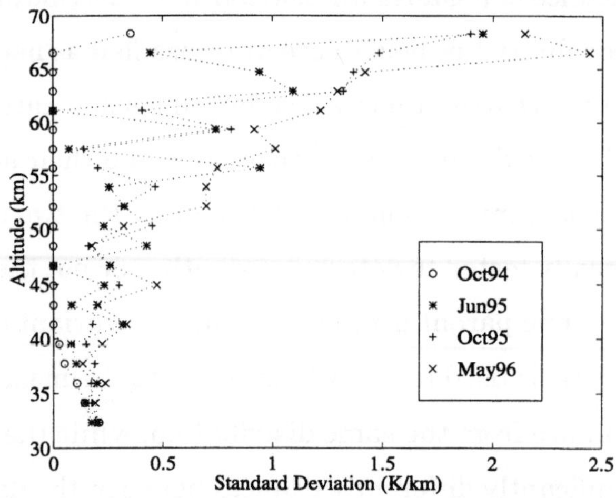


**Fig. 1. The difference between the standard deviations of the lapse rate measurements and the photocount noise for the four groups.**

all have residual variability. To further illustrate the difference between the Oct94 group and the other groups, the difference between the percentage of space and time with positive, negative and indeterminate lapse rate differences can be compared. The percent differences in time are shown in Figure 2. The number of negative regions is between

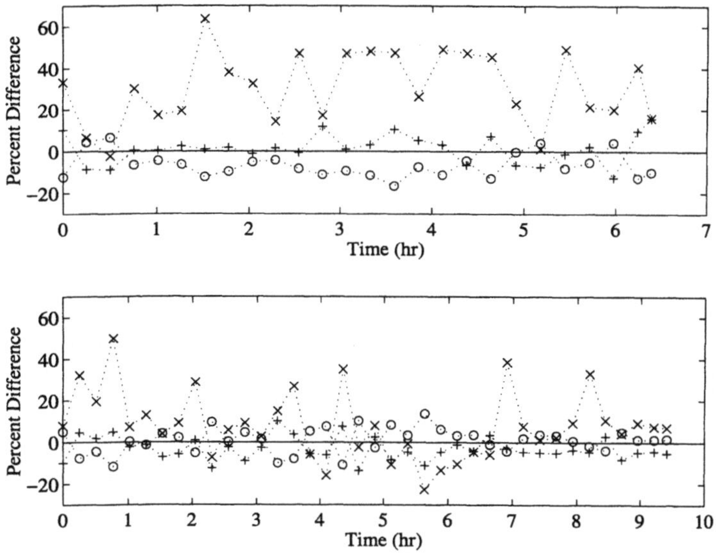


**Fig. 2. Percent difference in time between the lapse rate measurements and synthetic data for the variable pod (top) and the noise pod (bottom) for positive ('+'), negative ('x') and indeterminate ('o') differences.**

20-30% greater for the average of the Jun95, Oct95 and May96 groups (henceforth the "variable" pod) compared to the cloned data, while the indeterminate percent differences are about 10% less. The Oct94 group (henceforth the "noise" pod) has essentially the same number of positive, negative and indeterminate layers as the cloned data. The spatial percent differences are shown in Figure 3. In the stratosphere, below 40 km both pods show large differences from the cloned data. In the mesos-

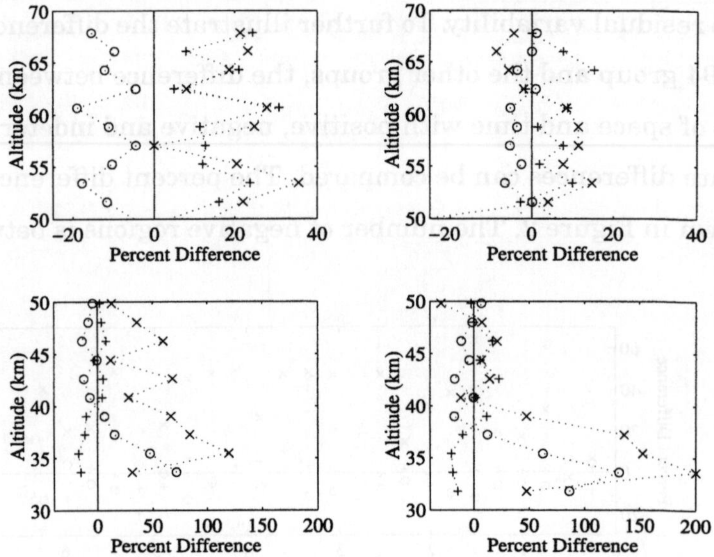


**Fig. 3. Percent difference in space between the lapse rate measurements and synthetic data for the variable pod (left) and the noise pod (right) for positive ('+'), negative ('x') and indeterminate ('o') differences. The height ranges have been plotted on difference scales for clarity.**

phere, the variable pod percent differences are consistent with a distribution with a larger variance than the photocount noise. The residual attributed to atmospheric variability (Fig. 1) is consistent with the comparison between the percent differences.

The spatial percentages of positive, negative and indeterminate layers are shown in Figure 4. Throughout most of the upper stratosphere the positive regions dominate, while the increase of the negative regions is roughly monotonically increasing with height. Above the stratopause, the number of indeterminate regions is almost always a few percent less than 60%, as compared to an average of about 63% for the indeterminate layers in the noise pod.

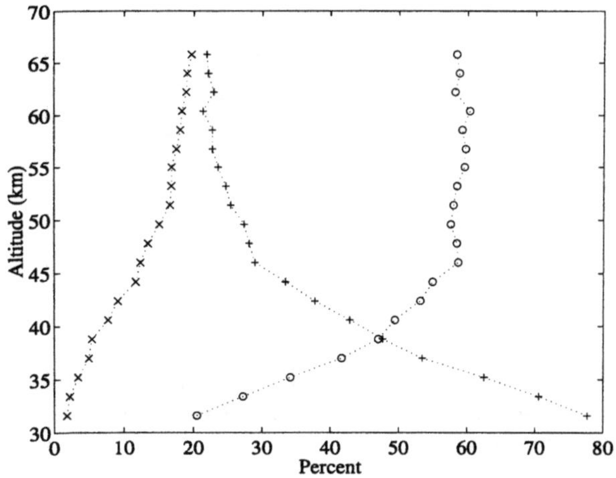


**Fig. 4. Percentage of positive ('+'), negative ('x') and indeterminate regions ('o') as a function of height for the variable pod.**

Hamilton (1984) computed the Richardson number for a large body of sounding rocket measurements. The nearest station to London, Ontario in his study is Wallops Island, Virginia, about 5° of latitude to the south. The percentage of regions with $Ri < 1$ for Hamilton's averages during equinox agrees well in the stratosphere with the negative percentages, but are about 40% lower in the mesosphere.

The coherence of the variable pod can be found in the manner described by Sica and Thorsley (1996a). The coherence, that is the distribution of contiguous positive, negative and indeterminate regions in space or time, is shown in Figure 5. The general distributions are similar to those on an individual night. The negative regions are extremely confined in space and time relative to the positive regions.

The individual lapse rate differences can be plotted as a function of height and time. In the mesosphere, these plots are similar from night to night. However, below 40 km, substantial variations in the amount and duration of layers is evident on different nights. Figure 6 shows the

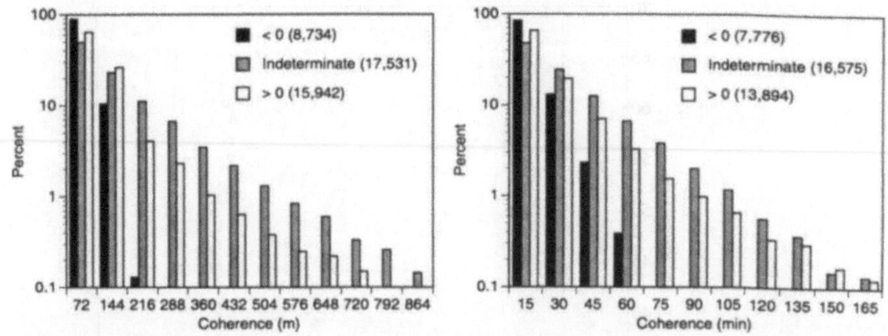


**Fig. 5. Spatial (left) and temporal (right) coherence for the variable pod. The numbers in brackets are the number of measurements in each distribution**

lapse rate differences on October 12, 1995. Below 40 km, substantial positive regions exist (e.g. from 0200 - 0300 hr; 0400-0500 hr, etc.). On

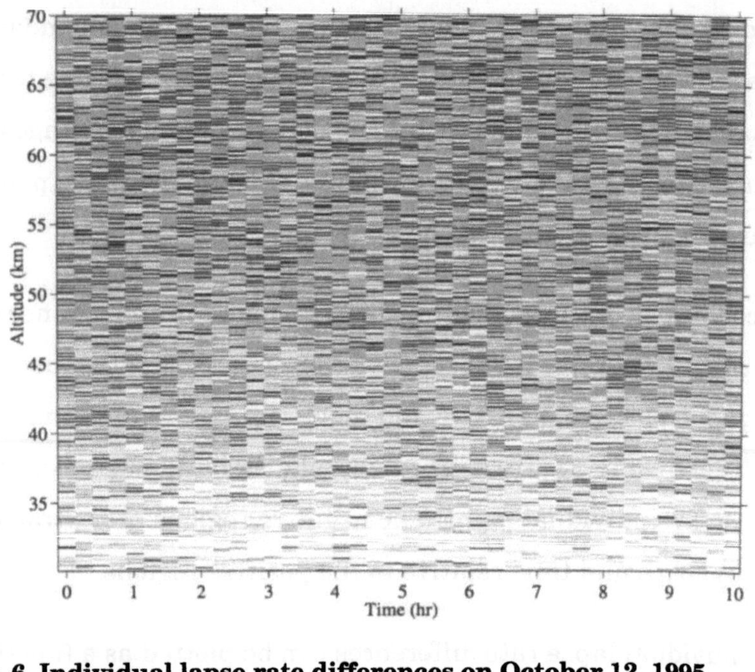


**Fig. 6. Individual lapse rate differences on October 12, 1995. Positive regions are white, indeterminate regions are gray and negative regions black. Zero time on the ordinate is 2024 EDT.**

October 17, 1995 few extended regions of stability exist compared to October 12 (Fig. 7). There are also substantially more negative layers below 40 km on this night.

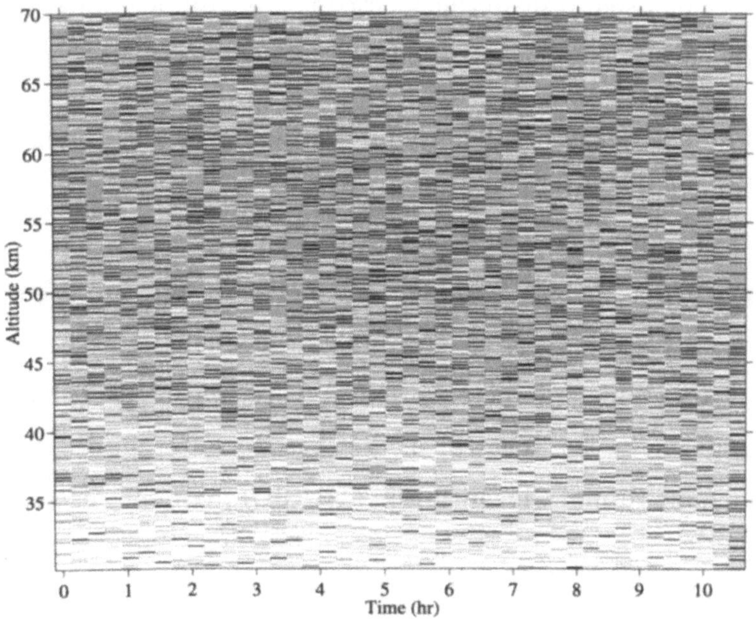


**Fig. 7. Individual lapse rate differences on October 17, 1995. Positive regions are white, indeterminate regions are gray and negative regions black. Zero time on the ordinate is 2009 EDT.**

## 4. Vertical Wavenumber and Temporal Spectrum

The material in this and the following section summarizes just a few of the main points reached in a detailed study by Sica (1996b; 1996c). The measurements used for this study were obtained on the night of August 30, 1994 commencing at 2319 EDT. The vertical (temporal) density fluctuations are temporally (spatially) filtered using finite impulse response filters. The data series were smoothed using either a Kaiser-Bessel filter with 40 dB of attenuation (in space) or by 3s and 5s (in time; Hamming 1977). The power spectral densities calculated have units of meters (spatial) or seconds (temporal); that is the angular units have been removed. From these natural units the gravity wave polarization equations can be used to scale the spectrum to a kinetic energy density or velocity spectrum (Gardner et al. 1989a).

Spectral analysis of the density fluctuations was performed using parametric models. Statistical procedures such as correlogram analysis require the autocorrelation sequence of a process to be set to zero outside a specified spectral window. Parametric models can use information outside of this range, which eliminates the sidelobes associated with window functions and increases spectral resolution. Specific models of use for gravity wave spectral studies are the autoregressive model and the Multiple SIgnal Classification (MUSIC) frequency estimator (Marple 1987).

The average vertical wavenumber spectrum using the MUSIC algorithm is shown in Figure 8. The arrow on the ordinate indicates the Hines parameter (Hines 1991). The spectrum in the regions of and to the right of the arrow are considered to be nonlinear. The conversion from density fluctuations to kinetic energy density should be treated with caution in these regions. For the average vertical wavenumber spectrum in the upper stratosphere on this night the majority of the tail region is linear (that is to the left of the arrow).

The average temporal spectrum in this height range shows a monotonic descent, without the shoulder often seen in correlograms (Fig. 8). A strong feature is present at $2.2 \times 10^{-3}$ $s^{-1}$. Closer analysis of the spectrum shows companion features to this strong feature at the second and

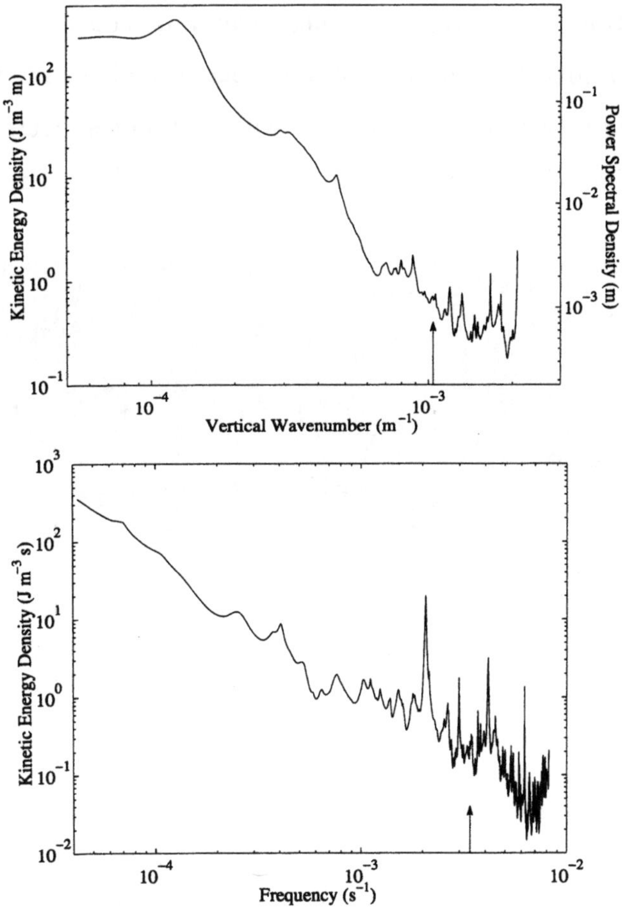


**Fig. 8. Average vertical wavenumber (top) and temporal (bottom) spectrum in the upper stratosphere on August 30, 1994. The arrow on the vertical wavenumber spectrum indicates the Hines parameter, while the arrow on the temporal spectrum indicates the measured buoyancy frequency.**

third harmonics. A detailed study of this event is in progress. The monotonic decrease of the spectrum continues down to the measured buoyancy frequency ($3.4 \times 10^{-3}\ s^{-1}$). Below this point additional energy appears to be present, before the spectral energy rapidly decreases above $7 \times 10^{-3}\ s^{-1}$.

The signal-to-noise ratio of the measurements coupled with the MUSIC algorithm allow the evolution of the spectrum with time to be examined. The individual one minute vertical wavenumber spectra can be shown as a function of time and wavenumber (Fig. 9). Note the coherence of

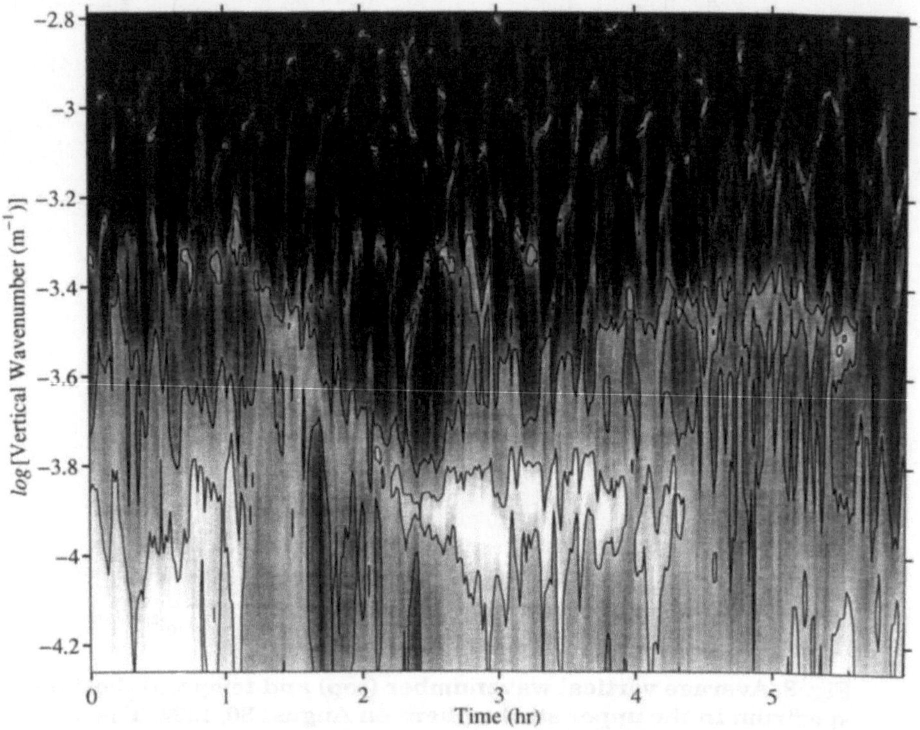


**Fig. 9. Contour "image" for the individual MUSIC spectra in the upper stratosphere on the night of August 30, 1994. The ordinate is time while the abscissa is the logarithm of the kinetic energy density. The individual contours are separated by 10 dB, with each gray-scale variation corresponding to a 10 db decrease in kinetic energy density (white 0 dB, black -60 dB). The peak value is 2950 J/ $m^3$ m. Zero hours corresponds to 2319 EDT.**

many features well beyond the bandwidth of the temporal filtering (7 min). At small wavenumbers, periods of high power exist between 0000 to 0100 hr and 0230 to 0415 hr, while the peak power decreases -20 to

-30 dB at times during the interim period. During this time, the regions of highest power shift to higher wavenumbers (e.g. the period around 0200 hr at -3.8 = $\log_{10}[1/(6310 \text{ m})]$). The power in the middle wavenumber bands shifts to lower wavenumbers at 0200 hr, just prior to an enhancement of the power around $1 \times 10^{-4}$ $m^{-1}$. At the highest wavenumbers several "islands" of power enhancement occur in the $1 \times 10^{-3}$ $m^{-1}$ region. Some of these events (e.g. at -3.2 = $\log_{10}[1/(1580 \text{ m})]$] and 0450 hr) are extremely intense, with peaks as high20 dB above the typical background.

## 5. Calculation of Energy Dissipation and the Eddy Diffusion Coefficient

The energy dissipation can be found from the product of the root mean square velocity and the standard deviation of the time rate of change of the velocity (Sica 1996c).The energy dissipation, which varies from a minimum of 1.5 mW/kg to 80 mW/kg near the stratopause, has a mean value of 11 mW/kg. The energy dissipation can also be calculated for individual waves by Prony's method, which allows the frequency, amplitude, growth rate and phase to be estimated from the measurements (Marple 1987). Once a fit to the data series has been computed, specific frequencies, if and when they exist, can be tracked as a function of time and the phase velocity, which is assumed equal to the group velocity determined. Knowing the group velocity, the growth rate and the amplitude of the wave perturbation expressed as an average horizontal wind fluctuation, the energy dissipation can be calculated without assuming a form for the spectrum (Hines 1965). A wave with a wavenumber of about $1 \times 10^{-4}$ $m^{-1}$ and a monotonically decreasing phase is present from

0000 to 0040 hr. The energy dissipation estimated from Hines' formula is about 70 mW/kg, consistent with the spectral calculation.

The energy dissipation can be used to estimate the eddy diffusion coefficient. The eddy diffusion coefficient, $D$, is proportional to the energy dissipation and inversely proportional to the square of the angular buoyancy frequency:

$$D = \beta\frac{\varepsilon}{N^2} \tag{1}$$

where $\beta$ is traditionally taken as a constant. Hocking (1991) discusses the choices of $\beta$ typically employed, with values in the range between 0.2 and 1.0. However, it has recently been pointed out by McIntyre (1989) that $\beta$ actually varies as a function of the saturation of the vertical wavenumber spectrum. McIntyre calculates how the constant varies with the saturation of the spectrum. The question is how can the amount of saturation of the spectrum be determined from the measurements?

The parametric models independently estimate the driving noise variance of the input process. Hence, it seems reasonable that the amount of saturation, or gain, of the spectrum relates to the ratio of the input driving noise variance to the average power of the output power spectral density. When the input driving noise is less than the average power, the spectrum is not saturated; when the input driving noise exceeds the average power, the output is saturated with respect to the input. Using the vertical wave number spectra and McIntyre's calculations, a mean value of 0.074 is obtained for $\beta$.

The resulting eddy diffusion profile is shown in Figure 10. The eddy diffusion coefficient decreases from about 1 $m^2/s$ at 32 km to 0.3 $m^2/s$ around 40 km, before increasing to about 7 $m^2/s$ at the stratopause.

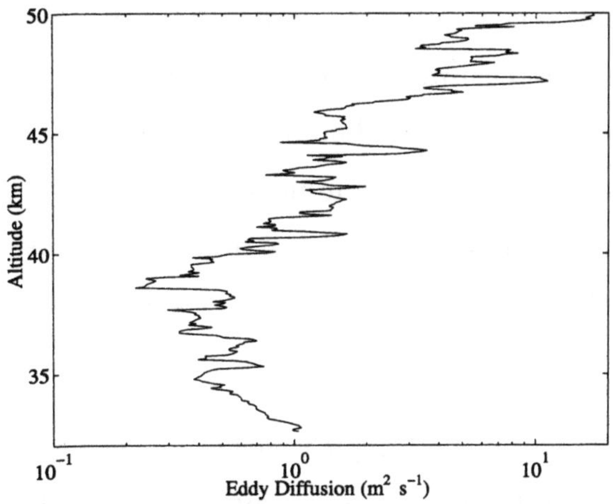


**Fig. 10. Eddy diffusion coefficient in the upper stratosphere using the β value of McIntyre (1989) determined from the estimated input driving noise variance of the vertical wavenumber spectrum.**

Some variability exists in the determination, in part due to instrumental noise and in part, perhaps in large part, to geophysical variations. The uncertainties are hard to quantify. The absolute value depends on the choice of β, which varied over time by about 50%. Much of the relative uncertainty is in the calculation of the variance of the acceleration fluctuation which amplifies the high frequency, low power components of the spectrum. Variations also exist in the buoyancy frequency which are greater than the statistical error, and these variations also contribute to the structure evident in the eddy diffusion profile.

## 6. Conclusions

The measurements of stable and unstable layers has been extended to 13 nights of measurements. The general percentages of layers and layer coherence appear to be similar on nights where the variance of the lapse rates exceed the photocount variance. Maps of the layers on individual nights appear similar (and random) above about 40 km; below these heights the amount of positive and negative layers can show significant structure and variability from night to night. Dr. Hines has suggested that the unstable layers are whitecaps on the crests of large amplitude gravity waves.

The spectral analysis of the Purple Crow Lidar measurements highlight the intermittency of the gravity wave dynamics. Sica (1996b) shows several other examples of the intermittency of the spectra, in addition to mesospheric spectra. Prony's method promises to be useful for exploring the temporal and spatial evolution of individual waves (see Sica (1996c) for details). The energy dissipation, as determined by two independent methods, is larger than suggested in previous studies. The energy dissipation is used to find an eddy diffusion profile. This profile is in reasonable agreement with previous estimates. The estimation of the eddy diffusion profile attempts to include the degree of saturation of the vertical wavenumber spectrum.

*Acknowledgments.* RJS has greatly benefited from scientific discussions with Professors C. Gardner, K. Hamilton, M. McIntyre, T. Shepherd and W. Hocking. Thanks also go to Team Purple Crow, in particular Dr. S. Argall and S. Beatty. This work was supported by grants from the

National Scientific and Engineering Research Council, the Canadian Space Agency and Canada's Atmospheric Environment Service.

## REFERENCES

Chanin, M. L. and A. Hauchecorne (1984) Lidar Studies of Temperature and Density Using Rayleigh Scattering, in *Handbook for MAP: Ground-Based Techniques,* **13**, ICSU Scientific Committee on Solar Terrestrial Physics, Urbana, Illinois, paper 7.

Gardner, C. S., M. S. Miller and C. H. Liu (1989a) Rayleigh lidar observations of gravity wave activity in the upper stratosphere at Urbana, Illinois. *J. Atmos. Sci.*, **46**, 1838-1854.

Gardner, C. S., D. C. Senft, T. J. Beatty, R. E. Bills and C. A. Hostetler (1989b) Rayleigh and Sodium Lidar Techniques for Measuring Middle Atmosphere Density, Temperature, and Wind Perturbations and their Spectra. *World Ionosphere / Thermosphere Study Handbook,* **2**, ICSU Scientific Committee on Solar Terrestrial Physics, Urbana, Illinois, Chapter 6.

Hamilton, K. (1984) A study of the occurrence of dynamically unstable conditions in the middle atmosphere, *Can. J. Phys.*, **62**, 963-967.

Hines, C. O. (1965) Dynamical heating of the upper atmosphere, *J. Geophys. Res.*, **70**, 177-183.

Hines, C. O. (1991) The saturation of gravity waves in the middle atmosphere. Part II: Development of Doppler-spread theory. *J. Atmos. Sci.*, **48**, 1360-1379.

Hocking, W. K. (1991) The effects of middle atmosphere turbulence on coupling between atmospheric regions. *J. Geomag. Geoelectr.*, **43**, Suppl., 621-636.

Marple, S. L. (1987) *Digital spectral analysis*. Prentice-Hall, Inc., Chapters 5-11.

McIntyre, ME (1989) On dynamics and transport near the polar mesopause. *J. Geophys. Res.*, **89**, 14617-14628.

Press, W. H., S. A Teukolsky, W. T. Vetterling, and B. P. Flannery (1992) *Numerical Recipes in FORTRAN, the Art of Scientific Computing*, 2nd edition. Cambridge University Press, 613-622.

Sica, R. J., S. Sargoytchev, P. S. Argall, E. F. Borra, L. Girard, C. T. Sparrow, and S. Flatt (1995) Lidar measurements taken with a large-aperture liquid mirror. 1. Rayleigh-scatter system. *Appl. Opt.*, **34**, 6925-6936.

Sica, R. J. and M. D. Thorsley (1996a) Measurements of superadiabatic lapse rates in the middle atmosphere. *Geophys. Res. Lett.*, submitted.

Sica, R. J. (1996b) Measurements of the Effects of Gravity Waves in the Middle Atmosphere Using Parametric Models of Density Fluctuations. 1: Spatial and Temporal Spectra. *J. Atmos. Sci.*, submitted.

Sica, R. J. (1996c) Measurements of the Effects of Gravity Waves in the Middle Atmosphere Using Parametric Models of Density Fluctuations. 2: Energy Dissipation and Eddy Diffusion. *J. Atmos. Sci.*, submitted.

# Climatology and hydrodynamic sources of internal gravity waves in the middle and upper atmosphere

Nikolai M. Gavrilov
Atmospheric Physics Department
Saint Petersburg University
Petrodvorets
Saint Petersburg, 198904
Russia

## Abstract

Results of studies of atmospheric internal gravity waves (IGW) at Saint Petersburg University are given. Height, seasonal and latitude variations of IGW parameters are studied from the data of ground-based radar networks. Theory of hydrodynamic IGW sources in the atmosphere is considered. Influence of mean wind and different mechanisms of IGW dissipation on the wave generation and propagation into the middle and upper atmosphere is discussed.

## 1. Introduction

Internal gravity waves (IGW) play an important role in the formation of general circulation, thermal regime and compositional structure of the middle and upper atmosphere. Progress in the theoretical models of IGW influence on the atmosphere requires the development of reliable mathematical methods and algorithms for study IGW climatology from experimental data, and mathematical modeling and parameterization of wave generation, propagation and effects of waves emitted by tropospheric sources. This paper describes some results of the study of IGW climatology in the middle and upper atmosphere from the data of incoherent scattering (IS), medium frequency and MST radars.

A comparison is made of the intensities of ion temperature perturbations having periods less than 6 hours at altitudes 100 – 400 km, observed at high-latitude and low-latitude IS radars. Vertical, seasonal and latitudinal variations of the intensity are studied using multi-year records of ion temperature measurements with the IS radars at Arecibo (18.3° N, 66.8° E); Millstone Hill (42.6° N, 71.5° W); Chatanika (65.1° N, 147.4° W); Sondrestrom (67.0° N, 51.0° W) and EISCAT (69.6° N, 19.2° E).

Height, seasonal and inter-annual variations of IGW intensities are studied using the MF radar observations at Saskatoon from 1979 to 1993. Inter-annual variations of monthly mean IGW intensities show an apparent negative correlation with solar activity, which is stronger for the longer-period component. Possible reasons and consequences of the observed IGW parameters variations and their correlation with solar activity are discussed.

NATO ASI Series, Vol. I 50
Gravity Wave Processes
Their Parameterization in Global Climate Models
Edited by Kevin Hamilton

In this paper we applied methods of calculation of IGW characteristics and adjust them to obtain all necessary parameters for each spectral harmonic of complicated wave noise existing in the atmosphere. The parameters of IGW harmonics are obtained using the measurements of wind velocity in the middle atmosphere with MU radar in five directions of the antenna beam (one vertical and four inclined to the north, west, south and east from zenith).

According to present-day knowledge, the main part of IGW propagating from below is originating in the troposphere. One of important sources of permanent generation of waves in the atmosphere is nonlinear interaction of different atmospheric motions. Scales of generated waves depend on the scales of atmospheric motions. Generation of IGWs in the troposphere may be caused mainly by Lighthill-type generation of waves by mesoscale turbulence. In this paper equations are obtained, which describe the IGW generation processes and calculations are made of characteristics of IGW propagating from tropospheric turbulent sources. Influence of the mean wind and different mechanisms of IGW dissipation on the IGW generation and propagation into the middle and upper atmosphere is considered.

## 2. Studies of IGW climatology

Studies of IGW climatology in Saint Petersburg university had begun from the development of statistical methods for the extraction of IGW parameters from ground-based observations of variations of nightglow characteristics (Gavrilov, 1982) and variations of wind velocity in the upper atmosphere (Gavrilov, 1984). These statistical methods have been applied to the study of IGW climatology from the observations of variations of OI 5577A night emission intensity in 49 points on the sky at Ashkhabad in 1964 – 1967 (Gavrilov and Shved, 1982), and from the observations of wind velocity in the meteor zone (Gavrilov et al, 1981; Gavrilov, 1987a). Study of IGW parameters in the stratosphere from observations of short-period variations of total ozone has been made (Semenov and Gavrilov, 1985).

Recently the study of IGW climatology in the thermosphere from the data of CEDAR Incoherent Scattering Radar Data Base, collected at National Center for Atmospheric Research, Boulder, U.S.A. was fulfilled (Gavrilov et al., 1994; Gavrilov, 1995). Height and seasonal structure of IGW parameters in the middle atmosphere was studied from 16-year continuous observations with medium frequency radar at Saskatoon, Canada (Gavrilov et al., 1995 ). Also, statistical study of IGW parameters is made using Japanese Middle and Upper Atmosphere (MU) radar at Kyoto (Gavrilov et al., 1991, 1996 ).

### *2.1 IGW intensity structure in the thermosphere*

IS radar observations are usually limited to daylight hours at altitudes below 200 km. These reasons make difficulties for the application of conventional spectral analysis to the study of short period ion temperature variations (see Gavrilov et al., 1994). To extract information about short-period fluctuations of ion temperature, first, we obtain the low-frequency component, composed of the mean value and 24-, 12-, 8- and 6-hr tidal harmonics:

$$\overline{T}_i(t) = T_{i0} + \sum_{k=1}^{4} C_k \cos(2\pi kt/24 - \phi_k), \tag{1}$$

where parameters $T_{i0}$, $C_k$ and $\phi_k$ are calculated using the least square fit method. The example of $\overline{T}_i(t)$ one can find in *Gavrilov et al.* (1994). Then the low-frequency component is subtracted from the initial data. Inclinations $T_i(t_j) - \overline{T}_i(t_j)$ contain information about variations having periods from 0.5 up to 6 hr. After the subtraction of (1) one can average squares of relative $[T_i(t_j) - \overline{T}_i(t_j)]^2/\overline{T}_i^2(t_j)$ residual values for each day and month over the entire multi-year period of observations with each IS radar.

The method of analysis described above is applied to the measurements of ion temperature with the network of IS radars listed above, and collected in the CEDAR data base of the National Center for Atmospheric Research, Boulder, USA. Fig.1 shows the altitude variations of intensity of ion temperature perturbations at low- and high-latitude IS radars. One can see from Fig.1 that the intensity has clear maxima 15 – 17 % in the altitude region 150 – 180 km at low latitudes (Arecibo radar). At high latitudes (Chatanika, Sondrestrom, EISCAT) the intensity maxima are absent or very small. Therefore, the values of maxima of the intensity at 150 – 180 km have tendency to decrease in the direction from the equator to high latitudes.

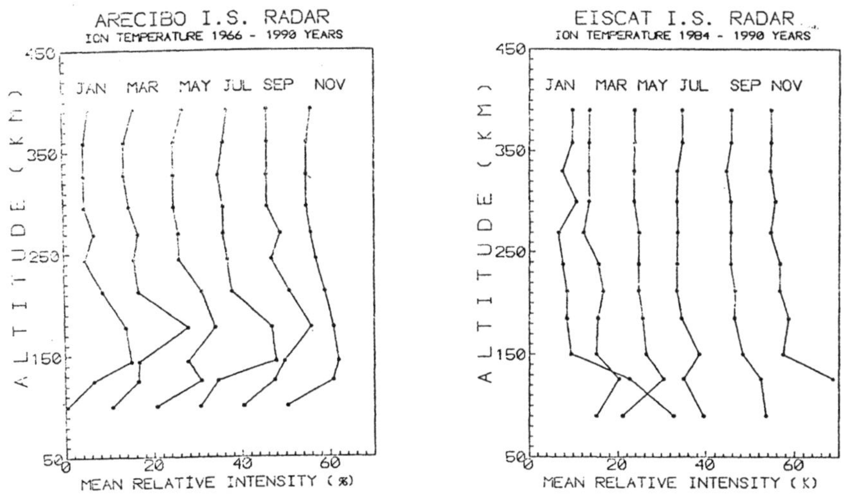


Fig.1. Vertical profiles of the relative intensity of short-period perturbations of ion temperature in thermosphere as measured with Arecibo and EISCAT IS radars. Lines are separated by adding of 10 % to each consecutive data set.

One may suppose that the observed perturbations of ion temperature at high latitudes are the complicated mixture of direct responses to ionospheric currents, particles precipitation, variations of magnetospheric electric fields and to the atmospheric waves, propagated from below, and generated in the thermosphere by the same sources.

Fig. 2 shows latitudinal variations of the intensity of ion temperature perturbations for different IS radars. In Fig. 2 the data for each altitude is plotted for two month intervals, separated by six months. One can see from Fig.2, that the character of latitudinal variations is different above and below 180 – 200 km. In the low thermosphere in January and in July the intensity of ion temperature perturbations has a maximum near the winter pole and a minimum – near the summer pole. At high altitudes the latitudinal distribution of the intensity is more symmetric. It has maxima near the poles and at the equator, and this distribution persists throughout the year.

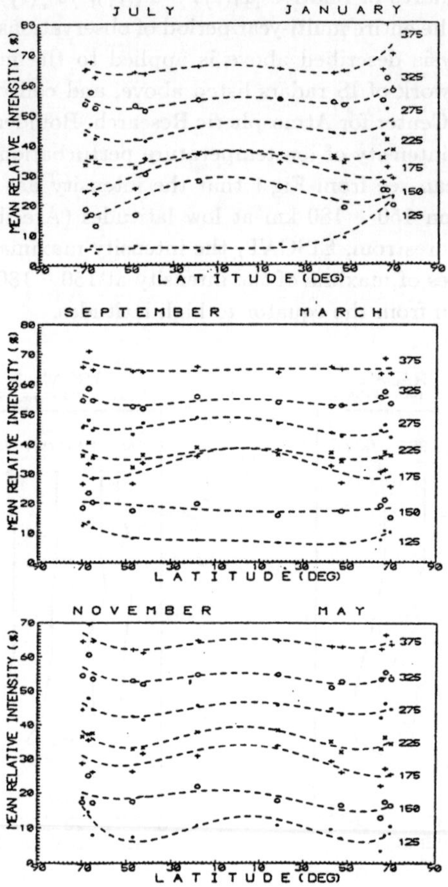


Fig.2. Dependence of the intensity of short-period relative variations of ion temperature on latitude at different altitudes (shown with numbers) in different months. To separate lines 10 % is added to each set.

Fig. 2 shows that the the influence of the electric fields plus waves generated in the auroral zone could dominate over the IGWs coming from lower atmosphere and could make the increased ion temperature perturbations at high latitudes in all seasons. More complicated is to explain the maximum of the intensity of the ion temperature perturbations at low latitudes in the thermosphere (see Fig.2). One possible reason for it could be the generation of IGWs by atmospheric tides, having largest amplitudes near equator.

### *2.2 Seasonal variations of IGW intensity*

Systematic study of multi-year averaged climatological characteristics, seasonal and inter-annual variations of the parameters of wind perturbations (attributed to IGW) in the upper part of the middle atmosphere were made with MF radar at Saskatoon, Canada. We used three different band-pass filters to separate the entire IGW intensity into parts that can be attributed to the waves having periods 0.2 -2.5; 1.5 - 6 and 2 - 10 hr respectively for filters called WH, HD and RS. These filters may be useful tools to investigate the relative variations, climatology and inter-annual variations of wind perturbations intensity in each different part of the IGW frequency spectrum.

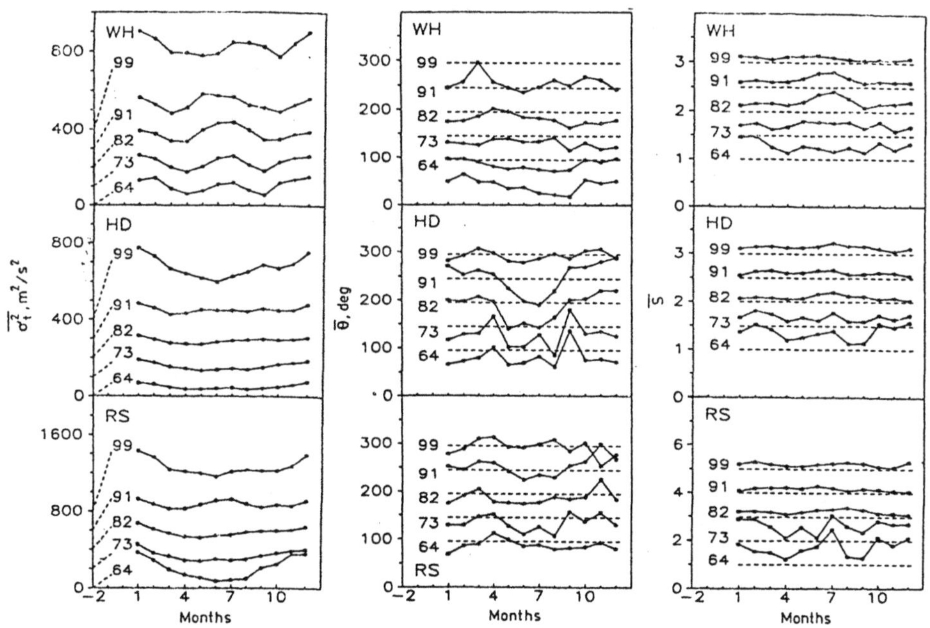


Fig. 3. Plots from left to right show multi-year averaged values of IGW intensity $\overline{\sigma_t^2}$, oval tilt angle $\overline{\theta}$, and oval axes ratio $\overline{S}$ for WH (top), HD (middle) and RS (bottom) data. Dashed curves show values $\overline{\theta} = 90°$, $\overline{S} = 1$ and and also places on the $\overline{\sigma}^2$ axis, which correspond to $\overline{\sigma_t^2} = 0$ at each altitude (numbers near curves).

When one makes an averaging of wind perturbations over a long period (e.g. one month), polarization ellipses of many single IGWs passing during this period overlap. If the wave process is isotropic, i.e. numbers and energies of IGWs going in any direction are the same, one expects the shape of curve representing variance $\sigma^2$ of wind component in the direction with azimuth $\varphi - \sigma^2(\varphi)$ (see (2) ) to be circular. So, oval shapes of $\sigma^2(\varphi)$ indicate that either the number or intensity of IGWs going in some predominant direction is larger than that for other directions. The tilt angle $\theta$ above gives the azimuth of this predominant direction, and the ratio of the oval axes $\sigma^2_{max}/\sigma^2_{min}$ characterizes the anisotropy of IGW intensity (see (Gavrilov et al., 1995)).

Monthly averaged WH, HD and RS data described above are calculated for each month of the entire cycle of MF radar measurements during December 1979 to May 1993. To obtain climatological characteristics, values for each month of the year are averaged over all years of observations.

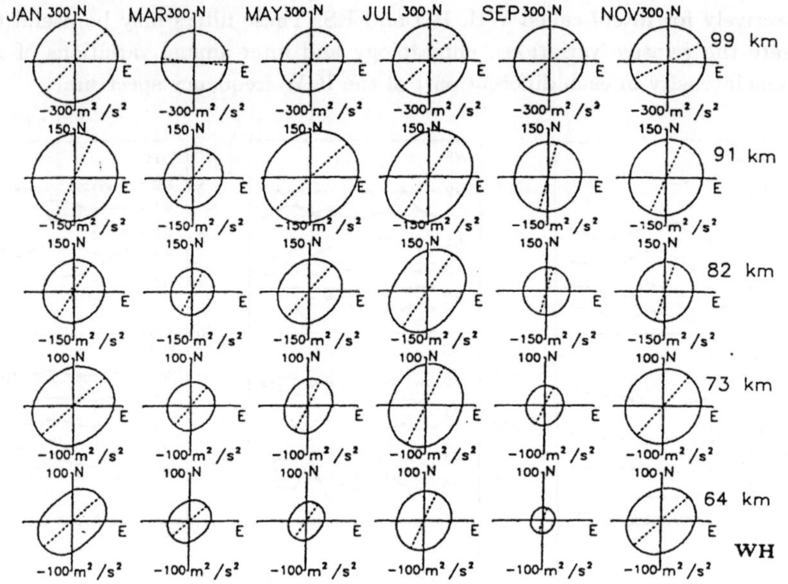


Fig. 4. Multi-year averaged distributions of $\overline{\sigma^2(\varphi)}$ (perturbations ovals) for WH data in different months at different altitudes.

Seasonal variations of $\sigma_t^2$ in Fig.3 are different for different perturbation frequencies. For WH data at most altitudes $\overline{\sigma_t^2}$ has two maxima - in winter and in summer (see Fig.3). Sometimes summer maximum is comparable or larger than the winter one. For HD and RS data in Fig.3 summer maximum of $\overline{\sigma_t^2}$ is less common and always is smaller than the winter maximum.

Fig.4 and Fig.5 show multi-year averaged polarization ovals for different months and altitudes for WH and RS data. One can see near zonal ovals orientation in most cases for RS data and closer to meridional orientation for WH data.

The dimensions of the ovals in Fig.4 show very clearly the semiannual variation of the intensity of WH data. For RS data in Fig.5 the semiannual variation of intensity is smaller, but there is a semiannual variation of tilt angle $\theta$ at altitudes 70 - 80 km as seen in fig.3. Ovals for HD data have shapes and orientations closer to RS data.

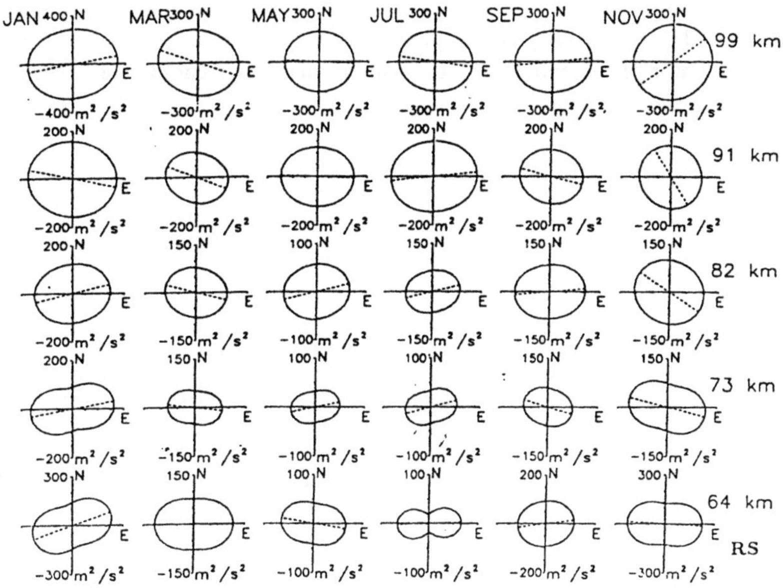


Fig. 5. The same as Fig. 4, but for RS data.

### *2.3 A trend of IGW characteristics versus solar activity*

The cycle of wind measurements at Saskatoon from 1979 to 1993 covers the main parts of 21 and 22 cycles of solar activity. A wide choice of solar and geomagnetic indices is available (Solar-Geophysical data, 1979 - 1993). Solar radio flux at wavelength 10.7 cm, $F_{10.7}$, and geomagnetic $A_p$ indices have been chosen for comparison with the variations of IGW parameters.

The variation of monthly differences of IGW intensity $\sigma_t^2 - \overline{\sigma_t^2}$ versus monthly mean values of $F_{10.7}$ and taken from Solar-Geophysical data (1979 - 1993) are shown in Fig. 6. The regression of IGW intensity upon $F_{10.7}$ and $A_p$ can be approximated with straight line at high level of statistical confidence.

Considering Fig. 6 one can find good evidence of negative correlation of intensity of RS and HD data on solar $F_{10.7}$. For WH data the probability of such correlation is smaller. Also good enough evidence exists for the dependence of polarization oval axes ratios and IGW predominant azimuth on $F_{10.7}$. Similar regression exists of IGW parameters on geomagnetic $A_p$ index.

One of the reasons for the negative correlation of IGW intensity with solar activity in the upper atmosphere could be variations of neutral and electron density during solar cycle. Modeling (Gavrilov et al., 1994) shows that increase in neutral density and ion drag during high solar activity leads to decreases in the amplitudes of IGW propagated to the upper atmosphere from troposphere. But this reason can explain the dependence of IGW intensity on solar activity only above 90 - 100 km.

Below 90 km there is no systematic data about the dependence of atmospheric density, temperature and electron density upon solar activity. The negative correlation of IGW intensity with solar activity obtained in section 5 at altitudes as low as 60 - 70 km, may be an indicator of a response of wave sources or wind structure of lower layers of the atmosphere to the variations of solar activity.

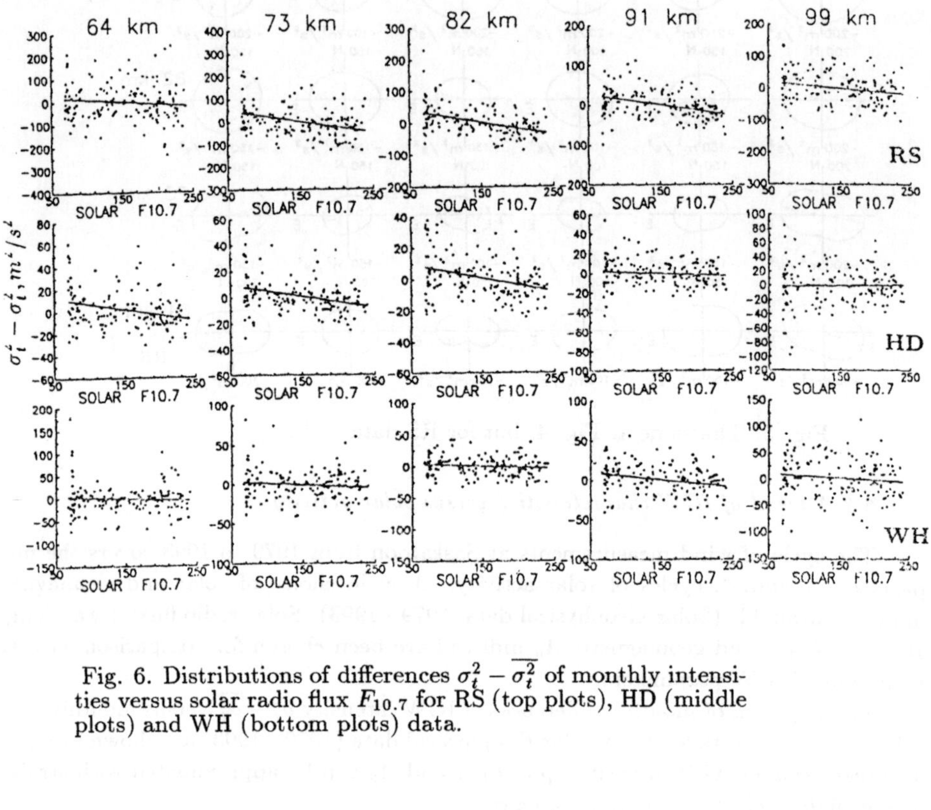


Fig. 6. Distributions of differences $\sigma_i^2 - \overline{\sigma_i^2}$ of monthly intensities versus solar radio flux $F_{10.7}$ for RS (top plots), HD (middle plots) and WH (bottom plots) data.

### *2.4 Statistical characteristics of IGW harmonics*

A statistical procedure is developed to obtain parameters of IGWs using measurements of Doppler velocity in the middle atmosphere with the MU radar in five directions of the antenna beam. The method is applied to find sets of IGW parameters for the winter and summer seasons of 1987 – 1988 for 6 km thick layers centered at 70, 75 and 80 km at Shigaraki, Japan (34.9°N, 139.4°E).

The statistical method includes spectral analysis of the data for the five directions of the antenna beam, then calculation of the axes and the tilt angle of a velocity hodograph ellipse, amplitudes and momentum flux, vertical and horizontal wave numbers, azimuths of wave vector and momentum flux for each spectral harmonic.

Developed procedure of calculation and selection of IGW harmonics is applied to measurements of wind velocity in the altitude region 60 – 90 km, obtained with the MU radar at Shigaraki (34.9° N, 139.4° E) during the winter (December – February) and summer (June – August) of 1987 – 1988. The numbers of selected IGW harmonics having periods from 5 min up to 6 hr vary between 360 and 2000 depending on the season and altitude.

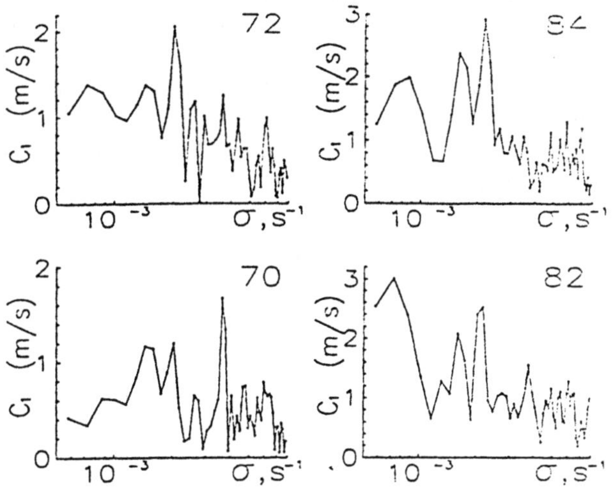


Fig. 7. Amplitude spectra of wind variations observed with the MU radar in the vertical direction on 3 February 1987.

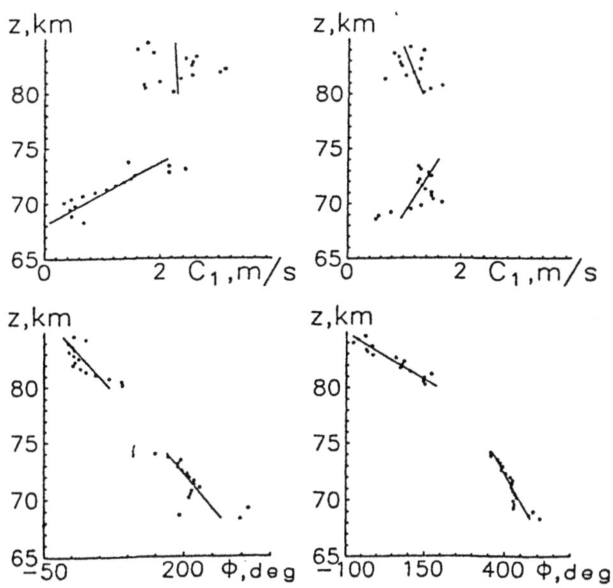


Fig. 8. Altitude variations of amplitudes (top plots) and phases (bottom plots) of the main spectral harmonics shown in Fig. 7.

Fig. 7 and Fig. 8 shows examples of the variations of spectra, amplitudes and phases of wave harmonics versus altitude. One can see that the IGW spectra on Fig 7, 8 are very complicated. Criteria are formulated for the selection of the spectral harmonics attributed to IGWs: a) local spectral maximum of amplitude or momentum flux; b) straight line fit of the harmonic phase versus altitude; c) equal or opposite azimuths of the momentum flux and wave vector; d) requirement of vertical wavenumber to be real. This gives a limitation on the value of horizontal phase velocity; e) requirement that the horizontal length of the IGW harmonic be much larger than the horizontal distances between the regions of atmosphere observed by different antenna beams (about 30 km at altitudes near 80 km).

Fig. 9 represents wave amplitudes and momentum fluxes averaged over frequency intervals for IGWs with energy flux directed upwards and downwards in the winter season at different altitudes.

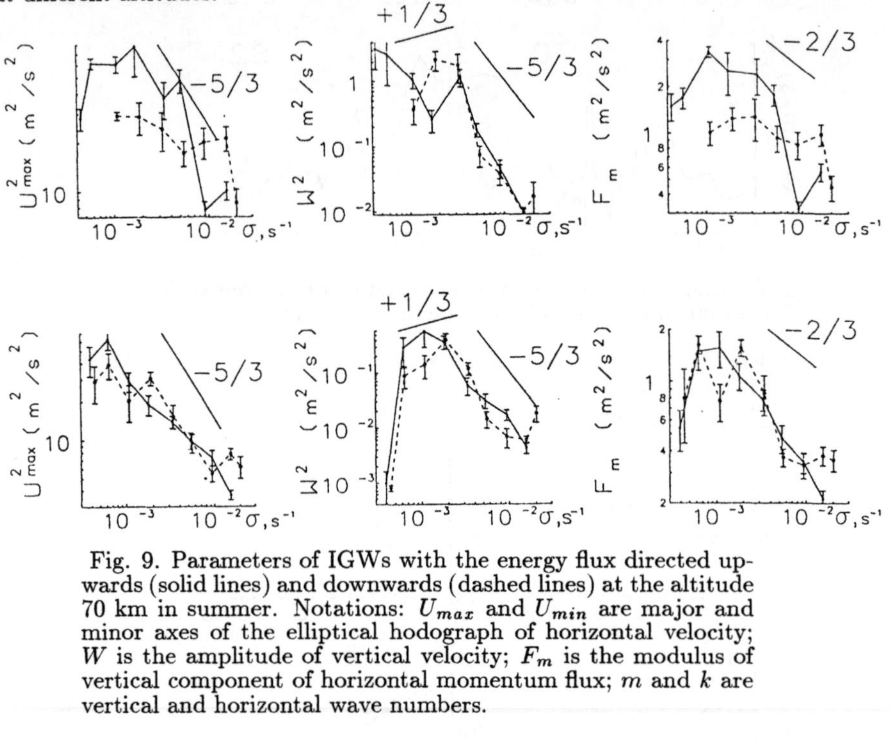


Fig. 9. Parameters of IGWs with the energy flux directed upwards (solid lines) and downwards (dashed lines) at the altitude 70 km in summer. Notations: $U_{max}$ and $U_{min}$ are major and minor axes of the elliptical hodograph of horizontal velocity; $W$ is the amplitude of vertical velocity; $F_m$ is the modulus of vertical component of horizontal momentum flux; $m$ and $k$ are vertical and horizontal wave numbers.

One can see in Fig. 9 that usually amplitudes and modulus of momentum flux are larger for upcoming IGW, especially at lower frequencies. This confirms the results of (*Fukao et al.*, 1985; *Lintelman and Gardner*, 1994) showing that IGWs propagated from below should be important for the dynamics of the middle and upper atmosphere.

A comparison of the parameters of IGWs propagated upwards in winter and summer shows that in summer the IGWs have larger vertical and horizontal wavelengths (smaller $m$ and $k$) and comparable or larger amplitudes and momentum fluxes than in winter.

## 3. Hydrodynamical IGW sources

Sources of internal waves in a flow might be subdivided into *surface sources* (caused by the interactions of the flow with rigid walls), and the *internal sources*, generating waves inside the flow (Lighthill, 1978). The internal wave sources might be in turn subdivided into two categories. The sources of ***first kind*** are described by linearised wave theory, and create only the amplification of primary waves, existing in the flow. Here are, for example, different kinds of instabilities in the atmosphere. Wave sources of ***second kind*** lead to the origin of new wave harmonics nonexisting in the flow before. Mainly, they are nonlinear interactions of different components of the flow.

The atmosphere has a broad spectrum of wave motions from acoustic waves with frequencies of tens kilohertz up to planetary waves. Big varity of the surface wave sources (orography), and internal sources of the first kind (instabilities) does not allow to explain observed variety of IGW manifestations in the atmosphere. For example, mountains and jet streams exist not everywhere and not everytime while wave motions are observed in the middle and upper atmosphere practically permanently (Vincent, 1984 Gavrilov *et al.*, 1995, 1996). To settle the contradiction one should take into account another kind of wave sources: dynamical motions of all scales. The importance of those nonlinear sources is determined by their permanent operation because hydrodynamic motions are inherent feature of the atmosphere.

The generation of sound by small-scale turbulence was studied by Lighthill (1952, 1978). The generation of long-period inertia-gravity waves by synoptic scale motions is described by the theory of quasi-geostrophic ajustment that was developed at the dynamical meteorology starting from (Rossby, 1937; Obukhov, 1949; Kibel, 1955). According to the theory wave motions appear as result of permanent competition between the violation of equilibrium of pressure and velocity and the tendency of the atmosphere to establish the quasi-geostrophic balance of those fields (Pedlosky, 1982; Gill, 1982; Gavrilov, 1988a; Fritts and Luo, 1992; Luo and Fritts, 1993; Medvedev and Gavrilov, 1995). The emission of mesoscale IGWs is provided by mesoscale turbulent motions (Stein, 1967, 1968; Townsend, 1965; Drobyazko and Krasilnikov, 1985; Gavrilov, 1987b, 1992).

### *3.1. Mathematical equations*

Turbulent motions of different scales permanently occur in the atmosphere. Fluctuations having time scales up to several minutes are attributed to microscale turbulence, from minutes to several hours form the mesoscale one, and stochastic regime of synoptic scale irregular motions might be regarded as macro-scale turbulence (Monin and Yaglom, 1965). Microscale turbulent motions are near isotropic and don't response to the gravity and Coriolis forces acting in the Earth atmosphere. Gravity is essential for the mesoscale turbulence, also eddies with time scales from several hours are influenced by the Coriolis force, which is very important for macroscale turbulent vortexes.

On the other hand the atmosphere has a broad spectrum of wave motions from acoustic waves with frequencies of tens kilohertz up to planetary waves having periods of several days. Because atmospheric hydrodynamic equations are nonlinear, turbulent and wave motions are not independent: the energy exchange exists that leads in particular to the generation of waves by turbulence. To describe the latter process let's consider a conventional set of hydrodynamic equations for a plane rotating atmosphere (see Gill, 1982):

$$\partial\rho/\partial t + \nabla(\rho \mathrm{v}) = 0;$$
$$\rho d\mathrm{v}/dt + 2\rho f \mathrm{e}_z \times \mathrm{v} + \nabla p - \rho g \mathrm{e}_z = 0; \qquad (2)$$
$$dp/dt - c_s^2 d\rho/dt = 0,$$

where $p$, $\rho$, v are pressure, density, and velocity; $c_s$ is the speed of sound; g is the acceleration due to gravity; $f$ is the Coriolis parameter; $\mathrm{e}_z$ is a unit vertical vector; $d/dt = \partial/\partial t + \mathrm{v} \cdot \nabla$.

Let's consider motionless atmosphere with background values of $p_0$, $\rho_0$ varying in altitude $z$ according to the barometric law, and $\mathrm{v}_0 = 0$. Dynamic fluctuations $\mathrm{v}_d$, $\rho_d = \rho - \rho_0$, $p_d = p - p_0$ are usually small, so that $|\rho_d/\rho_0| \ll 1$, $|p_d/p_0| \ll 1$ and $|\mathrm{v}_d|/c_{s0} \ll 1$. Let Mach's number $M$ be a measure of largest value of $|\mathrm{v}_d|/c_{s0}$. When $M$ is small, one should search for the solution to the equations (1) as series expansions with respect to $M$:

$$p = p_0 + Mp_t + M^2 p' + ...; \quad \rho = \rho_0 + M\rho_t + \rho' + ...;$$
$$\mathrm{v} = M\mathrm{v}_t + M^2\mathrm{v}'; \; c_s^2 = c_{s0}^2 + Mc_{st}^2 + M^2 c_s'^2 + ... \qquad (3)$$

Substituting (3) into (2) and equating coefficients at the same powers of $M$ to zero one may have the following set of equations for the components of hydrodynamic fields of the order of $M^2$:

$$d_0\rho'/dt + \nabla(\rho_0 \mathrm{v}') = \rho_0 q;$$
$$\rho_0 d_0 \mathrm{v}'/dt + 2\rho_0 f \mathrm{e}_z \times \mathrm{v}' + \nabla p' - \rho' g \mathrm{e}_z = \rho_0 \mathrm{r}; \qquad (4)$$
$$\partial p'/\partial t + \mathrm{v}' \cdot \nabla p_0 - c_{s0}^2(\partial\rho'/\partial t + \mathrm{v}' \cdot \nabla\rho_0) = \rho_0 h,$$

Right hand parts of the equations (4) include the hydrodynamic sources of mass $q$, momentum $\vec{r}$ and energy $h$ that describe the generation of the second order motions by the motions of the first order with respect of $M$:

$$q = -\rho_0^{-1}\nabla(\rho_t \mathrm{v}_t);$$
$$\mathrm{r} = -\rho_0^{-1}(\rho_t d_0 \mathrm{v}_t/d_0 t + \rho_0 \mathrm{v}_t \cdot \nabla \mathrm{v}_t); \qquad (5)$$
$$h = -\rho_0^{-1}(\mathrm{v}_t \cdot \nabla p_t - c_{s0}^2 \mathrm{v}_t \cdot \nabla\rho_t) + c_{st}^2(d_0\rho_t/dt + \mathrm{v}_t \cdot \nabla\rho_0).$$

At the periods of acoustic-gravity waves main contribution into the first order components make irregular turbulent motions, which may be treated as three-dimensional and solenoidal at small scales, and become two-dimensional and quasi-geostrophic at medium and large scales (Drobyazko and Krasilnikov, 1985; Gavrilov, 1988a,b; Medvedev and Gavrilov, 1995).

To study IGW generation by hydrodynamic sources we use here an approach based on the consideration of the equation of wave energy. It is known that the specific density of wave energy $E_w$ consists of tree parts: kinetic $E_{wk}$, elastic $E_{we}$ and thermobaric $E_{wt}$ that are given by the following expressions (see Dikiy, 1969):

$$E_w = E_{wk} + E_{we} + E_{wt}; E_{wk} = \rho_0 < |\mathrm{v}'|^2 >; \tag{6}$$
$$E_{we} =< p'^2 > /2\gamma p_0; E_{wt} =< (p' - c_{s0}^2\rho')^2 > /2\gamma b p_0,$$

where the sign <> denotes averaging over a period of wave harmonic; $\gamma = c_p/c_v$; $b = (\gamma - 1) + g^{-1}\partial c_{s0}^2/\partial z$; $c_p$ and $c_v$ are the specific heat capacities at constant pressure and volume respectively. Multiplying equations (3) by $\rho'$, $\mathrm{v}'$ and $p'$ respectively, one may derive the equation for $E_w$ having the form of

$$d_0 E_w/dt = -\partial F_{w\alpha}/\partial x_\alpha + \rho_0 Q; \tag{7}$$

where $F_{w\alpha}$ is the total wave energy flux (see Gavrilov, 1990); $Q$ is the specific rate of wave energy production by hydrodynamic sources of momentum, mass and heat:

$$Q = \rho_0^{-1}\langle \rho_0 \mathrm{r} \cdot \mathrm{v}' + qp' + h(p'(b+1)/(c_{s0}^2 b) - \rho'/b)\rangle \tag{8}$$

The formula (8) for the specific rate of the wave energy production $Q$ may be simplified using IGW polarization relations, which are most simple for the short low-frequency IGW harmonics having $m^2 \gg 1/4H^2$ and $\omega^2 \ll N^2$, where $m$ and $\omega$ are vertical wave number and intrinsic frequency respectively; $N$ is the Brunt-Vaisala frequency; $H$ is atmospheric scale heights. If one directs the axis $x$ parallel to the horizontal wave vector of a plane monochromatic IGW harmonic, he may obtain from the formulae by Gavrilov *et al.* (1996) the following relations for mentioned IGWs:

$$\frac{v'}{u'} \simeq i\frac{f}{\omega}; \frac{w'}{u'} \simeq -\frac{\omega\delta}{N}; \frac{p'}{p_0 u'} \simeq \frac{\gamma\omega\delta^2}{kc_{s0}^2}; \frac{\rho'}{\rho_0 u'} \simeq i\frac{N\delta}{g}; \tag{9}$$

where $\delta = (1 - f^2/\omega^2)^{1/2}$; $u$, $v$, $w$ are the velocity components along horizontal axes $x$, $y$ and vertical axis $z$ respectively. The expression (8) for $Q$ shows that if $u'$ depends on time in monochromatic IGW harmonic as

$$u' = U\cos\sigma t, \tag{10}$$

then $Q$ is not equal to zero only when frequency spectra of the sources r, $q$ and $h$ also contain harmonics with observable frequency $\sigma$:

$$r_\alpha = r_{\alpha m}\cos(\sigma t + \varphi_\alpha); \quad \alpha = x, y, z;$$

$$q = q_m\cos(\sigma t + \varphi_q); \quad h = h_m\cos(\sigma t + \varphi_h), \tag{11}$$

where $r_{\alpha m}$, $q_m$, $h_m$ and $\varphi_\alpha$, $\varphi_q$, $\varphi_h$ are the phase shifts of respective values relatively to the variations of $u'$. Putting (9) – (11) into (8) we have

$$Q = sU; \quad s = 1/2\big(r_{xm}\cos\varphi_x + r_{ym}(f\sin\varphi_y)/\omega - \tag{12}$$

$$r_{zm}\frac{\omega\delta}{N}\cos\varphi_z - h_m\frac{N\delta}{bg}\sin\varphi_h + q_m\frac{\gamma\omega\delta^2}{kc_{s0}^2}\cos\varphi_q\big).$$

One can see that the rate of wave energy production is governed by the value of parameter $s$, which has the dimension of force acting upon unit of atmospheric gas mass. According to evaluations presented by Drobyazko and Krasilnikov (1985) and Gavrilov (1987), the main contribution to the IGW generation comes from the first term in square brackets of (12), which is connected with quasi-two-dimensional mesoscale motions, so $s \sim \overline{r}_{xm} \sim u_s \partial u_s / \partial s$, where $u_s$ and $\partial u_s / \partial s$ are the scales of horizontal velocity and its horizontal gradient, respectively. In accordance with the tables of characteristics for the tropospheric meteorological values (Belov, 1975), we put $u_s \sim 10^1$ m s$^{-1}$; $\partial u_s / \partial s \sim 10^{-5} - 10^{-4}$ s$^{-1}$, and therefore $\overline{r}_m \sim 10^{-5} - 10^{-3}$ m s$^{-1}$.

Medvedev and Gavrilov (1995) have made a numerical modeling of meteorological fields, and have calculated the characteristic values of $r_x$ using the scheme of short-term weather forecast. They have calculated the space distributions of $r_x$ for the real meteorological situations during the period from 15.04.1981 to 06.08.1981. The maximum value of $r_{xm}$ has been found for each day. the main range of values $r_{xm}$ is $(4 - 10) \cdot 10^{-3}$ m s$^{-2}$, but sometimes the values of $r_{xm}$ larger than $2 \cdot 10^{-2}$ m s$^{-2}$ were obtained. It must be noted that the maximum values of $r_{xm}$ are realized in limited regions of increased wave generation, but over the main part of the considered territory the values of $r_x$ are not more than above mentioned estimates of $\overline{r}_{xm}$ (see also Gavrilov, 1992).

Analysis of vertical profiles of $r_x$ calculated by Medvedev and Gavrilov (1995) shows that maxima of $r_{xm}$ are mainly concentrated at altitudes 9 – 12 km near the tropopause. Calculations of the meteorological fields evolution using real initial data (see Medvedev and Gavrilov, 1995) show that the regions of high values of $r_x$ have much smaller dimensions than macro-scale motions creating the wave generation. These regions are situated in the central parts of cyclones, anti-cyclones, and jet streams. More often the wave sources are associated with deep cyclones involving the entire troposphere, jet stream curves, and closeness of two regions with high and low pressure, which may produce "saddle points".

### *3.2. Results of calculations*

The essential features of the vertical distributions of the strength of tropospheric wave sources (see above) may be approximated by the following formula:

$$s(z) \;=\; s_m \exp(-(z - z_m)^2/d^2), \tag{13}$$

where $z_m$ and $d$ are the height of maximum and the halfwidth of the layer in which the generation of IGWs takes place.

In realistic atmosphere background values have complicated dependence upon altitude, which requires a numerical solution to (7). For our calculations we use the vertical profiles of the mean temperature, wind, and the rates of IGW dissipation due to radiative heat exchange and ion drag according to Gavrilov and Yudin, (1987). The coefficient of turbulent viscosity is represented by the following sum:

$$K \;=\; K_0 \;+\; K_i \;+\; K_n, \tag{14}$$

where $K_0 \sim 10^0$ m$^2$ s$^{-1}$ is the background value, $K_i$ is the contribution from the shear and dynamical instability of the IGW profiles of temperature and wind. The value of $K_i$ is equal to zero up to the level of IGW breaking, and it is calculated using semi-empirical model by Gavrilov and Yudin (1992) above the breaking altitude. The term $K_n$ in (14) represents the contribution from the alternative mechanism of the turbulization of the atmosphere, which is due to nonlinear destruction of the primary IGW into secondary wave harmonics (Rozenfeld (1983)).

Fig. 10 shows the vertical profiles of amplitudes and $K$ produced by an IGW harmonic propagated from tropospheric hydrodynamic source in an isothermal windless atmosphere for different mechanisms of the turbulizing of the atmosphere by IGWs. Dashed line in Fig. 10 shows that without dissipation the amplitude grows exponentially with altitude. The nonlinear destruction of the IGW into secondary harmonics leads to the wave energy dissipation and formation of the amplitude maximum near $z \sim 100$ km (see thin lines in Fig. 10). The parameter of IGW supersaturation $a = U/|c - u_0|$ is about $a \sim 2$ here, which is much larger than the values of $a \sim 1$ usually appeared in the atmosphere

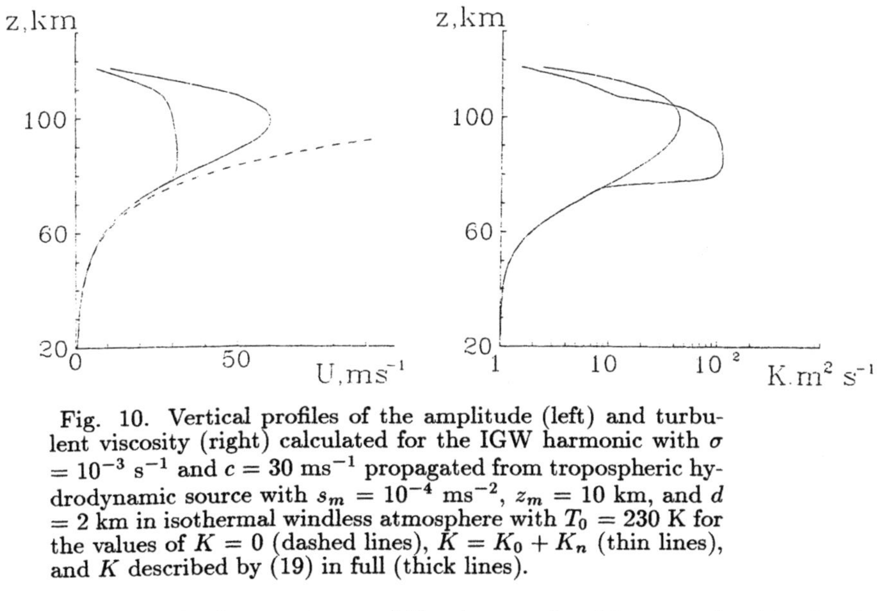


Fig. 10. Vertical profiles of the amplitude (left) and turbulent viscosity (right) calculated for the IGW harmonic with $\sigma = 10^{-3}$ s$^{-1}$ and $c = 30$ ms$^{-1}$ propagated from tropospheric hydrodynamic source with $s_m = 10^{-4}$ ms$^{-2}$, $z_m = 10$ km, and $d = 2$ km in isothermal windless atmosphere with $T_0 = 230$ K for the values of $K = 0$ (dashed lines), $K = K_0 + K_n$ (thin lines), and $K$ described by (19) in full (thick lines).

The reason for the decrease in $a$ could be the wave breaking near the altitude where $a \sim 1$ become. The IGW instability gives rise to the generation of turbulence and to the sharp increase in the term $K_i$ in (14) (compare thick and thin lines in Fig. 10). Above the breaking level the IGW amplitude in Fig. 10 become practically constant, and then the amplitude decreases at high altitudes due to increase in the molecular viscosity and heat conductivity, and ion drag.

The presence of the mean wind makes differences for IGWs propagated upstream and downstream. Fig. 11 depicts the parameters of such IGW harmonics propagated in the isothermal atmosphere with constant values of $u_0$. One may see that at low altitudes amplitude, wave energy flux, and other parameters in Fig. 11 are larger for the IGW propagated in the direction of mean flow. At high altitudes the IGW going opposite to the mean flow becomes more intensive. To understand such behaviour one should take into account the dual action of the mean wind on the IGWs: the wave-mean flow interaction, and the change of the vertical wave number.

It is widely accepted that IGWs produce the acceleration of the mean flow $a_w = -\rho_0^{-1}\partial(\rho_0\overline{u'w'})/\partial z$ (see Gavrilov, 1990), which becomes nonzero at any change in $\rho_0\overline{u'w'}$. In the region of action of the tropospheric wave source near $z = z_m \sim 10$ km the modulus $|\rho_0\overline{u'w'}|$ increases that causes the appearance of the wave accelerations. The consideration of the directions of the accelerations shows that the interaction with the mean flow makes an increase in the energy of IGWs propagated in the direction of the mean flow, and the decrease in energy of upstream IGWs. Thus, at the same strength $s_m$ the wave source generates more effectively downstream waves than upstream IGWs. This difference is expressed mathematically by the appearance of the factor $\sigma/\omega$ in the right hand part of the equation (14), and explain the dominance of downstream IGWs at low altitudes near the wave source in Fig. 11.

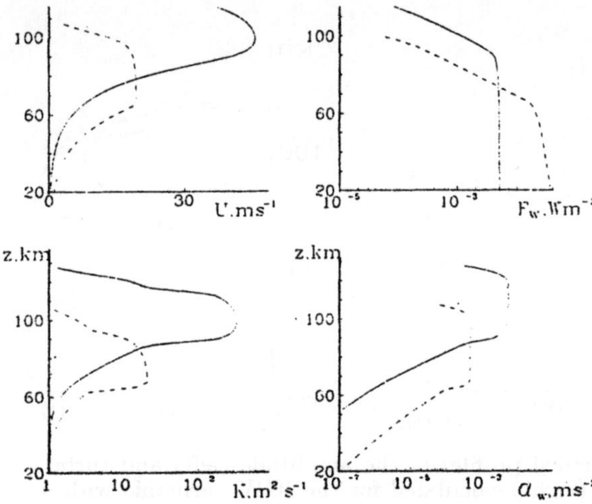


Fig. 11. Vertical profiles of IGW parameters calculated for the same model as Fig. 1, but including constant mean wind $u_0 = 10$ ms$^{-1}$ for the waves propagated in the direction of the mean flow (dashed lines), and opposite to it (solid lines).

On the other hand, above the region of the wave source action the influence of $u_0$ leads to the increase in $\omega$ for the upstream IGW, and to the decrease in $\omega$ for the downstream wave. According to the IGW dispersion equation larger $\omega$ corresponds to smaller value of the vertical wave number $m$. Thus, the downstream IGW has larger value of $m$, which corresponds to larger dissipation of wave energy. Also, the downstream IGW breaks and become saturated at smaller altitude than the upstream wave. The larger dissipation of the downstream IGW may explain the dominance of the upstream wave in the upper part of Fig. 11. In the region of strong dissipation the wave acceleration of the mean flow has the same direction as the wave horizontal phase speed. So, the dominance of the upstream wave in Fig. 11 means the dominance of the drag of the mean flow by IGWs.

## 4. Conclusion

In this paper we consider the mechanism of IGW generation by mesoscale turbulence in the troposphere. The equations are derived that describe the generation of waves by hydrodynamic sources of momentum, heat and mass. Calculations are made of amplitudes, wave energy fluxes, turbulent viscosities, heating rates, and accelerations of the mean flow caused by IGWs generated in the troposphere. A comparison of different mechanisms of turbulizing of the atmosphere by IGWs shows that after the IGW breaking and saturation the main source of turbulence is the convective and dynamical instability of the wave. Nonlinear destruction of a primary IGW into a spectrum of secondary waves may be important for nonsaturated stable IGWs.

The mean wind increases the effectiveness of generation as well as larger dissipation of IGWs propagated in the direction of the wind. Competition of both effects may lead to the formation of eastward wave accelerations in summer, and westward accelerations in winter near $z \sim 100$ km.

*Acknowledgment.* This work was partly supported by the Russian Basic Research Foundation and the International Association for Cooperation with New Independent States (INTAS).

## References

Belov, P. N., (1975) Numerical methods of weather forecast. Hydrometeoizdat press, Leningrad

Drobyazko, I. N., and V. N. Krasilnikov (1985) Acoustic-gravity wave generation by atmospheric turbulence. Radiophysics, Izvestia of VUZ of USSR 28:1357–1365

Fritts, D. C., and Z. Luo (1992) Gravity wave excitation by geostrophic adjustment of the jet stream, Part 1: Two-dimensional forcing. J. Atmos. Sci. 49:681–697

Fukao, S., Y. Maekawa, T. Sato, and S. Kato (1985) Fine structure in mesospheric wind fluctuations observed by the Arecibo UHF Doppler radar. *J. Geophys. Res. 90*:7547–7556

Gavrilov N. M. (1982) Methodical aspects of the identification of wave variations of the nocturnal sky airglows. In *Problems of Atmospheric Physics*, Leningrad University Press, Leningrad 17:31–66

Gavrilov N. M. (1984) Methodical aspects of studying internal gravity waves using the meteor radar method. In *Meteor Studies*, Geophys. committee of USSR Acad. Sci. press, Moscow 9:106–123

Gavrilov N. M. (1987a) Study of internal gravity waves in the meteor zone. Handbook for MAP 25:153–166

Gavrilov, N. M. (1987b) On the nonlinear generation of wave motions in the atmosphere by quasi-geostrophic and eddy motions. In *Meteor studies*, Geophys. committee of USSR Acad. Sci. press, Moscow 13:49–71

Gavrilov N. M. (1992) Internal gravity waves in the mesopause region: Hydrodynamical sources and climatological patterns, Adv. Space Res. 12:(10)113–(10)121

Gavrilov N. M. (1995) Distributions of the intensity of ion temperature perturbations in the thermosphere. J. Geophys. Res. A100, Paper No. 94.0400, in press

Gavrilov N. M., S. Fukao, T. Nakamura, T. Tsuda, M. D. Yamanaka, and M. Yamamoto (1996) Statistical analysis of gravity waves observed with the MU radar in the middle atmosphere: 1. Method and general characteristics. J. Geophys. Res. 101:in press

Gavrilov N. M., B.V. Kalchenko, B.L. Kashcheev and G.M. Shved (1981) Detection of the relationship between the intensity of internal gravity waves in the upper atmosphere and the tidal phase. Atmos. and Oceanic Phys., Izvestia USSR Acad. Sci. 17:680–689

Gavrilov, N. M. (1988a) Wave generation by vortex and quasi-geostrophic motions in the atmosphere. In *Investigations of dynam. processes in the upper atmos.*, Hydrometeoizdat press, Moscow 69–74

Gavrilov, N. M. (1988b) On the generation of internal gravity waves in the atmosphere by mesoscale turbulence. In *Investigations of dynam. processes in the upper atmos.*, Hydrometeoizdat press, Moscow 74–77

Gavrilov N. M., A. H. Manson, and C. E. Meek (1995) Climatological monthly characteristics of middle atmosphere gravity waves (10 min – 10 hr) during 1979 – 1993 at Saskatoon. Ann. Geophys. 13:285–295

Gavrilov N. M., A.D. Richmond, F. Bertin, and M. Lafeuille (1994) Investigation of seasonal and interannual variations of internal gravity wave intensity in the thermosphere over Saint Santin. J. Geophys. Res. 99:6297–6306

Gavrilov N. M. and G.M. Shved (1982) Study of internal gravity waves in the lower thermosphere from isophots of the nocturnal sky airglow. Atmos. and Oceanic Phys., Izvestia USSR Acad. Sci. 18:8–17

Gavrilov N. M., A.D.Richmond, J.S.H.M.Van Baelen, T.Tsuda, S.Kato, S.Fukao and M.Yamamoto (1991) Investigation of Internal Gravity Wave Motions in the Troposphere and Stratosphere with the MU-radar. In *Atmospheric Waves.* Preprint Volume of the Eighth Conference on Atmospheric and Oceanic Waves and Stability, October 14 - 18, 1991:308–311

Gavrilov N. M. and V.A. Yudin (1987) Numerical modelling of the IGW propagation from nonstationary tropospheric sources. itref Atmos. Oceanic Phys., Izvestia of USSR Acad. Sci. 23:241–248

Gavrilov N. M. and V.A.Yudin (1992) Model for coefficients of turbulence and effective Prandtl number produced by breaking gravity waves in the upper atmosphere. J. Geophys. Res. 97:7619 - 7624

Gill, A. (1982) Atmosphere-ocean dynamics. Academic press New York – Toronto

Kibel, I. A. (1955) On the adaptation of air motion to geostrophic one. Rept. of USSR Acad. Sci. 104:60–63

Lighthill, M. J. (1952) On sound generated aerodynamically, 1. General theory. Proc. Roy. Soc. London A211:564–587

Lighthill, M. J. (1978) Waves in fluids, Cambridge Univ. Press

Lintelman, S. A., and C. S. Gardner (1994) Observation and interpretation of spectra of atmospheric gravity wave perturbations with upward and downward phase progression. J. Geophys. Res., 99:16959–16971

Luo, Z., and D. C. Fritts (1993) Gravity-wave excitation by geostrophic adjustment of the jet stream, Part II: Three-dimensional forcing. J. Atmos. Sci. 50:104 – 115

Medvedev, A. S., and N. M. Gavrilov (1995) The nonlinear mechanism of gravity wave generation by meteorological motions in the atmosphere. J. Atmos. Terr. Phys.. 57:1221 -1231

Monin, A. S., and A. M. Yaglom (1965) Statistical hydromechanics. Nauka press, Moscow

Obukhov, A. M. (1949) On the problem of geostrophic wind. Geophys. Geograph., Izvestia of USSR Acad. Sci. 13:281–306

Pedlosky, J. (1982) Geophysical fluid dynamics. Springer-Verlag New York Berlin

Rossby, C. G. (1937) On the mutual adjustment of pressure and velocity distribution in certain simple current systems. J. Mar. Res. 1:15–28

Rozenfeld, S. H. (1983) On the damping of internal gravity waves in the atmosphere due to generation of secondary harmonics. Atmosph. Oceanic Phys., Izvestia of USSR Acad. Sci. 19:1011–1019

Semenov V.K. and N.M. Gavrilov (1985) On the characteristics of intradiurnal variations of the total ozon content. Atmos. and Oceanic Phys., Izvestia of USSR Acad. Sci. 21:791-795

Stein, R. S. (1967) Generation of acoustic and gravity waves by turbulence in an isothermal stratified atmosphere. Solar Phys. 2:285–432

Stein, R. S. (1968) Waves in the solar atmosphere: 1. The acoustic energy flux. Astrophys. J. 154:297–306

Townsend, A. A (1965) Excitation of internal waves by a turbulent boundary layer. J. Fluid. Mech. 22:241–252

# Observational Studies of Gravity Waves Associated with Convection

**Kaoru Sato**
Department of Geophysics
Faculty of Science
Kyoto University
Kyoto 606–01, Japan

**ABSTRACT:** Using twice-daily radiosonde data from Singapore taken over a 10 year timespan, equatorial short-period (1–3 day) waves are examined in terms of the quasi-biennial oscillation (QBO). Power spectra of temperature ($T$) and zonal wind ($u$) fluctuations have year-to-year variation synchronized with the QBO. In both QBO phases, the short-period waves have large variance, comparable to that of long-period (5–20 day) Kelvin waves in westerly shear. Synchronization with the QBO is also clear for cross spectra of $T$ and $u$ and particularly for the co-spectra. A new method is considered to estimate positive and negative momentum fluxes associated with oppositely propagating gravity and equatorial waves, separately. This method uses both quadrature and co-spectra of $T$ and $u$. Results of an analysis of the Singapore data suggest that positive and negative momentum fluxes associated with short-period waves are largely cancelled and the summation of absolute values of the momentum flux are 3–10 times larger than that associated with long-period Kelvin waves. The importance of gravity waves with periods shorter than 1 day is also discussed.

## 1 Introduction

Many theoretical and observational studies show that gravity waves play an important role in the global atmospheric circulation, redistributing momentum in the atmosphere through their generation, propagation and breaking. Although the majority of studies have emphasized the role of gravity waves in the global circulation of middle and high-latitude regions, some theoretical studies suggest the importance of gravity waves for the mechanism of the quasi-biennial oscillation (QBO) and semi-annual oscillation (SAO) of the mean zonal wind in the equatorial stratosphere and mesosphere. Gravity waves in the equatorial region are considered to be mainly excited by convection though various generation mechanisms. Long-term observational data need to be analyzed to elucidate the relation between gravity waves and such global-scale oscillations in circulation of the equatorial middle atmosphere.

## 2 Equatorial 1–3 day waves in long-term observational data

### 2.1 Characteristics of power and cross spectra

Interannual variation of short-period waves near the equator using twice-daily rawinsonde data taken at Singapore (7.6S, 112.7E) over a ten year period was discussed by Sato et al.(1994). Figure 1 shows the power and cross spectra of the temperature ($T$) and zonal wind ($u$) fluctuations. Dominant peaks are observed in the power spectra in the short-period range of 1–3 days during both phases of the QBO and around 10 day period (corresponding to long-period Kelvin waves) in the westerly shear phase of the QBO, indicating that the activity of short-period waves

NATO ASI Series, Vol. I 50
Gravity Wave Processes
Their Parameterization in Global Climate Models
Edited by Kevin Hamilton

is strong in both phases of the QBO. The variance of 1–3 day waves is comparable to that of long-period Kelvin waves.

It is shown from the thermodynamic equation that quadrature spectra of $T$ and $u$ are proportional to $-\overline{u'w'}$ (e.g. Maruyama, 1994). Thus, large negative values around the 10 day period in the quadrature spectra correspond to positive $\overline{u'w'}$ associated with long-period Kelvin waves. This bias is not clear in the short-period range. On the other hand, an interesting and mysterious feature is observed in co-spectra of $T$ and $u$ (Figs. 1c and 1d), that is clearly synchronized with the QBO across the whole range of frequencies. The variation cannot be explained by the previous theory of equatorial waves in a uniform background wind, which predicts that co-spectra should be almost zero.

Using a theoretical and numerical approach, Dunkerton (1995) analyzed the covariance of $T$ and $u$ components for two-dimensional gravity waves in a background wind having vertical shear. For slowly varying, conservative, incompressible waves, the covariance is proportional to the vertical shear and vertical flux of horizontal momentum. The sign of the covariance is determined by the vertical shear, independent of horizontal and vertical direction of gravity-wave propagation, which is consistent with the results of Sato et al.(1994). An extended theory to three-dimensional equatorial waves suggests that the relation of the covariance and vertical shear does not change significantly (Dunkerton 1996).

## 2.2 Estimates of momentum fluxes

Maruyama (1994) estimated momentum flux associated with equatorial 1–3 day waves from the covariance of $\partial T/\partial t$ (temporal variation of $T$) and $u$ fluctuations (corresponding to quadrature spectra of $T$ and $u$) observed at Singapore, assuming that they are due to short-period Kelvin waves. He showed that the momentum flux is large and comparable to that of long-period Kelvin waves. However, it is also possible that the short-period fluctuations are due to a mixture of eastward and westward propagating waves which have positive and negative momentum fluxes, respectively. Thus, the net momentum flux estimated by Maruyama may be residual after some cancellation between positive and negative momentum fluxes.

A new indirect method is proposed to estimate momentum fluxes based on Dunkerton's theory. The indirect estimates obtained from co-spectra correspond to the summation of absolute values of momentum flux associated with each wave, whereas direct estimates from quadrature spectra give the summation of momentum flux:

$$\begin{cases} \sum \overline{u'w'} = -\dfrac{g}{\overline{T}N^2}\displaystyle\int_{\omega_1}^{\omega_2} Q_{Tu}(\omega)\omega d\omega, & (1) \\ \sum |\overline{u'w'}| = \dfrac{2g}{N\overline{T}\overline{U}_z}\displaystyle\int_{\omega_1}^{\omega_2} C_{Tu}(\omega)\omega d\omega. & (2) \end{cases}$$

From these two equations, we can estimate positive and negative momentum fluxes, separately.

An analysis was made for the rawinsonde data at Singapore used by Sato et al. (1994). The direct estimate for Kelvin waves (5–20 day components) is 2–9 $\times 10^{-3}$m$^2$s$^{-2}$and accords with the indirect estimate within the estimation error. This result supports use of the indirect method.

Results of the indirect and direct estimates assuming 2-dimensional gravity waves for 1–3 day waves are shown in Fig. 2. It should be noted that the direct and indirect estimates of momentum flux from (1) and (2) are possible only when the mean wind is sufficiently weak so as not to cause severe Doppler effects. The vertical profiles shown in Fig. 2 were obtained for periods of weak mean wind (and large shear associated with the QBO), and the time of observation depends on altitude. Thus the displayed vertical profiles of momentum flux do not provide the wave drag from their convergence.

| | Mode | $c$ (m s$^{-1}$) | $\lambda_x$ (km) | $\lambda_z$ (km) |
|---|---|---|---|---|
| | n=–1 Kelvin waves | $> 6.7$ | $> 870$ | $> 2.0$ |
| (a) | n=1 eastward IGW | 7.7–19.6 | 1000–2500 | 2.0–4.2 |
| | n=3 eastward IGW | 9.4 | 1200 | 2.0 |
| | n=1 westward IGW | –25 to –31.8 | 3200–4100 | 5.7–6.7 |
| | n=–1 Kelvin waves | $> 15$ | $> 1900$ | $> 2.7$ |
| | n=1 eastward IGW | 15–44.1 | 1900–5700 | 3.4–6.6 |
| (b) | n=3 eastward IGW | 15–20.6 | 1900–2700 | 2.6–3.1 |
| | n=1 westward IGW | –7.2 to –61.8 | 930–8000 | 2.0–9.3 |
| | n=3 westward IGW | –8.6 to –23.8 | 1100–3100 | 2.0–3.6 |
| | n=5 westward IGW | –11.7 to –14.7 | 1500–1900 | 2.0–2.2 |

**Table 1**: Rough estimates of phase velocity $c$, zonal ($\lambda_x$) and vertical ($\lambda_z$) wavelengths of possible equatorial modes in the westerly shear (a) and easterly shear (b).

The direct estimate for 1–3 day waves in westerly shear is 0–4 $\times 10^{-3}$m$^2$s$^{-2}$which is almost half of the momentum flux associated with long-period Kelvin waves. This result is consistent with that of Maruyama (1994). The direct estimate for easterly shear is about zero. On the other hand, the indirect estimate for westerly (easterly) shear is 20–60 (7–30) $\times 10^{-3}$m$^2$s$^{-2}$based on the theory of two-dimensional gravity waves, which is larger than the direct estimate by one order of magnitude. For equatorial waves, the numeral factor in the numerator on the right hand side of equation (2) depends on the mode structure, and hence the indirect estimate is modified if we assume equatorial waves for 1–day waves. Thus, we inferred possible equatorial modes using the following three criteria. First, the minimum observable vertical wavelength is about 2 km because the vertical resolution of radiosonde observations is about 1 km. Secondly, it is deduced from the power spectra of $T$ and $u$ that the ratio between zonal kinetic energy and potential energy should be 1–4. Thirdly, the phase velocity should not equal the mean wind below the shear since the waves are considered to originate from the troposphere. Parameters of possible equatorial modes satisfying the criteria are listed in Table 1. The reduction of the indirect estimate under the assumption of possible equatorial waves is about 30–70%. Therefore, the discrepancy between direct and indirect estimates is still large for any possible equatorial waves, indicating a large cancellation of positive and negative momentum fluxes for both westerly and easterly shear.

In the above estimate we assumed that the observed frequency is the true wave frequency, i.e. no aliasing was present. Since rawinsonde observation at Singapore is made twice a day, aliasing of waves with periods shorter than 1 day is possible. Such very short-period waves are commonly observed when vigorous convection exists as discussed in the next section. Since the direct and indirect estimates of momentum flux are proportional to the wave frequency, the true values may be larger than the estimates given here.

The indirect estimate for westerly shear is almost twice as large as that for easterly shear. The characteristics of waves near the source in the troposphere are thought to be independent of the QBO in the stratosphere, so that the difference in wave activity could be attributed to the differing characteristics of wave propagation under the strong QBO shear (Sato and Dunkerton, 1996, submitted to J. Geophys. Res.).

## 3 Very short-period (< 1 day) waves associated with convection

Convection has the ability to generate very short-period (< 1 day) gravity waves with small horizontal scales of $O(10)$–$O(100)$ km. Recent studies of aircraft observations in the tropics showed that such small-scale gravity waves are associated with convection and large momentum fluxes were observed (Pfister et al., 1993; Alexander and Pfister, 1995). Observational studies with a VHF clear-air Doppler radar in the middle latitude also showed that a large amount of gravity wave energy at high frequencies (up to the Brunt-Väisälä frequency) are generated by cumulus convection. Sato (1992) and Sato et al.(1995) examined gravity waves and turbulence associated with cumulus convection which appears daily in the afternoon of mid-summer using VHF/UHF clear-air Doppler radars in Shigaraki, Japan (35N, 136E). The momentum flux associated with the gravity waves is about 50 $\times 10^{-3} m^2 s^{-2}$. Sato (1993) reported that gravity waves associated with a typhoon have large momentum fluxes ($\sim$200 $\times 10^{-3} m^2 s^{-2}$), comparable to those of severe mountain waves. Although their occurrence may be sporadic, the relative importance of such very short-period gravity waves should be clarified through further observational and numerical studies.

## Acknowledgments

The author wishes to thank Prof. J. P. Mathews for his comments on the manuscript.

## References

Alexander M.J., and L. Pfister, 1995: Gravity wave momentum flux in the lower stratosphere over convection. Geophys. Res. let., **22**, 2029.

Dunkerton, T.J., 1995: Horizontal buoyancy flux of internal gravity waves in vertical shear. J. Meteor. Soc. Japan, **73**, 747.

Dunkerton, T.J., 1996: Covariance of zonal wind and temperature for equatorial waves in vertical shear. J. Meteor. Soc. Japan, in preparation.

Maruyama, T., 1994: Upward transport of westerly momentum due to disturbances of the equatorial lower stratosphere in the period range about 2 days — A Singapore data analysis for 1983–1993. J. Meteor. Soc. Japan, **72**, 423.

Pfister, L., K. R. Chan, T. P. Bui, S. Bowen, M. Legg, B. Gary, K. Kelly, M. Proffitt, and W. Starr, 1993: Gravity waves generated by a tropical cyclone during the STEP tropical field program: A case study. J. Geophys. Res., **98**, 8611.

Sato, K., 1992: Vertical wind disturbances in the afternoon of mid-summer revealed by the MU radar. Geophys. Res. Let., **19**, 1943.

Sato K., 1993: Small-scale wind disturbances observed by the MU radar during the passage of Typhoon Kelly. J. Atmos. Sci., **50**, 518.

Sato K., and T. J. Dunkerton, 1996: Estimates of momentum flux associated with equatorial Kelvin and gravity waves. Submitted to J. Geophys. Res. (CADRE special issue).

Sato, K., F. Hasegawa, and I. Hirota, 1994: Short-period disturbances in the equatorial lower stratosphere. J. Meteor. Soc. Japan, **72**, 423.

Sato, K., H. Hashiguchi, and S. Fukao, 1995: Gravity waves and turbulence associated with cumulus convection observed with the UHF/VHF clear-air Doppler radars. J. Geophys. Res., **100**, 7111.

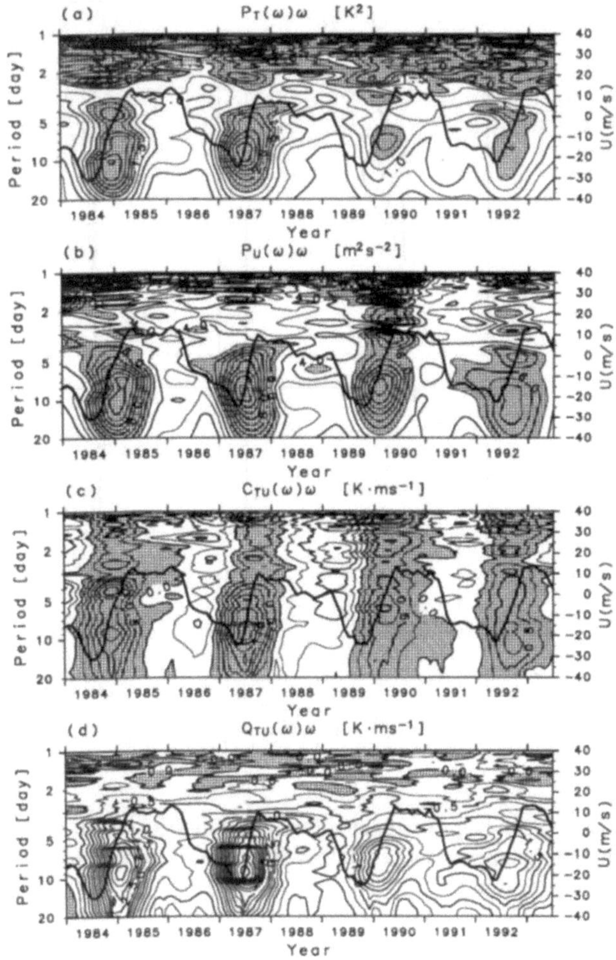


**Fig. 1**: Dynamic power spectra of (a) temperature fluctuations $T$ and (b) zonal wind fluctuations $u$, in energy content form, for 1984–1993, averaged over the height range 20–25 km at Singapore. Co-spectra and quadrature spectra of $T$ and $u$ are also shown in (c) and (d), respectively, in flux content form. Temporal smoothing was applied using a lowpass filter with cutoff of 20 months in order to display the relation with the QBO more clearly. Thick solid curves show the monthly mean zonal wind averaged over the same height range; the scale is shown on the right axis. Contour intervals are 0.25 $K^2$ in (a), 1 $m^2s^{-2}$ in (b), and 0.25 Km $s^{-1}$in (c) and (d). Shading indicates values larger than 1.5 $K^2$ in (a), larger than 5 $m^2s^{-2}$ in (b), and positive values in (c) and (d).

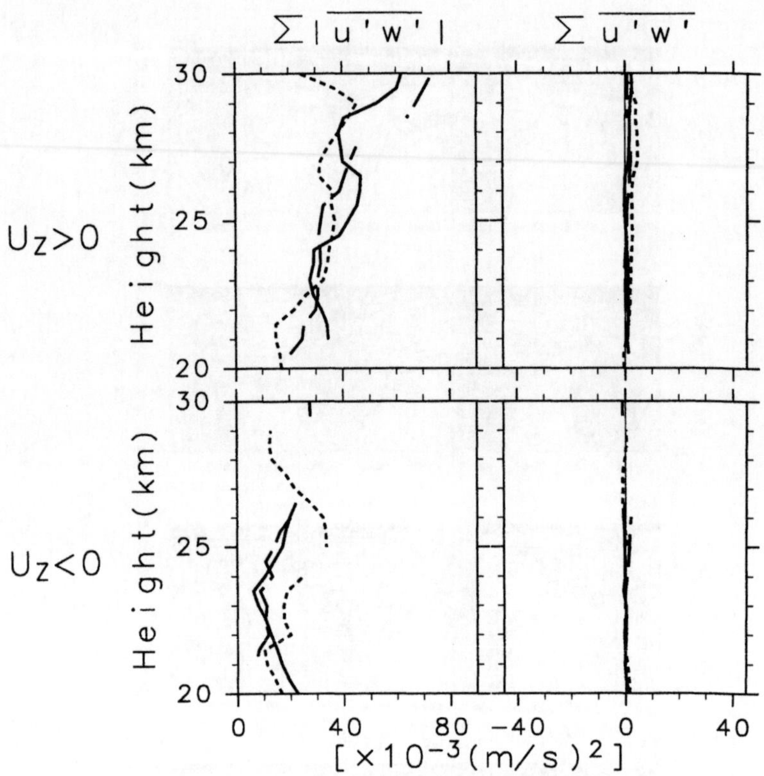


**Fig. 2**: Indirect (left) and direct (right) estimate of momentum flux $\overline{u'w'}$ as a function of height for 1–3 day waves during time periods when the vertical shear is greater than $5\times10^{-3}s^{-1}$ (top) and less than $-5\times10^{-3}s^{-1}$ (bottom).

# Experimental Constraints on Gravity Wave Parameterization from in Situ Measurements of Temperature and Turbulence

Franz-Josef Lübken
Bonn University, Department of Physics
Nussallee 12
53115 Bonn, Germany

**ABSTRACT.** The thermal structure in the high latitude mesosphere and thermosphere has presumably created the strongest stimulation for a realistic parameterization of gravity wave breaking in models. The meridional temperature reversal and the very cold mesopause temperatures in summer of ~130 K have put a serious constraint on any theoretical description of the physical processes acting in the upper atmosphere. In this paper mesospheric temperature measurements at high latitudes are summarized. The thermal structure in summer exhibits a remarkable and persistent seasonal and interannual repeatability with a rather abrupt change from summer to winter conditions in mid August. At typical noctilucent cloud altitudes (82 km) the mean temperature is again and again observed to be in the range 150 ± 2 Kelvin ("equithermal submesopause"). The second part of this paper deals with insitu measurements of turbulent energy dissipation rates. Turbulent heating is strongest around the mesopause region in summer (10 to 20 K/d). During winter, the heating rates are considerably smaller and turbulence is distributed over a wider altitude range. It is shown that intense radar echoes observed in the mesopause region during summer ("polar mesosphere summer echoes") are not necessarily caused by strong neutral air turbulence.

## 1 Introduction

Stimulated by remote sensing of noctilucent clouds (NLC) and meteor winds *Kellog and Schilling* [1951] speculated about a meridional temperature reversal in the mesosphere and lower thermosphere (MLT) more than 40 years ago. Unambiguous experimental proof for a 'warm' and 'cold' mesopause in summer and winter, respectively, was later provided by the rocket grenade measurements over Fort Churchill [*Stroud et al.*, 1960], which showed that the mesopause in summer is colder by as much as 50–60 K compared to winter (actually, the mesopause in winter was above the instrumental altitude range for the grenade technique). Theoretical studies accompanying these observations showed that there exists a radiative surplus at the summer mesopause which could be compensated for by a meridional pole to pole circulation [*Murgatroyd and Singleton*, 1961]. The cooling (heating) is achieved by the adiabatic vertical upward (downward) motions in the summer (winter) hemisphere accompanying these meridional circulation as a consequence of mass conservation. The problem with this explanation of the temperature reversal was that the meridional flow of some meters per second causes a large acceleration of the zonal flow due to the Coriolis torque: A meridional flow of 1 m/s experiences a zonal acceleration of ~10 m/s per day. It was realized quickly that frictional forces could decelerate the zonal flow

NATO ASI Series, Vol. I 50
Gravity Wave Processes
Their Parameterization in Global Climate Models
Edited by Kevin Hamilton

and prevent it from growing to extensively large values [*Haurwitz*, 1961; *Leovy*, 1964]. Although there were some speculations about the possible source of the friction earlier, it was *Houghton* [1978] who first proposed that gravity waves could be the cause of the drag. A major step forward was achieved by *Lindzen* [1981] who parameterized the drag produced by breaking gravity waves by assuming that the wave saturation amplitudes are limited by convective instabilities which in turn depend on the actual phase speeds of the waves, and on the background winds. Lindzen's parameterization is still used successfully in various models [*Holton*, 1983; *Schoeberl et al.*, 1983; *Ebel*, 1984; *Garcia and Solomon*, 1985; *McIntyre*, 1989; *Fritts and Luo*, 1995].

The deposition of momentum required to decelerate and even reverse the zonal flow stems from instabilities of gravity waves when their amplitude grows to extensively large values and breaking occurs. These instabilities result in turbulent motions which lead to a flux divergence of momentum and to a dissipation of kinetic and potential energy. The energy cascades from larger to smaller scales and is finally dissipated into heat due to molecular viscosity. Apart from being a heat source, turbulence also acts as an effective heat sink since the vertical exchange of air parcels in a stably stratified atmosphere transports heat downwards. The question whether turbulence is effectively heating or cooling the upper atmosphere has been discussed controversially in the literature [*Chandra*, 1980; *Gordiets et al.*, 1982; *Ebel et al.*, 1983; *Gärtner and Memmesheimer*, 1984; *Zimmerman and Keneshea*, 1986; *Lübken et al.*, 1993a].

From this brief historical overview it is evident that measurements of the thermal structure in the MLT region have caused a strong stimulation for a realistic parameterization of gravity wave breaking in models. Any new scheme of gravity wave breaking and its parameterization must finally be tested versus the measurements of the thermal structure in the mesopause region with special emphasis on its seasonal variation. In addition, turbulence plays the most important role in the transfer of momentum and energy from the gravity waves to the mean flow. Gravity wave parameterizations must therefore be consistent with a reliable measurement of turbulent parameters. It is the purpose of this paper to present an updated status report on insitu measurements of temperatures and turbulent energy dissipation rates, including some new results from rocket campaigns performed in recent years.

## 2 Thermal Structure of the Mesopause Region

### 2.1 Insitu measurements of density and temperature

The largest record of density and temperature measurements in the mesosphere stems from the so called 'passive falling spheres' (FS), which give accurate data below ~90–95 km, depending on the mean density in the MLT [*Jones and Peterson*, 1968; *Schmidlin*, 1991]. Temperatures are deduced from the measured densities assuming hydrostatic equilibrium. A recent comparison of FS data with results obtained from ground based lidars has shown general agreement between both techniques taking into account the error bars and the spatial and temporal differences in the data set [*Lübken et al.*, 1994a]. A systematic overview of all existing FS flights at high latitudes is published in *Lübken and von Zahn*

[1991]. Since then a number of FS launchings were performed at high latitudes, most of them in conjunction with sounding rocket flights and ground based observations (see Table 1).

Table 1: Falling sphere flights from the Andøya Rocket Range (69°N,16°E) from 1992 until 1995.

| Campaign | period | # FS |
|---|---|---|
| AEROSOL-I | April 23, 1992 – June 3, 1992 | 21 |
| SCALE | July 27, 1993 – August 1, 1993 | 5 |
| ECHO | July 28, 1994 – August 17, 1994 | 12 |
| SCIFER | January 25, 1995 | 1 |
| AEROSOL-II | April 2, 1995 – May 31, 1995 | 11 |

The seasonal distribution of all FS flights from Andøya is shown in Figure 1. A total of 171 flights were performed in the period 1983–1995. All seasons are covered, however, there are only few flights available around the equinoxes.

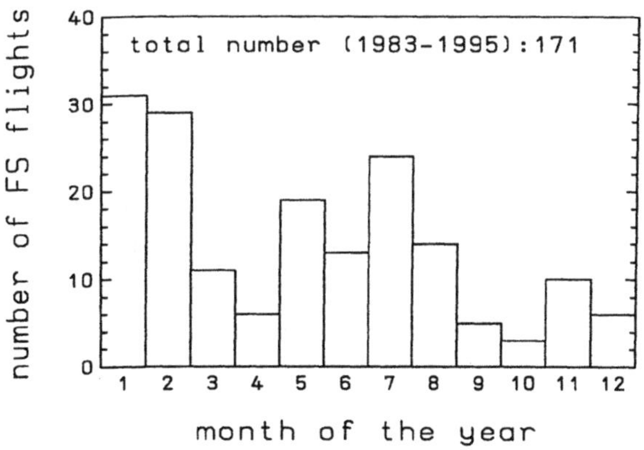


Figure 1: Seasonal distribution of falling sphere launchings from the Andøya Rocket Range (69°N) in the period 1983 – 1995. A total of 171 flights were performed in this period.

Absolute total number densities were obtained recently in the altitude range from ~115–90 km by means of the TOTAL and CONE instruments on board the TURBO rocket. Both the TOTAL and CONE sensors basically consist of an ionization gauge with very high time and altitude resolution of approximately 1 ms and 1 m, respectively [*Hillert et al.*, 1994; *Lübken*, 1996b] (the main difference between TOTAL and CONE is the geometrical configuration of the electrodes within the ionization gauge). Relative density fluctuations at very small scales are deduced from these measurements in the altitude range from 115 to 65 km. The TOTAL and CONE launchings are listed in Table

2 (there were 3 more TURBO launchings from Biscarosse, Southern France, which are not considered here since they were not performed at high latitudes).

Table 2: Campaigns with launches of the TURBO rocket. The geographical coordinates of Andøya and Kiruna are (69°N,16°E) and (68°N,21°E), respectively.

| Campaign Label | Campaign Period | | Location | Number of Flights | Ionization Gauge |
|---|---|---|---|---|---|
| DYANA | January – March | 1990 | Andøya | 6 | TOTAL |
| NLC | July – August | 1991 | Kiruna | 2 | TOTAL |
| METAL | September – October | 1991 | Andøya | 6 | TOTAL |
| SCALE | July – August | 1993 | Andøya | 2 | CONE |
| ECHO | July – August | 1994 | Andøya | 3 | CONE |

## 2.2 Mean temperatures and thermal variability as a function of season

In Figure 2 all individual FS profiles measured during three summer months (June, July, and August) and during two winter months (January and February) are shown, respectively. The summer/winter difference in the thermal structure of the mesosphere is evident from these profiles: The mesopause is very cold in summer (~130 K) and is located at an altitude of approximately 88 km. In winter, the upper mesosphere is significantly warmer by as much as 60 to 70 Kelvin (actually, the mesopause in winter is above the altitude range of the FS technique). These Figures are in fact very similar to our earlier compilation [*Lübken and von Zahn*, 1991], except that the number of profiles has significantly been increased since then.

Apart from the seasonal dependence of the mean temperature, there is an apparent difference in the variability of the profiles from winter to summer: Whereas the deviation from the mean (RMS) is on the order of 10 to 15 Kelvin in winter, it is significantly smaller in summer (~6–7 K). The repeatability in summer is remarkable, considering that profiles from 5 different years and from a period of nearly three months contribute to this compilation (please note that the individual profiles from August where all obtained in the first half of the month).

The monthly mean profiles (Figure 3) demonstrate the remarkable persistence of the thermal structure in summer: The mean from June, July, and August are practically identical. This Figure also emphasizes the grouping into winter and summer season, with a rather short transition period around the equinoxes (the transition from winter to summer is not clearly as established due to a rather poor statistics around the spring equinox).

Figure 4 summarizes the seasonal variation of the thermal structure in the upper stratosphere and mesosphere. The transition from summer to winter conditions occurs around mid August and is rather rapid with a change of more than 40 Kelvin in 3 weeks at summer mesopause altitudes. A rapid transition is also evident in the data reported by *Theon and Smith* [1971] who have summarized a series of temperature soundings by acoustic grenades from Pt. Barrow, Alaska (71°N).

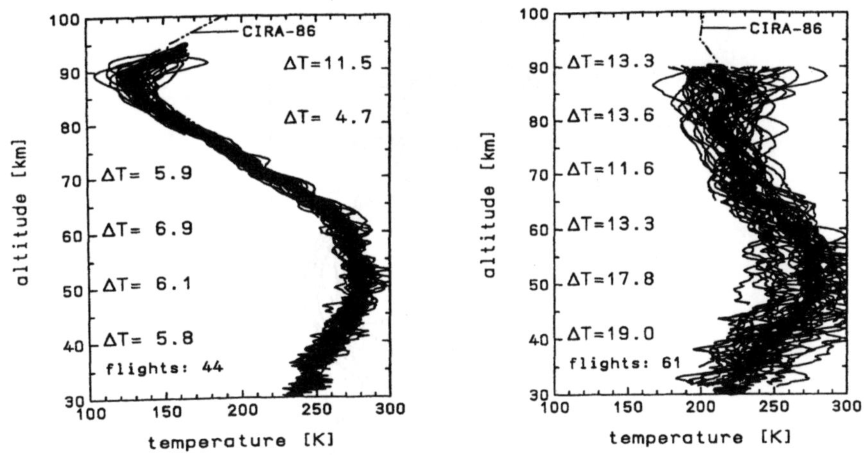


Figure 2: All FS profiles measured from the Andøya Rocket Range during three summer months (June, July, and August; left panel), and during two winter months (January and February; right panel). The numbers in the plot indicate the variability (RMS) of the temperatures at a given altitude.

## 2.3 The "equithermal submesopause" in summer

The thermal structure of the polar summer mesopause has attracted significant attention since is has been speculated that an anthropogenically caused increase of greenhouse gases could change the thermal structure in this region much stronger compared to other parts of the atmosphere [*Roble and Dickinson*, 1989; *Rind et al.*, 1990; *Thomas*, 1996; *Portmann et al.*, 1995]. The increase of NLC occurrence rate over the last 30 years is attributed to such a temperature trend [*Gadsden*, 1990]. However, it was noted by *Thomas et al.* [1989] that an increase of water vapor concentration caused by a rise of methane abundance could have a similar effect and would also increase the probability of ice particle formation, thus of NLC occurrence rate.

There have been only very few simultaneous measurements of NLCs and temperatures in previous years. The main reason for this deficiency is that noctilucent clouds cannot be observed visually at stations located within the polar circle due to the permanently bright sky. The situation has changed since lidars are capable of detecting NLCs, and temperatures were measured simultaneously by FS from the same location. A compilation of such measurements is presented in *Lübken et al.* [1996] where it is shown that NLCs appear significantly below the mesopause, but still at altitudes where the temperatures are below ~150 Kelvin. The lidar data show that the majority of the NLCs is located around

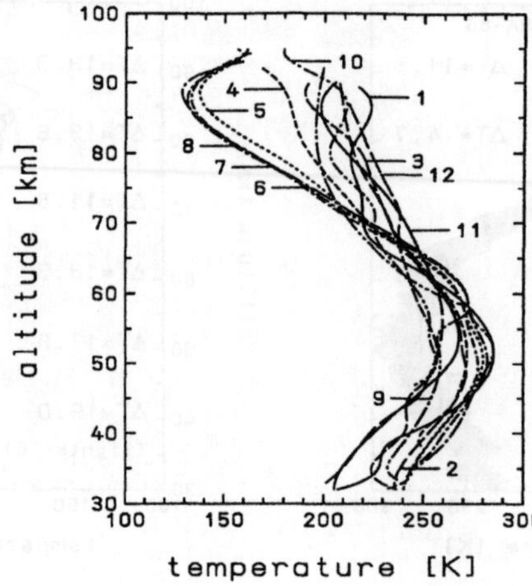


**Figure 3: Monthly mean profiles from all FS flights performed from the Andøya Rocket Range in the period 1983 – 1995. The numbers at the profiles indicate the month. The mean profiles were smoothed by a spline fit.**

82 km which is surprisingly close to the very first height determinations more than 100 years ago [*Jesse*, 1896]. A compilation of all temperature measurements available from the polar mesopause region (with or without observations of NLCs) shows a remarkable repeatability of the thermal structure ever since the first measurements were performed more than 30 years ago: At the lower edge of the NLC heights (82 km) the mean temperature is again and again observed to be in the range 150 $\pm$ 2 Kelvin (the variability within each data set is only a few Kelvin). Such an "equithermal submesopause" in summer puts a serious constraint on any model prediction of secular changes of temperatures in the upper mesosphere as well as on the parameterization of gravity wave breaking in the upper summer mesosphere.

## 3 Turbulence in the Mesosphere and Lower Thermosphere

We have noted earlier that turbulence is presumably the most important physical process by which gravity waves damp their momentum and their energy to the background atmosphere. Unfortunately, the measurement of turbulent wind fluctuations is not possible in the upper atmosphere due to experimental limitations. Instead, one measures the fluctuations of a passive tracer on a mean gradient caused by turbulent velocity fluctuations. Tracers such as chemoluminescent clouds, and chaff clouds have been released from a rocket and their spatial and temporal evolution has been studied in order to infer

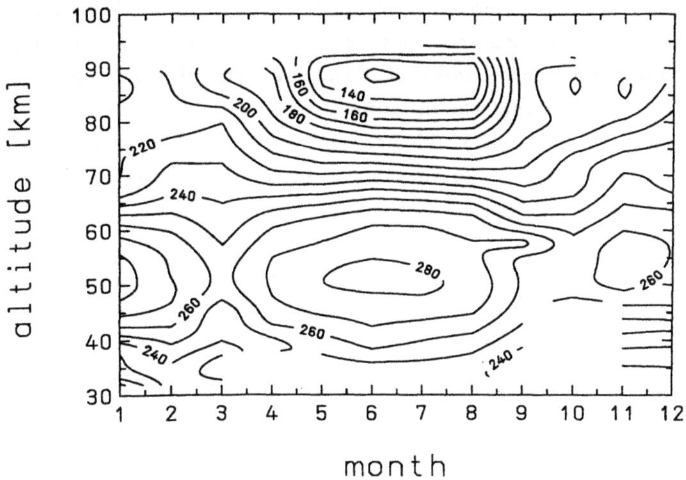


Figure 4: Contour plot of the monthly mean profiles shown in Figure 3. Contour lines are shown in degree Kelvin. Please note that the data base rather limited in the months April, September, and October.

turbulent diffusion [*Blamont and DeJager*, 1961; *Wu and Widdel*, 1989]. The problem with these techniques is that non-turbulent effects can disperse the clouds, and that the spatial resolution is only some tens of meters which is not sufficient to detect the smallest scales in the turbulent field.

A common technique to obtain information about turbulence is to study its effect on the backscattered signal of a radar. Both the signal strength and the broadening due to turbulent motions have been used [*Gage and Balsley*, 1984; *Hocking*, 1985; 1996]. This technique allows for a nearly continuous time coverage of turbulent processes in the upper atmosphere, and may therefore be used to qualitatively map the variation with season. However, when deducing quantitative results, uncertainties are introduced since it is difficult to appropriately correct for non-turbulent effects on the backscattered signal. Furthermore, background parameters needed to deduce quantitative results are normally not measured simultaneously and have to be taken from reference atmospheres. We will come back to the relationship between neutral air turbulence and radar echo strength in section 3.2.

### 3.1 Rocketborne measurements of turbulent parameters

More recently, neutral and plasma density fluctuations measured from a rocket have been used as passive tracers for turbulence [*Thrane and Grandal*, 1981; *Blix et al.*, 1990; *Lübken et al.*, 1994b]. The idea behind these measurements is that any vertical excursion of an air parcel due to turbulent motions results in a difference between the number density (neutral or plasma) in the air parcel and that of the environment. The main advantage of these *in situ* measurements is that the instruments are capable of detecting fluctuations down to spatial scales of a few meters. This capability is crucial since it allows to detect the 'break' in the spectrum of the fluctuations which occurs at the transition between the classical $k^{-5/3}$ power low in the inertial subrange and a much steeper dropoff in the viscous subrange [*Lübken*, 1992].

As mentioned earlier, the CONE and TOTAL instruments were installed on rockets in order to measure neutral density fluctuations down to very small scales of a few meters (see Table 2). During several flights small scale fluctuations were observed in layers of a few kilometer thickness. Subsequent spectral analysis indicated that these variations have been caused by turbulent motions. As is described in more detail in *Lübken et al.* [1993a] a theoretical model is fitted to the measured spectrum. We have applied a model first presented by *Heisenberg* [1948] which exhibits the classical $k^{-5/3}$ power law in the inertial subrange and a $k^{-7}$ dependence in the viscous subrange. Between these subranges a smooth transition takes place. The frequency spectrum is given by:

$$W(\omega) = \frac{\Gamma(5/3)\ sin(\pi/3)}{2\pi v_R} \cdot C_n^2 \cdot f_\alpha \cdot \frac{(\omega/v_R)^{-5/3}}{\left[1 + \{(\omega/v_R)/k_\circ\}^{8/3}\right]^2} \tag{1}$$

where $\Gamma(5/3)$=0.90167 ; $v_R$ is the rocket velocity ; $C_n^2 = a^2 \cdot N_n/\epsilon^{1/3}$ is the structure function constant ; $\epsilon$ is the energy dissipation rate ; $N_n$ represents the amount of inhomogeneity which disappears per unit time due to molecular diffusion; $f_\alpha = 2$ and $a^2 = 1.74$ are numerical constants. The normalization of the frequency spectrum in equation 1 is such that $\sigma^2 = \int_{-\infty}^{\infty} W(\omega)d\omega$, where $\sigma$ is the variance of the fluctuations. The Heisenberg model 'breaks' at a wavenumber $k_\circ \equiv \omega_\circ/v_R$ which is the intersection of the asymptotic forms of $W(\omega)$ in the inertial and viscous subrange, respectively. The corresponding spatial scale $\ell_\circ^H \equiv (2\pi)/k_\circ$, also called 'inner scale', is given by (see *Lübken et al.*, 1993a for more details):

$$\ell_\circ^H/\eta = 9.90 \quad ; \quad \eta = \left(\frac{\nu^3}{\epsilon}\right)^{1/4} \tag{2}$$

$\nu$ is the kinematic viscosity, and $\eta$ is called the 'Kolmogoroff microscale'. As can be seen from these equations a measurement of $\ell_\circ^H$ is synonymous with an observation of $\epsilon$ ($\nu$ is deduced from the insitu measurements of density and temperature).

Towards larger scales the inertial subrange is limited by the so called 'outer scale' which is given by [*Weinstock*, 1981]:

$$L_B = 9.97\sqrt{\frac{\epsilon}{\omega_B^3}} \tag{3}$$

($\omega_B$ ist the Brunt-Väisälä frequency). In Figure 5 a spectrum deduced from neutral density fluctuations at 85 km is shown together with the best fit Heisenberg model. The observed spectrum is nicely represented by the model with an inner scale of $\ell_\circ^H$=21.0 m.

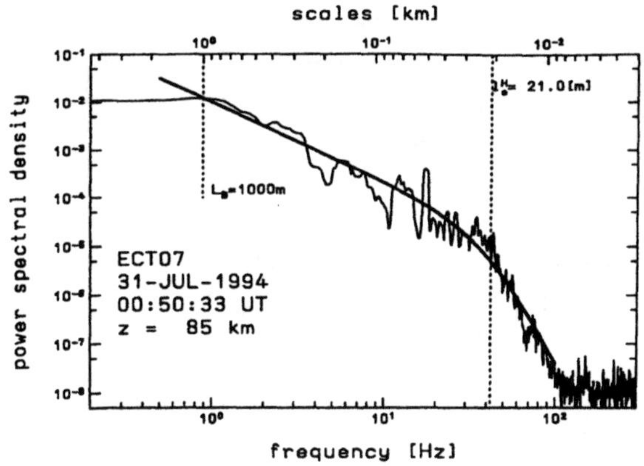


Figure 5: Spektrum of the neutral density fluctuations measured during flight ECT07 of the ECHO campaign in the altitude bin 85±2.5 km. The best fit Heisenberg spectrum is also shown. From this fit the following turbulent parameters are obtained: $\ell_0^H$=21.0 m ; $\epsilon$=107 mW/kg ; heating rate: 9.2 K/d ; $N_n = 5.5 \times 10^{-7}$ 1/s. The mean kinematic viscosity is $\nu$ = 1.29 $m^2/s$. The vertical lines indicate the 'inner scale' ($\ell_0^H$) and the 'outer scale' ($L_B$), defined by equations 2 and 3, respectively.

Applying equation 2 and using the measured kinematic viscosity of $\nu$=1.29 $m^2/s$ results in an energy dissipation rate of 107 mW/kg (corresponding heating rate: 9.2 K/d). This turbulent heating is comparable with other heat sources in the upper mesosphere, for example due to absorption of solar radiation, or due to chemical heating [*Mlynczak*, 1996].

Several examples of spectra similar to that in Figure 5 have been published in the literature; they demonstrate that the spectra observed in turbulent layers are appropriately fitted by theoretical models such as the Heisenberg model. [*Lübken et al.*, 1993a; *Hillert et al.*, 1994; *Lehmacher and Lübken*, 1995; *Lübken*, 1996a]. These spectra result in reliable and unambiguous values for the turbulent energy dissipation rate $\epsilon$.

In Figure 6 a spectrum is shown which was observed at higher altitudes (100 km) during the summer flight SCT06. No significant fluctuations are observed at small scales. For comparison, three theoretical model profiles are shown: One for $\epsilon$=600 mW/kg, and one each where $\epsilon$ is increased and decreased by a factor of 5, respectively. A comparison between the model spectra and the observed spectrum in Figure 6 clearly demonstrates that turbulence is very weak (if present at all) at this altitude. Moreover, the turbulent intensity is much weaker than the theoretical model of *Fritts and Luo* [1995] which predicts values around 500 mW/kg in the 100 km height region.

The two examples shown in these Figures are typical for our summer flights: Turbulence is rather strong around the polar summer mesopause and is practically absent above 95 km. A summary of all TOTAL and CONE measurements of turbulent parameters is

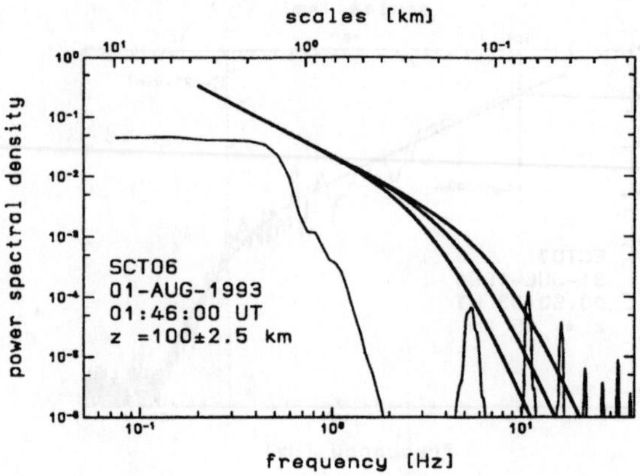


**Figure 6: Spektrum of the fluctuations measured during flight SCT06 of the SCALE campaign in the altitude range 100±2.5 km. The spectrum does not show any significant fluctuations at small scales. For comparison the theoretical Heisenberg type spectra for three $\epsilon$ values are shown: 600 mW/kg, 600/5 mW/kg, and 600*5 mW/kg.**

presented in *Lübken* [1996a]. The mean energy dissipation rates deduced from all 6 summer flights are shown in Figure 7. Turbulence is significantly heating the polar summer mesopause region by typically 10 to 20 Kelvin per day. Practically no turbulence is observed above 95 km and below 80 km. In contrast, the mean of all 12 winter measurements (also shown in Figure 7) shows a rather different height structure: Turbulent heating is small in the entire atmosphere (below 1–2 K/d), and is distributed over a larger altitude range. It is interesting to note that this seasonal variation is in qualitative agreement with the calculations from *Alexander* [1996] who has determined the energy dissipation rate profiles for breaking gravity waves which are generated in the troposphere by a frontal zone.

## 3.2 Are strong radar echoes indicative of intense turbulence?

In the early 1980s it was realized that the backscattered signals from VHF radars at high latitudes show a strong and persistent modulation with season [*Ecklund and Balsley*, 1981]. In particular, the signals from around the summer mesopause were extremely intense compared to other times of the year. In the literature this seasonal dependence of echo strength and height is sometimes deliberately taken as a direct indication for neutral air turbulence, despite the fact that other factors controlling the backscatter process (e. g., ionization by solar radiation) also vary with season. Moreover, it was realized that the summer echoes are too strong to be explainable by classical backscatter from turbulent structures within the plasma. These strong echoes were later called "polar mesosphere summer echoes" (PMSE) [*Röttger et al.*, 1988; *Cho and Kelley*, 1993].

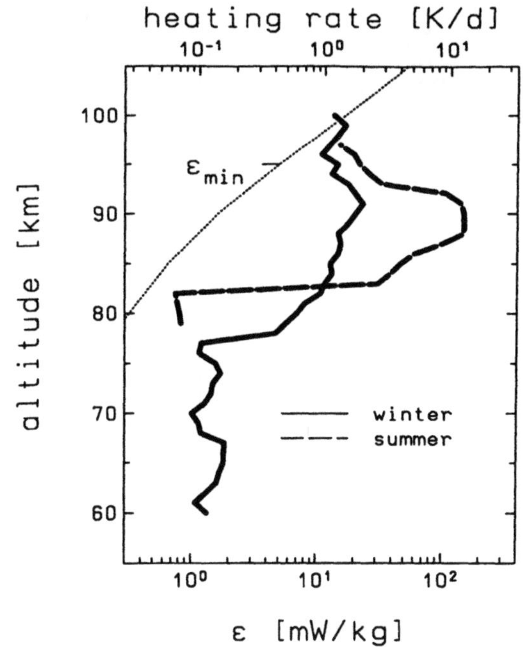


Figure 7: Mean summer and winter profiles for the turbulent energy dissipation rate $\epsilon$. A total of 7 and 12 flights contributed to the summer and winter means, respectively (see Table 2). The dotted profile shows an estimate of the theoretical lower limit of $\epsilon$. The profiles were smoothed by calculating a 5 point running mean.

It was suggested by *Kelley et al.* [1987] that massive charged aerosols (presumably present around the summer polar mesopause) could reduce the molecular diffusivity of the plasma relative to the neutrals. This would allow fluctuations in the plasma to exist at significantly smaller scales compared to the Kolmogoroff microscale (representing the smallest fluctuations in the neutral number density). The ratio of plasma to neutral diffusivities is called 'Schmidt number' *Sc*. The Schmidt number is greater than unity if charged aerosols with large masses are present in the atmosphere.

During the SCALE campaign in 1993, the first insitu measurements of electron, ion, and neutral number density fluctuations were measured in the vicinity of a PMSE layer which exhibited a double layered structure [*Lübken et al.*, 1993b]. In the upper layer both neutral and plasma fluctuations were detected and, indeed, the electron fluctuations extended to smaller scales compared to the neutrals. A Schmidt number of 6.5 was deduced from a comparison of the spectra. However, in the lower layer no neutral air turbulence was observed which lead us to conclude that at least two different mechanisms can create PMSE, one of which does not require neutral air turbulence.

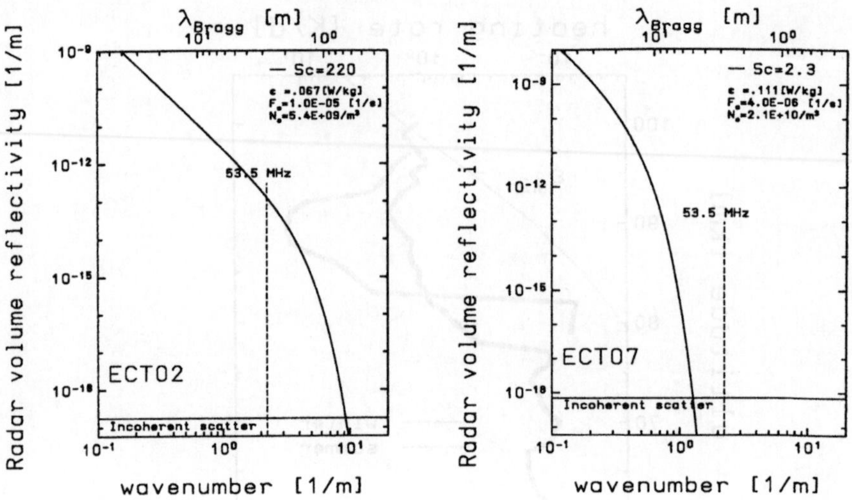


**Figure 8: Radar volume reflectivities (solid line) deduced from insitu measurements of the energy dissipation rate and the Schmidt number indicated in the plot. For comparison the reflectivity simultaneously measured by a VHF radar in the vicinity of a PMSE is shown (dashed line). The VHF radar transmits at a frequency of 53.5 MHz which corresponds to a Bragg scale of 3 meters. The left panel is from flight ECT02 (around 88 km) ; the right panel is from flight ECT07 (around 85 km).**

**During the ECHO campaign in 1994 these measurements were repeated with launchings of the TURBO rocket and simultaneous observations of PMSE by the ALOMAR/SOUSY VHF radar [*Bremer et al.*, 1996]. In Figure 8 the radar volume reflectivities derived from the insitu measurements of $\epsilon$ and $Sc$ are shown together with the reflectivity measured by the radar during the rocket flight (see *Giebeler* [1995] for more details). During flight ECT02 a relatively large Schmidt number of 220 was obtained. Together with the measured $\epsilon$ value this results in a reflectivity at the Bragg scale of 3 m which nicely agrees with the ground based observations (the frequency of the ALOMAR/SOUSY is 53.5 MHz which corresponds to a scale of $\sim$ 3m). The small difference is negligible considering the uncertainties introduced by the assumption about the horizonal extent of the turbulent layer.**

**During flight ECT07, however, neutral turbulence is too weak to create any significant backscatter (see Figure 8, right panel). The corresponding volume reflectivity at a scale of 3 m is too small by several orders of magnitude in order to account for the observed radar signal.**

**These measurements show that the relationship between neutral air turbulence and**

plasma fluctuations causing the radar echoes is not established at the moment. We have seen that at least two mechanisms can create PMSE, one of which is not related to neutral air turbulence. We conclude that strong VHF radar echoes cannot simply be taken as an indication of intense neutral air turbulence, at least not under the special conditions present around the polar mesopause in summer.

## 4 Concluding Remarks

The following observational features of the thermal and small scale dynamical structure of the upper atmosphere at high latitudes have to be taken into account in any parameterization of gravity wave breaking:

1. Our measurements confirm and reinforce the remarkable seasonal variation of the mean temperatures and their variability in the MLT region.
2. The mesospheric thermal structure is very persistent throughout the summer season with a rather abrupt change from summer to winter conditions in mid August.
3. At typical NLC altitudes of 82 km the temperature in summer is again and again measured to be 150±2 K with very little interannual variability ("equithermal submesopause").
4. Turbulence is rather patchy and occurs in layers with a thickness of a few kilometers.
5. The dissipation of turbulent kinetic energy is strongest around the summer mesopause with mean values of 10–20 Kelvin per day. During summer, we have not observed any significant turbulent heating above 95 km and below 80 km. During winter, the turbulent energy dissipation rates are generally smaller and turbulence is distributed over a larger altitude range.

**Acknowledgments.** I thank Dr. Jochen Giebeler for his support for the TOTAL and CONE measurements and the comparison with the radar results. The ALOMAR/SOUSY reflectivities were kindly provided by Dr. Singer from Kühlungsborn. The excellent work by the crew of the Mobile Raketenbasis (Germany), and the Andøya Rocket Range (Norway) is gratefully acknowledged. The rocket flights were supported by the Bundesministerium für Bildung, Wissenschaft, Forschung und Technologie under DARA grants No. 01OE88027 and 01OE89201.

## 5 References

Alexander, M. J., A simulated spectrum of convectively generated gravity waves: Propagation from the tropopause to the mesopause and effects in the middle atmosphere, *J. Geophys. Res.*, *101*, 1571–1588, 1996.

Blamont, J.-E., and C. DeJager, Upper atmospheric turbulence near the 100 km level, *Ann. Geophys.*, *17*, 134–144, 1961.

Blix, T. A., E. V. Thrane, and O. Andreassen, In situ measurements of fine scale structure and turbulence in the mesosphere and lower thermosphere by means of electrostatic positive ion probes, *J. Geophys. Res.*, *95*, 5533–5548, 1990.

Bremer, J., P. Hoffmann, W. Singer, C. Meek, and R. Rüster, Simultaneous PMSE observations with ALOMAR-SOUSY and EISCAT-VHF radar during the ECHO-94 campaign, *Geophys. Res. Lett.*, *23*, 1075–1078, 1996.

Chandra, S., Energetics and thermal structure of the middle atmosphere, *Planet. Space Sci.*, *28*, 585–593, 1980.

Cho, J. Y. N., and M. C. Kelley, Polar mesosphere summer radar echoes: observations and current theories, *Rev. Geophys.*, *31*, 243–265, 1993.

Ebel, A., H. J. Jakobs, and P. Speth, Turbulent heating and cooling of the mesopause region and their parameterization, *Ann. Geophys.*, *1*, 359–370, 1983.

Ebel, A., Contributions of gravity waves to the momentum, heat and turbulent energy budget of the upper mesosphere and lower thermosphere, *J. Atmos. Terr. Phys.*, *46*, 727–737, 1984.

Ecklund, W. L., and B. B. Balsley, Long-term observations of the arctic mesosphere with the MST Radar at Poker Flat, Alaska, *J. Geophys. Res.*, *86*, 7775–7780, 1981.

Fritts, D. C., and Z. Luo, Dynamical and radiative forcing of the summer mesopause circulation and thermal structure, I: mean solstice conditions, *J. Geophys. Res.*, *100*, 3119–3128, 1995.

Gadsden, M., A secular change in noctilucent cloud occurrence, *J. Atmos. Terr. Phys.*, *52*, 247–251, 1990.

Gage, K. S., and B. B. Balsley, MST radar studies of wind and turbulence in the middle atmosphere, *J. Atmos. Terr. Phys.*, *46*, 739–753, 1984.

Garcia, R. R., and S. Solomon, The effect of breaking gravity waves on the dynamics and chemical composition of the mesosphere and lower thermosphere, *J. Geophys. Res.*, *90*, 3850–3868, 1985.

Gärtner, V., and M. Memmesheimer, Computation of the zonally-averaged circulation driven by heating due to radiation and turbulence, *J. Atmos. Terr. Phys.*, *46*, 755–765, 1984.

Giebeler, J., In-situ Messungen zur Untersuchung der Rolle von Turbulenz bei der Erzeugung von Radarechos in der polaren Mesosphäre im Sommer, Dissertation, Bonn University, BONN-IR-95-28 1995.

Gordiets, B. F., Y. N. Kulikov, M. N. Markov, and M. Y. Marov, Numerical modelling of the thermospheric heat budget, *J. Geophys. Res.*, *87*, 4504–4514, 1982.

Haurwitz, B., Frictional effects and the meridional circulation in the mesosphere, *J. Geophys. Res.*, *66*, 2381–2391, 1961.

Heisenberg, W., Zur statistischen Theorie der Turbulenz, *Z. Physik*, *124*, 628–657, 1948.

Hillert, W., F.-J. Lübken, and G. Lehmacher, TOTAL: a rocket-borne instrument for high resolution measurements of neutral air turbulence during DYANA, *J. Atmos. Terr. Phys.*, *56*, 1835–1852, 1994.

Hocking, W. K., Measurement of turbulent energy dissipation rates in the middle atmosphere by radar techniques: a review, *Radio Sci.*, *20*, 1403–1422, 1985.

Hocking, W., An assessment of the capabilities and limitations of radars in measurements of upper atmosphere turbulence, *Adv. Space Res.*, *17, No.11*, (11)37 - (11)47, 1996.

Holton, J. R., The influence of gravity wave breaking on the general circulation of the middle atmosphere, *J. Atmos. Sci.*, *39*, 2497–2507, 1983.

Houghton, J. T., The stratosphere and mesosphere, *Quart. J. R. Met. Soc.*, *104*, 1–29, 1978.

Jesse, O., Die Höhe der leuchtenden Nachtwolken, *Astron. Nachr.*, *140*, 161–168, 1896.

Jones, J., and J. W. Peterson, Falling sphere measurements, 30 to 120 km, *Meteorol. Monogr.*, *8*, 176–177, 1968.

Kelley, M. C., D. T. Farley, and J. Röttger, The effect of cluster ions on anomalous VHF backscatter from the summer polar mesosphere, *Geophys. Res. Lett.*, *14*, 1031–1034, 1987.

Kellog, W. W., and G. F. Schilling, A proposed model of the circulation of the upper stratosphere, *J. Meteorol.*, *8*, 222–230, 1951.

Lehmacher, G., and F.-J. Lübken, Simultaneous observation of convective adjustment and localized turbulence production in the mesosphere, *Geophys. Res. Lett.*, *22*, 2477 - 2480, 1995.

Leovy, C., Simple models of thermally driven mesospheric circulation, *J. Atmos. Sci.*, *21*, 327–341, 1964.

Lindzen, R. S., Turbulence and stress owing to gravity wave and tidal breakdown, *J. Geophys. Res.*, *86*, 9707–9714, 1981.

Lübken, F.-J., On the extraction of turbulent parameters from atmospheric density fluctuations, *J. Geophys. Res.*, *97*, 20,385–20,395, 1992.

Lübken, F.J., Seasonal variation of turbulent parameters at high latitudes as determined by insitu measurements of neutral density fluctuations, *J. Geophys. Res.*, *(revised)*, 1996a.

Lübken, F.-J., Aerodynamical effects in number density measurements in the lower thermosphere with the CONE instrument, *Adv. Space Res.*, *(in press)*, 1996b.

Lübken, F.-J., and U. von Zahn, Thermal structure of the mesopause region at polar latitudes, *J. Geophys. Res.*, *96*, 20,841–20,857, 1991.

Lübken, F.J., W. Hillert, G. Lehmacher, and U. von Zahn, Experiments revealing small impact of turbulence on the energy budget of the mesosphere and lower thermosphere, *J. Geophys. Res.*, *98*, 20,369–20,384, 1993a.

Lübken, F.-J., G. Lehmacher, T. Blix, U.-P. Hoppe, E. Thrane, J. Cho, and W. Swartz, First in-situ observations of neutral and plasma density fluctuations within a PMSE layer, *Geophys. Res. Lett.*, *20*, 2311–2314, 1993b.

Lübken, F.-J., W. Hillert, G. Lehmacher, U. von Zahn, M. Bittner, D. Offermann, F. Schmidlin, A. Hauchecorne, M. Mourier, and P. Czechowsky, Intercomparison of density and temperature profiles obtained by lidar, ionization gauges, falling spheres, datasondes, and radiosondes during the DYANA campaign, *J. Atmos. Terr. Phys.*, *56*, 1969–1984, 1994a.

Lübken, F.-J., W. Hillert, G. Lehmacher, U. von Zahn, T. Blix, E. Thrane, H.-U. Widdel, G. A. Kokin, and A. K. Knyazev, Morphology and sources of turbulence in the mesosphere during DYANA, *J. Atmos. Terr. Phys.*, *56*, 1809–1833, 1994b.

Lübken, F.-J., K.-H. Fricke, and M. Langer, Noctilucent clouds and the thermal structure near the arctic mesopause, *J. Geophys. Res.*, *101*, 9489 – 9508, 1996.

McIntyre, M. E., On dynamics and transport near the polar mesopause in summer, *J. Geophys. Res.*, *94*, 14,617–14,628, 1989.

Mlynczak, M. G., Energetics of the middle atmosphere: theory and observation requirements, *Adv. Space Res.*, *17(11)*, (11)117–(11)126, 1996.

Murgatroyd, R. J., and F. Singleton, Possible meridional circulations in the stratosphere and mesosphere, *Quart. J. R. Met. Soc.*, *87*, 125–135, 1961.

Portmann, R. W., G. E. Thomas, S. Solomon, and R. R. Garcia, The importance of dynamical feedbacks on doubled $CO_2$-induced changes in the thermal structure of the mesosphere, *Geophys. Res. Lett.*, *22*, 1733–1736, 1995.

Rind, D., R. Suozzo, N. Balachandran, and M. Prather, Climate change and the middle atmosphere. part I: the double $CO_2$ climate, *J. Atmos. Sci.*, *47*, 475–494, 1990.

Roble, R. G., and R. E. Dickinson, How will changes in carbon dioxide and methane modify the mean structure of the mesosphere and thermosphere?, *Geophys. Res. Lett.*, *16*, 1441–1444, 1989.

Röttger, J., C. LaHoz, M. C. Kelley, U.-P. Hoppe, and C. Hall, The structure and dynamics of polar mesosphere summer echoes observed with the EISCAT 224 MHz radar, *Geophys. Res. Lett.*, *15*, 1353–1356, 1988.

Schmidlin, F. J., The inflatable sphere: a technique for the accurate measurement of middle atmosphere temperatures, *J. Geophys. Res.*, *96*, 22673–22682, 1991.

Schoeberl, M. R., D. F. Strobel, and J. P. Apruzese, A numerical model of gravity wave breaking and stress in the mesosphere, *J. Geophys. Res.*, *88*, 5249–5259, 1983.

Stroud, W. G., W. Nordberg, W. R. B. F. L. Bartman, and P. Titus, Rocket-grenade measurements of temperatures and winds in the mesosphere over Churchill, Canada, *J. Geophys. Res.*, *65*, 2307–2323, 1960.

Theon, J.S., and W. Smith, The meteorological structure of the mesosphere including seasonal and latitudinal variations, , S. 131–146, D. Reidel Publ. Co., Dordrecht, 1971.

Thomas, G. E., J. J. Olivero, E. J. Jensen, W. Schröder, and O. B. Toon, Relation between increasing methane and the presence of ice clouds at the mesopause, *Nature*, *338*, 490–492, 1989.

Thomas, G., Is the polar mesosphere the miner's canary of global change ?, *Adv. Space Res.*, *18(3)*, 149–158, 1996.

Thrane, E. V., and B. Grandal, Observations of fine scale structure in the mesosphere and lower thermosphere, *J. Atmos. Terr. Phys.*, *43*, 179–189, 1981.

Weinstock, J., Energy dissipation rates of turbulence in the stable free atmosphere, *J. Atmos. Sci.*, *38*, 880–883, 1981.

Wu, Y.-F., and H.-U. Widdel, Turbulent energy dissipation rates and eddy diffusion coefficients derived from foil cloud measurements, *J. Atmos. Terr. Phys.*, *51*, 497–506, 1989.

Zimmerman, S. P., and T. J. Keneshea, Turbulent heating and transfer in the stratosphere and mesosphere, *J. Atmos. Terr. Phys.*, *48*, 491–507, 1986.

# Analysis of Intermittency in Aircraft Measurements of Velocity, Temperature and Atmospheric Tracers using Wavelet Transforms

Julio T. Bacmeister
Naval Research Laboratory,
Washington D.C., 20375

Stephen D. Eckermann
Computational Physics, Inc.
Fairfax, VA, 22031

Lynn Sparling
Hughes STX
Lanham, MD, 20706

K. Roland Chan and Max Loewenstein
NASA Ames Research Center
Moffet Field, CA, 94035

Michael H. Proffitt
CIRES University of Colorado
Boulder, CO 80303

*Abstract.* Wavelet transforms of geophysical data are a promising new technique which may extend traditional analyses employing Fourier transforms. Here we will employ wavelet transforms to examine velocity, potential temperature, and trace gas mixing ratio measurements from high-altitude ER-2 aircraft (altitude≈20 km). Probability density functions and quantities similar to structure functions are calculated from wavelet transforms. This analysis reveals differences between the variability of meteorological quantities such as velocity and potential temperature, and that of passive atmospheric tracers. Our analysis suggests that trace gas variability contains intermittent episodes of high variability, while velocity and temperature flucuations are more uniformly distributed. We propose a simple model consisting of two sets of variability with different degrees of "roughness" and continuity. This "bi-fractal" model is used to interpret structure functions obtained for velocity and $N_2O$. Finally, we show a case of gravity wave-tracer filament interaction in the ER-2 data. We argue that such interactions may increase cross-isentropic mixing of trace gases by gravity waves.

## 1. Introduction

The nature of mesoscale atmospheric variablity is still a subject of intense debate. Studies of the power spectral density (PSD) of atmospheric kinetic energy as a function of horizontal wavenumber ($k$) have generally revealed power law spectra with indices in the range $-5/3$ to $-3$ at horizontal scales between 1000 km and 1 km. These include extensive analyses of Global Atmospheric Sampling Program (GASP) data [e.g., *Nastrom et al.* 1986], as well as analyses of more restricted measurement campaigns [e.g., *Lilly and Lester*, 1974]. Theories for the observed spectra invoke either quasi 2D turbulence or a spectrum of inertia-gravity waves or some comination of both. Quasi 2D turbulence may produce either a $k^{-3}$ in the downscale enstrophy cascade, or a $k^{-5/3}$ spectrum the inverse

NATO ASI Series, Vol. I 50
Gravity Wave Processes
Their Parameterization in Global Climate Models
Edited by Kevin Hamilton

energy cascade [*Lilly*, 1983]. Thus, the relevance of quasi 2D turbulence to atmospheric energy spectra depends on the scale of actual energy inputs to atmosphere. *Lilly* [1983] has argued that the frequently observed $-5/3$ slope at scales of 10 to 500 km, is the result of strong energy injection at smaller scales by convective motions. Gravity wave theories also predict horizontal wavenumber spectra of kinetic energy with spectral indices (slopes in log-log space) in the observed range [e.g. *Gardner et al.* 1993]. A recent study by *Bacmeister et al.* [1996] argues that gravity wave theories give a good fit to observed spectra of horizontal and vertical velocity and temperature in the stratosphere. A key prediction of the gravity wave theory is a change of slope in the horizontal velocity spectra, with slopes around -1.5 or -2.0 at horizontal scales longer then 12km and slopes between -2.5 and -3.0 for scales shorter than 6km. The observed spectra of horizontal velocity exhibit this behavior. The ratio of horizontal and vertical kinetic energy is also well predicted by the gravity wave theory.

Any explanation of velocity and temperature variability in the atmosphere must also account for the observed variablity of passive tracers such as $N_2O$ and Ozone. Horizontal wavenumber power spectra of trace gas mixing ratios generally exhibit spectral indices close to -5/3 [e.g. *Nastrom et al.*, 1986; *Murphy*, 1989; *Bacmeister et al.*, 1996]. Bacmeister et al. argued that mixing ratio spectra of trace gases in the stratosphere differ significantly from those of velocities and temperature. In contrast to stratospheric velocity spectra, no transition to steeper slopes at short horizontal scales is present in stratospheric spectra of trace gas mixing ratio. Generally speaking, theories for the mesoscale power spectrum of trace gas mixing ratios in the atmosphere are unsatisfactory. A well established theoretical prediction is available only for the inertial subrange of 3D isotropic turbulence, where the $k^{-5/3}$ Obukhov-Corrsin spectrum for tracers is predicted, and for linear gravity waves, where tracer PSD's are predicted to have the same shape as horizontal velocity and potential temperature PSD's. For 2D turbulence, theoretical predictions for the variance spectrum of a passive tracer are still unclear. The simplest arguments based on a correspondence of the tracer field and the vorticity field predict a $k^{-1}$ spectra in the enstrophy cascading portion of the spectrum. However, it has been pointed out that this $k^{-1}$ tracer variance spectrum has rarely if ever been observed [*Pierrehumbert*, 1995]. Pierrehumbert argues that $k^{-1}$ is approached only in the limit of zero dissipation, and that in the presence of dissipation the tracer variance spectrum will lie between $k^{-1}$ and $k^{-2}$. In any case, neither turbulence nor gravity wave theories appear capable of simultaneously explaining the observed spectra of velocities, temperatures and passive tracers.

Power spectra of atmospheric variability are an incomplete description of a sequence of data. Simple mathematical models can yield sharply dissimilar data sequences with identical power spectra [e.g., *Davis et al.*, 1994]. Qualities of the data such as intermittency are not accessible via straightforward Fourier techniques. However, such information can be obtained from wavelet transforms with little more effort than that required to obtain

PSD's. By construction, wavelet transforms yield information about the distribution of variability at particular scales. Quantities analogous to the structure functions commonly used in turbulence research are also easily calculated from wavelet transforms [e.g, *Katul et al.*, 1994]. Structure functions are generalizations of power spectra, and reveal information about the degree of intermittency in the variables examined. The bulk of the work employing structure functions to analyze flow data has focussed on long data sets collected in fully developed 3D turbulent flows rather than meteorological data. However, recently *Davis et al.* [1994] examined atmospheric liquid water content data using higher order structure functions. Their analysis indicated the presence of complex multifractal structure in the liquid water data.

This study extends the analyses of ER-2 data presented in *Bacmeister et al.* [1996]. That study focussed on the shapes of horizontal wavenumber PSD's of velocity, temperature and tracer data. In this study we will present preliminary results addressing the issue of intermittency in the same data set. We will rely heavily on discrete wavelet tranforms (DWT's) [*Daubechies,* 1988, 1992; *Mallat*, 1989] which will be described briefly in Section 2. In Section 2 we will also define quantities similar to higher order structure functions which are easily calculated from DWT's. In Section 3 these quantities as well as other DWT diagnostics will be used to define objective measures of intermittency in both the dynamical quantities and trace gas mixing ratio. Results will be interpreted in terms of a simple bifractal model for trace gas variablity. We will also show what appears to be a clear example of strong coupling between horizontal tracer variation and vertical gravity wave displacments.

## 2. Data and Methods

### *2.1. Description of Data*

The data used in this study are high-resolution measurements of horizontal velocities, potential temperatures, and ozone and $N_2O$ mixing ratios obtained in the mid-stratosphere during recent aircraft campaigns employing NASA's ER-2 aircraft. Velocity, temperature and pressure data were measured from the ER-2 by the Meteorological Measurement System (MMS) [*Scott et al.*, 1990]. MMS makes *in situ* measurements of the three-dimensional vector wind velocity, as well as temperature and pressure, at a sampling frequency of 5 Hz. However, in this analysis we use coarse grained MMS data on a 1 Hz time grid. Ozone measurements were obtained by the NOAA Dual-Beam Ozone Photometer [*Proffitt et al.*, 1989] at a sampling frequency of 1 Hz. Mixing ratios of $N_2O$ were obtained from the Airborne Tunable Laser Absorption Spectrometer (ATLAS) [*Loewenstein et al.*, 1989]. ATLAS measurements were also recorded at 1 Hz. The measurement campaigns in which the data used here were obtained are; the second Arctic Airborne Stratospheric Expedition (AASE II), the Stratospheric Pho-

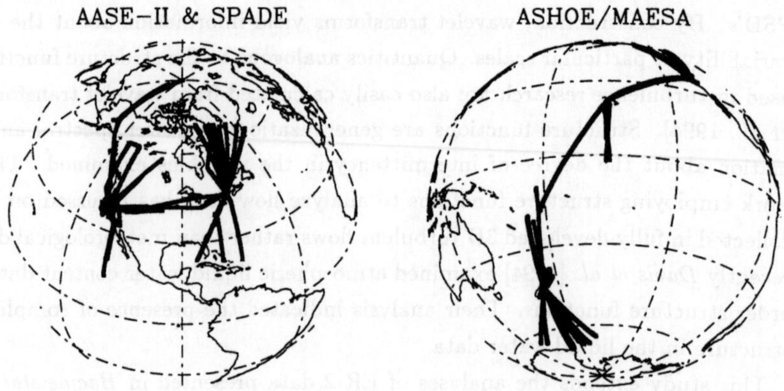


Figure 1:
**Map showing ER-2 flight tracks during AASE II, SPADE, and ASHOE/MAESA measurement campaigns.**

tochemistry, Aerosol and Dynamics Experiment (SPADE), and the Airborne Southern Hemisphere Ozone Experiment/Measurements for Assessing the Effects of Stratospheric Aircraft (ASHOE/MAESA). The dates and locations of these and three earlier aircraft campaigns employing the ER-2 described more completely in *Bacmeister et al.* [1996] and references therein.

For convenience a map showing ER-2 flight tracks during AASEII, SPADE, and ASHOE/MAESA is displayed in Figure 1. The data set has a definite bias towards wintertime polar and mid-latitude conditions, since the primary objective of 2 of the 3 missions was to examine polar ozone chemistry. However, during AASE II and SPADE equatorward flights were conducted regularly, and during ASHOE/MAESA 2 out of every 6 deployments took place from Hawaii. Thus, approximately one third of the flights examined here sampled tropical air. The ER-2's cruising altitude is close to 20 km or 60,000 feet, 20,000–30,000 feet higher than the commercial airliners used in the GASP investigations. Pressures below 70 mb are reached routinely. Approximate air speed of the ER-2 during cruise is about 200 m $s^{-1}$. This air speed will be used to convert temporal scales in the data time series to spatial scales. The validity of this transformation is discussed and demonstrated by *Bacmeister et al.* [1996].

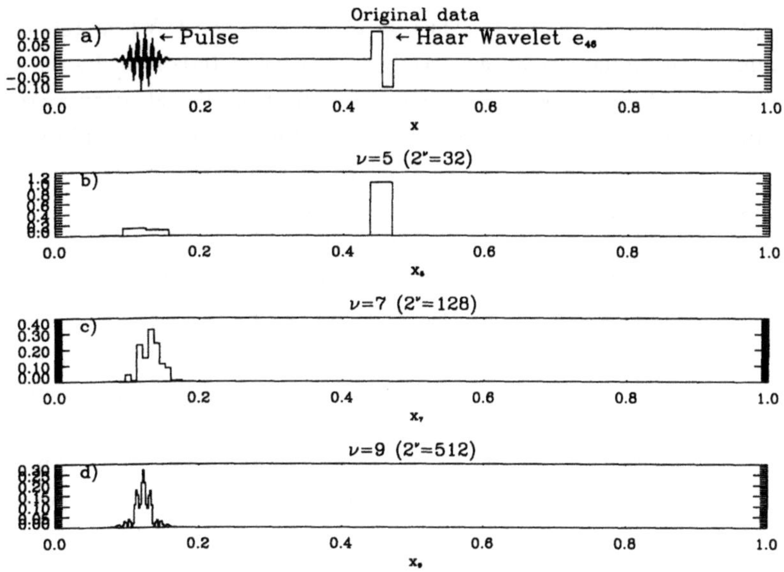


Figure 2:
(a) Artificial 4096 element data set consiting of a modulated high-frequency pulse near $x$=0.1 and a Haar Wavelet of scale $l \sim 2^{-5}$ positioned near $x$=0.425. (b-d) Three different "octaves" of Haar wavelet transform of data in (a). Since the scale of the Haar wavelet in (a) is $2^{-5}$ it shows up in (b) as a discreet peak near the middle of the 5-th octave of the DWT. The pulse projects power onto this and higher octaves.

### *2.2. Wavelet Transforms and Structure Functions*

Orthonormal wavelets are families of orthogonal functions which, unlike sines and cosines, are localized in both the spatial and wavenumber domains [*Daubechies*, 1989, 1992; *Mallat*, 1989]. Due to the localization of the basis functions, DWT spectra contain information about the location of variability in a time series as well as its frequency. Unfortunately, for a data sequence with $2^N$ elements, only $N$ independent frequencies or "octaves" are obtained from a DWT. Adjacent octaves represent frequencies separated by factors of two. Thus, the price for the spatial information contained in the DWT is a significant loss of frequency resolution when compared to an FFT. However, in the following analyses we will be most interested in variability at logarithmically spaced length scales, so that the degraded frequency resolution of the DWT's is not a serious drawback.

The DWT of a data sequence $X(s)$ with $2^N$ elements can written schematically as

$$X(s) \rightarrow \mathsf{DWT} \rightarrow [D_{X,0}(i_0), D_{X,1}(i_1), ..., D_{X,N-1}(i_{N-1})] = [D_{X,\nu}(i_\nu)], \tag{1}$$

where at each octave $D_{X,\nu}$ is a $2^\nu$ element vector. The $D_{X,\nu}(i_\nu)$ give the variability at a characteristic wavenumber $\sim 2^\nu$ in the neighborhood of $s = i_\nu(2^{(N-\nu)}ds)$ where $ds$

the sampling interval in the data. Thus, as $\nu$ increases, the DWT at this octave gives information about variability at succesively shorter characteristic scales on increasingly fine grids. The DWT is most easily understood when individual octaves are compared to the original data sequence. Figure 2 shows a synthesized data set and several octaves from its DWT. At each octave the DWT shows variability at a characteristic frequency or scale as a function of location. An obvious relationship exists between the peaks in the DWT spectrum and the localized variability in the data. There is also a correspondence between the scale of the variability and the octave in which its signature appears.

The greatest practical limitation of fast DWT's [e.g., *Press et al.*, 1992] is that the data vector must $2^n$ elements long, where $n$ is an integer. In the present study we perform a single DWT for each quantity measured during a flight. A typical flight is around 25,000 s long and therefore each data vector is 25,000 elements long. Zero padding is used to bring the length of each vector to 32,768 or $2^{15}$ elements. A fast DWT is performed on the padded data series. At each octave, the portion of the DWT corresponding to the zero padding is zero to machine accuracy, and is thus easily identified and tagged. Edge effects can be accounted for by tagging DWT components within a wavelet-filter length of the zero padding. The "good" or untagged DWT elements from each octave are pooled and used to calculate probability density functions. Averages of DWT amplitudes, for power spectrum or structure function calculations, are simply taken over the set of "good" elements. This processing is somewhat sloppier than that used in *Bacmeister et al.* [1996] in that there will not be an exact correspondence between the time interval covered in each octave (i.e., near the edges a number of DWT elements from higher octaves will be kept even if at some lower octave the corresponding DWT element is eliminated). Also, the spectra in *Bacmeister et al.* [1996] represent log-averages of subspectra from 1024 s flight segments, whereas in this study only simple averages are used. In any case, it will be seen that this simpler processing yields results equivalent to those in *Bacmeister et al.* [1996]. Finally, we mention again that in this study, spatial scales are obtained by simply multiplying time scales in the data by the ER-2 air speed of 200 m $s^{-1}$. Thus, time scales of 1 sec are assumed to represent horizontal scales of 200 m.

Many families of orthonormal, compactly supported wavelets have been identified [*Daubechies*, 1992]. These functions are exactly zero outside a finite interval. Unfortunately, only one member of the class of compact orthonormal wavelets has any simple symmetries. This is the Haar basis, which is the simplest wavelet basis known. An example of a Haar wavelet is shown in Figure 2. Haar wavelets are easy to apply, but tend to introduce high frequency components into power spectra due to their discontinuous shape. Other, higher-order, compactly supported wavelets yield power spectra in nearly exact agreement with FFT power spectra [e.g., *Bacmeister et al.*, 1996]. However, these higher order orthonormal wavelets do not possess simple symmetries and may not be appropriate for the structure function analyses to be described below. It is not obvious that asymmetric wavelets cannot provide analogous kinds of information. In fact comparison of DWT results using various bases may prove informative. However, in the interest of brevity and compatability with previous work, we will use the Haar basis in this preliminary analysis.

*2.3. Structure Functions*

Given a series of measurements of some fluid quantity $X(s)$, where $s$ can be either time or distance, p-th order structure functions are defined according to:

$$S_X(l,p) = \langle |X(s+l) - X(s)|^p \rangle \tag{2}$$

where $\langle\rangle$ represents some average over the data (i.e., over all $s$), and $l$ is a characteristic length scale. Classical turbulence theory gives predictions for the behavior of the these quantities as a function of $l$. In fact, if $X$ is velocity $u$ and $p = 2$, the structure function (2) is directly related to the kinetic energy spectrum according to $E(l) \sim l\ S_u(l,2)$. Turbulence theory predicts $S_u(l,2) \sim l^{2/3}$, which is simply a restatement of the familiar $E(k) \sim k^{-5/3}$ power law for kinetic energy in 3D inertial range turbulence.

The wavelet transform is essentially a convolution of a function like the Haar wavelet in Figure 2 with a sequence of data. Numerous researchers have argued that DWT's using simple antisymmetric wavelets such as the Haar basis are essentially differencing operators at logarithmically spaced characteristic scales [e.g *Katul*, 1994]. They conclude that DWT's using simple antisymmetric wavelets can be used to calculate quantities resembling structure functions. An approximate correspondence between the structure function (2) and the DWT (1) is proposed to be;

$$S_X(l,p) \sim \langle |D_{X,\nu}|^p \rangle \, (2^{(N-\nu)/2} ds)^{-p} \tag{3}$$

where the characteristic scale of the structure function is taken to be $l = 2^{(N-\nu)} ds$ and $\langle\rangle$ is an average over the elements of $D_{X,\nu}$ [*Katul et al.* 1994]. In the present analysis we will calculate the average $\langle\rangle$ over the set of "good" DWT amplitudes as described in Section 2.2, or over subsets thereof.

## 3. Results

*3.1. Previous Results for Horizontal Wavenumber Spectra*

The most familiar characterization of variability that can be derived for a sequence of data is the power as a function of frequency or in this case, as a function of horizontal wavenumber $k$. Horizontal wavenumber PSD's have been derived for a variety of atmospheric data sets [e.g. *Nastrom et al.*, 1986], including ER-2 data like that examined in this study [e.g., *Murphy*, 1989; *Bacmeister et al.*, 1996]. Figure 3 shows horizontal wavenumber PSD's for horizontal wind $u$, potential temperature $\theta$, ozone mixing ratio $R_{O3}$ and $N_2O$ mixing ratio $R_{N2O}$ from the ER-2 data set [adapted from Bacmeister et al.

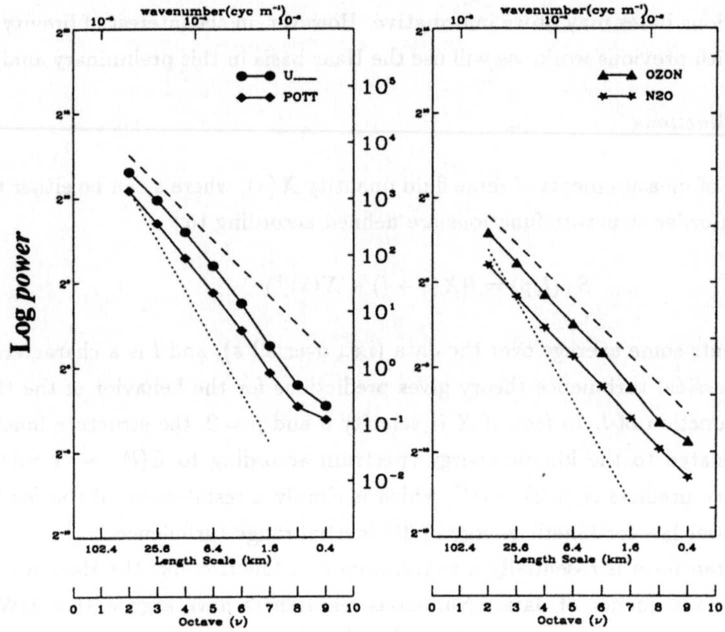


Figure 3:
**Power spectral densities (PSD's) of zonal velocity and potential temperature (left panel), and ozone and $N_2O$ mixing ratios (right panel) as a function of horizontal scale. Units of power spectral density are arbitrary. Average PSD's have been calculated using all data from AASE II, SPADE and ASHOE/MAESA campaigns using Daubechies 6-element wavelets and log-averages of subspectra as described in *Bacmeister et al* [1996]**

1996]. The $u$ and $\theta$ spectra nearly parallel everywhere except at the longest and shortest scales in the analysis. Between 0.8 km and 51.2 km the spectra exhibit a curved shape with slopes between –1.6 and –2.0 at longer scales and steeper slopes close to –2.5 at shorter scales. At scales longer than 50 km the $\theta$ spectrum retains a slope of around –2.0 while the $u$ spectrum appears to flatten. Parallel spectra for $u$ and $\theta$ are consistent with a gravity wave interpretation for the dynamical variability. Furthermore, as discussed in detail in Bacmeister et al. the shapes of $k-$spectra for the dynamical quantities (including vertical velocity, not shown here) are remarkably consistent with the separable vertical wavenumber- intrinsic frequency, $(m, \omega)-$spectra of *Gardner et al.* [1993]. In this comparison transformations based on linear gravity wave theory were employed to transform the $(m, \omega)-$spectra of horizontal velocity into $k-$spectra of other wave quantities. The ratio of vertical-to-horizontal velocity variance in ER-2 data was also found to be in good agreement with theoretical predictions.

However, the ozone and $N_2O$ mixing ratio PSD's in Figure 3 appear to be significantly different from the dynamical spectra. They are generally flatter between 0.8 and 51.2 km and do not display the curved shape of the dynamical spectra. The slope of the tracer spectra at scales longer than 1.6 km is around –1.7. Some evidence of a break

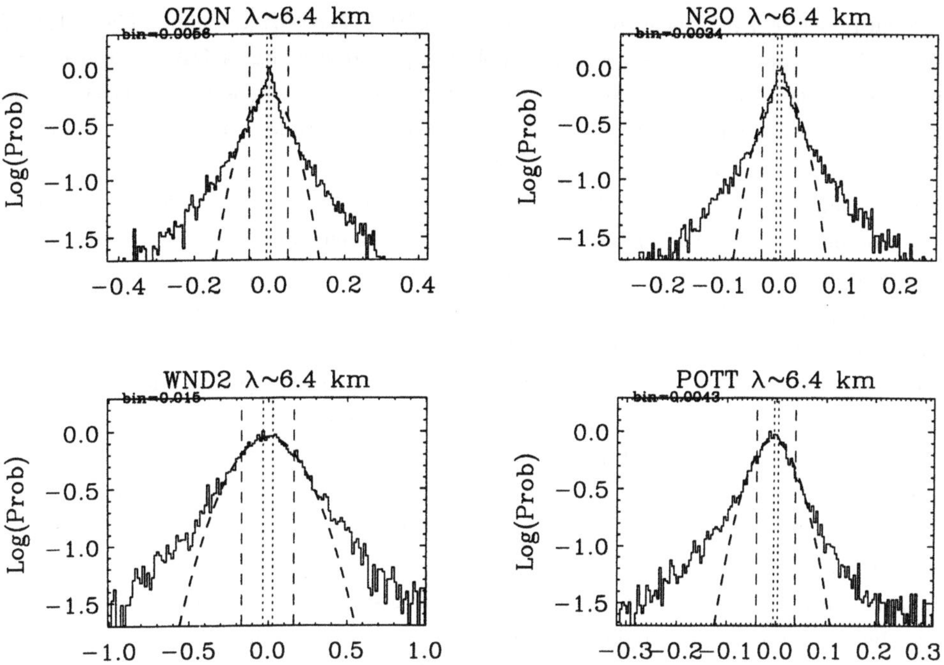


Figure 4:

Probability density functions (PDF's) for DWT amplitudes. PDF's for fluctuations with a charcteristic scale of 6.4 km are shown. (Top left) PDF for ozone mixing ratio fluctuations; (Top right) $N_2O$; (Bottom left) zonal velocity and; (Bottom right) potential temperature. Dashed parabolas show Gaussian fit to middle 66% of the data in each case. Vertical long-dashed lines enclose 50% of the data. Vertical short-dashed lines enclose 10% of the data.

to a steeper spectral slope is evident between 3.2 km and 1.6 km for ozone. Simple linear gravity wave theory would predict tracer spectra shaped like those of horizontal velocity and potential temperature. Current 2-D turbulence theories predict a wide range of slopes for tracer spectra. For a long-lived tracer in a nondiffusive 2-D enstrophy cascade $k^{-1}$ is expected, but diffusion may steepen the spectrum to something between $k^{-1}$ and $k^{-2}$ [*Pierrehumbert* 1995]. However, regardless of the present uncertainties in theoretical predictions for the tracer spectra, the basic conclusion remains that the tracer variance spectra and the dynamical energy spectra in Figure 3 are not presently consistent with a single theoretical explanation.

### *3.2. Probability Density Functions (PDF's)*

The most obvious way to obtain further information about variability from DWT's is to simply examine the distribution of DWT amplitudes at particular scales. Figure 4 shows histograms or probablity density functions (PDF's) of DWT amplitudes at a scale of 6.4 km for $u$, $\theta$, $R_{O3}$, and $R_{N2O}$. The PDF's shown in Figure 4 are taken over the entire set of ER-2 measurements for the ASHOE/MAESA mission. The value of the PDF at a particular position $x$ is simply the probability that a DWT amplitude in particular octave will fall between $x-0.5\Delta x$ and $x+0.5\Delta x$, where $\Delta x$ is chosen to be about 0.05 of the mean absolute value of the DWT amplitudes. The PDF's in Figure 4 have been normalized. The portion of each PDF shown represents 95% of all the "good" elements in the data. The plots are log-linear, so that the tails of the PDF's are prominent. The dashed parabolic curves in each panel show least-squares fits of the central 66% of the data to a Gaussian (or normal) distribution, which appears parabolic in log-linear plots. The difference between the tracer PDF's (ozone and $N_2O$) and the dynamical PDF's ($u$ and $\theta$) is striking. The tracer PDF's disagree with the 66% Gaussian fit everywhere, even near the center of the PDF's. In contrast, the dynamical PDF's are in excellent agreement with the 66% Gaussian fit in the middle of the distribution. Thus, the majority of dynamical fluctuation "events" are normally distributed. Good agreement with a Gaussian distribution is found for upto 80% of the dynamical fluctuations. Similar PDF's are obtained for all 4 quantities for a range of fluctuation scales spanning 1.6 km to 51.2 km. The occurrence of non-Gaussian statistics for the tracer fluctuation amplitudes suggests variability dominated by a number of large disturbances. Non-Gaussian statistics have been observed for a number of "turbulent" flows including the atmospheric boundary layer [*Katul et al.* 1994], High-Re wind tunnel [*Yamada and Ohkitani*, 1991] and in the Solar wind [*Marsch and Tu*, 1994].

*3.3. Structure Functions*

The PDF's shown in Figure 4 contain useful information about variability that is not contained in traditional Fourier power spectra. However, scale dependent information about variability is not easily diagnosed using PDF's. Analysis of structure functions (3) can recover some of the information about intermittency contained in the PDF analysis, as well as some of the scale dependent information contained in Fourier power spectra.

For many geophysical data sets it is well known that the power spectrum of variability is described approximately by a power law, i.e, $E(k)\sim k^{-\alpha}$. The most familiar example in fluid mechanics is the Kolmogorov spectrum for inertial-range, isotropic, 3-D turbulence, where $\alpha=5/3$. When the power spectrum is described by a power law the same will be true for the second order structure function, i.e.,

$$E(k) \sim k^{-\alpha} \rightarrow S(l,2) \sim h^{\zeta(2)}$$

where the scaling exponents $\alpha$ and $\zeta(2)$ are straightforwardly related by $\alpha = \zeta(2)+1$ [e.g., *Davis et al.* 1994]. Much of the information in structure function analysis is contained in the way the scaling exponent $\zeta(p)$ for the structure functions varies with $p$. The scaling exponents $\zeta(p)$ are determined by performing linear least squares fits to the curves $\log_2 S(l,p)$ vs. $\log_2 l$ for different values of $p$. The fits should be performed over a range

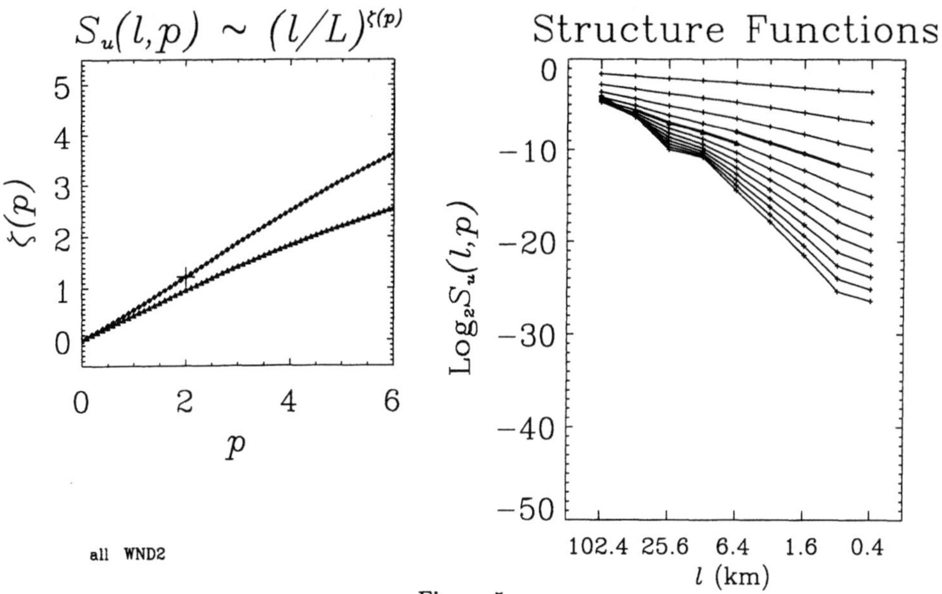


Figure 5:
(Right) Structure functions $S(l,p)$ as functions of length scale for zonal velocity; and (left) scaling exponents $\zeta(p)$ vs $p$ derived from linear fits to $S(l,p)$. The two $\zeta(p)$ vs $p$ curves correspond to different scale ranges used in linear fit to structure functions. The steeper $\zeta(p)$ vs $p$ curve corresponds to the scale range 0.8 to 6.4 km. The shallower $\zeta(p)$ vs. $p$ curve corresponds to the scale range 6.4 to 51.2 km. The $\log_2 S(l,p)$ vs. $\log_2 l$ curves on the right are shown for $p$=0.5, 1.0, ...., 6.0.

of scales where the energy or variance spectrum is well described by a power law [*Davis et al.* 1994]. For variables with normally distributed fluctuation amplitudes at all scales and energy or variance spectra steeper than $k^{-1}$, the $\zeta(p)$ vs. $p$ curve will be a straight line.

Figures 5 and 6 show structure functions and scaling exponents for ASHOE/MAESA zonal velocity and $N_2O$ data. As discussed above, the determination of $\zeta(p)$ involves choosing a scale range over which to perform the linear fit to the structure functions. For the zonal velocity structure functions, we performed two linear fits; one over the scale range from 0.8 km to 6.4 km and another from 6.4 km to 51.2 km. This was done since our earlier analysis of zonal velocity power spectra [*Bacmeister et al.*, 1996] suggested a change in the spectral index somewhere between 3.2 and 12.8 km. This change in spectral index is reflected in the structure functions for zonal velocity, which exhibit an apparent break at 12.8 km. Nevertheless, consistent with the Gaussian PDF's for zonal velocity fluctuations shown in Figure 4, the scaling exponent $\zeta(p)$ for horizontal velocity is approximately linear in $p$ over both scale ranges.

In contrast the $\zeta(p)$ vs $p$ curve for $N_2O$ (Figure 6) is decidedly nonlinear. At best, the $\zeta(p)$ vs $p$ curve for $N_2O$ can be decomposed into two approximately linear regions; one for

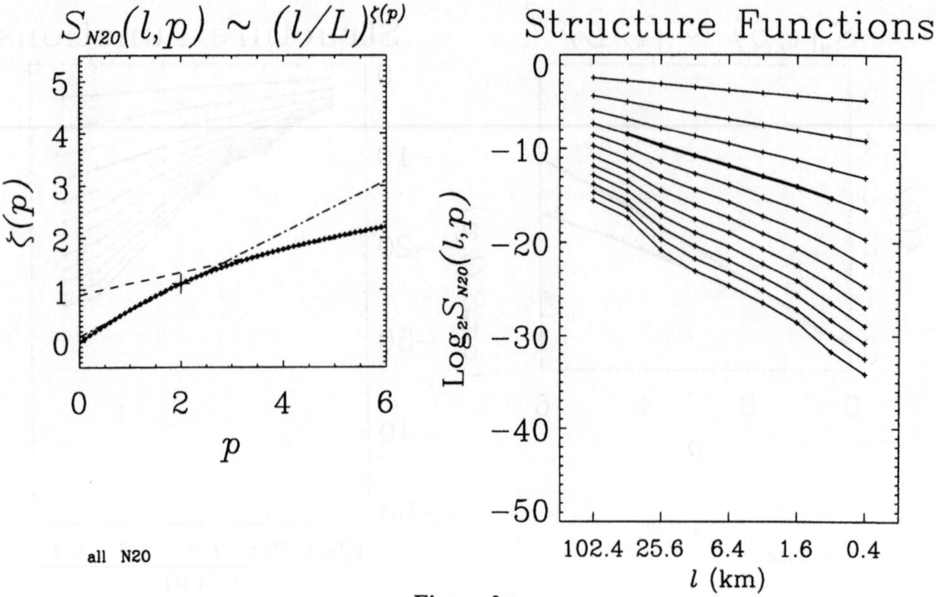


Figure 6:
**As in Figure 5 but for $N_2O$ mixing ratios. In this case, the fits to determine $\zeta(p)$ are made over a single scale range from 0.8 km to 51.2 km. The straight lines plotted over $\zeta(p)$ vs $p$ curve have slopes of 0.5 and 0.22. The y–intercept of the shallower line occurs at around 0.9. This suggests the presence of highly discontinuous or intermittent variability in the $N_2O$ data (see text).**

$p<2$, where $\zeta(p)$ has a slope of around 0.5; and a second for $p>3$, where the slope of $\zeta(p)$ is much shallower, close to 0.22. Between, $p=2$ and $p=3$ the $\zeta(p)$ gradually changes from one slope to the other. This is clearly an oversimplification of the shape of the $\zeta(p)$ curve for $N_2O$, but it is one that leads to a simple interpretation in terms of "fractal" sets.

*Bi-Fractal Interpretation.* Here we outline an interpretation of the structure functions and exponent curves in Figures 5 and 6 in terms of variability with different degrees of "roughness" [see *Davis et al.*, 1994] and with different degrees of continuity. The argument presented follows one given in *Aurell et al.* [1992] for structure functions arising in connection with Burger's Equation. Assume a certain kind of variablility produces fluctuations of scale $l$ which have a characteristic amplitude $H(l)$. Also, assume that $N(l)$ such fluctuations typically occur per unit length. The contribution to the $p$-th order structure function at scale $l$ from this variability will be,

$$S(l,p) \sim H(l)^p \; N(l) \; l$$

The simplest kind of variability is simply a smooth, continuous curve. In this case the number of fluctuations at a given scale $N(l)$ varies as $l^{-1}$ while their amplitude $H(l)$ varies as $l$. The structure function is then simply $S(l,p) \sim l^p$ and the $\zeta(p)$ vs. $p$ curve will simply be a straight line of slope 1. If we want continuous but "rougher" variability we keep $N(l) \sim l^{-1}$ but allow $H(l) \sim l^{\alpha}$ where $1>\alpha>0$. In other words the amplitude of fluctuations decreases more slowly than $l$ as $l$ decreases. The structure function for this variability becomes $S(l,p) \sim l^{\alpha p}$ and the $\zeta(p)$ vs. $p$ curve is a straight line of slope $\alpha$. This describes the horizontal velocity results in Figure 5 for a choice of $\alpha$=0.8 or 1.1. If we now also allow the number of fluctuations to vary as $l^{-d}$ where $d$ is less than one, we can describe variability that is not continuous, or in the context of a 1-D data record, sets of variability whose "dimension" is less than one. If such a discontinuous set with $N(l) \sim l^{-d}$ and $H(l) \sim l^{\alpha_d}$ is added to continuous variability with $N(l) \sim l^{-1}$ and $H(l) \sim l^{\alpha_c}$ we will get a $p$-th order structure function,

$$S(l,p) \sim l^{\alpha_c p} + l^{1-d} l^{\alpha_d p}$$

In order to determine the scaling exponent $\zeta(p)$ (that is the slope of the $\log S(l,p)$ vs $\log l$ curve) we let $l \to 0$. This gives,

$$\zeta(p) = \mathrm{MIN}\left[\alpha_c p,\ 1-d+\alpha_d p\right]$$

In other words for this simple "bifractal" model the $\zeta(p)$ vs. $p$ curve becomes the minimum envelope of two lines. Furthermore, the y-intercept of the shallower of the two lines has a clear interpretation as the "dimension" of the set of discontinuous variability. In the case of ASHOE $N_2O$ the shallow $\zeta(p)$ vs. $p$ curve at $p>3$ implies the existence of highly discontinuous variability with $d$ near or below 0.1. The preceeding is highly oversimplified, and glosses over important questions about the PDF's of fluctuations at each scale. However, it gives a simple interpretation for the $N_2O$ structure functions that can be used to test ideas. More complete discussions of "bifractal" and "multifractal" data sets are given for example in *Meneveau and Sreenivasan* [1991] *Davis et al.* [1994], and *Frisch* [1995] (Chap. 8.5).

## 4. Discussion

Analysis of atmospheric data in terms of PDF's and structure functions is in its infancy, and the work presented here is preliminary. However, it seems safe to draw several conclusions from the preceeding analysis. First of all, the PDF's and structure functions shown here reinforce earlier tentative statements in *Bacmeister et al.* [1996] that dynamical quantities such as velocities and potential temperature are subject to different sources of variability than tracer mixing ratios. Both structure functions and PDF's suggest a richer character for tracer variability. Tracer variability appears to be strongly multifractal, containing a continuous background of "rough" variability as well as intermittent events of high amplitude. In contrast, dynamical variability appears to be dominated by continuous, normally distributed background variability. Velocity and potential temperature PDF's exhibit non-Gaussian tails but these do not appear to represent a significant fraction of the variability. However, it should be emphasized that the ER-2 sampling is somewhat biased against periods of intense dynamical variability at small scales, such as large mountain waves or severe turbulence (for obvious safety reasons).

In their analysis of ER-2 data *Bacmeister et al.* [1996] argued that the observed horizontal wavenumber spectra of velocities and potential temperature dynamical were consistent with a gravity wave interpretation. The same was not true for the tracer variance spectra. A simple gravity wave interpretation based on vertical displacements of a smooth background tracer field would predict tracer variance spectra similar in shape to those of horizontal velocity or potential temperature. Similar fluctuation PDF's and structure functions for horizontal velocities, potential temperature, and tracers would also be expected from a linear gravity wave interpretation. Instead, we find non Gaussian PDF's and tracer variance spectra near $k^{-1.7}$ over a wide range of scales. Such statistics may be characteristic of a wide class of 2D, area-preserving maps applied to a tracer field [e.g. *Pierrehumbert*, 1995]. Thus, the results of Section 3 appear to be consistent with the idea that a significant fraction of tracer variability in the stratosphere is associated with quasi-isentropic motions which stretch and fold large scale tracer gradients into small-scale "filaments". This has long been the dominant interpretation among researchers investigating stratospheric tracer transport, for small scale features observed in trace gas data from the ER-2 and other sources [e.g., *Schoeberl et al.*, 1993; *Waugh and Plumb*, 1994] This interpretation has been buttressed by the the success of passive advection models using global wind analyses in predicting the location of filaments observed by the ER-2 [e.g., *Schoeberl et al.*, 1993].

### *4.1. Gravity Wave-Tracer Filament Interaction*

It appears likely that quasi-isentropic filamentation is responsible for a significant proportion of observed tracer variability. However, it is also clear that the background gravity wave field in the atmosphere must also generate some degree of tracer variability. This is all the more probable since the same process of differential advection, which can generate small scale tracer structures in the horizontal, can also lead to the formation of small-scale features in the vertical [e.g. *Reid et al.*, 1994]. Gravity wave vertical displacements can be expected to interact strongly with thin filaments of tracer. Clear examples of such interactions have been observed. Figure 7 shows ER-2 measurements of potential temperature, horizontal wind, as well as ozone mixing ratio taken on December 14, 1991 over northern Quebec, Canada. The ozone field shows a series of sharp, almost square-wave like, variations between 62000 and 65000 UT. The ozone mixing ratio appears to jump between two envelope curves, the higher curve around 2.8 ppmv and the lower curve around 2.0 ppmv. During the same period, the potential temperature, as well as the horizontal velocity, exhibited enhanced variability. However, the enhancement in dynamical variability appears to relatively small compared to the enhanced tracer variability. Figure

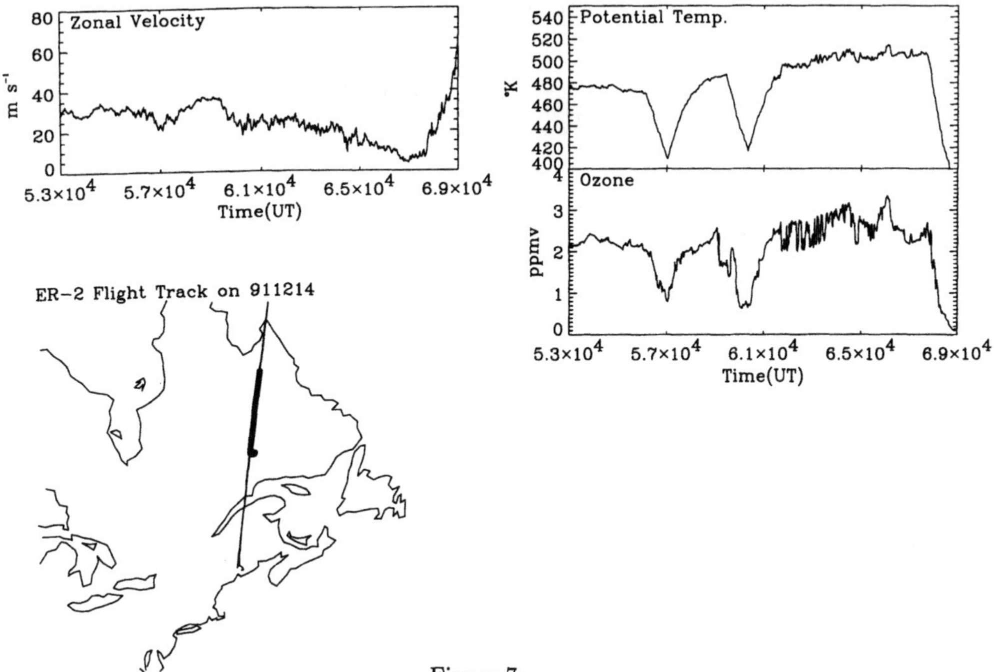


Figure 7:
Map in lower left hand panel shows ER-2 flight track on Dec. 14, 1991. Other 3 panels show flight data recorded during flight. Upper left shows zonal velocity. Upper right shows potential temperature and lower right shows ozone mixing ratio. Highlighted portion of track indicates the portion of flight occurring between 62000 and 65000 UT sec. During this time interval strong variability was observed in ozone mixing ratio (see lower right panel).

8 shows ozone mixing ratio as a function of potential temperature for the same portion of the flight track. Adiabatic gravity wave vertical displacements should not perturb the $O_3$–$\theta$ relationship. However, between 490°K and 510°K large deviations from the apparent background $O_3$–$\theta$ curve are evident. The ozone mixing ratio in the background curve appears to increase almost linearly from around 2.7 ppmv at 490°K to 3.0 ppmv at 510°K. The filament exhibits sharply reduced ozone mixing ratios with values between 1.9 and 2.1 ppmv in the same $\theta$ range. Such a reduction of ozone mixing ratio along a $\theta$ surface at this time and location suggests the intrusion of an air mass from further equatorward. It is clear in this case that the presence of a filament drastically enhances the ozone variability produced by wave-induced vertical displacements. In this example there did not appear to be significant wave-induced turbulence in the vicinity of the filament, based on a cursory examination of vertical and horizontal velocity DWT's. However, repeated superpositions of thin filaments and wave-induced turbulence could significantly enhance climatological cross-isentropic mixing of tracers and vorticity. This could play an important role in producing the diffusion required in the study of *Pierrehumbert* [1995] to produce tracer spectra steeper than $k^{-1}$.

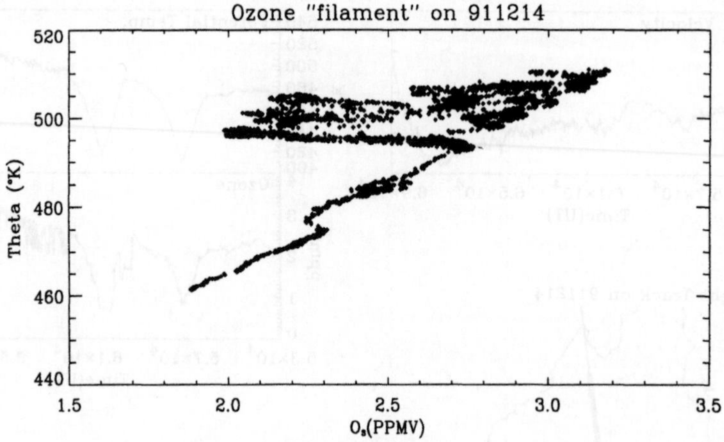


Figure 8:
**Ozone mixing ratio as a function of potential temperature for 62000 to 65000 UT on Dec. 14, 1991. Note sharply reduced ozone between 490°K and 510°K suggesting a thin intrusion of air from further south (equatorward).**

## 5. Summary and Conclusions

This study presented a wavelet-based analysis of PDF's and structure functions for velocity, potential temperature, and trace gas ($N_2O$ and ozone) mixing ratio data measured by a high-altitude ER-2 aircraft in the lower stratosphere. Clear differences between the tracer variability and the dynamical variability were detected. Tracer variability was dominated by large intermittent events, while dynamical variability was dominated by relatively continuous nearly Gaussian fluctuations. This was evident in both PDF and structure function analyses. The behavior of the structure functions for $N_2O$ was interpreted in terms of a simple bifractal model. It was argued that the differences between tracer and dynamical variability may be due to the different roles of gravity wave motions vs. quasi-isentropic stretching and folding in producing the observed variability in each case. Finally, an example of strong interactions between a tracer "filament" and gravity wave motions was presented.

The interpretation of these data is still far from clear. However, it appears safe to conclude that tracer variability in the stratosphere is quantifiably more intermittent than dynamical variability. The intermittent character of tracer variability is likely to be related to quasi-isentropic stretching and folding by large scale motions, which yields for example sudden narrow filaments. The same stretching and folding would not significantly enhance dynamical variability at small scales. An earlier study of ER-2 data suggested that dynamical variability was largely consistent with a spectrum of gravity wave motions. The lack of intermittency in the dynamical fields may therefore place constraints on the nature of wave-wave interactions which might occur in the atmosphere. However, before such implications can be fully examined in the ER-2 data set, questions of ER-2 sampling biases (e.g. its avoidance of intense dynamical variances), as well as questions of resolution (e.g., can MMS resolve typical turbulent length scales) must be carefully considered.

*Acknowledgments.* This work was partially funded by NASA contract W-18,566

## References

Aurell, E., U. Frisch, J. Lutsko, and M. Vergassola, On the multifractal properties of the energy dissipation derived from turbulence data. *J. Fluid Mech., 238,* 467-486, 1992.

Bacmeister, J. T., S. D. Eckermann, P. A. Newman, L. R. Lait, K. R. Chan, M. Loewenstein, M. H. Proffitt and B. L. Gary, Stratospheric horizontal wavenumber spectra of winds, potential temperature and atmospheric tracers observed by high-altitude aircraft. *J. Geophys. Res.*, 101, 9441-9470, 1996.

Daubechies, I., Orthonormal bases of compactly supported wavelets, *Comm. Pure Appl. Math., 41*, 909-996, 1988.

Daubechies, I., *Ten lectures on wavelets*, 357 pp., CBMS-61, SIAM, Philadelphia, 1992.

Davis, A., A. Marshak, W. Wiscombe and R. Cahalan, Multifractal characterizations of nonstationarity and intermittency in geophysical fields: Observed, retrieved, or simulated, *J. Geophys. Res., 99*, 8055-8072, 1994.

Frisch, U., Turbulence: the legacy of A. N. Kolmogorov, 296 pp., Cambridge Univ. Press., New York, 1995.

Gardner, C. S., C. A. Hostetler, and S. J. Franke, Gravity wave models for the horizontal wave number spectra of atmospheric velocity and density perturbations, *J. Geophys. Res., 98,* 1035-1049, 1993a.

Katul, G. G., M. B. Parlange, C. R. Chu, Intermittency, local isotropy, and non-gaussian statistics in atmospheric surface layer turbulence, *Phys. Fluids, 6,* 2480-2492, 1994.

Lilly, D. K., Stratified turbulence and the mesoscale variability of the atmosphere. *J. Atmos. Sci., 40,* 749-761, 1983.

Lilly, D. K., and P. F. Lester, Waves and turbulence in the stratosphere, *J. Atmos. Sci., 31,* 800-812, 1974.

Loewenstein, M., J. R. Podolske, K. R. Chan, and S. E. Strahan, Nitrous Oxide as a dynamical tracer in the 1987 Airborne Antarctic Ozone Experiment.,*J. Geophys. Res., 94*, 11,589-11,598, 1989.

Mallat, S. G., A theory for multiresolution signal decomposition: The wavelet representation, *IEEE Trans. Pattern Anal. Machine Intell.* 11, 674-693, 1989.

Marsch, E., and C. Y. Tu, Non-Gaussian probability distributions of solar wind fluctuations, *Ann. Geophysicae, 12*, 1127-1138, 1994.

Meneveau, C., and K. R. Sreenivasan, The multifractal nature of turbulent energy dissipation, *J. Fluid Mech., 224*, 429-484, 1991.

Murphy, D., Time offsets and power spectra of the ER-2 data set from the 1987 Airborne Antarctic Ozone Experiment, *J. Geophys. Res., 94*, 16737 - 16748, 1989.

Nastrom, G. D., W. H. Jasperson, and K. S. Gage Horizontal spectra of atmospheric tracers measured during the Global Atmospheric Sampling Program. *J. Geophys. Res., 91,* 13201-13209, 1986

Pierrehumbert, R. T., Tracer microstructure in the large-eddy dominated regime. in *Chaos applied to fluid mixing*, ed. H. Aref, M. S. El Naschie; Pergamon/Elsevier, 347-365, 1995.

Press, W. H., S. A. Teukolsky, W. T. Vetterling, and B. P. Flannery, *Numerical Recipes in FORTRAN*, 2nd ed., 963 pp. Cambrige University Press, Cambridge, UK, 1992.

Proffitt, M. H., M. J. Steinkamp, J. A. Powell, R. J. McLaughlin, O. A. Mills, A. L. Schmeltekopf, T. L. Thompson, A. F. Tuck, T. Tyler, R. H. Winkler, and K. R. Chan, In situ ozone measurements within the 1987 ozone hole from a high-altitude ER-2 aircraft, *J. Geophys. Res., 94,* 16547-16556, 1989.

Reid, S. J., G. Vaughan, N. J. Mitchell, I. T. Pritchard, H. J. Smit, T. S. Jorgensen, C. Varostos, and H. de Becker, Distribution of ozone laminae and the possible influence of inertia-gravity waves, *Geophys. Res. Lett., 21,* 1479-1482, 1994.

Schoeberl, M. R., A. R. Douglass, R. S. Stolarski, P. A. Newman, L. R. Lait, D. W. Toohey, L. M. Avallone, J. G. Anderson, W. Brune, D. W. Fahey, and K. Kelly, The evolution of ClO and NO along air parcel trajectories, *Geophys. Res. Lett., 20,* 2511-2514, 1993.

Scott, S. G., T. P. Bui, K.R. Chan, and S.W. Bowen, The meteorological measurement system on the NASA ER-2 aircraft, *J. Atmos. Ocean. Tech., 7*, 525-540, 1990.

Waugh, D. W., and R. A. Plumb, Contour advection with surgery: A technique for investigating fine scale structure in tracer transport, *J. Atmos. Sci., 51,* 530-540, 1994.

Yamada, M., and K. Ohkitani, Orthonormal wavelet analysis of turbulence, *Fluid. Dyn. Res., 8*, 101-115, 1991.

# Observations of Gravity Waves with the UARS Microwave Limb Sounder

Dong L. Wu and Joe W. Waters
Mail Stop 183-701
Jet Propulsion Laboratory, California Institute of Technology
4800 Oak Grove Drive
Pasadena, California 91109
U.S.A.

## 1. Introduction

Geographical and spectral distributions of gravity waves are crucial for large-scale circulation and local mixing in the atmosphere. Lack of global gravity wave (GW) climatology makes it difficulty to quantify the total momentum and energy forcings contributed by the small-scale eddies. Observations of the GW distributions have been provided previously by various techniques such as radar [Meek et al. 1985; Vincent and Fritts 1987; Fukao et al. 1994], lidar [Wilson et al. 1991], balloon [Allen and Vincent 1995], rocket [Hirota 1984], aircraft [Nastrom and Gage 1985] and satellite [Fetzer and Gille 1994]. However, each of these techniques only measures waves of certain spatial and temporal scales. Observations from radar, lidar, balloon and rocket yield good temporal and vertical resolutions usually at one geographical location while aircraft observations provide good horizontal resolution but for a short period of time. It is difficult in general for space-borne sensors to obtain the same resolutions, but observations of GWs at somewhat larger scales are feasible, for example using saturated radiances from the Upper Atmosphere Research Satellite (UARS) Microwave Limb Sounder (MLS) [Wu and Waters 1996].

Observations from UARS MLS can provide geographical distributions and seasonal climatology of small-scale GWs in the middle atmosphere. The gravity waves to which MLS is sensitive are those with relatively long (>10 km) vertical wavelengths and therefore are of importance to the momentum budget in the mesosphere and lower thermosphere. The UARS MLS, in operation since 12 September 1991, was designed to measure profiles of molecular abundances ($O_3$, ClO, and $H_2O$), temperature and pressure in the middle atmosphere using thermal emission features near 63, 183 and 205 GHz [Waters 1993; Barath et al. 1993]. We recently made an extraordinary use of the saturated radiances from the 63-GHz radiometer channels, which measure atmospheric temperatures at different altitude layers, for small-scale gravity wave study [Wu and Waters 1996]. This study is benefited from a good global coverage ranging from 34° latitude in one hemisphere to 80° in the other, because the MLS field of view is 90° from the UARS orbital velocity and the orbit is 598 km high with 57° inclination. The UARS makes 10 yaw maneuvers each year allowing alternating views of high latitudes in the two hemispheres with a periodicity of ~36 days. Section 2 provides some examples of MLS raw radiance measurements where gravity waves are evident as coherent patterns among the measurements at different altitudes. In section 3 sampling issues and temperature weighting functions are discussed. Section 4 describes a variance analysis technique for routine MLS limb-scanning observations which is used to map GW activity on a

NATO ASI Series, Vol. I 50
Gravity Wave Processes
Their Parameterization in Global Climate Models
Edited by Kevin Hamilton

monthly basis. Section 5 discusses the GW variance maps obtained for January and July periods and a climatology for 1992/93. Summary and conclusions are in section 6.

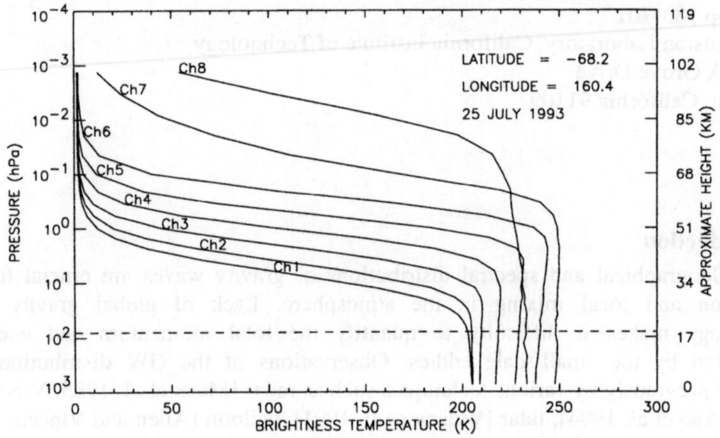


**Figure 1**. An example of MLS $O_2$ radiance profiles near 63 GHz as the instrument step-scans from the mesosphere to the surface. Fluctuations in the saturated radiances are generally large for center channels but small for wing channels due to atmospheric temperature variability and instrument noise.

## 2. MLS 63-GHz Radiance Measurements

Two general scanning mechanisms have been implemented in the MLS operation for atmospheric sampling: limb-scanning and limb-tracking. In routine limb-scanning mode the instrument step-scans the atmospheric limb in ~65 seconds at 2 seconds/step from ~90 km to the surface producing increments of ~5 km in the mesosphere and 1-3 km in the stratosphere and troposphere. Figure 1 shows the radiance profiles of a single scan measured by the 63 GHz radiometer, which resolves an $O_2$ emission line into 15 spectral channels for retrieving atmospheric tangent pressure and temperature [Fishbein et al. 1996]. All radiances are saturated when the instrument views tangent heights below ~18 km because of strong line absorption. Radiances near line-center (channel 8) saturate at higher altitudes than those near the line-wing (channels 1 and 15 that are symmetric about channel 8). The saturated radiance is a good measure of atmospheric temperature and depends little on the tangent height of pointing. Thus, as the satellite moves along, fluctuations in the saturated radiances actually reflect atmospheric temperature variations in the horizontal direction. Nevertheless, although the radiances below ~18 km are saturated, they still depend weakly and nearly linearly on pointing. By removing this weak linear trend, one can detect fluctuations in the saturated radiances due to horizontal variations in atmospheric temperature.

In limb-tracking mode the instrument tracks the limb, usually at ~18 km tangent height, and makes measurements with the same 2-second sampling frequency as in limb-scanning mode (equivalent to ~15 km horizontal resolution along the suborbital track). Since all 63 GHz channel radiances are optically thick in the 18-km-limb-tracking observation, MLS in this mode measure atmospheric temperature continuously with a good horizontal resolution. Such a data set is quite useful for GW studies. Figure 2 illustrates the sampling pattern for the limb-tracking mode, showing the MLS suborbital tracks on 28 December 1994.

centers separated by ~15 km, while a larger gap in the middle is caused by a 6-second instrument calibration. The limb-tracking mode was used nearly continuously during 23-30 Dec. 1994, 1 Feb.-20 Mar. and 7-15 Apr. 1995, and normally scheduled for every-third-day operation since then while MLS is on. (Because of degradation in the UARS power system, the instruments on board are now operated in a time-sharing mode). Radiance measurements from channels 1-2 and 14-15 are not shown here because these channels are not fully saturated at high latitudes with the 18-km-limb-tracking mode and pointing variations may contaminate the radiance variances.

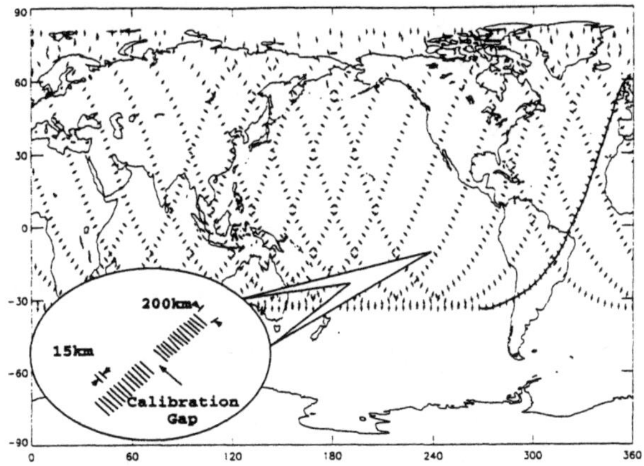


**Figure 2**. UARS/MLS sampling tracks on 28 December 1994 marked by the first measurement of each major frame (65.5 seconds). The inset details the set of individual measurements in a single major frame with the short lines indicating the orientation of the temperature weighting functions (see text). On this day MLS was preferentially observing the Northern Hemisphere. The ascending portion of orbits 1 is highlighted with solid lines.

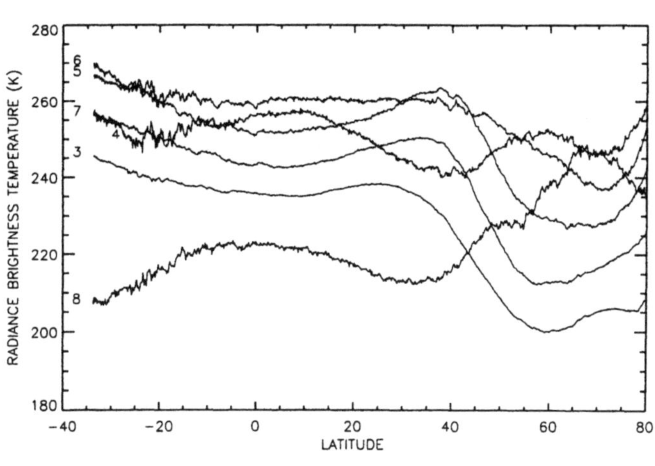


**Figure 3**. Channel 3-8 radiance measurements from the ascending part of orbit 1 on 28 December 1994.

Figure 3 gives an example of limb-tracking radiance measurements taken on 28 December 1994, when the radiance brightness temperatures observed have large gradients near the vortex edge (~40-50°N) for channels 3-5. The weaker latitudinal gradients in channel 6-8 radiances suggest that the vortex is weakening at altitudes above 50 km. The radiance gradient differs from orbit to orbit because the vortex was not zonally symmetric at that time in the Northern Hemisphere. Coherent variations can be seen in the radiance fluctuations at different channels (or altitudes) with scales from hundreds to thousands of kilometers. Caution should be given to the interpretation of large-scale radiance variations from channels 7-8. These channels and channel 9 (not shown in Figure 3), close to the $O_2$ line center, can vary by 3-8 K due to the Zeeman effect associated with Earth's magnetic field. However, these variations generally change slowly along the orbital track except above 70°N and below 25°S (during this observing day) where the viewing angle changes rapidly with respect to the magnetic field lines.

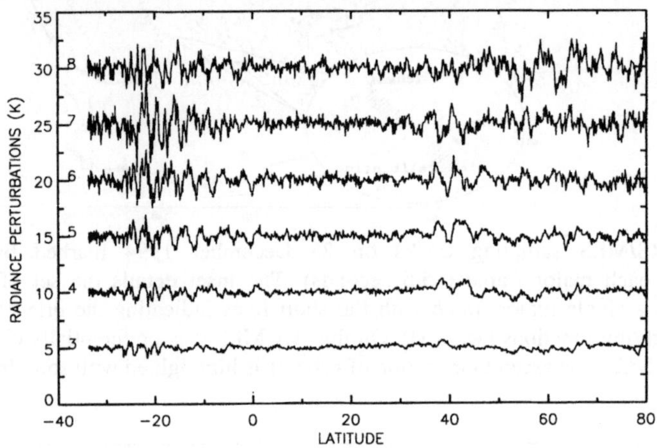


**Figure 4**. Radiance fluctuations derived from Figure 3 with large-scale (>1000 km) variations removed. Radiance fluctuations of each channel are displaced by 5 K with the channel number indicated at the left of each measurement series. There are more samples near the turning latitudes at 34°S and 80°N than elsewhere. In the middle portion of the orbit, a 10° latitude band corresponds to a distance of ~1300 km.

The small-scale fluctuations can be seen more readily in Figure 4 where large-scale (>1000 km) variations are removed from the radiances in Figure 3. The filtered data in Figure 4 are obtained by differencing the raw and smoothed data (i.e., averages over ~1000 km). The radiance fluctuations from the first ascending orbit show a magnitude of 0.5-3 K at 10°-30°S over South America where deep convection is known to be strong during this period. Oscillations of 1-2K are apparent in the mid- and high-latitude Northern Hemisphere. Phase coherence and amplitude growth with height are clearly seen at some latitudes, suggesting the presence of vertically propagating waves. The 0-30°N region is relatively quiet where the fluctuations are mainly instrument noise.

It should be noted that the MLS radiance fluctuations are expected to largely underestimate atmospheric temperature variations mainly because of the broad vertical and horizontal weighting functions. Atmospheric temperature variances are typically 1-5 $K^2$ in the lower stratosphere according to radiosonde measurements [Allen and Vincent, 1995] and 1-10 $K^2$ in the upper stratosphere from rocket observations [Hirota, 1984; Eckermann et al., 1994]. Moreover, the gravity wave spectra can vary largely with time, height, and place, which adds complexity to interpretation of the radiance fluctuations. The observed magnitude of the radiance fluctuations is a result of convolutions of wave spectra, wave propagating directions, instrument weighting functions, and sampling patterns. Aliasing between wave amplitudes and propagating directions may give problems in directly relating the radiance fluctuations to GW parameters. The quantitative GW information in the MLS measurements is therefore to some extent as limited as what can be obtained from other techniques. Before we further discuss the radiance fluctuations, the weighting functions and the instrument spatial resolution need to be described.

## 3. Temperature Weighting Functions

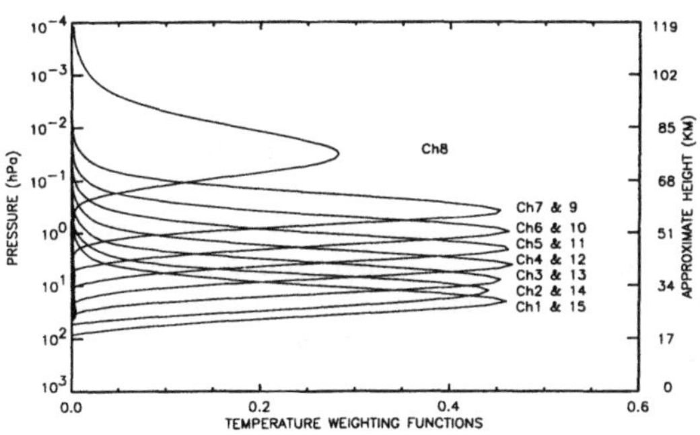


**Table 1**. The 63 GHz channel parameters

| Ch. | Apprx. Height (km) | Layer Thickness (km) | Noise (K) | Elevation Angle (degree) |
|---|---|---|---|---|
| 1, 15 | 28 | 10 | 0.07 | 3.2 |
| 2, 14 | 33 | 10 | 0.08 | 3.9 |
| 3, 13 | 38 | 10 | 0.12 | 4.5 |
| 4, 12 | 43 | 10 | 0.18 | 5.1 |
| 5, 11 | 48 | 10 | 0.26 | 5.6 |
| 6, 10 | 53 | 10 | 0.37 | 6.0 |
| 7, 9 | 61 | 10 | 0.49 | 6.7 |
| 8 | 80 | 15 | 0.45 | 8.0 |

**Figure 5** Temperature weighting functions of channels 1-15 for the MLS 63 GHz radiometer viewing the limb at 18 km (calculated by W. G. Read).

Instrument spatial resolution and sampling patterns are key to sensing GW-scale disturbances in the atmosphere. The MLS has temperature weighting functions and sampling schemes that are suitable for observing some small-scale gravity waves. Figure 5 describes the MLS temperature weighting functions for 18 km tangent height radiances, showing eight altitude layers (~10-15 km) where temperature is measured by the saturated radiances of different channels. Because the MLS line-of-sight (LOS) direction is perpendicular to the orbit velocity, horizontal averagings are ~100-300 km cross-track (perpendicular to suborbit path) due to radiative transfer through the limb path, and ~30 km along-track (parallel to suborbit path) due to the antenna field-of-view (FOV) smearing. The vertical and horizontal averagings can substantially reduce the magnitudes observed from actual atmospheric temperature fluctuations, but they are still detectable due to low radiometer noise (varying from 0.07 to 0.5 K). Table 1 summarizes the key parameters of the temperature weighting functions and the 63 GHz channel noise.

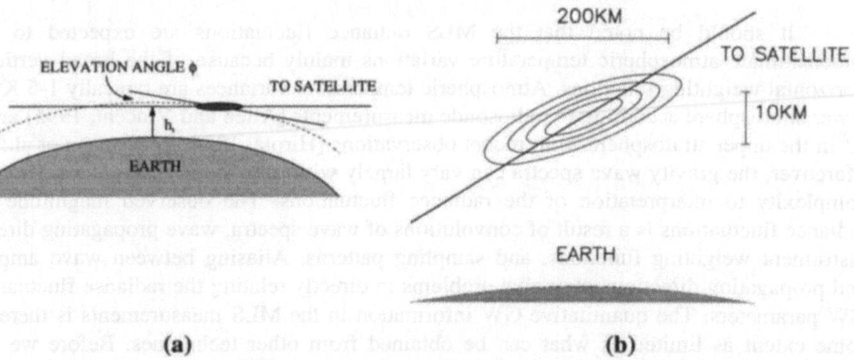


**Figure 6** (a) Observing geometry for MLS saturated radiances. (b) Detailed weighting functions in the saturation region where the line of sight is indicated by the straight line. The weighting function in (b), normalized by its peak, is contoured at 0.2, 0.4, 0.6 and 0.8 from edge to center, while the line of sight is indicated by the straight line.

One must consider the 3-dimensional nature of the temperature weighting functions in order to understand what waves can be observed with MLS. Atmospheric waves propagate vertically as well as horizontally, and therefore the amplitude of observed radiance fluctuations depends on orientations of the weighting function relative to wave propagation direction. Figure 6 illustrates the MLS observing geometry where the temperature weighting function has an asymmetric shape in the vertical plane, showing slight tilting from the local horizontal plane. Because of the spatial asymmetry of the temperature weighting functions, the magnitude of radiance fluctuations observed depends on the angle between the LOS direction and wave vector. The observed magnitude is larger if the LOS direction is aligned more along with wavefronts.

The horizontal projection of the weighting functions is also asymmetric with a width of ~20 km and a length 150-300 km. As illustrated in Figure 7, the instrument is more sensitive to the waves propagating in the *y* (along-track) direction due to larger averaging in the *x* (LOS) direction: waves are filtered differently in the *x* and *y* directions. As shown in Figure 8 the response to horizontal wavenumber in the *x*-direction ($k_x$) is cut off at high positive numbers because of the long (~200 km) weighting function. Little responses at small and negative wavenumbers are due to the angle preference of the tilted weighting function which produces severe averaging on the waves with wavefronts not parallel to the LOS direction. The response cutoff at small wavenumbers in the *y*-direction ($k_y$) is due to the data point truncation

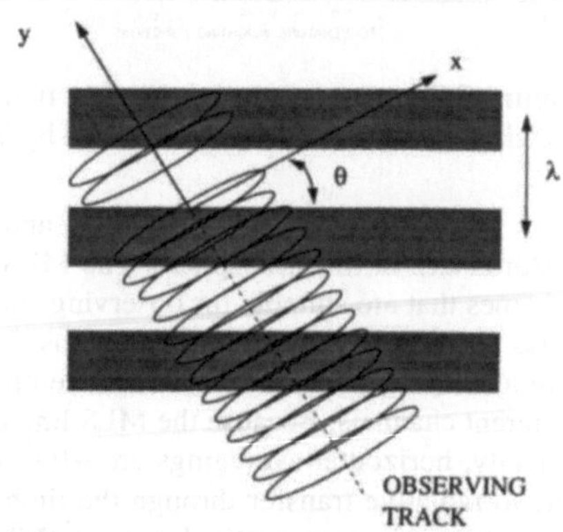


**Figure 7.** Temperature weighting functions of the saturated radiances projected on the horizontal plane. The footprints of the weighting function move across wavefronts (shaded) with an angle θ, and $k = 2\pi\lambda$.

that any analysis method must include. The low response for high $k_x$ is caused by the spatial smearing (~15 km) during the 2-second integration and the width of the temperature weighting function (~20 km).

The combination of the spectral filters in the $x$ and $y$ directions yields an optimal viewing angle and an optimal wavelength for MLS to observe atmospheric waves. This is simply because the response functions in Figure 8 provide spectral constraints in two orthogonal directions. The two orthogonal wavenumbers are related by $k_x = k\cos\theta$ and $k_y = k\sin\theta$, where $k$ and $\theta$ are defined in Figure 7. Figure 9(a) shows the calculated radiance response as a function of horizontal wavelength and observing angle. For waves with a 10 km vertical wavelength, the optimal observing angle is ~30° between the instrument FOV and wavefronts, while the most observable horizontal wavelength is ~100 km. Because the UARS orbit has an inclination of 57°, the MLS is more sensitive to meridionally-propagating waves near the equator and more sensitive to zonally-propagating waves near the orbit turning latitudes (Figure 2). Figure 9(b) gives the maximal radiance response as a function of vertical wavelength, where poorer sensitivity in smaller vertical wavelengths is the direct effect of the MLS broad vertical weighting functions. There should be a vertical wavelength that is most observable if the GW spectrum and dispersion relation are considered. However, the wave spectrum and its climatology are not well known at present, although some theoretical and observational studies suggest that short vertical wavelengths (<10 km) may dominate the waves in the lower atmosphere (< ~50 km) while long (>10 km) waves, due to atmospheric filtering, are more important in the upper atmosphere [Meek et al. 1985; Smith et al. 1985].

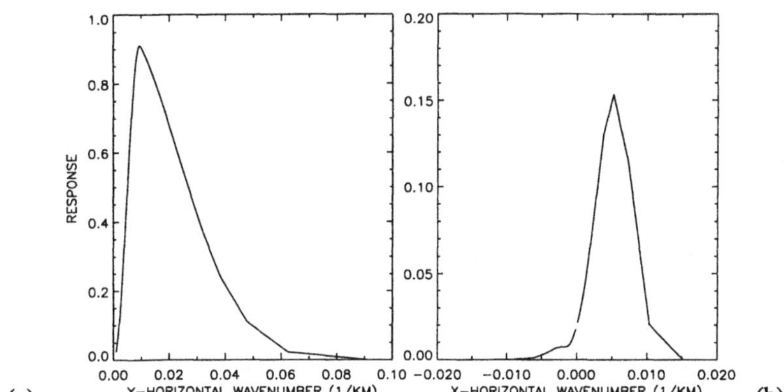


**Figure 8**. Radiance responses to the horizontal wavenumbers (a) along and (b) cross the track. The response function in (a) is a simple convolution of a 6-point (~90 km) truncation functions and along-track smearing, while the response function (b) is calculated by convolving the instrument weighting function (Figure 6b) with a 1 K monochromatic wave that has a 10 km vertical wavelength. For a given vertical wavelength, positively propagating waves have wavefronts more parallel to the MLS field of view direction.

The minimal observable wavelength can be determined from Figure 9 if the minimal detectable radiance amplitude or variance is specified, which depends on the analysis method used. Section 4 describes a simple technique of variance analysis for MLS limb-scanning data, with which the minimal observable vertical wavelength is ~7 km in a monthly average. The minimal detectable variance can be shown to depend on the number of data points averaged and the noise level.

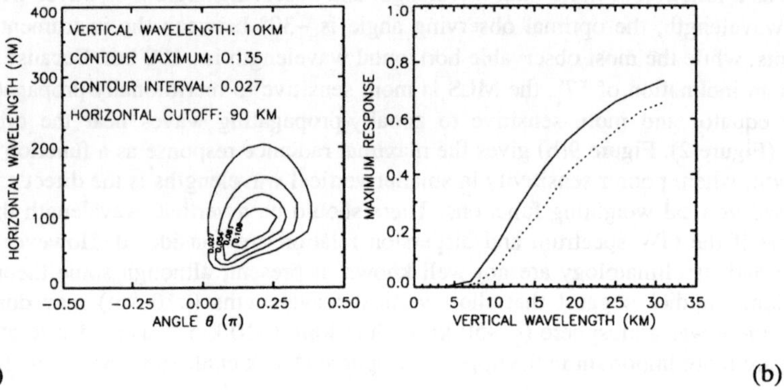


**Figure 9**. MLS radiance response as a function of (a) horizontal wavelength and observing angle θ and (b) vertical wavelength. The radiance responses in (b) correspond to the peak value in (a), where the solid (dashed) line is for channel 1 (8).

## 4. Variance Analysis

The method discussed here is a radiance variance analysis that can be used to derive the atmospheric temperature variance from the saturated radiances measured with the MLS limb-scan mode (like one in Figure 1). There is a maximum of 6 consecutive data points in the bottom of each scan where all channels can meet the saturation requirements. A radiance variance for each channel can be calculated from these 6 measurements. The radiance variance calculated for a give channel contains instrument noise and atmospheric temperature variations from a specified altitude layer as illustrated in Figure 2. To calculate the radiance variance, a linear variation is firstly removed from the 6 radiance measurements, which removes effects of the weak tangent pressure dependence and large-scale wave modulations. Subsequently, the estimated total radiance variance for a given channel, $\tilde{\sigma}^2$, similar to the definition of reduced chi-square, is obtained by

$$\tilde{\sigma}^2 = \frac{1}{4}\sum_{i=1}^{6}(y_i - a - bz_i)^2 \tag{1}$$

where $y_i$ and $z_i$ denote, respectively, radiance measurements and tangent heights. Parameters $a$ and $b$ are determined from the linear least squares fit to the 6 measurements, and 4 is the degree of freedom. As we discussed above, this total radiance variance, $\sigma^2$, is mainly due to atmospheric temperature fluctuations, instrument noise and other possible error sources such as non-linear terms in the pressure dependence, namely,

$$\sigma^2 = \sigma_{GW}^2 + \sigma_N^2 + \sigma_{NL}^2 \qquad (2)$$

where $\sigma_N^2$ is the variance due to instrument noise for the given channel and is known from instrument calibrations in each limb scan. The non-linear pressure contribution, $\sigma_{NL}^2$, is small and only important for channels 1/15 and 2/14, and can be reasonably estimated from radiance models (when these channels are used in the analysis). As a result, the atmospheric fluctuation, $\sigma_{GW}^2$, can be derived by subtracting $\sigma_N^2$ from the estimated total radiance variance $\tilde{\sigma}^2$. We now interpret the atmospheric temperature fluctuations contributing to the radiance variance as a manifestation of upward propagating GWs [Hines 1960]. The same procedure is repeated for all the radiance channels, and the results are averaged for the channel pairs that are symmetric about the line center. Other fluctuation sources, such as the antenna pointing, are either insignificant or very occasional and, therefore, neglected in the analysis here.

The uncertainty in the estimated total radiance variance is the fundamental limit for detecting weak gravity wave signals, and this depends on the number of data points averaged and the instrument noise. To reduce this uncertainty, we generally average measurements over a month for each latitude-longitude grid. The statistical uncertainty of the averaged radiance variance, therefore, is reduced significantly and given by $|\tilde{\sigma}^2 - \sigma^2| \approx \sqrt{2/m}\sigma^2$, where $m$ is the total number of data averaged within the grid. For example, an 80-variance average will make a wave variance of $10^{-3}$ $K^2$ statistically significant in the channel 2 radiances.

## 5. Gravity Wave Variances

The variance maps presented in this section are 40-day averages for two periods near solstices: January (20 December 1992 - 29 January 1993) and July (18 June to 28 July 1993), centered on UARS yaw days. Figure 10 shows the resulting maps at seven altitudes, and striking features in these maps are large amplitudes associated with the stratospheric polar vortex in the winter hemisphere and subtropical land masses in the summer hemisphere. These features evolve with height and change remarkably above the stratopause.

Background winds are expected to play a major role in determining the GW variance amplitudes observed with MLS. Theoretical studies [Schoeberl and Strobel 1984; Miyahara et al. 1987] show that a strong background wind is a favored condition for GWs to propagate vertically because of the large intrinsic phase speed (i.e. difference between horizontal wave phase speed and the background wind) that prevents the waves from breaking. One may interpret the enhanced variance associated with the stratospheric polar jet as the result of vertically propagating GWs as well. This interpretation is consistent with some aircraft observations in the lower stratosphere [Hartmann et al. 1989], where a positive correlation was found between small-scale static stability and wind speed. It is the selective filtering effect of the jetstream that acts to reshape the wave spectrum by allowing upstream propagating GWs to grow more efficiently with height than others, and likely causes the variance enhancement observed in the jetstream. However, horizontal finestructures near the vortex can also contribute to the variance observed by MLS, but we cannot quantify this contribution at present. As another result of background wind filtering, the variances in the subtropical summer hemispheres show larger amplitudes at the latitudes (10°S-30°S in January and 10°N-30°N in July) where winds are stronger. The distribution of these summer GW variances is consistent with that of the large-wavenumber momentum flux calculated from the GFDL SKYHI high-resolution general circulation model [Miyahara et al. 1987], both enhanced over Madagascar, Australia, South Pacific and Brazil during January.

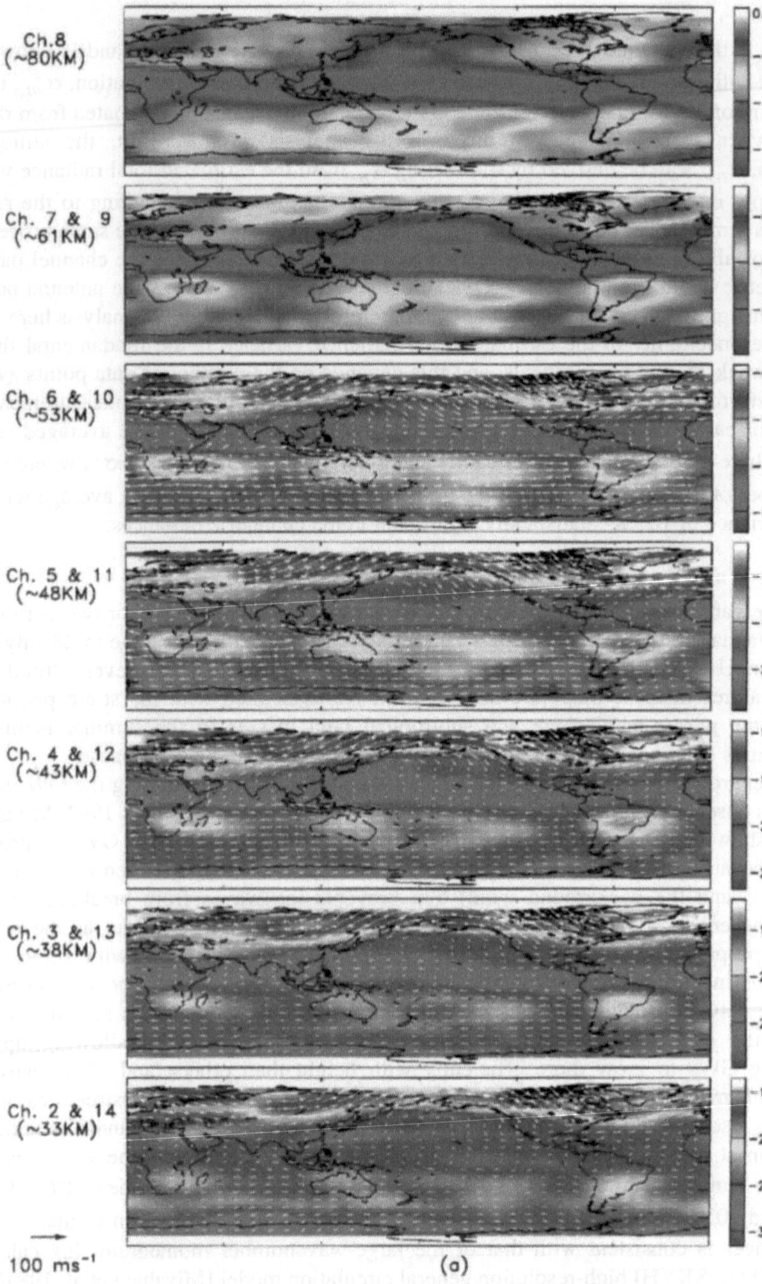


**Figure 10(a).** Maps of gravity wave variances for January. Latitude and longitude bins are respectively 5° and 10° with more than 40 measurements in each grid point. The variances are in a unit of $K^2$ and colored in a logarithmic scale, i.e., $\log_{10}(\sigma^2_{GW})$. Winds (up to ~1 hPa) are derived from the US National Meteorological Center data [Manney et al. 1994] and averaged over the same periods.

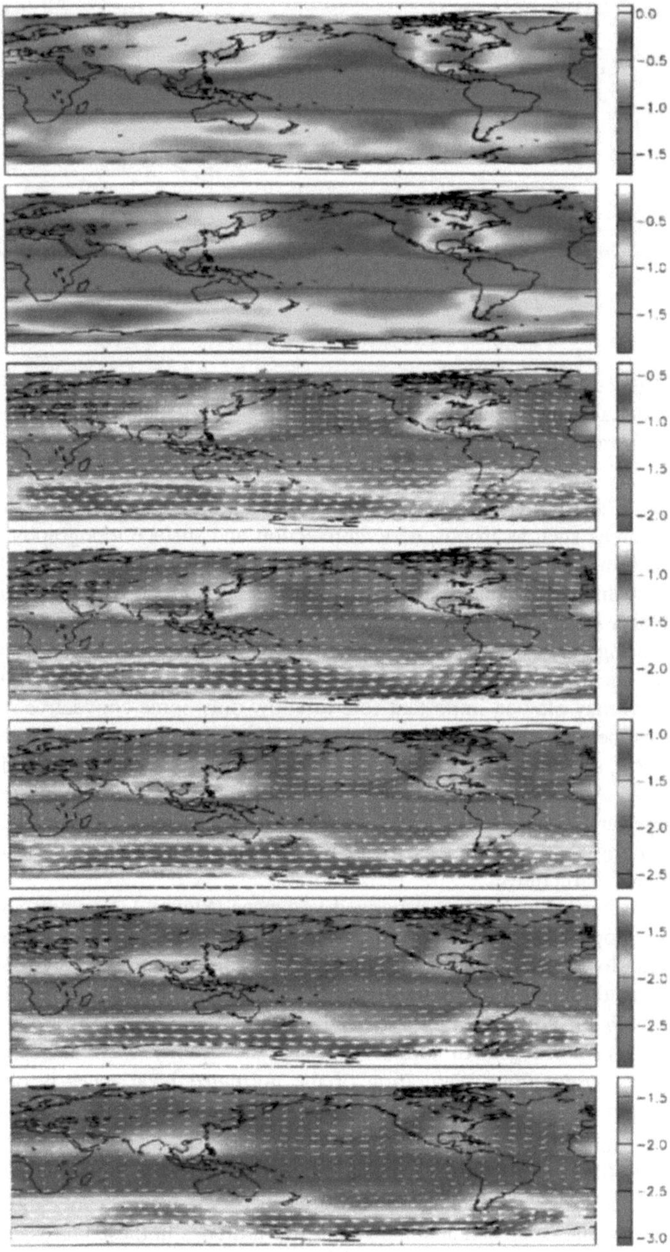


**Figure 10(b).** As in Figure 10(a) but for July.

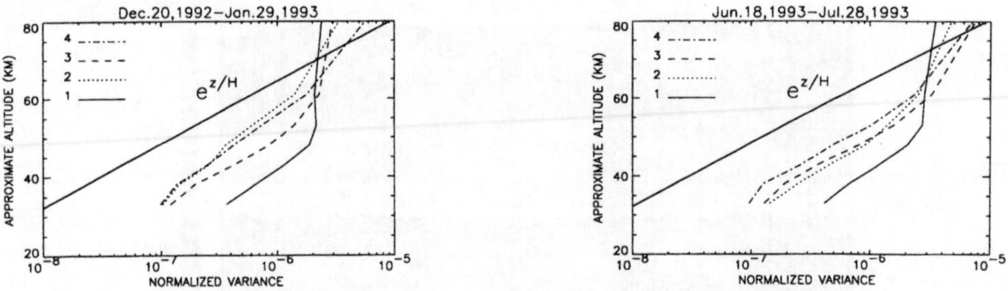


(a) (b)

**Figure 11.** Variance growth at different latitudes during (a) January and (b) July compared with the exponential growth expected for non-breaking GWs. The variances are normalized by the squared mean radiance brightness temperature.

The spatial distribution of variances shown in Figure 10 also provide information on gravity wave sources such as tropospheric convection and surface topography. Tropospheric cumulus convection, frequently occurring in the summer over tropical and subtropical land masses, is most likely responsible for the large stratospheric variances observed near Madagascar, North Australia and Brazil during January, and South Asia, Central America and North Africa during July. However, not all convectively-generated GWs can propagate into the stratosphere because of background wind structures. As discussed above, the prevailing subtropical winds may play an important role to allow only some convective disturbances to reach the stratosphere. Topography-related variances can be seen clearly in the maps from channels 8 and 7/9, showing the enhancements over Europe, Asia and America during January and July, and over Andes during July. The zonal asymmetries of the wintertime jetstreams in the stratosphere are generally believed due to differences in surface topography and tropospheric forcings between the two hemispheres, and GWs may provide a considerable contribution to such structures.

The height variation of the normalized variances (Figure 11) reveals some important aspects of the propagating nature of the observed perturbations. Despite very different amplitudes in the lower stratosphere, these variances exhibit approximately the same growth rate with height throughout the stratosphere, which is consistent with the theoretical exponential growth for non-breaking GWs and rocket observations [Hirota and Niki 1985]. This property in variance growth further supports the gravity wave interpretation of the radiance variance observations. Saturation of the normalized variances is observed in the mesosphere, implying wave breakdown/saturation and momentum drag at these altitudes [Fritts 1984]. In Figure 12, zonal means of the normalized variances show that the saturation occurs lower over the vortices (50°N-70°N in January and 50°S-70°S in July) and subtropical land masses (10°S-30°S in January and 10°N-30°N in July). This suggests that the stratospheric jets would be closed at somewhat lower altitudes in these places, consistent with the structure of climatological mean zonal winds during these periods [Fleming et al. 1990].

Moreover, dynamic heating due to the GW breaking/saturation in the jetstreams may reverse the temperature lapse rate remarkably and create temperature inversion layers in the mesosphere. Recent maps of temperature inversions in the mesosphere [Leblanc et al. 1995] show a distribution supported by the MLS wave variance observations.

Alexander and Holton [this volume] have simulated the GW variances that would be observed by MLS, using a quasi-linear model with a broad wave spectrum input at the lower boundary and convolving the predicted temperature variations at higher altitudes with the weighting functions similar to those in Figure 2. In their simulations the GW forcing at the lower boundary was set to be uniform in latitude and longitude, left only with mean zonal winds as variables affecting the wave spectra. As a result, a non-uniform distribution of the temperature variance is obtained and strikingly similar to the variance maps observed by MLS, very much catching the first-order variability. The non-uniform distribution resulted from a uniform forcing again suggest the importance of background winds in gravity wave propagation in the middle atmosphere. It is also suggested by their model calculations that the variances observed by MLS are much likely due to the atmospheric GWs and hence useful for validating some GW parameterization schemes.

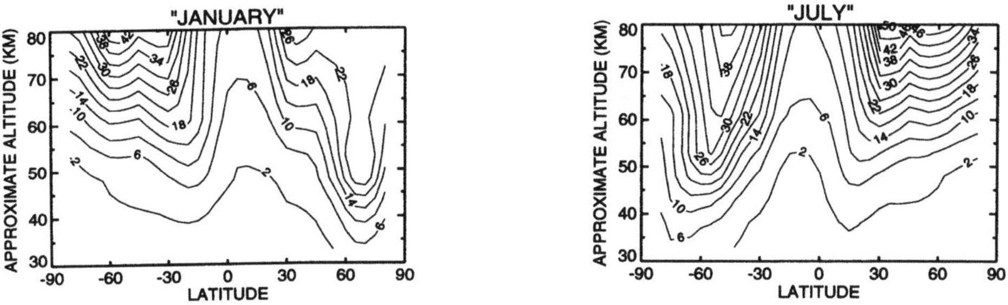


(a) (b)

**Figure 12.** Zonal mean normalized GW variances for (a) January and (b) July. Contours are in units of $10^{-7}$.

Figure 13 provides time series of daily averaged wave variances for the period of October 1992-October 1993. Ascending-descending differences are evident in the averaged wave variances, implying complicated propagation behaviors of the GWs. Let us focus on the features in the 0-40° latitude summer hemisphere which show a strong annual variation in the stratosphere and a somewhat semiannual variation in the mesosphere. These variances, as discussed above, they are likely associated with the GWs generated by tropospheric deep convection and reach the stratosphere with aid of strong westward winds. During the winter/summer months the variances in the stratosphere are very sensitive to the MLS viewing geometry, showing a large difference before and after the yaw days. Since the variances generally vary slowly with time within a UARS month (i.e., the time period between two adjacent yaw maneuvers) for both ascending and descending orbits, the sudden decreases/increases in the variances after a yaw maneuver are due to the changes of MLS viewing geometry with respect to propagating GWs.

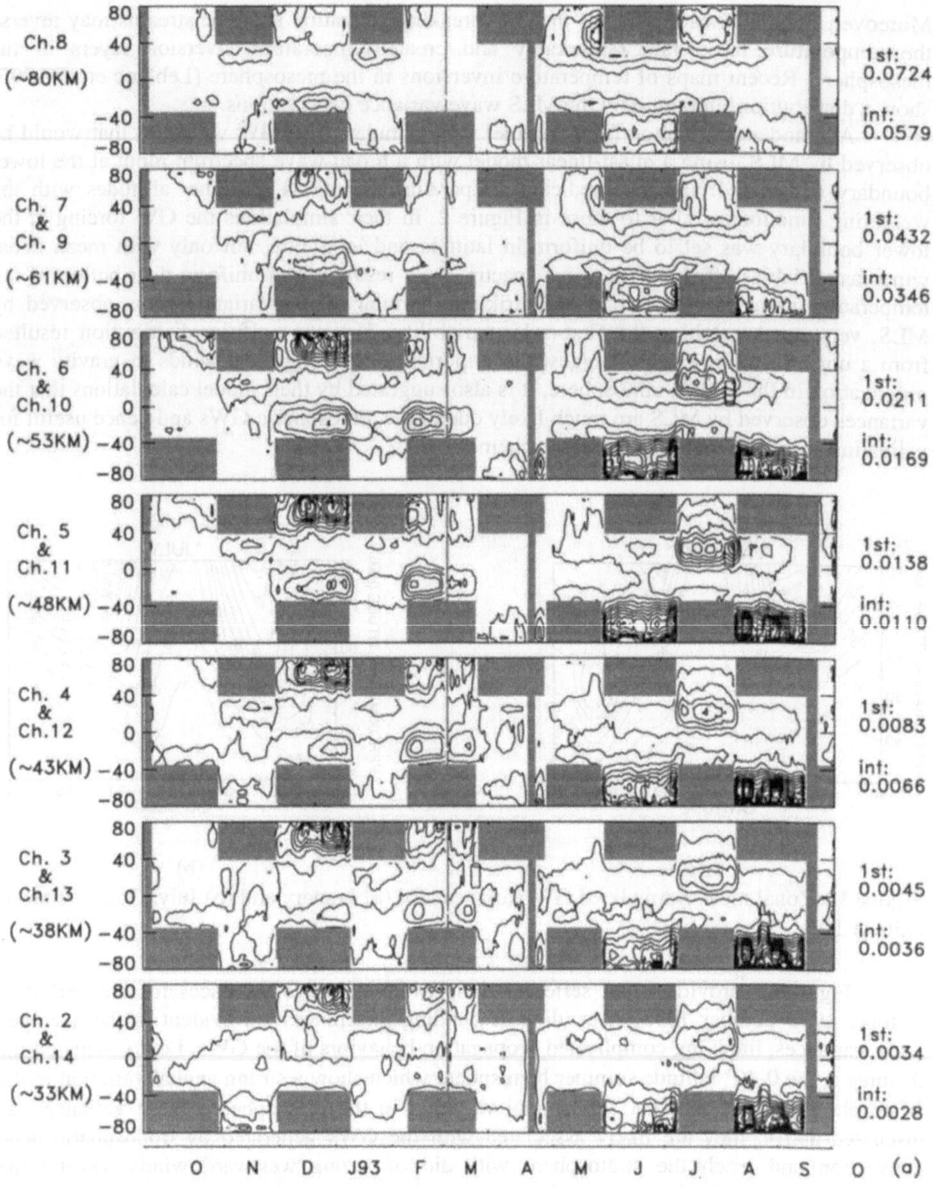


**Figure 13(a).** Time-latitude plots of daily GW variances for ascending measurements. The first contour and contour interval, in units of $K^2$, are noted on the right of each panel. A 5-day smoothing is employed for each latitude band with data gaps highlighted by shaded areas. The large data gaps alternated at high latitudes are caused by the UARS yaw maneuvers. At a given latitude the MLS looking angle with respect to the atmosphere is approximately the same for an entire UARS month.

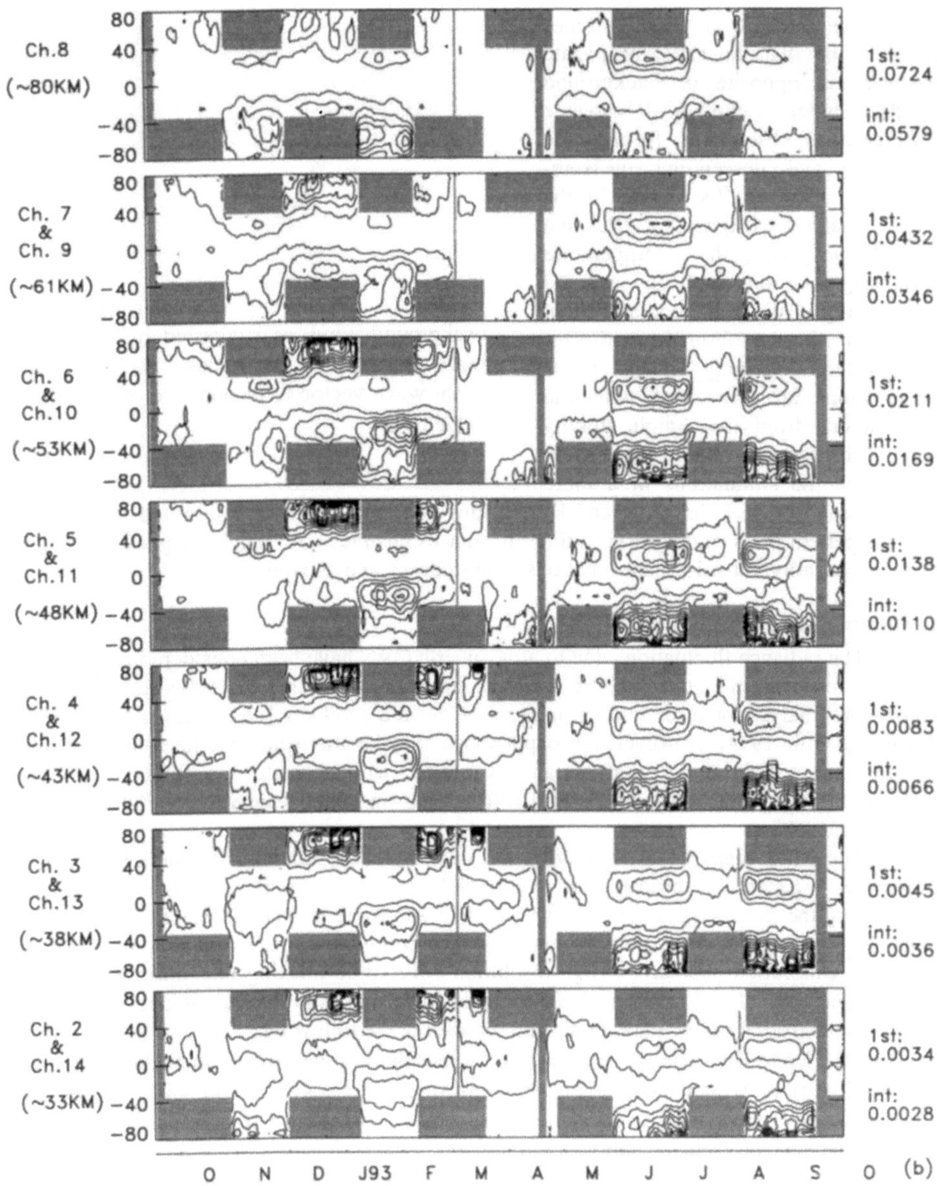


**Figure 13(b)**. As in Figure 13(a) but for descending measurements.. As a result of the yaw maneuver, the MLS viewing direction changes at a given latitude, providing an opportunity for observing wave propagation directions. To the first-order approximation, the variance differences between ascending and descending measurements are caused by the angle variations between the instrument FOV and wave vectors.

The scenario at these latitudes needs to be that the GWs propagate eastward in opposite of background winds in order to explain all variance differences observed from the ascending and descending measurements [Figure 14]. As shown in the early analysis [Figures 8-9], the convolution of the temperature weighting functions with an eastward propagating GW can generate large differences in the observed variances with different viewing angles. These observations are consistent with the expectations from the simple gravity wave propagation theory [Lindzen, 1981] and the simulations in a more realistic atmosphere [Alexander and Holton, this volume]. It can be also noted in Figure 13 that the pattern of ascending-descending differences changes with height, suggesting that the wave propagation directions are functions of altitude as well as latitude.

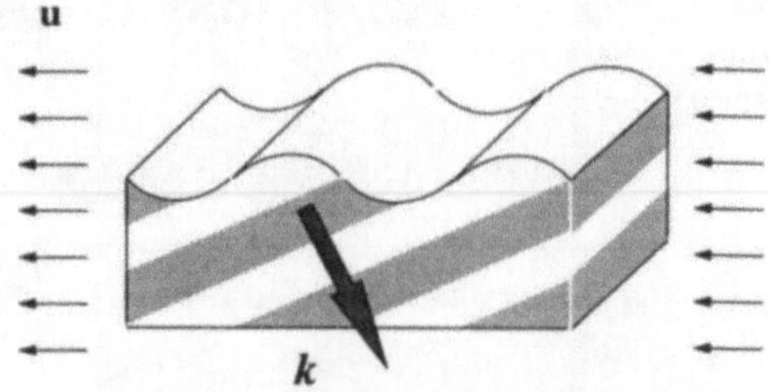


**Figure 14**. Gravity wave propagation in a background wind, as suggested by the MLS wave variance observations, showing the zonal wave vector $k_\lambda$ in opposite of the wind velocity **u**.

**6. Summary and Conclusions**

The radiance fluctuations due to atmospheric GWs have been considered as "noise" to the MLS retrieval, in fact, to most remote sensing techniques. This "noise" is reflected in the Chi-square analysis or error budget analysis of desired products [i.e., Fishbein et al. 1996; Fetzer and Gille 1994]. However, this paper shows that the atmospheric temperature variances can be extracted from the total radiance variances with accurate and frequent on-board calibration, and therefore provide useful information on small-scale wave variability in the middle atmosphere.

The MLS maps of GW activity can contribute substantially to our knowledge of gravity wave generation, propagation, and breakdown, which provide an observational basis for refining the parameterization schemes used in atmospheric modelings. The advantage of using satellite measurements for GW studies is their continuous spatial and temporal coverage with one calibration standard. High-resolution MLS limb-tracking and normal limb-scan observations have clearly shown some detectable wave signals in the saturated radiances. The major results from this new data set are summarized in the followings:

- The MLS 63 GHz channels can measure atmospheric temperature fluctuations of small (~100 km) horizontal and large (>10 km) vertical scales in 30-80 km altitudes. For instance, a variance of $10^{-3}$ $K^2$ can be statistically significant in a 10°x5° monthly averaged map for channel 2 and 14 saturated radiances.
- The wave variances calculated from the saturated radiances reveal some interesting new features associated with the stratospheric polar vortices, tropospheric deep convection zones, and surface topography. Much of the observed variance enhancements can be explained with gravity waves propagating in strong background winds.
- The normalized variances show the expected exponential amplitude growth with height in the stratosphere and saturation in the mesosphere.

- The wave variance is dominated by an annual variation in the stratosphere and a semiannual variation in the mesosphere.
- Separate analyses of the ascending and descending measurements show that the variances are sensitive to wave propagation directions, and suggest that the subtropical variances associated with deep convection are likely caused by the gravity waves that propagate upward and eastward in the prevailing westward stratospheric wind.

Further study of this data set will be focused on the gravity wave spectrum and the structures of the strong radiance perturbations in the stratospheric polar vortex. More difficult questions such as, to what extent the vortex finestructures contribute to the MLS radiance variances, and how gravity and planetary waves interact with each other, need to be answered. The limb-tracking data are particularly useful for addressing these questions and need to be fully explored in the future. More advances in the GW observations are anticipated while the UARS MLS continues collecting data with the limb-tracking mode.

The technique described in section 4 for variance analysis is for general purposes and can be used for measurements from the MLS 183 GHz channels as well. Similar to the 63 GHz radiometer, the 183 GHz radiances will saturate to the atmospheric temperature of various altitude layers. The differences, however, are their temperature weighting functions and a narrower beamwidth for the 183 GHz channels. These differences allow us to compare the variances calculated from the two radiometers so as to gain more knowledge about the height variation of temperature fluctuations.

**Acknowledgments.** We thank Joan Alexander, Bill Read, Evan Fishbein, Jim Holton, Gloria Manney and Lee Elson for valuable discussions, and Robert Jarnot, Dennis Flower and Richard Lay for operating the instrument. This work was performed at the Jet Propulsion Laboratory, California Institute of Technology, under contract with the National Aeronautics and Space Administration.

**References**

Alexander, M. J., and J. R. Holton (1996), Non-stationary gravity wave forcing of the stratospheric zonal mean wind. *This volume.*

Allen, S. J., and R. A. Vincent (1995), Gravity wave activity in the lower atmosphere: Seasonal and latitudinal variations. *J. Geophys. Res.*, **100**, 1327-1350.

Barath, F.T., et al. (1993), The Upper Atmosphere Research Satellite Microwave Limb Sounder instrument, *J. Geophys. Res.* **98**, 10,751-10,762.

Eckermann, S., I. Hirota, and W. K. Hocking (1994), Gravity wave and equatorial wave morphology of the stratosphere derived from long-term rocket soundings, *Q. J. R. Meterol. Soc.*, 121, 149-185.

Fetzer, E.J., and J.C. Gille (1994), Gravity wave variance in LIMS temperatures. Part I: Variability and comparison with background winds. *J. Atmos. Sci.*, **51**, 2461-2483.

Fishbein, E. F., R.E. Cofield, L. Froidevaux, R.F. Jarnot, T. Lungu, W.G. Read, Z. Shippony, J.W. Waters, I.S. McDermid (1996), Validation of UARS MLS temperature and pressure measurements, *J. Geophys. Res.*, **101**, 9983-10,016.

Fleming, E. L., S. Chandra, J. J. Barnett, and M. Corney (1990), Zonal mean temperature, pressure, zonal wind and geopotential height as functions of latitudes. *Adv. Space Res.*, 10, 11-59.

Fritts, D.C. (1984), Gravity wave saturation in the middle atmosphere: A review of theory and observations. *Rev. Geophys.*, 22, 275-308.

Fukao, S., M.D. Yamanaka, N. Ao, W.K. Hocking, T. Sato, M. Yamamoto, T. Nakamura, T. Tsuda, and S. Kato (1994), Seasonal variability of vertical eddy diffusivity in the middle atmosphere, 1. three-year observations by the middle and upper atmosphere radar. *J. Geophys. Res.*, **99**, 18,973-18,987.

Hartmann, D. L., K. R. Chan, B. L. Gary, M. R. Schoeberl, P. A. Newman, R. L. Martin, M. Loewenstein, J. R. Podolske, and S. E. Strahan (1989), Potential vorticity and mixing in the south polar vortex during spring. *J. Geophys. Res.*, **94**, 11,625-11,640.

Hines, C. O. (1960), Internal gravity waves at ionospheric heights. *Can. J. Phys.*, **38**, 1441-1481.

Hirota, I. (1984), Climatology of gravity waves in the middle atmosphere. *J. Atmos. Terr. Phys.*, **46**, 767-773.

Hirota, I., and T. Niki (1985), A statistical study of inertia-gravity waves in the middle atmosphere. *J. Meteor. Soc. Japan*, **63**, 1055-1066.

Leblanc, T., A. Hauchecorne, M.-L. Chanin, C. D. Rodgers, F. W. Taylor, and N. J. Livesey (1995), Mesospheric temperature inversions as seen by ISAMS in December 1991. *Geophys. Res. Lett.*, **22**, 1485-1488.

Manney, G. L., R. W. Zurek, M. E. Gelman, A. J. Miller and R. Nagatani (1994), The anomalous Arctic lower stratospheric polar vortex of 1992-1993. *Geophys. Res. Lett.*, **21**, 2405-2408.

Meek, C. E., I. M. Reid, and A. H. Mason (1985), Observations of mesospheric wind velocities 2. Cross sections of power spectral density for 48-8 hours, 8-1 hours, and 1 hour to 10 min over 60-110 km for 1981. *Radio Sci.*, **20**, 1383-1402.

Miyahara, S., Y. Hayashi and J. D. Mahlman (1987), Interactions between gravity waves and planetary-scale flow simulated by the GFDL SKYHI general circulation model. *J. Atmos. Sci.*, **43**, 1844-1861.

Smith, S. A., D. C. Fritts, and T. E. VanZandt (1985), Comparison of mesospheric wind spectra with a gravity wave model. *Radio Sci.*, **20**, 1331-1338.

Schoeberl, M. R., and D. F. Strobel (1984), Nonzonal gravity wave breaking in the winter mesosphere. *Dynamics of the Middle Atmosphere*, J.R. Holton and T. Matsuno, Eds. (Terra, Tokyo, 1984), pp.45-64.

Vincent, R.A., and D.C. Fritts (1987), A climatology of gravity wave motions in the mesopause region at Adelaide, Australia. *J. Atmos. Sci.*, **44**, 748-760.

Waters, J. W. (1993), Chap.8 in *Atmospheric Remote Sensing Microwave Radiometry*, M.A. Janssen, Ed., Wiley, New York, pp.383-496.

Wilson, R., M.L. Chanin, and A. Hauchecorne (1991), Gravity waves in the middle atmosphere observed by Rayleigh Lidar: 2. Climatology. *J. Geophys. Res.*, **96**, 5169-5183.

Wu, D. L. and J. W. Waters (1996), Satellite observations of atmospheric gravity waves, *J. Res. Lett.*, in press.

# Maintainance of the Tropical Oscillations of Zonal-Mean Winds in the Middle Atmosphere by Convectively-Forced Gravity Waves.

Fabrizio Sassi[1] and Rolando R. Garcia
National Center for Atmospheric Research [2]
P.O. Box 3000
Boulder, 80307, Colorado - USA

General Circulation Models do not produce sufficient westerly momentum source in the upper stratosphere, yielding a weak penetration of the westerly winds of the Semiannual Oscillation below the stratopause during equinox. Since the observed spectrum of meteorological quantities is "red" in frequency and zonal wavenumber, it is generally assumed that a large contribution to the momentum forcing in the middle atmosphere comes from the planetary scales and that additional momentum forcing can be supplied by small-scale gravity waves. The latter, being beyond the resolution of current numerical models, need to be parameterized.

In this paper, we propose that high-frequency gravity waves at resolvable zonal scales can provide sufficient zonal force in the upper stratosphere and mesosphere. These waves are generated in troposphere by tropical convection and, although they have small amplitude at the source level, they may attain substantial magnitudes at higher levels. We investigate the propagation of such waves in the middle atmosphere with an equatorial $\beta$-plane model that incorporates a tropospheric heating whose spatial and temporal characteristics are based on analysis of tropical cloud radiances.

## 1. Introduction

The Semiannual Oscillation (SAO) of the zonal-mean zonal wind is the salient climatological feature of the tropical upper stratosphere and mesosphere (e.g., Hirota 1978, 1980 and references therein). The period of the oscillation is set by the seasonal cycle at stratopause level, where the summer to winter meridional circulation exerts an easterly body-force during solstices; on the other hand, the descent of the westerly winds is mantained by dissipating waves which are generated in the troposphere and carry westerly momentum upward (Dunkerton 1979). In the mesosphere, both phases of the SAO descend with time; the winds in the upper mesosphere have the opposite sign of the ones at stratopause level. The reason for

---

[1] Also at University of L'Aquila, L'Aquila, Italy

[2] The National Center for Atmospheric Research is sponsored by the National Science Foundation

NATO ASI Series, Vol. I 50
Gravity Wave Processes
Their Parameterization in Global Climate Models
Edited by Kevin Hamilton

the out of phase relationship is that the stratopause winds filter out the waves propagating in the same direction of the background zonal-mean zonal winds (Dunkerton. 1982).

Although the physical mechanism of the SAO seems to be well understood, numerical simulations of the oscillation have not been successful. In particular, General Circulation Models (GCMs) are able to reproduce the easterly phase of the SAO at the stratopause level, but the strength and descent of the westerly winds below 50 km remain poorly modelled (Sassi et al. 1993, Hamilton et al., 1995). A common aspect of several modelling efforts of the stratopause SAO is the lack of sufficient westerly momentum source at stratopause. Since most of the forcing appears to be at the planetary scales and little at other resolvable scales, it has been suggested (Hitchman and Leovy, 1988) that westerly momentum may be supplemented by small-scale gravity waves which need to be parameterized since they are beyond the resolution of current numerical models.

Another possibility has emerged recently from the analysis of cloud radiances made available by the International Satellite Cloud Climatology Project (ISCCP) at temporal and spatial scales not observed before (Salby et al., 1991). Using the spectrum of radiances as a proxy for tropical convection, Bergman and Salby (1994) showed that the momentum fluxes emerging at tropopause level contain a variety of temporal and spatial scales. In particular at phase speeds relevant to the SAO ($\sim$ 40-50 $\text{ms}^{-1}$), wave-activity is substantial over a broad range of wavenumbers (up to $k \simeq 20$) and frequencies (well beyond the diurnal). Since polar-orbiting satellite observations can resolve up to $k \simeq 6$ and frequencies up to the diurnal, a substantial fraction of the spectrum is left unresolved and it is not possible to verify directly whether the high-frequency wave activity associated with tropical convection is capable of supplying sufficient momentum forcing in the upper stratosphere and mesosphere. In this paper, we tackle this issue using an equatorial $\beta$-plane model which is forced in the troposphere by introducing a heating perturbation. The perturbation has a frequency-spectrum that resembles Salby et al's (1991) analysis, while the spatial structure is based on the evidence provided by Bergman and Salby (1994). We will show that, with acceptable total forcing, it is possible to obtain an SAO at stratopause and mesopause levels that compares well with observations. We will also investigate the role of the planetary-scale waves and of the diurnal oscillation in forcing the zonal-mean circulation in the middle atmosphere. The paper is organized as follows: in Section 2 we briefly describe the model; in Section 3 we present some results obtained with different tropospheric forcings; Section 4 gives the final remarks.

## 2. Model Description

In this Section we provide a brief description of the model and of the implementation of the tropospheric forcing. A more complete discussion can be found in Sassi and Garcia (1996).

The model used in this study is a three-dimensional equatorial $\beta$-plane model; the governing equations (zonal and meridional momentum, and thermodynamics) are based on Andrews et al (1987; Section 4.7). Waves are not allowed to interact with each other but they do interact with the zonal-mean state through the convergences/divergences of heat and momentum. Since this is an equatorial channel model, it is necessary to impose a source of easterly momentum at the equatorial stratopause, in order to reproduce easterly winds at solstice. The strength and location of the easterly forcing are adjusted seasonally to obtain a meridional profile of zonal-mean zonal winds in accord with observations.

An important component of the model is the tropospheric forcing. An unsteady heating is introduced in the thermodynamic equation for each zonal wave:

$$\frac{\partial T'}{\partial t} + ... = \mathcal{H}(x, y, t)F(z) \tag{2.1}$$

where $T'$ is the temperature wave amplitude, $\mathcal{H}$ and $F$ determine how the heating is distributed in space and time; $x$,$y$,$z$ and $t$ are longitude, latitude, height and time, respectively. The vertical structure $F$ is a sinusoidal function of $z$, which makes the total heating non-zero only between the ground and 12 km. In order to determine the horizontal and spatial structure of heating, we followed Salby et al (1991) who showed that it is possible to distinguish between an oceanic contribution with negligible power at high frequencies and a continental component that contains a diurnal oscillation; the zonal structure was chosen following Bergman and Salby who showed that the high-frequency forcing is localized to the regions of the strongest convection. With this in mind, for the oceanic component of the forcing, we let:

$$H_o(x, y, t) = S_o(x)T_o(t)Y(y) \tag{2.2.1}$$

where $T_o$ is a stochastic process (Markov autoregressive process of the second order):

$$T_o(t) = r \cdot T_o(t - \Delta t) + (1 - r) \cdot \mu + \epsilon \tag{2.2.2}$$

$r$ is the correlation coefficient ($r$=0.5 at lag 1 day); $\Delta t$ is the time step; $\mu$ is the mean of the stochastic process ($\mu = 0$); $\epsilon$ is a random process with variance $\langle \epsilon^2 \rangle = 1 - r^2$. These parameters have been chosen in order to make the resulting spectrum resemble Salby et al's (1991; see their Fig. 8). The zonal structure is

$$S_o(x) = A \exp\left(\frac{-x^2}{2L_A^2}\right) \tag{2.2.3}$$

where $A$ and $L_A$ are the amplitude and the zonal half-width of the oceanic forcing. Similarly, the continental component of the forcing is specified as follows:

$$H_c(x, y, t) = S_c(x)T_c(t)Y(y) \tag{2.3.1}$$

where $T_c$ is defined as

$$T_c(t) = r \cdot T_c(t - \Delta t) + (1 - r) \cdot \mu + \epsilon + \sqrt{2}\cos\Omega t \tag{2.3.2}$$

$\Omega$ is the frequency corresponding to period 1 day ($\Omega = 2\pi/86400$ rad day$^{-1}$) and the other parameters have the same values as in (2.2.2). The zonal structure is

$$S_c(x) = B \exp\left(\frac{-x^2}{2L_B^2}\right) \tag{2.3.3}$$

where $B$ and $L_B$ are the amplitude and the zonal half-width of the continental forcing. In all cases, $S_c$ and $S_o$ are modulated by a meridional Gaussian envelope $Y$ which has a half-width of 850 km. The envelope's center changes position in the meridional plane according to the season; it is centered at 1100 km (~10 degrees of latitude) at solstice and at the equator at equinox.

In order to determine the amplitudes ($A$ and $B$) and zonal scales ($L_A$ and $L_B$) of the forcing, we used some guidance from observations. Since Bergman and Salby showed that, for the continental heating source, about one-half of the total power is contained in the high frequency components, it may be expected that the oceanic component accounts for the generation of the planetary scales. We also know from satellite observations (Salby et al, 1984; Hitchman and Leovy, 1988; Canziani et al, 1994) that time-mean wave amplitudes in the tropical upper stratosphere are about 1 K at zonal wavenumber 1 and substantially less at zonal wavenumbers $\geq$4. Several authors (Hitchman and Leovy, 1988; Canziani et al, 1994) have noted that, for time-mean amplitudes much below 1 K, the wave structures are difficult to recognize in the presence of measurement noise. With these facts in mind, we ran the model with only the oceanic forcing (2.2.1) and we tuned the parameters ($A$ and $L_A$ in 2.2.1) to obtain stratospheric wave amplitudes at $k$=1 of 1 K and substantially less at $k$=4. We found that planetary wave amplitudes consistent with observations could be obtained with $A$=13 K day$^{-1}$ and $L_A$=2000 km. Nevertheless, it will be shown in the following discussion that, with oceanic forcing alone, the model does not produce a realistic SAO. It turns out that it is also necessary to force the continental component (2.3.1) with $B$=20 K day$^{-1}$ and $L_B$=200 km to obtain a satisfactory simulation.

Most of the simulations described in the following used 15 zonal waves, from $k$=1 up to $k$=15; meridional resolution is 300 km (lateral boundaries are at 4050 km); vertical resolution is 1 km, extending from the ground up to 120 km (an absorbing layer is present in the uppermost 20 km); the time step required for a stable simulation is 5 minutes. In the next Section we present results of a set of simulations obtained forcing the model with the unsteady heating described by (2.2.1) and (2.3.1). We show first the simulation we obtained with the complete spectrum; then we present the results from a simulation with only the planetary-scale waves, and finally a simulation in which the continental forcing (2.3.1) has no diurnal component.

## 3. Results

*3.1 Simulation with the complete spectrum.*

In Fig. 1 we show the equatorial zonal-mean zonal wind *(a)* and equatorial Eliassen-Palm (EP) flux divergence *(b)* for one year of a simulation with the

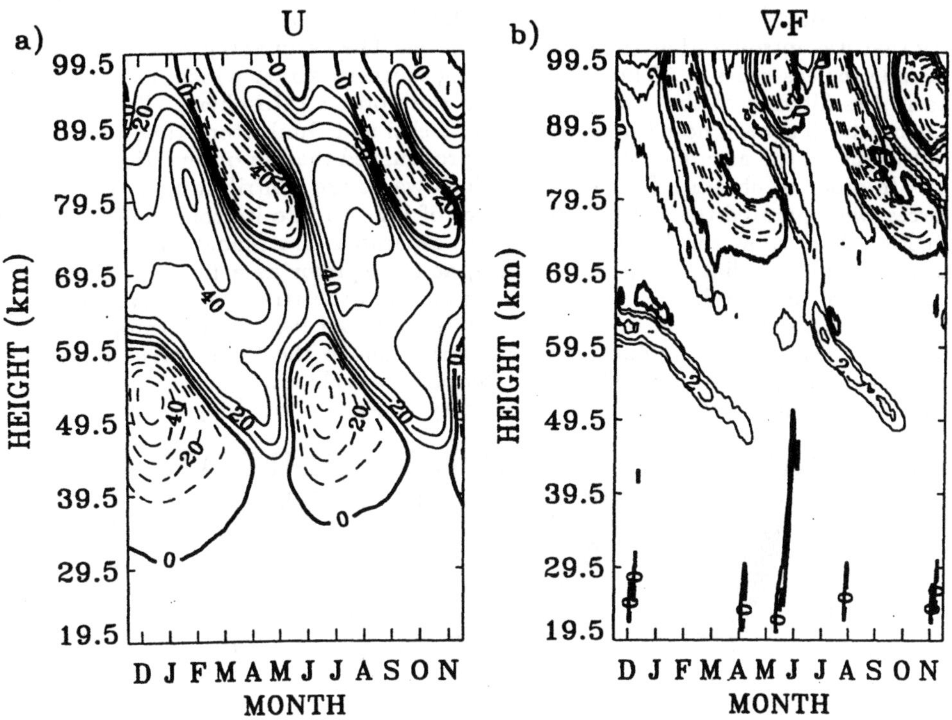


Fig. 1. *a*) Equatorial zonal-mean wind during the first year of the simulation, as a function of height (in km). Time is expressed as the month of the simulation, with the tags located in the middle of each month. Contour interval is 10 $ms^{-1}$. *b*) The EP flux divergence; contours are $\pm 1$, $\pm 2$, $\pm 4$, $\pm 8$, $\pm 10$ $ms^{-1}day^{-1}$.

complete forcing (2.2.1) and (2.3.1). The zonal-mean zonal wind (Fig. 1a) has a stratopause oscillation with easterlies at solstice and descending westerlies at equinox; in the upper mesosphere, there is another oscillation with a period of six months, but out of phase with the stratopause winds (no attempt was made to simulate the upper mesospheric time-mean easterly winds); the two SAO are separated by time-mean westerlies in the the lower mesosphere. The stratopause westerlies penetrate into the upper stratosphere reaching 45 km; maximum winds are about 30

$ms^{-1}$ at the equinox near 50 km. The mesospheric SAO has both phases descending with time: maximum westerlies occur at solstice, reaching 60 $ms^{-1}$; easterlies at equinox attain –40 $ms^{-1}$. The westerly forces exerted by all waves (Fig. 1b) at the stratopause level reach up to 2 $ms^{-1}day^{-1}$; weaker forces are present in the upper stratosphere (1 $ms^{-1}day^{-1}$). In the upper mesosphere, intense easterly forces occur at equinox, up to –10 $ms^{-1}day^{-1}$ in September and October. The easterly force in the upper mesosphere during fall equinox is somewhat larger than at spring equinox ( –10 $ms^{-1}day^{-1}$ versus –8 $ms^{-1}day^{-1}$): this is the result of a seasonal asymmetry that was introduced at the stratopause level. The asymmetry was obtained by making the easterlies at 50 km weaker during southern winter which results in Fig. 1a in less filtering of the easterly waves in the upper stratosphere. The seasonal asymmetry has been noted previously (e.g., Delisi and Dunkerton, 1988); however, in this model the asymmetry is much less pronounced than in observations.

In order to obtain information on which waves are contributing to the zonal forces, we separated the EP flux divergence of Fig. 1b into its zonal components. In Fig. 2 we show the EP flux divergence separated into planetary scales ($k$=1 to 3) and intermediate scale ($k$=4 to 15), at the equator as a function of height and time of the simulation. It appears that around the stratopause level the two zonal components are comparable ($\sim$ 1 $ms^{-1}day^{-1}$) but below the stratopause the planetary-scale waves provide a larger contribution to the zonal forces than the intermediate scales. Above 50 km, the intermediate-scale waves provide more zonal force (2 $ms^{-1}day^{-1}$ at 60 km), than planetary scales (1.5 $ms^{-1}day^{-1}$ at 55 km).

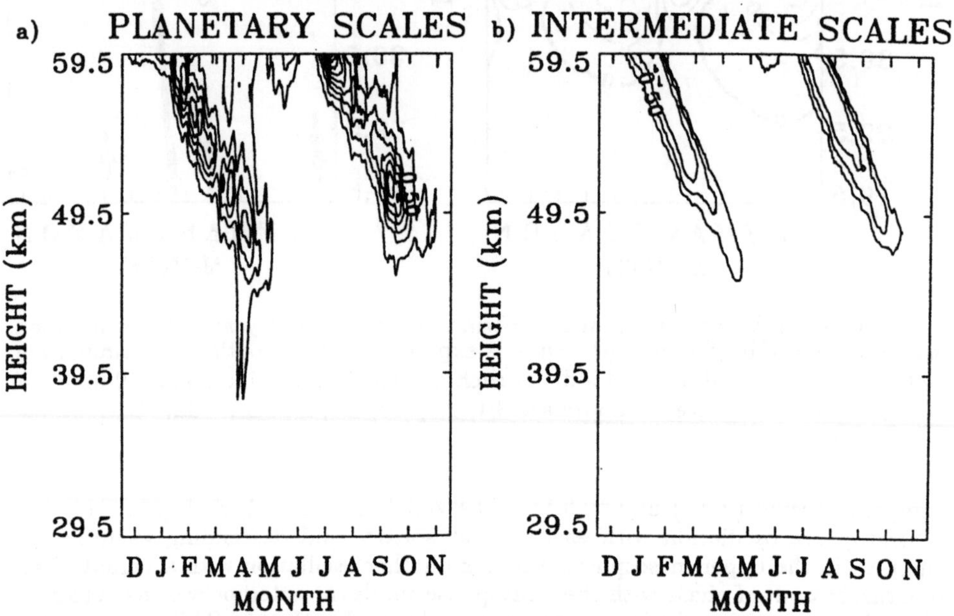


Fig. 2. $a$) The force exerted by the planetary-scale waves ($k$=1 to 3) and $b$) by intermediate-scale waves ($k$=4 to 15). Contour interval is 0.25 $ms^{-1}day^{-1}$ in $a$) and contours in $b$) are ±0.25, ±0.5, ± 1, ±2, $ms^{-1}day^{-1}$; no zero contour line is plotted.

In Fig. 3 we show the total EP flux divergence separated in each zonal component at the stratopause level. Westerly forces occur prevalently at equinox at all zonal scales. The largest westerly forces are located at the planetary scales (at $k$=2, the zonal force is $\sim$ 0.6 $ms^{-1}day^{-1}$ in late equinox); beyond $k$=4, the magnitudes decrease rapidly although there are forces of about 0.05 $ms^{-1}day^{-1}$ at the highest zonal wavenumbers. In fact, Bergman and Salby suggested that up to $k$=20 there is substantial westerly momentum flux at the tropopause level available at

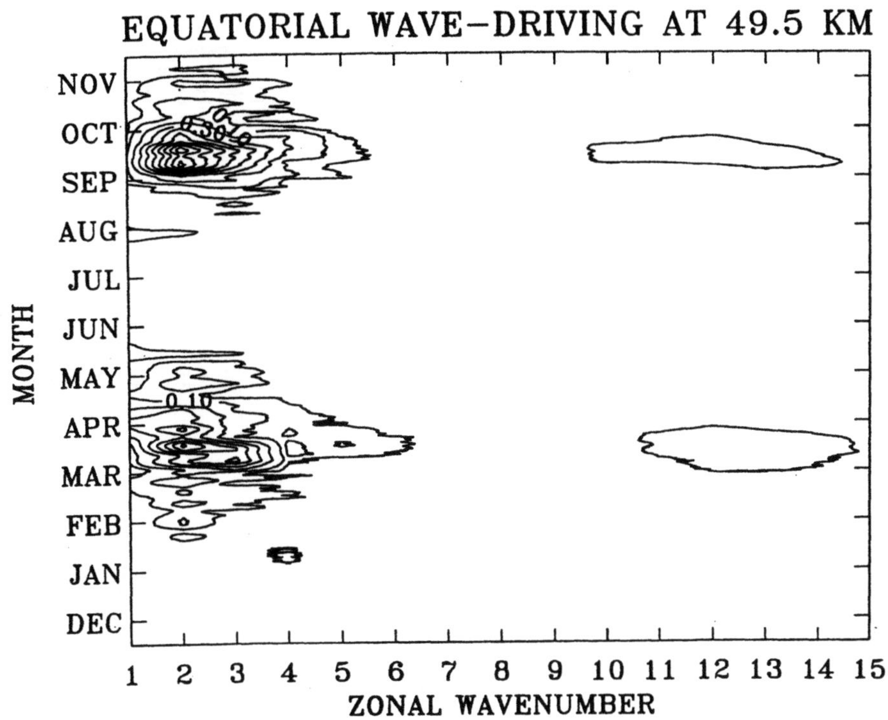


Fig. 3. EP flux divergence at 49.5 km, as function of zonal wavenumber and time of the simulation. Contour interval is 0.1 $ms^{-1}day^{-1}$; the 0.05 contour is also plotted, and no zero contour is drawn.

zonal phase speeds relevant to the SAO (see Fig. 1 in Sassi and Garcia, 1994). Our results indicate that at stratopause level the zonal force at intermediate scales maximizes at $k$=12-13, decreasing at higher zonal wavenumbers; however, had we run the model with more zonal waves, we could have obtained a larger contribution to the westerly zonal force from the intermediate-scale waves. In fact, what seems relevant is the total force on a broad range of zonal scales, rather than an intense forcing in a limited spectral region.

In Fig. 4 we show the EP flux divergence as a function of zonal wavenumber

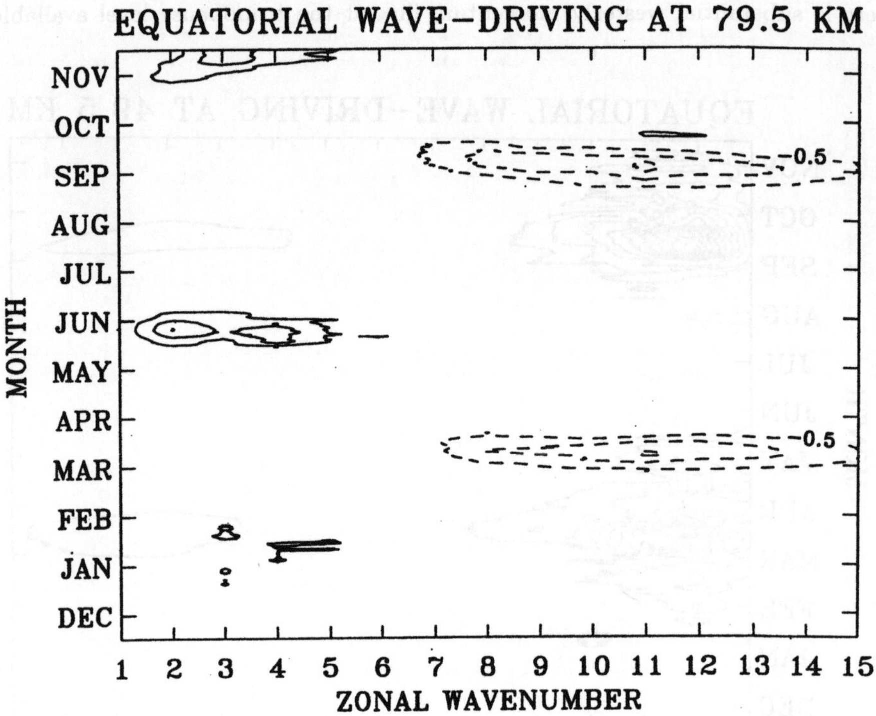


Fig. 4. As in Fig. 3, but at 79.5 km. Contour interval is 0.5 $ms^{-1}day^{-1}$, with no zero line plotted.

and time at the 79.5 km level. At this level, there are westerly forces at the planetary scales and easterly forces only at the intermediate scales. The planetary scales produce westerly zonal forces at all times throughout the middle atmosphere (compare Fig. 3 to Fig. 4). This fact suggests that Kelvin waves dominate at these zonal scales. On the other hand, the presence of easterly zonal forces at intermediate scales (alternating at other times and levels with westerly forces) indicates that inertio-gravity waves dominate at these scales. Easterly forcing at intermediate scales can attain magnitudes of −1.5 $ms^{-1}day^{-1}$, while the planetary scales can provide westerly a force up to 1 $ms^{-1}day^{-1}$. It should be noted that the forces at

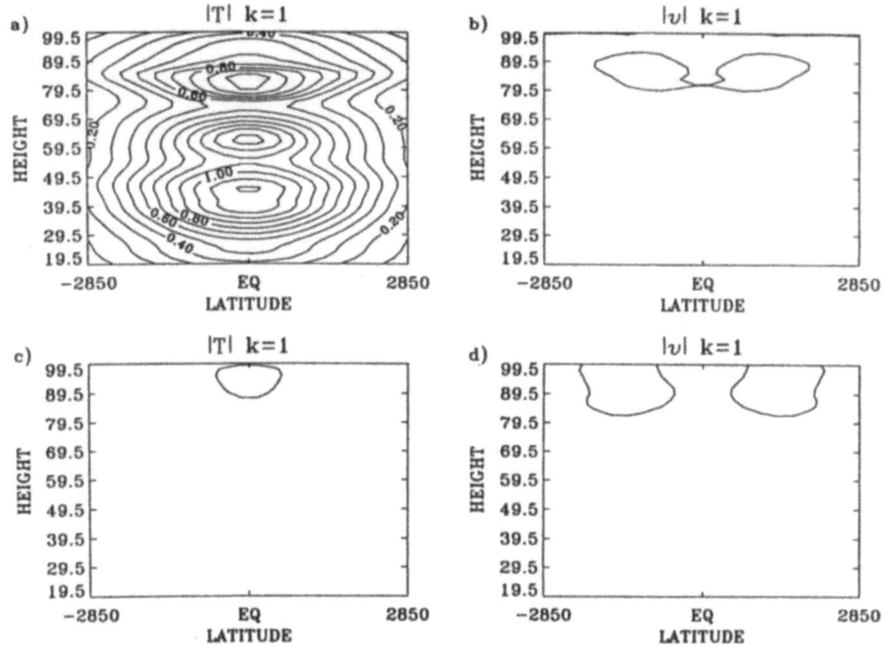


Fig. 5. Symmetric temperature amplitude at zonal wavenumber 1 for eastward (*a*) and westward (*c*) propagating waves, as a function of height and distance from the equator (in km). Contour interval is 0.1 K. Antisymmetric meridional velocity at zonal wavenumber 1 for eastward (*b*) and westward (*d*) propagating waves; contour interval is 0.2 ms$^{-1}$.

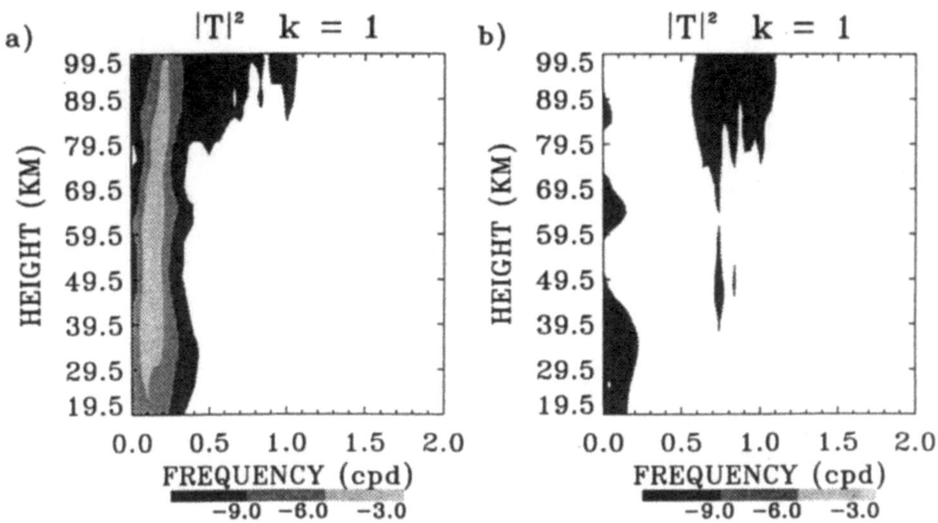


Fig. 6. Frequency spectrum of the log($T^2$) as a function of height at zonal wavenumber 1 for the symmetric mode: *a*) for the eastward and *b*) for the westward component.

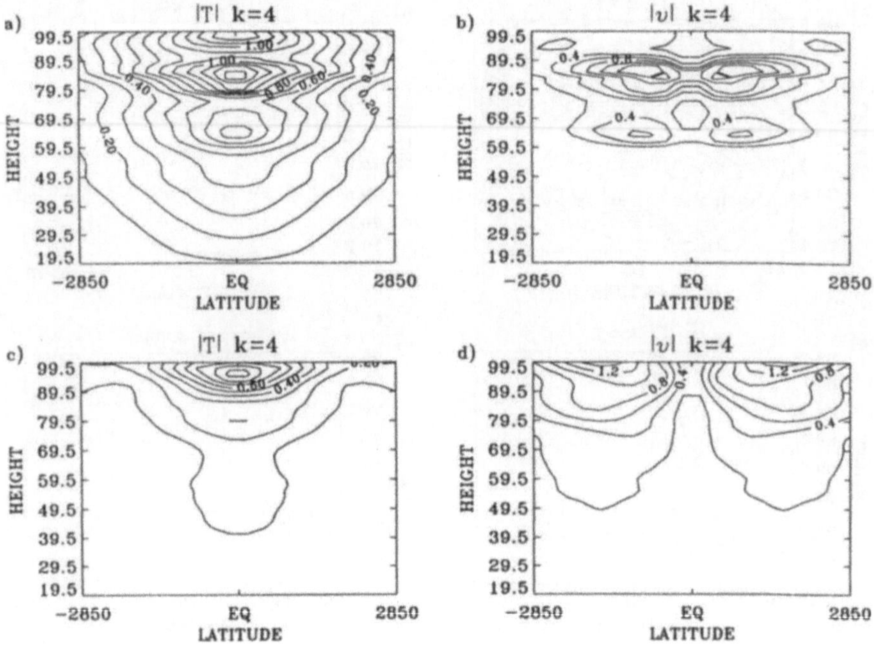


Fig. 7. As in Fig. 5, but for zonal wavenumber 4.

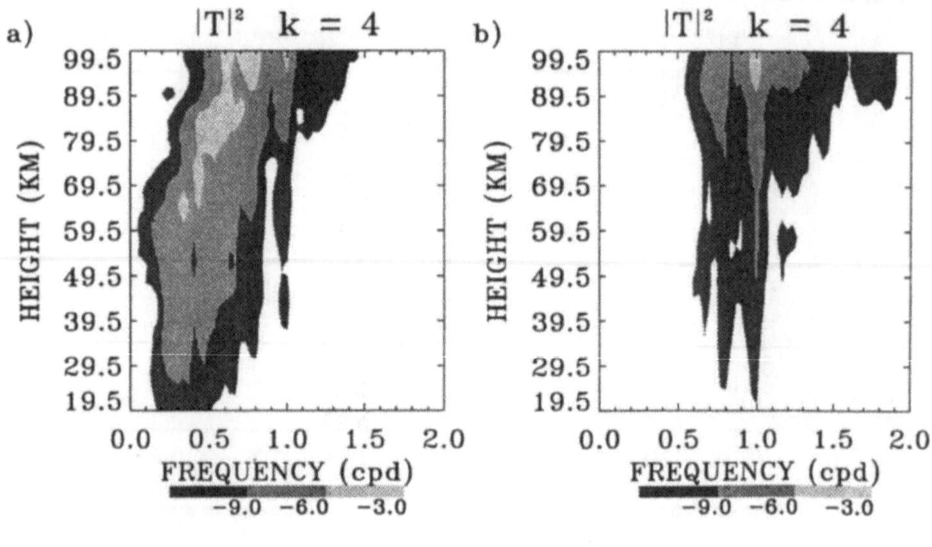


Fig. 8. As in Fig. 6, but for zonal wavenumber 4.

intermediate scale occupy a broad range of zonal wavenumbers, up to the current model's resolution.

It is well known that the meridional structure of the wave modes in an equatorial $\beta$-plane model can be separated into symmetric and antisymmetric modes, where symmetry is intended about the equator and it is always referred to the temperature field. In the remainder of this Section we will briefly examine the meridional structure of the symmetric modes obtained by analysing the model results; antisymmetric modes are not presented here, although they are of of some relevance in the upper mesosphere (see, Sassi and Garcia, 1996). The meridional structure of the symmetric waves at zonal wavenumber 1 is illustrated in Fig. 5. This Figure was obtained by Fourier analyses of the wave field in the middle of the second solstice of Fig. 1. The easterly and westerly waves were separated according to the method of Hayashi (1971) and the amplitude was plotted at each level and latitude. Most of the amplitude at $k$=1 is in the westerly components (Fig. 5a,b); the lack of a visible meridional velocity, along with a strong signal in the temperature field, indicates that Kelvin waves are predominant at this zonal scale. The largest temperature is 1.2 K at the stratopause, well in accord with observations (Salby et al, 1984; Canziani et al, 1994); however, the temperature in the lowermost part of the stratosphere is about twice as large as expected. No visible wave amplitudes appear in the easterly component (Fig. 5c,d). It is worth noting that the meridional extent of the wave field increases with height; since the equatorial Rossby radius of deformation is proportional to the square-root of the zonal phase velocity, meridonal broadening of the wave with height is a result of the selective filtering of the slowest component by the underlying winds.

The frequency spectrum of temperature at $k$=1 is shown in Fig. 6 as a function of height. Most of the power is in the westerly waves, in accord with the results of Fig. 5. In the lower stratosphere, the dominant period is ~10 days, while in the upper mesosphere is ~3-4 days. The spectrum shown in Fig. 6a is typical of a packet of Kelvin waves whose slowest components are progressively dissipated.

In Fig. 7 we show the meridional structure at zonal wavenumber 4. The westerly components (Fig. 7a,b) are still predominant in the stratosphere and appear as Kelvin waves. Throughout the stratosphere, the wave amplitudes are much smaller than at $k$=1; however, it should be remembered from Fig. 3, that the zonal force exerted at $k$=4 is not negligible. In the mesosphere, inertia-gravity waves are present. The amplitudes are much larger than in the stratosphere, with temperature attaining amplitudes of 1.4 K and meridional velocities of 0.8 $ms^{-1}$. The easterly components (Fig. 7c,d) are negligible in the stratosphere, but they reach visible values in the upper mesosphere; the meridional structures are suggestive of inertia-gravity waves.

In Fig. 8 we show the frequency spectrum of temperature at $k$=4, for westerly (Fig. 8a) and easterly (Fig. 8b) waves. Contrary to the case in Fig. 6, there is visible power over a broad spectral region, even up to 1 cpd in the mesosphere. At higher levels, power at frequencies in excess of 1 cpd is evident, suggesting that satellite retrievals around the mesopause level could be contaminated by these high frequency components.

A more detailed analysis of the wave modes and their spectra is presented in Sassi and Garcia (1996). In the next Sections we will address how different components of the tropospheric forcing affect the simulation in the middle atmosphere.

*3.2 Simulation with planetary-scale waves only.*

It was mentioned in the Introduction that current GCM simulations of the stratopause SAO show insufficient westerly momentum forcing at 50 km; in fact it appears that in some models the planetary-scale waves provide most of the zonal force generated at the stratopause (Sassi et al, 1993) but they do not seem to supply sufficient westerly force in the upper stratosphere (Hitchman and leovy, 1988). With this model we can verify the contention that the planetary scales alone cannot provide sufficient zonal forcing of the SAO. Therefore, we have run the same model as in the previous Section but with only three waves, $k$=1 to 3. In Fig. 9 we show the zonal-mean zonal wind of a one year long simulation. The stratopause SAO has weakend substantially; in fact, westerly winds descending from the lower mesosphere are not capable of breaking through the solstical easterlies. Time-mean westerlies are still present throughout the mesosphere, yielding a complete disappearance of the mesopause SAO. From the discussion of the previous Section, it appears that the planetary scales are mostly Kelvin waves that carry only westerly momentum; therefore, it is not a surprise to find no easterlies at any time in this simulation. The weakening of the stratopause SAO is also associated with the insufficient zonal force that planetary-scale Kelvin waves can exert.

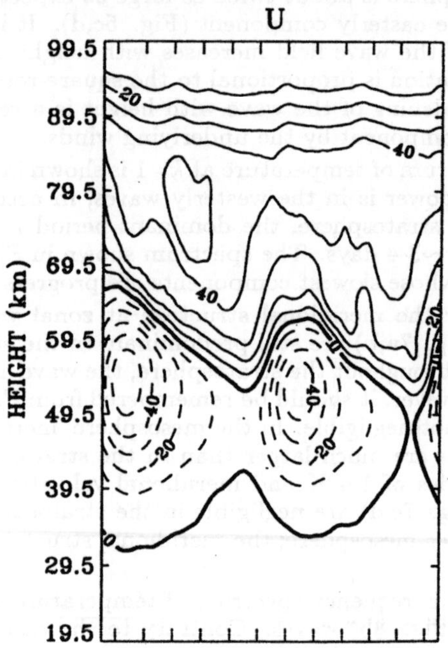


Fig. 9. As in Fig. 1a, but for the simulation with only planetary-scale waves.

*3.3 Simulation without diurnal forcing.*

Salby et al (1991) showed that the spectrum of cloud radiances over continents is similar to the spectrum over the oceans, except that the former has a strong diurnal oscillation and stronger forcing at high frequencies. Sassi and Garcia (1994) addressed the issue of the importance of high-frequency gravity waves in forcing the SAO at the stratopause and the mesopause; they used a one-dimensional model forced at a single zonal wavenumber ($k$=12). In order to verify their contention regarding the role of the diurnal forcing, in this Section we show results of a simulation with the same model as in Section 3.1, but without the diurnal forcing in (2.3.2). Therefore the frequency spectra of convection over the ocean and the continent are identical in this case. In Fig. 10 we show the equatorial zonal-mean zonal wind (Fig. 10a) and equatorial EP flux divergence (Fig. 10b) as a function of height and time of the simulation. The stratopause SAO is present but it is not as strong as in the simulation with the complete spectrum: the westerly winds descend only to 50 km, barely breaking through the solsticial easterlies. Time-mean westerlies are present throughout most of the mesosphere, although weak easterlies appear at the uppermost levels. The EP flux divergence shows magnitudes substantially smaller than the early case with the complete spectrum. At the stratopause, the zonal force is 1 $ms^{-1}day^{-1}$ and reaches 2 $ms^{-1}day^{-1}$ in the mesosphere. Easterly forcing appears only above 90 km, but it is quite weak (–2 $ms^{-1}day^{-1}$).

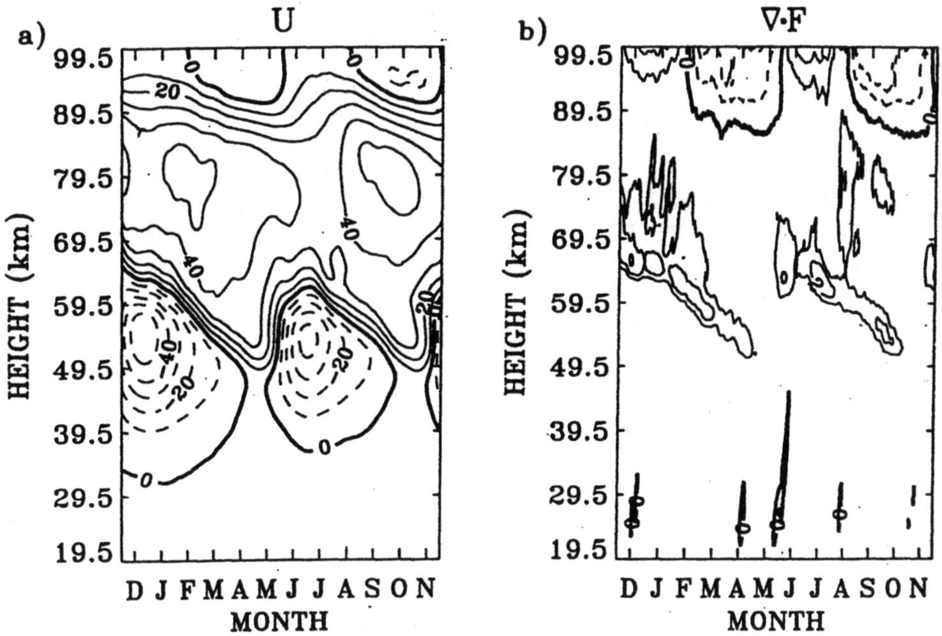


Fig. 10. As in Fig. 1, but for the simulation without the diurnal forcing.

The lack of the mesopause SAO and the weakness of the stratopause oscillation are indications that the diurnal forcing is instrumental in the SAO at both levels. This fact is evident in Fig. 11, which shows the planetary (Fig. 11a) and

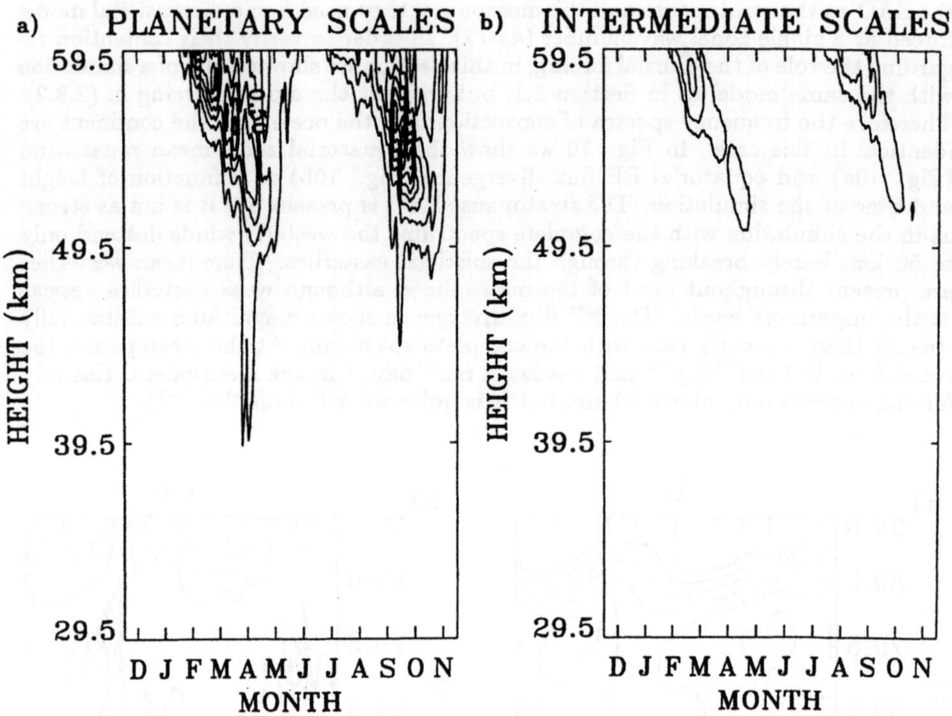


Fig. 11. As in Fig. 2, but for the simulation without the diurnal forcing.

intermediate (Fig. 11b) scale contributions to the total EP flux divergence. The planetary-scale zonal force is similar to the earlier case with the complete forcing, although a slight time lag and somewhat smaller magnitudes are present because of the slower and shallower descent of the westerly winds at the stratopause. On the other hand the intermediate scales are much less intense and hardly manifest themselves at the stratopause. In this case, the planetary scales contribute solely to the zonal force exerted at 50 km, since the lack of the diurnal oscillation makes the intermediate scale much weaker. At higher levels (not shown), most of the zonal force is still exerted by the planetary-scale waves, with negligible contributions from the intermediate scales.

## 4. Conclusions

It emerged from some recent studies (Salby et al 1991; Bergman and Salby 1994) that tropical convection produces wave-activity rich in spatial and temporal scales. Unfortunately, the question of which waves generated by tropical convection can be instrumental in supplying momentum to the equatorial middle atmosphere cannot be fully addressed by analysis from current polar-orbiting satellites since some of the most important features occur beyond the attainable spatial and temporal resolutions. Therefore, we have presented a numerical simulation with an equatorial $\beta$-plane model of the stratopause and mesopause SAO. The model has been forced in the troposphere by unsteady heating, whose temporal and spatial structures have been specified according to the analysis of the cloud radiances of Salby et al (1991). Realistic SAO at stratopause and mesopause levels are simulated by the model with reasonable forcing. We presented a budget of the zonal momentum in the upper stratosphere and lower mesosphere, which showed that waves at zonal wavenumbers other than the planetary are important for generating and mantaining the SAO. The planetary-scale waves show a structure in the meridional plane typical of Kelvin waves; temperature at the stratopause is comparable to satellite observations. At higher zonal wavenumbers the amplitude decrease; in the mesosphere, inertio-gravity waves are dominant at frequencies at and beyond the diurnal. This latter fact points to the possibility that satellite analysis at around the mesopause may be contaminated by high frequency waves that are not resolved by single polar-orbiters.

In a simulation in which the model was run with only the planetary scales, the results are quite unsatisfactory, since these waves are not able to provide sufficient westerly force in the middle atmosphere. This fact is in accord with evidence from satellite observations (Hitchman and Leovy, 1988) that the planetary scales alone cannot provide sufficient momentum in the upper stratosphere in order to balance the zonal mean momentum. In this case, the mesopause SAO is non-existent, since planetary-scale waves are predominantly Kelvin waves which carry only westerly momentum.

We also ran the model without diurnal forcing and the contribution to the zonal-mean momentum from waves at intermediate scales ($k$=4 to 15) was somewhat reduced yielding a weaker stratopause SAO. Further, the mesopause SAO is almost non-existent, since intermediate scale waves are the only source of easterly momentum in the mesosphere and without the diurnal forcing they do not attain visible amplitudes below the mesopause.

It appears from this calculations that a description of the momentum budget in the middle atmosphere cannot be reduced to the planetary scales and parameterized small-scale gravity waves; in fact, tropical convection produces waves over a variety of spatial and temporal scales that require further investigation in order to ascertain their role in forcing the zonal-mean zonal winds in the stratosphere and mesosphere.

## REFERENCES


Andrews, D.G., J.R. Holton, C.B. Leovy, 1987: *Middle Atmosphere Dynamics.* Academic Press Inc., 489 pp.

Bergman, J.W. and M.L. Salby, 1994: Equatorial Wave activity generated by fluctuations in observed convection. *J. Atmos. Sci.,* **51,** 3791-3806.

Canziani, P.O., J.R. Holton, E. Fishbein, L. Froidevaux, and J. Waters, 1994: Equatorial Kelvin waves: a UARS MLS view. *J. Atmos. Sci.,***51,** 3053-3076.

Dunkerton, T., 1979: On the role of the Kelvin wave in the westerly phase of the Semiannual Zonal wind Oscillation. *J. Atmos. Sci.,* **36,** 32-41.

—, 1982: Theory of the mesopause semiannual oscillation, *J. Atmos. Sci.,* **39,** 2681-2690.

Delisi, D.P. and T. Dunkerton, 1988: Seasonal variation of the Semiannual Oscillation. *J. Atmos. Sci.,* **45,** 2772-2787.

Hamilton, K., R.J. Wilson, J.D. Mahlman, and L.J. Umscheid, 1995: Climatology of the SKYHI troposphere-stratosphere-mesosphere general circulation model. *J. Atmos. Sci.,* **52,** 5-43.

Hayashi, Y., 1971: A generalized method for resolving disturbances into progressive and retrogressive waves by space Fourier and time cross-spectral analyses. *J. Meteor. Soc. Japan,* **49,** 125-128.

Hirota, I., 1978: Equatorial waves in the upper stratosphere and mesosphere in relation to the semiannual oscillation of the zonal mean wind. *J. Atmos. Sci.,* **35,**714-722.

—, 1980: Observational evidence of the semiannual oscillation in the tropical middle atmosphere: A review. *Pure Appl. Geophys.,* **118,** 217-238.

Hitchman, M.H. and C.B. Leovy, 1988: Estimation of Kelvin wave contribution to the Semiannual Oscillation. *J. Atmos. Sci.,* **45,** 1462-1475.

Salby, M.L., D.L. Hartman, P.L. Bailey, and J.C. Gille, 1984: Evidence of equatorial Kelvin modes in Nimbus-7 LIMS. *J. Atmos. Sci.,* **41,** 220-235.

—, H.H. Hendon, K. Woodberry, and K. Tanaka, 1991: Analysis of global cloud imageries from multiple satellites. *Bull. Americ. Meteor. Soc.,* **72,** 467-480.

Sassi, F., R.R. Garcia, and B.A. Boville, 1993: The stratopause Semiannual Oscillation in the NCAR Community Climate Model. *J. Atmos. Sci.,* **50,** 3608-3624.

— and —, 1994: A one-dimensional model of the Semiannual Oscillation driven by convectively-forced gravity waves. *J. Atmos. Sci.,* **51,** 3167-3182.

— and —, 1996: The role of Convectively-Forced Gravity Waves in the Tropical Semiannual Oscillation. *J. Atmos. Sci.* (Submitted)

# NUMERICAL MODELLING OF INERTIA-GRAVITY WAVE EMISSION BY FRONTS AND JETS

Michael J. Reeder [1]
*National Center for Atmospheric Research* [2]
*Boulder, CO 80303*
*USA*

Morwenna Griffiths
*Department of Meteorology*
*University of Reading*
*Reading, Berkshire RG6 6AU*
*UK*

ABSTRACT. Idealized numerical-model calculations following the development of upper-level and low-level fronts and their attendant jets are presented. As the fronts and jets evolve, long-wavelength low-frequency inertia-gravity waves are emitted. The key components of the model basic-state are: a time-dependent cross-front geostrophic shear flow in thermal wind balance with the along-front potential temperature, and a time-independent height-independent confluent deformation wind field with its axis of dilation parallel to the front. When the cross-front geostrophic wind shear, $\Lambda$, is negative and consequently the upper-level jet advects cold air along the upper front, vertically propagating inertia-gravity waves are a prominent part of the model solution, especially in the lower stratosphere. Like those observed in the stratosphere, the modelled waves have frequencies close to the inertial frequency, horizontal wavelengths of order $400 - 1200\,km$, and vertical wavelengths in the range $2 - 10\,km$. Numerical experiments with $\Lambda \geq 0$ show little gravity wave activity. The time-space mean of the vertical flux of horizontal momentum in the stratosphere associated with the modelled waves is approximately $1.4 \times 10^{-4}\,N\,m^{-2}$, where the average is taken over a 24 $h$ time interval, and over a 5000 $km$ wide $\times$ 14 $km$ high box in the lower stratosphere, centred over the jet/front system.

---

[1]Permanent affiliation: Centre for Dynamical Meteorology and Oceanography, Monash University, Clayton, Victoria 3168, Australia

[2]The National Center for Atmospheric Research is sponsored by the National Science Foundation

NATO ASI Series, Vol. I 50
Gravity Wave Processes
Their Parameterization in Global Climate Models
Edited by Kevin Hamilton

## 1. Introduction

Observations suggests that internal gravity waves in the middle atmosphere are generated principally in the the troposphere, and that the chief generation sites include mountains, convective clouds and jet/front systems (*e.g.*, see the review article by Fritts, 1993 and the references therein). Observations show also that the most energetic waves in the middle atmosphere are those with long wavelengths and frequencies close to the inertial frequency (*e.g.*, see the review by Vincent, 1990). Despite these observations, relatively little theoretical or modelling work has focused on the origin of long-wavelength low-frequency waves in the middle atmosphere. The principal aim of this study is to investigate the emission of low-frequency inertia-gravity waves by intensifying fronts and jets.

There have been relatively few previous theoretical or modelling studies of wave generation associated with frontogenesis. Ley and Peltier (1978) were the first to examine the problem theoretically. The basic assumption underpinning their work is that waves are generated by cross-front parcel accelerations neglected in the balanced confluent-deformation frontogenesis theory of Hoskins and Bretherton (1972). More recently, Gall *et al.* (1988), Garner (1989) and Snyder *et al.* (1993) have investigated the behaviour of surface fronts as they collapse to form a discontinuity, and have examined the gravity waves generated during this process. These numerical studies are based on primitive-equation extensions of Hoskins and Bretherton's (1972) confluent-deformation model, although Snyder *et al.* (1993) also examine a primitive-equation version of Eady's baroclinic instability problem. Recently, O'Sullivan and Dunkerton (1995) examined the generation of interia-gravity waves within an idealized three-dimensional baroclinic wave. Several recent studies have dealt with gravity wave generation associated with the geostrophic adjustment of upper-level jets (Fritts and Luo, 1992; Luo and Fritts, 1993) but have not explicitly taken into account the evolution of the jet/front system.

The dynamics of fronts in the jet-entrance region have been investigated by Reeder and Keyser (1988). In the present study we use a slightly modified version of one of their two-dimensional numerical models to examine the problem of low-frequency gravity wave emission. The remainder of the article is arranged as follows. Section 2 discusses the basic state and the formulation of the numerical model including the initial conditions. Results from the model are presented in section 3. Here, as in most of the article, the emphasis is placed on the modelled stratospheric waves. The results from a version of the model that filters gavity waves are presented and discussed in Section 4. A knowledge of the filtered solution proves to be useful when interpreting calculations of vertical flux of horizontal momentum, presented Section 5. Finally, some conclusions are drawn in Section 6. The wave sources and the dynamical processes that generate the waves are discussed in the papers Griffiths and Reeder (1996) and Reeder and Griffiths (1996).

## 2. Numerical model

The atmosphere is modelled as a hydrostatic compressible fluid on an f-plane, bounded above and below by two rigid horizontal boundaries. Moist processes are ignored. Rectangular Cartesian coordinates are assumed throughout this study, with $x$, $y$ and $z$ measuring the cross-front distance, the along-front distance, and the height respectively. The flow is partitioned into a time-dependent basic state which is an exact solution of the full equations of motion, and a finite-amplitude departure having no structure in $y$ (the along-front direction).

The basic state, first introduced by Keyser and Pecnick (1985), is essentially the sum of two background flows commonly used in frontogenesis modelling, the combination being representative of the jet entrance region in a developing baroclinic wave. The first background flow is the much-studied confluent-deformation wind field, and has been used in most previous investigations of gravity wave emission by fronts (Ley and Peltier, 1978; Gall *et al.*, 1988; Garner, 1989; Snyder *et al.*, 1993). The second background flow is a horizontally uniform cross-front geostrophic wind with a linear variation with height and in thermal wind balance with a linear along-front temperature gradient. The basic-state is defined by

$$U(x,z,t) = -\alpha x + \Lambda Z e^{-\alpha t}, \tag{1}$$

$$V(y,z,t) = \alpha y - 2\Lambda \frac{\alpha}{f} Z e^{-\alpha t}, \tag{2}$$

$$W = 0, \tag{3}$$

$$B(y,z,t) = \frac{g}{\theta_0}\theta_a(z) - \Lambda^2(1-e^{-2\alpha t})Z - f\Lambda e^{-\alpha t} y, \tag{4}$$

$$\Phi(x,y,z,t) = f\alpha xy - \frac{\alpha^2}{2}(x^2+y^2) - f\Lambda e^{-\alpha t} yZ - \frac{\Lambda^2}{2}(1-e^{-2\alpha t})Z^2 + \frac{g}{\theta_o}\int_0^z \theta_a(z)dz, \tag{5}$$

where $U, V$ and $W$ are the components of the basic-state velocity, B is the basic-state buoyancy ($= g\theta_b/\theta_o$, where $\theta_b$ is the basic-state potential temperature), $\Phi = P/\rho_a$ where $P$ is the basic-state dynamic pressure and $\rho_a$ is a reference density profile, $g = 9.81\ ms^{-2}$ is the acceleration due to gravity, $f = 10^{-4}\ s^{-1}$ is the Coriolis parameter, and $\theta_0 = 273.13$ $K$ is a characteristic potential temperature. Reference profiles of density and potential temperature $\rho_a(z)$ and $\theta_a(z)$ are defined in Reeder and Keyser (1988). Z=$z$-$z_r$ where $z_r$ is a reference height chosen such that the jet-front system is almost stationary relative to the model coordinate system. The magnitude of the horizontal deformation is $2\alpha$, and $\Lambda$ is the vertical shear of the cross-frontal geostrophic wind at the initial instant. Alternatively, $\Lambda$ is a measure of the initial along-front temperature gradient on account of the thermal wind relation

$$\frac{\partial B}{\partial y} = -f\Lambda e^{-\alpha t} = -f\frac{\partial U}{\partial z}. \tag{6}$$

When $\Lambda = 0$ the along-front temperature gradient vanishes and the basic state reduces to a confluent-deformation wind field. Note also how the factor $e^{-\alpha t}$ appears in the basic-state equations, particularly in combination with $\Lambda$. This factor reflects the advection of the basic-state along-front temperature field by the basic-state confluent-deformation wind field. Since $\alpha > 0$ in the present study, the along-front temperature gradient and cross-front geostrophic wind shear weaken exponentially with time (Eq. 6).

The equations governing the model are derived by adding a two-dimensional finite-amplitude perturbation to the basic-state equations (Eqs. 1-5) and then substituting these expressions into the full anelastic equations of motion. If $u$, $v$, and $w$ are the perturbation velocity components, and $b = (g/\theta_o)\theta$ and $\phi$ the perturbations from $B$ and $\Phi$ respectively with $\theta$ the the perturbation from $\theta_b$, then set of equations for the perturbation quantities is:

$$\frac{Du}{Dt} + U_x u + U_z w - fv = -\frac{\partial \phi}{\partial x} + \mathcal{D}(u), \tag{7}$$

$$\frac{Dv}{Dt} + V_y v + V_z w + fu = \mathcal{D}(v), \tag{8}$$

$$\frac{\partial \phi}{\partial z} - b = 0, \tag{9}$$

$$\rho_a \frac{\partial u}{\partial x} + \frac{\partial}{\partial z}(\rho_a w) = 0, \tag{10}$$

$$\frac{Db}{Dt} + B_y v + N_a^2 w = \mathcal{D}(b), \tag{11}$$

where

$$\frac{D}{Dt} = \frac{\partial}{\partial t} + (U + u)\frac{\partial}{\partial x} + w\frac{\partial}{\partial z},$$

and

$$N_a^2 = \left(\frac{g}{\theta_o}\right)\frac{d\theta_a}{dz} - \Lambda^2(1 - e^{-2\alpha t}).$$

Here $N_a^2$ is the buoyancy frequency of the basic-state. Note that from Eqs. (1) and (2), $U_x = -\alpha, U_z = \Lambda e^{-\alpha t}, V_y = \alpha$ and $V_z = -2\Lambda(\alpha/f)e^{-\alpha t}$. Thus the second and third terms in Eqs. (7), (8) and (11) physically represent the advection of the basic-state by the (finite-amplitude) perturbations. Horizontal dissipation is included through the term $\mathcal{D}(\mathrm{r}) = -\mathrm{K}_x\,(\Delta x)^2\partial^4 r/\partial x^4$ in order to control the high frequency noise, where $\Delta x$ is the horizontal grid resolution and $\mathrm{K}_x = 2.5 \times 10^5\ m^2 s^{-1}$. In the calculations presented here, the model domain is 7000 $km \times 70\ km$, with uniform horizontal and vertical grid spacing of 40 $km$ and 175 $km$ respectively. Sponge regions occupy the uppermost 40 $km$ of the model domain and the outermost 1000 $km$ adjacent to the lateral boundaries. A fuller account of the model can be found in Reeder and Keyser (1988), and Griffiths and Reeder (1996).

The same initial conditions are used for the along-front wind and cross-front potential temperature fields ($v$ and $\theta$) in each of the model integrations presented. These initial fields, shown in Fig. 1, comprise opposing upper and lower along-front jets in thermal wind balance with the potential temperature field. A cross-front ageostrophic circulation ($u$ and $w$) is added also (not shown), as is required for the system to evolve in an almost balanced fashion. The thick black line in Fig. 1 and in later figures represents the tropopause, defined here as that region wherein the potential vorticity lies between $1 \times 10^{-6}\ m^2 s^{-1} K kg^{-1}$ and $2 \times 10^{-6}\ m^2 s^{-1} K kg^{-1}$.

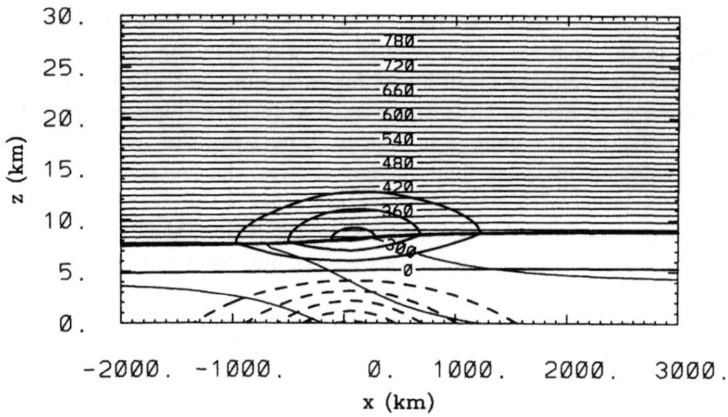


Figure 1. *Isotachs of along-front velocity, v (thick lines), and isentropes of total potential temperature, Θ (thin lines), at the initial time. Dashed lines denote negative values. Contour intervals are 5 $ms^{-1}$ and 15 K. The thickest black line represents the tropopause.*

## 3. Results

The governing equations described in the previous section are numerically integrated for three values of $\Lambda$ (the vertical shear of the cross-front geostrophic wind), the results of which are presented now. As discussed by Keyser and Pecnick (1985) and Reeder and Keyser (1988), the model solutions are characterized by the sign of $\Lambda$. When $\Lambda > 0$, the along-front wind advects warm air at upper levels and cold air at low levels; the converse is true for $\Lambda < 0$. When $\Lambda = 0$ the potential temperature gradient in the along-front direction vanishes.

Frontogenesis at the surface and at the tropopause are prominent features of the numerical solutions and are inextricibly connected to the generation of inertia-gravity waves in the model. The structure of the low-level and upper-level fronts that are developed by the model are, therefore, discussed briefly. Figure 2 shows the 36 hour isotachs of the along-front wind and the isentropes for the model runs with $\Lambda = 0\,s^{-1}$ (no shear), $\Lambda = 6\times10^{-3}\,s^{-1}$ (positive shear), and $\Lambda = -6\times10^{-3}\,s^{-1}$ (negative shear). For convenience these will be referred to as run 1, run 2 and run 3 respectively. In each case $\alpha$=$10^{-5}s^{-1}$.

Pronounced upper-level and low-level jet-front systems are common to each run, as are their associated tropopause folds. Keyser and Pecnick (1985) and Reeder and Keyser (1988) have discussed the dynamics of these systems in some detail. A comparison of Figs. 2b and 2c shows that a deep baroclinic region spanning the whole troposphere develops in run 2 whereas the upper-level and low-level fronts remain relatively distinct and structurally disconnected in run 3. Despite this, the tropopause fold extends further into the troposphere in run 3 than in run 2. The upper-level and low-level fronts and the tropopause fold formed in run 1 are disconnected and are weaker than those in runs 2 and 3.

In each case, frontogenesis is accompanied by a thermally direct cross-front circulation with subsidence in the cold air and ascent in the warm air. As an example, Fig. 4 shows isotachs of the vertical velocity at 36 hours for each run. Isotachs of the total cross-front wind, $U + u$, are shown also. The isotachs have a positive slope in run 2 but a negative slope in run 3, reflecting the change in sign of $\Lambda$ between the two runs (refer to Eq. 1). Although the evolution of these fields is not shown, the isotachs become more widely spaced and more vertical with time. The feature reflects the decreasing basic-state shear with time (see Eqs. 1 and 6). Away from the frontal region, the cross-front wind isotachs are vertical in run 1 since the basic state consists only of a confluent-deformation wind field.

Of particular relevance to this study is the prominent fan-like pattern of vertically propagating low-frequency inertia-gravity waves evident in the lower stratosphere in run 3 (Fig. 3c). The waves have relatively long horizontal and vertical wavelengths, in the range 400–1200 $km$ and 2–10 $km$ respectively, although the horizontal wavelength decreases and the vertical wavelength increases between with time. The modelled waves have frequencies in the range $.05 \times 10^{-4}$ $rad\ s^{-1}$ to $2 \times 10^{-4}$ $rad\, s^{-1}$ and intrinsic frequencies between $1 \times 10^{-4}$ $rad\, s^{-1}$ and $6 \times 10^{-4}$ $rad\, s^{-1}$. The phase lines in run 3 have a positive slope indicating that energy is radiating vertically and to the right relative to the air. Video-looping of the vertical velocity fields shows that the phase lines are slowly displaced downward and to the right with time. However, the horizontal position of the wave packet appears to be relatively steady with respect to the jet/front system. Evidently the wave packet reaches a position at which the horizontal component of the group velocity is opposed by the background flow. Perturbations in the cross-front wind isotachs, also shown in Fig. 3c, are in phase with the vertical velocity perturbations suggesting an upward transport of positive momentum. Note also, that only rightward propagating modes are seen in Fig. 3c because the leftward propagating modes strike a critical level in the lower stratosphere and are absorbed.

The inertia-gravity waves generated by the model in run 3 have many features in common to those observed in the stratosphere and mesosphere. Observations have shown that the largest contribution to the wave variance comes from waves with frequencies close to the inertial frequency f, and that the vertical wavelengths of such waves are typically about 2-4 $km$ in the lower stratosphere, increasing to around 20 $km$ near the mesopause (see *e.g.* Thompson, 1978; Hirota and Niki, 1986; Fritts and Chou, 1987; Eckermann and Vincent, 1989; Vincent, 1990; Hamilton, 1991; Allen and Vincent, 1995). Moreover, studies such as those by Sato (1994) and Fritts *et al.* (1988) have identified upper jets, and their associated fronts, as sources of low-frequency gravity waves. Similarly, Eckermann and Vincent (1993) observed that a broad spectrum of gravity waves accompanied the passage of fronts across southern Australia.

In contrast to run 3, the wave crests seen in run 2 (Fig. 3b) slope in the opposite direction indicating propagation upward and to the left relative to the air. Once again, the phase lines are tilted in the direction opposite to that of the cross-front wind in the stratosphere. In other words the wave crests slope against the flow in each case. In run 1, very weak amplitude waves are seen both to the left and to the right of the frontal region.

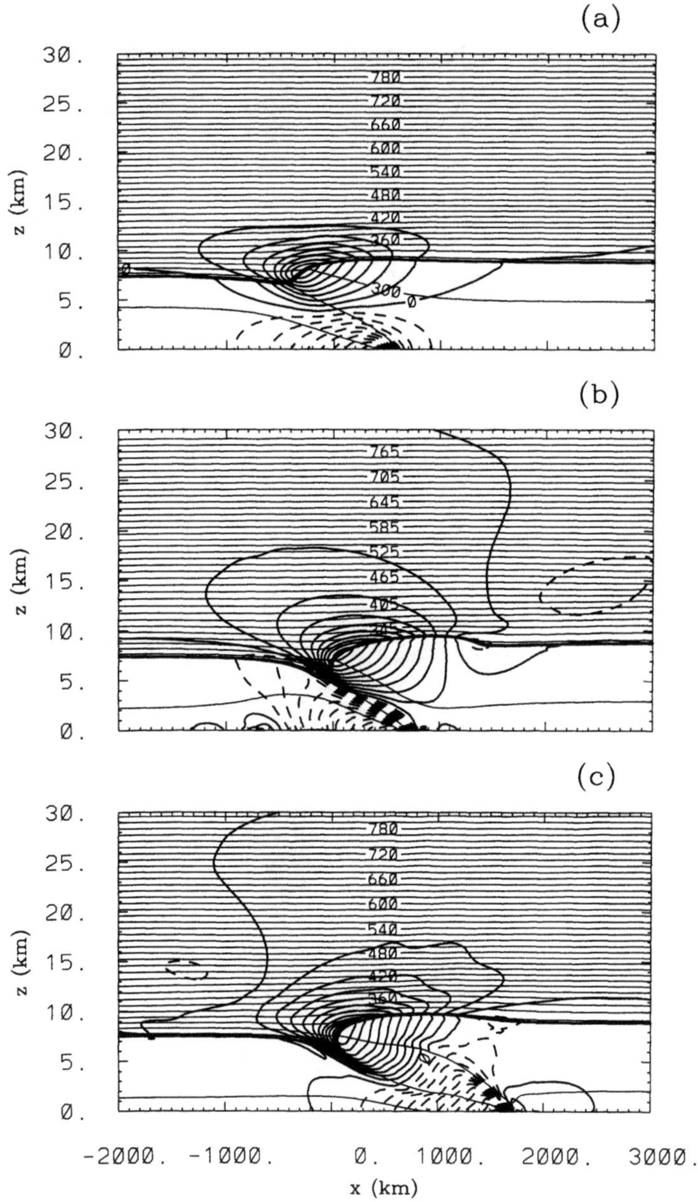


Figure 2. *Isotachs of along-front velocity, v (thick lines), and isentropes of total potential temperature, Θ (thin lines), at t=36 hours. Dashed lines denote negative values. Contour intervals are 5 $ms^{-1}$ and 15 K. The thickest black line represents the tropopause.* $\alpha = 1 \times 10^{-5}\ s^{-1}$. *(a) Run 1,* $\Lambda = 0\ s^{-1}$; *(b) run 2,* $\Lambda = 6 \times 10^{-3}\ s^{-1}$; *and (c) run 3,* $\Lambda = -6 \times 10^{-3}\ s^{-1}$.

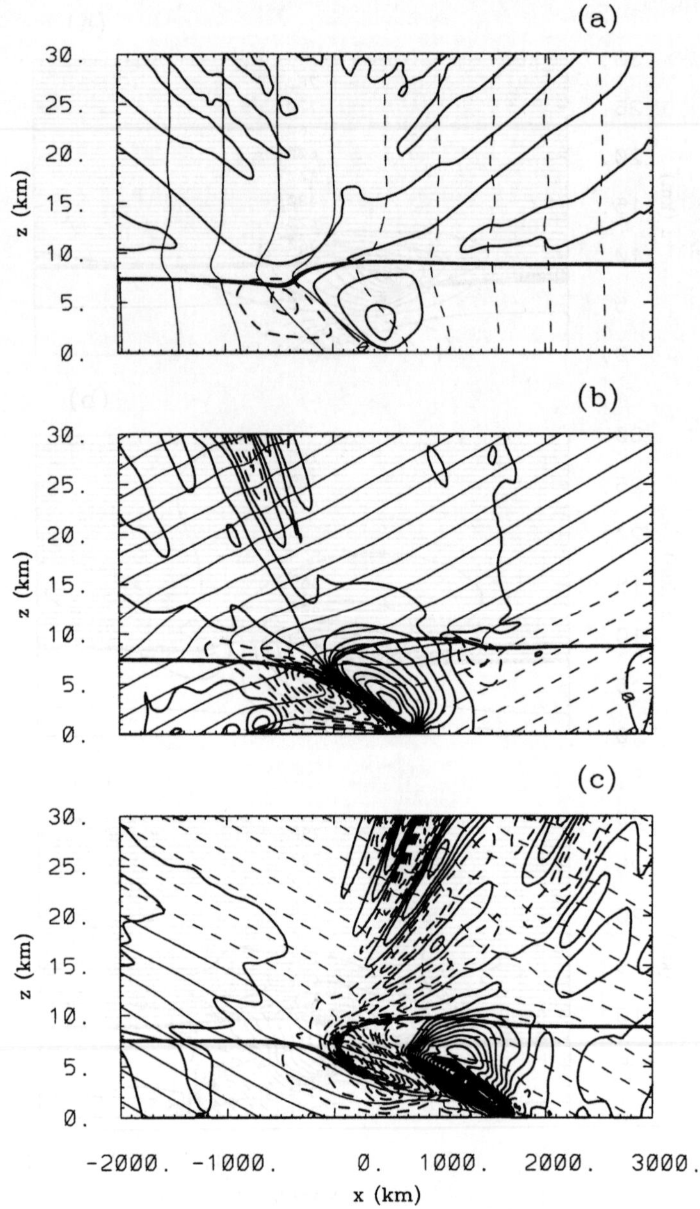


Figure 3. *Isotachs of vertical velocity, w (thick lines), and isotachs of total cross-front velocity, U + u (thin lines), at t=36 hours. Dashed lines denote negative values. Contour intervals are 0.6* $cms^{-1}$ *for vertical velocity and* $5\,m\,s^{-1}$ *for cross-front velocity. The thickest black line represents the tropopause.* $\alpha = 1 \times 10^{-5} s^{-1}$. *(a) Run 1,* $\Lambda = 0\ s^{-1}$*; and (b) run 2,* $\Lambda - 6 \times 10^{-3}\ s^{-1}$*, and (c) run 3,* $\Lambda = 6 \times 10^{-3}\ s^{-1}$.

The evolution of each system may be neatly summarized through time-series of the maximum potential temperature gradient of the upper-level and low-level fronts. These time-series are shown in Fig. 4. An important point to note is that initially the upper-level and low-level fronts develop very rapidly in run 3 before reaching a quasi-steady state. In contrast, the fronts in run 2 evolve more slowly than those in run 3 but eventually produce a considerably stronger surface temperature gradient and a slightly stronger temperature gradient near the tropopause. Run 1 shows a weaker development than run 3 at upper levels, however, at low levels the potential temperature gradient exceeds that of run 3 after about 45 hours of integration. As discussed by Reeder and Griffiths (1996), the degree of wave generation is not related to the strength of the fronts, but instead to the rate of frontogenesis.

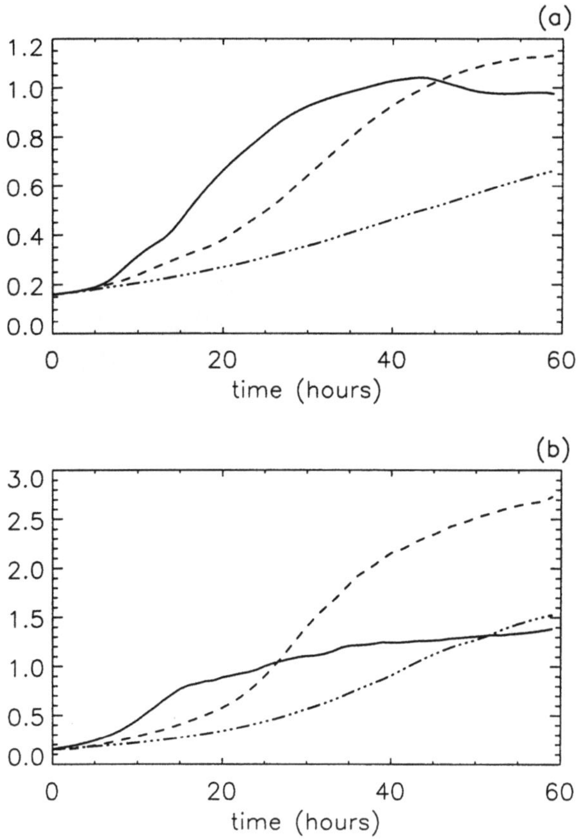


Figure 4. *Time-series of the maximum horizontal temperature gradient in the neighbourhood of the (a) upper and (b) lower front for each run of the model. Dot-dash lines represent run 1, dashed lines represent run 2, and solid lines represent run 3. Units are* $K\ (100\,km)^{-1}$. *Note the different scale used for the ordinates of (a) and (b).*

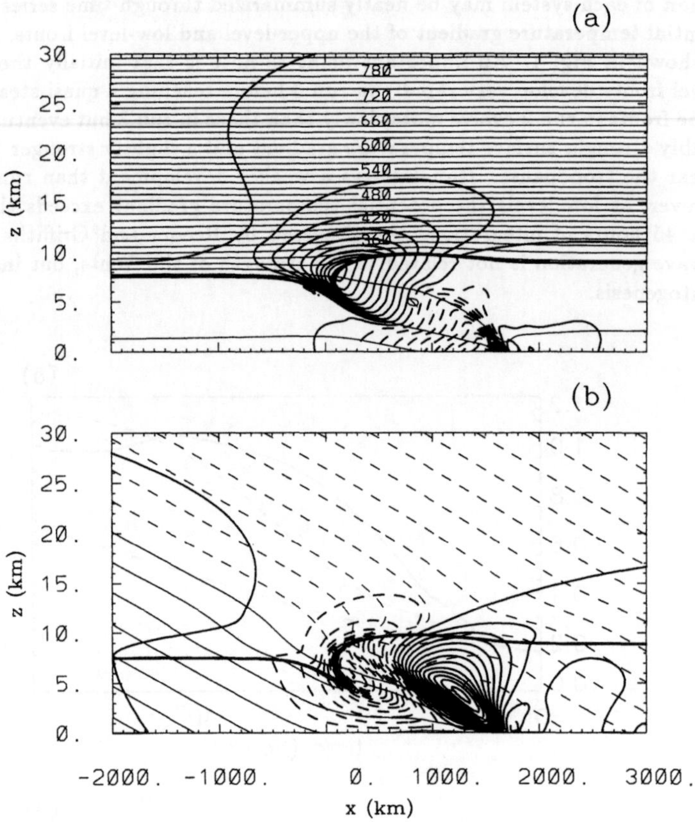


Figure 5. *The GM equivalent to run 3 at* $t = 36h$. $\alpha = 1 \times 10^{-5}\, s^{-1}$ *and* $\Lambda = -6 \times 10^{-3}\, s^{-1}$. *(a) Isotachs of along-front velocity,* $v_{gm}$ *(thick lines), and isentropes of total potential temperature,* $\Theta_{gm}$ *(thin lines); (b) isotachs of vertical velocity,* $w_{gm}$*, and total cross-front velocity,* $U + u_{gm}$*. Dashed lines denote negative values. Contour intervals are: (a)* 5 $ms^{-1}$ *for* $v_{gm}$ *and* 15 $K$ *for* $\Theta_{gm}$*, and (b)* 0.6 $cms^{-1}$ *for* $w_{gm}$ *and* 5 $ms^{-1}$ *for* $U + u_{gm}$*. The thick horizontally-oriented line represents the tropopause.*

## 4. Balanced model

We discuss now an approximation to the governing equations that filters gravity waves. These filtered solutions will prove useful when quantifying the upward flux of horizontal momentum by the gravity waves emitted by the fronts in run 3 (see next section).

Scale analysis, such as that developed by Hoskins and Bretherton (1972), suggests that the along-front wind is almost geostrophic, and that its vertical shear balances the cross-front temperature gradient. Such a simplification when applied to the governing equations is commonly called the Geostrophic Momentum (GM) approximation. An analysis by Reeder and Keyser (1988) showed that, provided $\alpha \ll f$ and $\Lambda \exp(-\alpha t) \leq N_a^2$, conditions satisfied in the present work, the GM approximation could be applied to the perturbation equations (Eqs. 7-11). The key simplification is that while the along-front wind is assumed to be strictly geostrophic (*i.e.*, $v = f^{-1}\partial\phi/\partial x = v_{gm}$), advection by the ageostrophic wind in the $x$-$z$ plane is retained (*i.e.*, $D/Dt$ is not approximated). An important consequence of the GM approximation is that, unlike the anelastic equations, they filter gravity wave solutions. Of course, the two systems also will exhibit differences in solution which have nothing to do with gravity wave radiation, but instead reflect the approximations underpinning the GM model.

The results from an integration of the GM model analogous to run 3 is shown in Fig. 5. An outline of the model and numerical methods used to solve the GM equations can be found in Reeder and Keyser (1988). Comparing Figs. 2c and 5a shows that the along-front wind and total potential temperature fields from the two model integrations are in close agreement. The most prominent differences between the solutions arise in the vertical velocity fields (*c.f.* Figs. 3c and 5b); in particular the GM solution fails to reproduce the stratospheric wave pattern (as expected) and the secondary updraft on the warm side of the front in the middle troposphere.

## 5. Momentum flux calculations

In general, radiating gravity waves induce a vertical flux of horizontal momentum. Problems concerning wave-mean-flow interactions are often handled by separating the flow into mean and departure quantities. Commonly, the mean flow is defined as the horizontal average of the flow and the 'waves' are defined as departures from the mean. In the present section this paradigm is followed and estimates of the generalized momentum flux, or Eliassen-Palm flux (Andrews and McIntyre, 1976), for the waves generated in run 3 are presented.

In this study, mean quantities are defined as

$$\langle r \rangle = \frac{1}{x_2 - x_1} \int_{x_1}^{x_2} r dx, \tag{12}$$

where $r$ is any given quantity, and $[x_1,x_2]$ defines the averaging interval. Consequently, departures from the mean, or 'waves', are defined as

$$r' = r - \langle r \rangle \, . \tag{13}$$

We note that the decomposition of the flow defined by Eqs. (12) and (13) is arbitrary and depends on the choice of $x_1$ and $x_2$.

Rewriting Eq. (7) in flux form and applying the averaging operator Eq. (12) yields

$$\frac{\partial \langle u\rangle}{\partial t}+\langle w\rangle \frac{\partial \langle u+U\rangle}{\partial z}+V_y\langle u\rangle - f\langle v_\star\rangle = -\frac{1}{\rho_a}\frac{\partial E}{\partial z}+A+R+\langle D\rangle . \tag{14}$$

Here

$$\langle v_\star\rangle = \langle v\rangle - \frac{1}{\rho_a}\frac{\partial}{\partial z}\left(\frac{\rho_a\langle v'\theta'\rangle}{\langle\theta_z\rangle}\right)$$

is the mean along-front residual circulation, and

$$E=\rho_a\langle (u'+U')w'\rangle - f\rho_a\frac{\langle v'\theta'\rangle}{\langle\theta_z\rangle}$$

is the vertical component of the generalized momentum flux (*i.e.*, the negative of the Eliassen-Palm flux). Further,

$$A=-\frac{1}{x_2-x_1}\,[\phi]_{x_1}^{x_2}$$

is the average pressure gradient, proportional to the difference in pressure at the left and right endpoints of the averaging interval, and

$$R=-\frac{1}{x_2-x_1}\left[u^2+2uU-\langle u+U\rangle u\right]_{x_1}^{x_2}$$

represents the terms in Eq. (14) describing fluxes at the endpoints of the averaging enterval. In the calculations that follow, the averaging interval is taken to be [$-2000$ *km*, $3000$ *km*] *i.e.*, the entire width of the model domain excluding the lateral sponge regions.

The mean values of the generalized momentum flux for run 3 and its GM equivalent, $\overline{E}^{zt}$ and $\overline{E}^{zt}_{gm}$, are $1.14\times10^{-3}$ and $9.6\times10^{-4}\ kg\,m^{-1}s^{-2}$ respectively. The overbar, $\overline{(\ )}^{zt}$, represents an average taken over the time interval 24 to 48 hours and over the height range 15 to 29 *km*. This height range was chosen to lie below the sponge, but well above the tropopause. The most significant point to note here is that $\overline{E}^{zt}$ and $\overline{E}^{zt}_{gm}$ are nearly equal even though the GM solution explicitly filters gravity waves. The 'waves' that contribute to $\overline{E}^{zt}_{gm}$ are not inertia-gravity waves but are simply departures from the mean attributable to the horizontal structure in the basic state and the penetration of the frontal circulation, albeit very weakly, into the stratosphere. For this reason, the magnitude of $\overline{E}^{zt}$ (and $\overline{E}^{zt}_{gm}$) is sensitive to the choice of averaging interval $[x_1,x_2]$. These results highlight the danger in identifying inertia-gravity waves as departures from an arbitrary mean state, and sound a note of caution when interpreting observations. An estimate of the generalized momentum flux associated with the inertia-gravity waves identified in run 3 is $\overline{E}^{zt}_{wave}\equiv\overline{E}^{zt}-\overline{E}^{zt}_{gm}=1.8\times10^{-4}\ kg\,m^{-1}\,s^{-2}$.

The table below lists the mean values for each term contributing to the momentum budget, Eq. (14), for both run 3 and its GM equivalent. The table shows that in each case the dominant balance is between the divergence of the generalized momentum flux and the difference in pressure at the lateral boundaries. The generalized momentum flux divergence attributable to inertia-gravity waves, $\rho_a^{-1}\partial(\overline{E}^{zt} - \overline{E}_{gm}^{zt})/\partial z$, and the unbalanced pressure difference at the lateral boundaries, $\overline{A}^{zt} - \overline{A}_{gm}^{zt}$, essentially balance one another. Note that the acceleration of the mean flow is four orders of magnitude smaller than either of these two terms. This result has implications for the parameterization of inertia-gravity waves.

The small momentum flux accompanying the modelled inertia-gravity waves is in accord with observations. While such waves contribute significantly to the variance of atmospheric wind and temperature fields, they appear to play only a minor part in the vertical transport of horizontal momentum; higher-frequency short-wavelength waves are believed to be the principal contributors to the vertical flux of horizontal momentum.

| | $\overline{\frac{\partial\langle u\rangle}{\partial t}}^{zt}$ | $\overline{\langle w\rangle \frac{\langle u+U\rangle}{\partial z}}^{zt}$ | $\overline{V_y \langle u\rangle}^{zt}$ | $\overline{f\langle v_\star\rangle}^{zt}$ |
|---|---|---|---|---|
| run 3 | $9.5 \times 10^{-7}$ | $1.4 \times 10^{-6}$ | $-1.3\times10^{-6}$ | $-1.3\times10^{-5}$ |
| GM run 3 | $-9.6\times10^{-7}$ | $1.8 \times 10^{-6}$ | $-1.3\times10^{-6}$ | $-1.3\times10^{-5}$ |

| | $\overline{\frac{1}{\rho_a}\frac{\partial E}{\partial z}}^{zt}$ | $\overline{A}^{zt}$ | $\overline{R}^{zt}$ | $\overline{\langle D\rangle}^{zt}$ |
|---|---|---|---|---|
| run 3 | $3.0 \times 10^{-3}$ | $-3.0\times10^{-3}$ | $2.9 \times 10^{-6}$ | $3.5 \times 10^{-10}$ |
| GM run 3 | $2.6 \times 10^{-3}$ | $-2.7\times10^{-3}$ | $2.7 \times 10^{-6}$ | $1.1 \times 10^{-11}$ |

Table. *Average values of the terms in Eq. (15) for run 3 and the GM version of run 3. Each term is averaged over the height range* $15 \le z \le 30$ *km and the time interval* $24 \le t \le 48$ *hours.*

## 6. Summary and conclusion

This study has addressed the problem of gravity wave generation during frontogenesis using a dry two-dimensional anelastic numerical model. The model basic state is a combination of two background flows commonly used in frontogenesis modelling; these two flows are a horizontal confluent-deformation wind field and a time-dependent cross-front geostrophic wind with a linear variation with height. All previous numerical studies concerned with wave generation during frontogenesis (Gall *et al.*, 1988; Garner, 1989; Snyder *et al.*, 1993) have used rigid-lid upper-boundary conditions, and none of these studies examined the possibility of wave generation associated with upper-level frontogenesis. Although several recent papers have dealt with gravity wave generation associated with the geostrophic adjustment of upper-level jets (Fritts and Luo, 1992; Luo and Fritts, 1993), these calculations are initial-value adjustment problems and do not explicitly address the evolution of the jet. The present paper builds on the studies cited above by: examining the role of upper-level frontogenesis in the generation of internal gravity waves; explicitly calculating the evolution of the jet/front system and allowing the internal dynamics to spontaneously generate gravity waves; allowing upward wave radiation into the stratosphere; and using the mixed confluent-deformation/horizontal-shear basic state developed by Keyser and Pecnick (1985).

The numerical model was initialized with opposing upper-level and lower-level along-front jets in thermal wind balance with the potential temperature field, together with the balanced vertical cross-front circulation. Numerical solutions for three cases have been presented. These cases are model runs for which $\Lambda = 0\,s^{-1}$ (run 1, no shear), $\Lambda = 6 \times 10^{-3}\,s^{-1}$ (run 2, positive shear), and $\Lambda = -6 \times 10^{-3}\,s^{-1}$ (run 3, negative shear). In each case fronts formed at the tropopause and at the surface.

Attention has been focused on the stratospheric wave field. Propagating inertia-gravity waves were generated in each model run, although the waves were by far most pronounced in run 3. Such waves are consistent with those observed in the lower stratosphere, having frequencies close to the inertial frequency and horizontal wavelengths on the order of $400 - 1200$ *km* and vertical wavelengths in the range $2 - 10$ *km*. In each case the phase lines of the modelled waves tilted into the wind. For instance, in run 3, the phase lines showed a positive slope indicating that energy was radiating vertically and to the right relative to the air.

It is significant that the strongest frontogenesis and the most intense upper-level and surface fronts formed in run 2, and not in run 3, the case which showed the most pronounced wave generation. On the other hand, time-series of the maximum potential temperature gradient showed that run 3 evolved most rapidly. As discussed in detail by Reeder and Griffiths (1996), the inertia-gravity waves were generated by the unsteady two-dimensional cross-front circulation. Thus, the relatively stronger wave emission in run 3 was a consequence of the more rapid evolution in this case.

The mean flux of horizontal momentum attributable to the inertia-gravity waves has been shown to be approximately $1.8 \times 10^{-4}\,kg\,m^{-1}\,s^{-2}$, where the flux was calculated as the difference between that for run 3 and that for its GM equivalent. The divergence of the generalized momentum flux associated with the wave field was shown to be balanced by the difference in pressure between the lateral boundaries of the model domain.

ACKNOWLEDGEMENTS. We should like to thank Steve Eckermann, David Karoly, Dan Keyser, Michael McIntyre and Joe Zehnder for their helpful criticisms and suggestions. This work was supported by grants from the Australian Research Council and the National Greenhouse Advisory Committee.

# References

Allen, S. J., and Vincent, R. A., 1995. Gravity wave activity in the lower atmosphere: seasonal and latitudinal variations. *J. Geophys. Res.*, **100**, 1327–1350.

Andrews, D. G., and McIntyre, M. E., 1976. Planetary waves in horizontal and vertical shear: the generalized Eliassen-Palm relation and the mean zonal acceleration. *J. Atmos. Sci.*, **33**, 2031–2048.

Eckermann, S. D., and Vincent, R. A., 1989. Falling sphere observations of anisotropic gravity wave motions in the upper stratosphere over Australia. *Pageoph.*, **130**, 509–532.

Eckermann, S. D., and Vincent, R. A., 1993. VHF radar observations of gravity wave production by cold fronts of southern Australia. *J. Atmos. Sci.*, **50**, 785–806.

Fritts, D. C., 1993. Gravity wave sources, source variability and lower and middle atmosphere effects. E. V. Thrane *et. al.*, Ed., *Coupling Processes in the Lower and Middle Atmosphere*, Kluwer Academic Publishers, 191–208.

Fritts, D. C., and Chou, H.-G., 1987. An investigation of the vertical wavenumber and frequency spectra of gravity wave motion in the lower stratosphere. *J. Atmos. Sci.*, **44**, 3610–3624.

Fritts, D. C., and Luo, Z., 1992. Gravity wave excitation by geostrophic adjustment of the jet stream. Part I: two-dimensional forcing. *J. Atmos. Sci.*, **49**, 681–697.

Fritts, D. C., Tsuda, T., Sato, T., Fukao, S., and Kato, S., 1988. Observational evidence of a saturated gravity wave spectrum in the troposphere and lower stratosphere. *J. Atmos. Sci.*, **45**, 1741–1759.

Gall, R., Williams, R. T., and Clarke, T. L., 1988. Gravity waves generated during frontogenesis, *J. Atmos. Sci.*, **45**, 2204–2219.

Garner, S. T., 1989. Fully Lagrangian numerical solutions of unbalanced frontogenesis and frontal collapse, *J. Atmos. Sci.*, **46**, 717–739.

Griffiths, M., and Reeder, M. J., 1996. Stratospheric inertia-gravity waves generated in a numerical model of frontogenesis. I: model solutions, *Quart. J. Roy. Meteor. Soc.*, **122**, 1153-1174.

Hamilton, K., 1991. Climatological statistics of stratospheric inertia-gravity waves deduced from horizontal rocketsonde and temperature data. *J. Geophys. Res.*, **96**, 20831–20839.

Hirota, I., and Niki, T., 1986. Inertia gravity waves in the troposphere and stratosphere observed by the MU radar. *J. Meteor. Soc. Japan*, **64**, 995–999.

Hoskins, B. J., and Bretherton, F. P., 1972. Atmospheric frontogenesis models: mathematical formulation and solution. *J. Atmos. Sci.*, **29**, 11–37.

Keyser, D., and Pecnick, M. J., 1985. A two-dimensional primitive equation model of frontogenesis forced by confluence and horizontal shear. *J. Atmos. Sci.*, **42**, 1259–1282.

Ley, B., and Peltier, W. R., 1978. Wave generation and frontal collapse. *J. Atmos. Sci.*, **35**, 3–17.

Luo, Z., and Fritts, D. C., 1993. Gravity wave excitation by geostrophic adjustment of the jet stream. Part II: three-dimensional forcing. *J. Atmos. Sci.*, **50**, 104–115.

O'Sullivan, D., and Dunkerton, T. J., 1995. Generation of inertia-gravity waves in a simulated life cycle of baroclinic instability. *J. Atmos. Sci.*, **52**, 3695–3716.

Reeder, M. J., and Griffiths, M., 1996. Stratospheric inertia-gravity waves generated in a numerical model of frontogenesis. II: wave sources and generation mechanisms. *Quart. J. Roy. Meteor. Soc.*, **122**, 1175-1195.

Reeder, M. J., and Keyser, D., 1988. Balanced and unbalanced frontogenesis. *J. Atmos. Sci.*, **45**, 3366–3386.

Sato, K., 1994. A statistical study of the structure, saturation and sources of inertio-gravity waves in the lower stratosphere observed with the MU radar. *J. Atmos. Terr. Phys.*, **50**, 755–774.

Snyder, C., Skamarock, W. C., and Rotunno, R., 1993. Frontal dynamics near and following frontal collapse. *J. Atmos. Sci.*, **50**, 3194–3211.

Thompson, R. O. R. Y., 1978. Observation of inertial waves in the stratosphere. *Quart. J. Roy. Meteor. Soc.*, **104**, 691–698.

Vincent, R. A., 1990. Gravity waves in the southern hemisphere middle atmosphere: a review of observations. A. O'Neill, Ed., *Dynamics, Transport and Photochemistry in the Middle Atmosphere of the Southern Hemisphere*, Kluwer Academic Publishers, 159–170.

# A Model of Non-Stationary Gravity Waves in the Stratosphere and Comparison to Observations

M. Joan Alexander
Department of Atmospheric Sciences
University of Washington
Seattle, WA

## Abstract

A model of gravity wave propagation and mean flow effects is compared to observations. Despite the absence of any gravity wave source variations in this model, monthly mean distributions in the gravity wave-driven zonal force and temperature variance show very realistic seasonal, vertical, and geographical variations in the middle atmosphere. A comparison to recent Upper Atmosphere Research Satellite observations by the Microwave Limb Sounder (MLS) suggests the observed patterns are not indicative of variations in gravity wave sources, but instead result from a convolution of two effects on the gravity wave spectrum: 1) Doppler-shifting by the background winds, and 2) vertical wavelength filtering by the observation technique itself. These results underscore the importance of variations in the background winds in interpreting observations of gravity waves and their effects on the middle atmosphere.

## Introduction

The existence of a zonal force due to gravity wave dissipation is now well established in the mesosphere, where this wave-driven force is fundamental to the understanding of the warm winter and cold summer mesopause temperatures. In the upper troposphere-lower stratosphere, stationary gravity wave drag associated with surface topography is considered an important component of global circulation models, slowing the eastward winds above the midlatitude tropospheric jet maximum and significantly affecting northern winter climate.

In the region between, which includes most of the stratosphere, we have had little constraining evidence for determining the sign and magnitude of the gravity wave-driven zonal force. Simple interpolation would lead to the conclusion that the force opposes the wind, as it does in the mesosphere and lowermost stratosphere. Rayleigh friction and stationary wave drag applications in models can only produce such drag forces, and so lead to the same conclusion. Planetary wave dissipation dominates the stratospheric forcing in the winter seasons so that the force due to gravity waves at these levels has been largely ignored.

Alexander and Rosenlof [1996] recently examined the role of non-stationary gravity wave-driven zonal-mean zonal forcing in the stratosphere, using data compiled from the Upper Atmosphere Research Satellite (UARS) and a theoretical model of gravity wave propagation and dissipation. This work was precipitated by the observation of Rosenlof [1996] that lower stratosphere temperatures in the southern summer season are 2.5-5 K warmer than northern summer, suggesting a stronger summer residual circulation

NATO ASI Series, Vol. I 50
Gravity Wave Processes
Their Parameterization in Global Climate Models
Edited by Kevin Hamilton

in the south. Differences in the inferred residual circulation could not be explained by differences in the planetary-scale wave driving as resolved in global assimilated data from either the National Meteorological Center (NMC) or the United Kingdom Meteorological Office (UKMO). Instead a force due to small-scale waves, not resolved in the global data, must be present.

A theoretical model of gravity wave propagation and dissipation [Alexander, 1996] was used in Alexander and Rosenlof [1996] to examine what role gravity waves play in generating this missing zonal force. This force is westward throughout most of the summer stratosphere, having the same sign as the stratospheric winds. Because this missing force has the same sign as the mean wind, it cannot be described by Rayleigh friction, by stationary wave drag, or by extension of mesospheric gravity wave drag down into the stratosphere. These all produce a force that opposes the winds. The lower stratosphere temperature effect noted above suggests that this missing zonal force may be significant to climate. Further, the fact that most previous gravity wave parameterization applications have applied a force of the wrong sign in the stratosphere, even if relatively small, may give significant effects in model results at these levels, and so it should not be ignored. Alexander and Rosenlof [1996] showed that initially isotropic globally uniform sources of gravity wave momentum flux interact very anisotropically with the mean winds to give seasonally varying patterns in the zonal mean zonal forcing that were similar to the observationally derived constraints. The conceptual approach is similar to theoretical work by Dunkerton and Butchart [1984] and Saravanan [1990] who demonstrated filtering effects of the background winds on initially uniform gravity wave sources in the troposphere.

The model also predicts global variations in gravity wave temperature variance like those deduced from observations by the UARS Microwave Limb Sounder (MLS) [Wu and Waters, 1996]. Geographical patterns in the MLS gravity wave variance have been presumed to be caused by variations in gravity wave sources. In this work, globally uniform gravity wave sources are allowed to propagate through both zonal mean and longitudinally varying monthly mean zonal wind and temperatures from the UKMO assimilated data [Swinbank and O'Neill, 1994] compiled for UARS. Vertical variations, seasonal variations, and global patterns in the model variance will all be shown to be strikingly similar to the patterns observed by MLS. The agreement between the model and the observations will help to confirm that the MLS technique is sensitive to gravity wave perturbations, however the patterns in variance observed with the MLS do not reveal variations gravity wave sources. Instead the patterns will be shown to result from a convolution of Doppler-shifting effects by the background winds on the gravity wave spectrum and vertical wavelength filtering inherent in the MLS observation technique. The model also relates the vertical profile of variance observed by MLS to the gravity wave-driven forcing estimated in Alexander and Rosenlof [1996] and provides an important confirmation of those results that has general implications for gravity wave parameterizations and models.

The constraints on gravity wave forcing in the stratosphere will be reviewed in the next section. Then the gravity wave model will be introduced and used to estimate the zonal mean zonal force in the middle atmosphere due to non-stationary gravity waves. Temperature variance maps, using the same model and a spectral filter designed to mimic the MLS observing technique, will then be presented and an interpretation of the MLS observations described, followed by discussion of these results and summary of the conclusions.

## Constraints on Gravity Wave Induced Forcing in the Stratosphere

Figure 1, reproduced from Alexander and Rosenlof [1996] shows the observationally derived constraint on small-scale wave-driven forcing that may be associated with gravity wave dissipation. In summertime, large-scale waves resolved in the UKMO data only contribute significantly to the total forcing in the lower stratosphere below ~30 hPa. Large-scale waves are also very important in the winter hemispheres at all altitudes. However in summertime above 30 hPa (~25 km), the force is almost entirely attributable to unresolved waves, and is likely associated with gravity wave dissipation. The required zonal force is westward in this region of summertime westward winds. The small scale force changes sign in both hemispheres near 2 hPa, marking the level where the gravity wave drag force in the mesosphere begins to appear. The altitude of this zero force line is used as a rough constraint on gravity wave amplitudes at the source level in the model presented below. It should be noted that the forcing estimates in Fig. 1 are highly derived quantities and have potentially large uncertainties associated with them. However, the model results in conjuction with the MLS comparison described below strongly support the general features shown here, in particular the sign of the force in the stratosphere, and the altitude of the zero force line occuring near 2 hPa (~50 km).

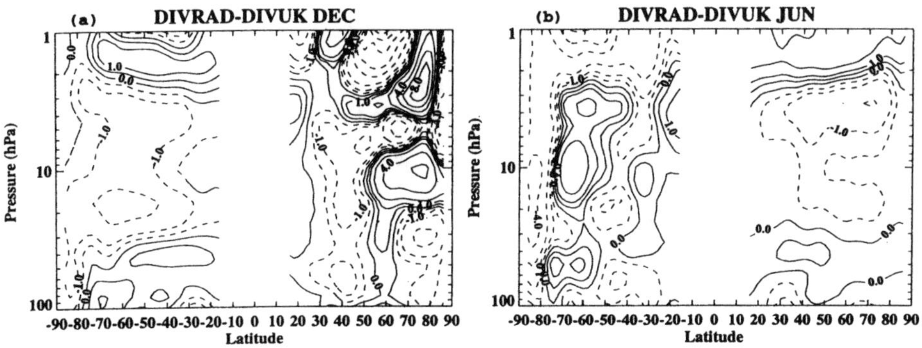


*Figure 1. Latitude-height maps of the small-scale zonal force for (a) December and (b) June. These show the difference of the total wave-driven force and the large-scale (resolved in the UKMO data) wave-driven force. Contours are 0, .5, 1, 2, 4 m $s^{-1}day^{-1}$, with dashed contours for negative values. (Reproduced from Alexander and Rosenlof [1996].)*

Another constraint will be applied on the magnitude of the force in the mesosphere, which has been estimated fairly consistently to be ~100-200 m $s^{-1}day^{-1}$ in numerous modeling and observation studies in the past [e.g. Holton, 1983; Garcia and Solomon, 1985; Fritts and Vincent, 1987; Fritts and Yuan, 1989; Brasseur et al., 1990]. This constraint will determine the value of the "efficiency factor" in the model, a scaling factor applied to the resulting force that represents the spatial and temporal intermittancy of the gravity wave sources.

### The Gravity Wave Model

The gravity wave propagation and dissipation model used in this study was described in Alexander [1996] and will be summarized here. The two-dimensional ray equations [e.g. Lighthill, 1978] are used to solve for the gravity wave propagation properties and amplitudes along the ray path. The dispersion relation applied is that derived from the linearized set of fundamental equations including nonhydrostatic and rotation effects in a stratified atmosphere:

$$\omega^2 = \frac{N^2k^2 + f^2(m^2 + \alpha^2)}{k^2 + m^2 + \alpha^2}$$

Here, $\omega$ = intrinsic frequency, $N$ = buoyancy frequency, $(k, m)$ = wavenumber vector, $f$ = Coriolis parameter, and $\alpha = 1/(2H)$, where $H$ = density scale height. $N, m, H$, and $\omega = (\omega_0 - Uk)$ all vary with height. The zonal wavenumber $k$ is assumed constant, therefore neglecting horizontal variations in the background wind $U$ along the ray. In general, the zonal distance travelled along a ray is small compared to the horizontal variations in $U$ except near critical levels, but dissipation rapidly attenuates the wave in those regions. Wave properties are specified at the source level chosen here to be 6 km. The wave spectrum is treated as a set of non-interacting and linearly propagating gravity waves. The amplitude growth of each spectral element with height follows from conservation of wave action flux:

$$F_A = \rho c_{gz} E/\omega = constant$$

where $\rho$ =density, $E$ =energy per unit mass, and $c_{gz} = \partial\omega/\partial m$ =vertical group velocity. $F_A$ is conserved unless the wave amplitude exceeds the convective instability limit,

$$E \leq \frac{N^2}{m^2}\frac{\omega^2}{\omega^2 - f^2}$$

This is the saturation condition. The polarization relations further relate energy density to momentum flux,

$$\rho\overline{u'w'} = \frac{\omega}{N}\left[\left(1 - \frac{\omega^2}{N^2}\right)\left(1 - \frac{f^2}{\omega^2}\right)\right]^{1/2} E$$

Critical levels occur where $\omega$ decreases to the minimum frequency $f$, and turning points (reflection) where $\omega$ exceeds the cutoff frequency $\omega_c$ defined by setting $m = 0$ in the dispersion relation [Marks and Eckermann, 1995].

For the purpose of evaluating mean flow effects, the ray paths are collapsed to a single vertical profile, momentum flux is integrated over the spectrum, and the gravity wave induced force per unit mass $D$ evaluated from the convergence of the momentum flux profile:

$$D = -\frac{1}{\rho}\frac{\partial}{\partial z}(\rho\overline{u'w'})$$

Waves that are reflected without breaking have no net effect on the mean flow, but do contribute to the total variance below the turning point.

Specification of the gravity wave spectrum at the source altitude is generally considered one of the biggest uncertainties in this type of model. Here, the sources are assumed to be isotropic in zonal phase speed and independent of geographic location and season. The spectrum includes 60 waves ranging in: horizontal wavelength 6.25-800 km; period 10 min - 4 hr; vertical wavelength 1-25 km; and zonal phase speed $\pm 70$ m $s^{-1}$. For these calculations, the winds and buoyancy frequency at the source level were modified so that $U_0 = 0$ and $N_0 \geq .01$ $s^{-1}$ everywhere so the sources specified in intrinsic frequency and horizontal wavenumber are truly uniform globally. The modified $U$ and $N$ are relaxed back to the UKMO data with an e-folding vertical scale of 3 km, so in the stratosphere the mean state is unaltered. This change from Alexander and Rosenlof [1996] was necessary because the larger variations in $U_0$ and $N_0$ in the UKMO data without zonal averaging would introduce significant and undesirable filtering effects on the spectrum at the source height. Each member of the spectrum is assigned the same momentum flux. (This choice may seem odd at first glance, but a flat momentum flux spectrum is associated with a decidedly red frequency spectrum in zonal wind and temperature.) The effects of these assumptions about the source spectrum will be discussed below.

## Zonal Mean Zonal Force

The results of the model using December and June zonal mean $U(\phi, z)$ and $N(\phi, z)$ as functions of latitude $\phi$ and height $z$ are shown in Figure 2. Below 55 km, 1993 UKMO background fields are used, while above 55 km the results using CIRA (Cospar International Reference Atmosphere; Fleming et al., [1990]) are shown. These maps represent the contribution to the zonal mean zonal force due to non-stationary gravity waves. The magnitudes and distributions in the stratosphere are similar to those in Figure 1.

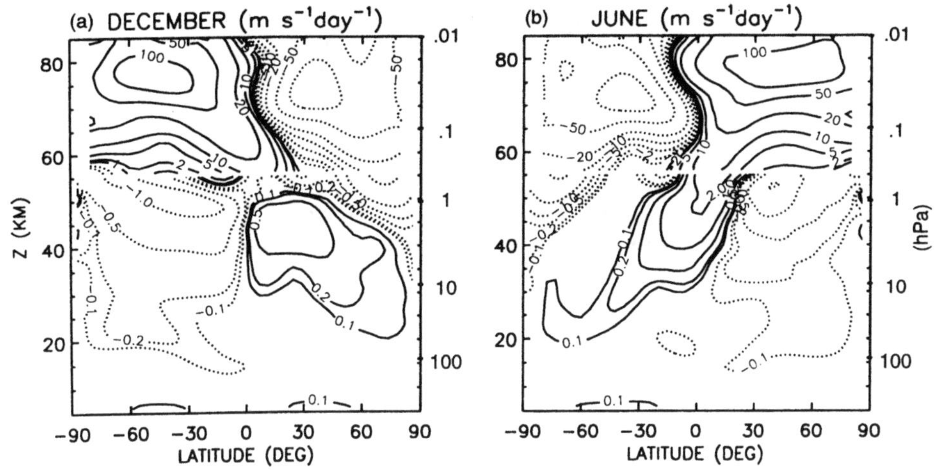


*Figure 2. Latitude-height maps of the gravity wave-driven zonal force dervied from the linear model with saturation and globally uniform gravity wave sources. Contours have semilog spacing, 0, .1, .2, .5, 1, 2, 5, 10, 20, 50, 100, 200 m $s^{-1}$ day$^{-1}$, with dashed contours for negative values.*

The details of the forcing distribution differ from Alexander and Rosenlof [1996] mainly because of differences between UKMO winds and the NMC winds used previously. The artificial modifications to the troposphere background fields described above affect the results below ~20 km. This region in general is the most sensitive to the details of the input source spectrum. Above ~20 km, the sign and magnitude of the stratospheric forcing is primarily sensitive to the average amplitudes of the waves at the source, and is fairly insensitive to other source spectrum details: source height, choice of discrete spectral frequencies, and shape of the source spectrum. The shape of the source spectrum in frequency or phase speed does affect the vertical profile of the force in the stratosphere. Redder spectra, emphasizing low frequencies or phase speeds, were also tested. These tended to give larger forcing at mid and lower stratosphere altitudes than Fig. 2, to exaggerate the asymmetry in the altitudes of the mesosphere summer/winter peak drag, and also shift the stratospheric forcing down to lower levels. It would be inappropriate to tune the input spectrum to try to exactly reproduce the observations in Fig. 1 given the potentially large uncertainty still associated with them, and given the lack of information on the spectral characteristics of gravity wave sources in nature.

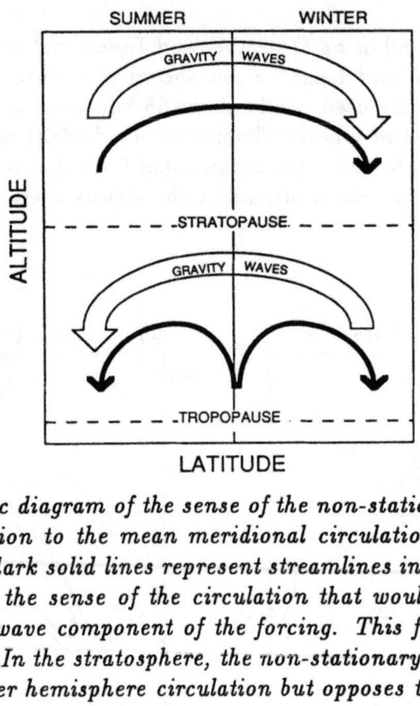


*Figure 3. Schematic diagram of the sense of the non-stationary gravity wave-driven circulation in relation to the mean meridional circulation in the mesosphere and stratosphere. The dark solid lines represent streamlines in the latitude-height plane. Open arrows show the sense of the circulation that would be induced by the non-stationary gravity wave component of the forcing. This forcing controls the mesosphere circulation. In the stratosphere, the non-stationary gravity wave driven force supports the summer hemisphere circulation but opposes the circulation in the winter hemisphere. The winter stratosphere is instead dominated by planetary waves and stationary gravity waves.*

These model results primarily suggest limits on the average amplitude of the waves in the spectrum. Average vertical velocity perturbations at the source height in the middle troposphere must not exceed ~0.5-1 m s$^{-1}$. If they are much higher, the mesospheric drag extends deeply into the stratosphere, and the sign of the stratospheric force can be reversed. This could occur without radical changes to the mesospheric forcing if the efficiency factor were simultaneously reduced with the increased amplitudes. The relationship between input wave amplitudes and the efficiency factor is however, not linear, so different choices for input spectra can imply very different gravity wave momentum fluxes at source levels. With the new constraints (Fig.1), the model can still accomodate a wide range of input momentum flux ($O \sim 10^{-3} - 10^{-2}$ kg m$^{-1}$s$^{-2}$), but this range is at least an order of magnitude smaller than the range this model can accomodate without the new stratospheric constraints.

The sign of the stratospheric forcing in Figs. 1 and 2 has implications for the mean meridional circulation in the stratosphere (see Fig. 3). The westward summertime force implies a poleward motion that would support the inferred Lagrangian mean circulation in the summer hemisphere stratosphere. On the other hand, the non-stationary gravity wave component of the zonal force in the winter hemisphere in Fig. 2 is eastward. This would imply an equatorward motion in wintertime that is opposite to the meridional motion induced by planetary and stationary gravity wave dissipation. So in the wintertime, the non-stationary gravity waves will work against the Lagrangian mean meridional circulation. These non-stationary gravity wave effects are sketched in Figure 3. The gravity wave component of the wave-driven forcing dominates in the mesosphere and is likely to be quite important in the summer stratosphere, but it is clearly only a minor component of the total force in the winter stratosphere.

## Patterns in Temperature Variance

The gravity wave model described above also predicts a temperature variance spectrum. Recent satellite observations by the UARS MLS [Wu and Waters, 1996; see also Wu, this book] provide the first global maps of monthly mean variance patterns associated with gravity waves. The MLS observation technique detects short horizontal wavelength perturbations (~30-100 km), but only those with long vertical wavelengths. Because the vertical wavelength of gravity waves evolves with height due to the influence of vertical variations in the the background winds through which they propagate, it will be proposed here that the MLS vertical wavelength filtering coupled to the zonal wind variations control the patterns in variance that Wu and Waters [1996] observe.

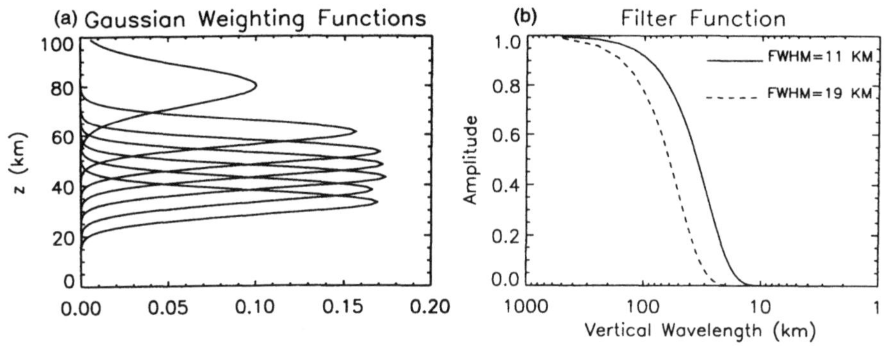


*Figure 4. (a) Gaussian weighting functions used to mimic the true MLS weighting functions. (Compare to Wu and Waters [1996].) (b) Amplitude vs vertical*

*wavelength describing the filtering effects of the MLS weighting functions in (a) on the gravity wave spectrum. An amplitude of 1.0 means the wave is unattenuated. The two curves represent the range of weighting function widths in (a). Only very long vertical wavelengths >20-30 km contribute much to the variance in the MLS observations.*

To quantify this effect, spectral filters based on the MLS vertical weighting functions are applied to the spectrum of waves in the model above. Figure 4a shows a set of Gaussian weighting functions with the same full width at half maximum (FWHM) and peak height as the true MLS weighting functions. The true weighting functions are not perfect Gaussians, but the differences will be trivial to the filtering operation. Most of the weighting functions in Figure 4a have FWHM ~11-12 km, except for the highest one that peaks near 80 km, which has FWHM of almost 19 km. Figure 4b shows the resulting filter function for Gaussian weighting functions having FWHM of 11 and 19 km. For FWHM=11 km, waves with vertical wavelength $\lambda_Z$=10 km are not visible. Waves with $\lambda_Z$=20 km are attenuated to 10% of their true amplitude. At $\lambda_Z$=30 km waves are attenuated to less than 40%, and at $\lambda_Z$=40 km still only 55% of the true amplitude will be observed. Thus only the very long vertical wavelength gravity waves contribute to the MLS variance observations. It is worth noting that the part of the $\lambda_Z$ spectrum to which MLS is sensitive is out of the range of detection in methods that analyze vertical profiles from radiosondes, rocket soundings, and lidar.

The same gravity wave source as that described in the previous section was specified at 6 km and the waves allowed to propagate through zonal wind $U(\lambda, \phi, z)$ and buoyancy frequency profiles $N(\lambda, \phi, z)$ that vary with longitude $\lambda$, latitude $\phi$, and height $z$. These are defined by January and July monthly mean UKMO data for 1993. Temperature variance is computed from the energy density using the polarization relations:

$$\left|\frac{T'}{T}\right|^2 = E\left(1 - \frac{f^2}{\omega^2}\right)\frac{N^2}{g^2}$$

Here $T'$ and $T$ are the perturbation and mean temperature respectively, and $g$ is the gravitational acceleration. The MLS-like filter functions (Fig.4) are applied at the observation heights and the variance is then integrated over the filtered spectrum. Performing this same calculation all over the globe on a $3.75 \times 2.5°$ longitude/latitude grid produces a map of variance associated with globally uniform gravity wave sources and observed monthly mean winds $U(\lambda, \phi, z)$ and temperatures $T(\lambda, \phi, z)$. The results are shown in Figure 5. Maps at altitudes 33, 38, 43, 48, and 53 km are shown. The UKMO data do not extend into the mesosphere, so comparisons at 61 and 80 km can't be performed.

Note that the magnitudes of modeled and observed variance are not easily compared. The MLS observations include filtering in horizontal wavelength with a continually varying geometry in the zonal/meridional plane throughout the MLS orbit. This horizontal wavelength filtering by the MLS has not yet been quantified, and our current understanding of gravity wave sources is too poor to estimate what fraction of the horizontal wavelength spectrum is retained. There may also be some temporal filtering in the observations because of the time it takes to assemble a map with global coverage. The variances in Fig. 5 are approximately an order of magnitude larger than the observations. Proper accounting of the horizontal and temporal filtering would naturally reduce this discrepancy. Relative values, however, should be directly comparable between model and data.

The patterns in Fig. 5 are remarkably similar to those derived from MLS observations [Wu and Waters, 1996]. Peak variance occurs in the wintertime jet. Secondary patchy maxima appear in the summer hemisphere subtropics. Moving upward in altitude, the summer variance becomes more comparable to the winter variance in magnitude. The longitudinal variability evident in both the winter and summer bands is also similar to the observations. The equatorial region appears extremely quiet with very weak variance compared to the higher latitudes. All of these same features appear in the MLS observations. Because the gravity wave sources are geographically uniform in this model, these patterns result solely from the zonal and meridional variations of the background atmosphere, $U(\lambda, \phi, z)$ and also $T(\lambda, \phi, z)$ to lesser extent.

The variance patterns generated with this model have a simple intuitive explanation. The MLS is sensitive only to the very long vertical wavelength waves. These are the waves propagating upstream against the background wind which have been Doppler-shifted to high intrinsic phase speed $(c - U)$ and long vertical wavelength $2\pi/m$. The effect is shown graphically in Eckermann [1995], and is most evident in the simplified hydrostatic form of the dispersion relation:

$$m^2 \simeq N^2/(c-U)^2$$

The agreement between the model and data is even more surprising considering the fact that the MLS should see both zonally and meridionally propagating waves, yet the model includes only the former. The good agreement likely results because meridional winds are relatively weak in the middle atmosphere, so meridionally propagating waves are not Doppler-shifted to such large vertical wavelengths as are the zonally propagating waves, and do not contribute that much to the observed variance.

Seasonal variations in the variance are shown in Figure 6 for the stratosphere and Figure 7 for the mesosphere. Background winds and temperatures were specified using the CIRA model for Fig.7. Seasonal variations in the zonal mean zonal wind are also shown. In the stratosphere (Fig.6), the winter winds are much stronger than in summer. The variance is thus much larger in winter than in summer, and an annual cycle in variance is the result. In the mesosphere (Fig.7), the summer and winter jets are much more comparable in strength. Variance then peaks in both summer and winter seasons with weaker values at equinoxes when the winds are weak, giving a semiannual cycle in the variance in the mesosphere. Figs. 6 and 7 underscore the fact that patterns in the variance of waves with long vertical wavelengths (to which the MLS is sensitive) are well correlated with the magnitude of the zonal winds.

Vertical variations in the model variance are shown in Figure 8. Solid and dotted curves are high latitude wintertime; dashed and dot-dashed are tropical profiles in January and July respectively. These latitudes and seasons are chosen for comparison to Wu and Waters [1996]. The modeled growth in variance with height is nearly identical to the MLS observations. The UKMO background winds do not extend high enough to model the transition to no growth that appears in the MLS observations between about 50-60 km altitude. Applying the CIRA winds, however, shows this transition occurs at very nearly the same height in the high latitude wintertime. (Note: the CIRA winds are quite unrealistic at tropical latitudes, so the tropical variances are not shown.) Above 70 km, the CIRA profiles decrease suddenly with height. This is a model artifact that arises due to the discrete gravity wave source spectrum and severe critical level filtering that begins to wipe out the spectrum. A broader phase speed spectrum should be included in future work to study the mesopause and thermosphere region.

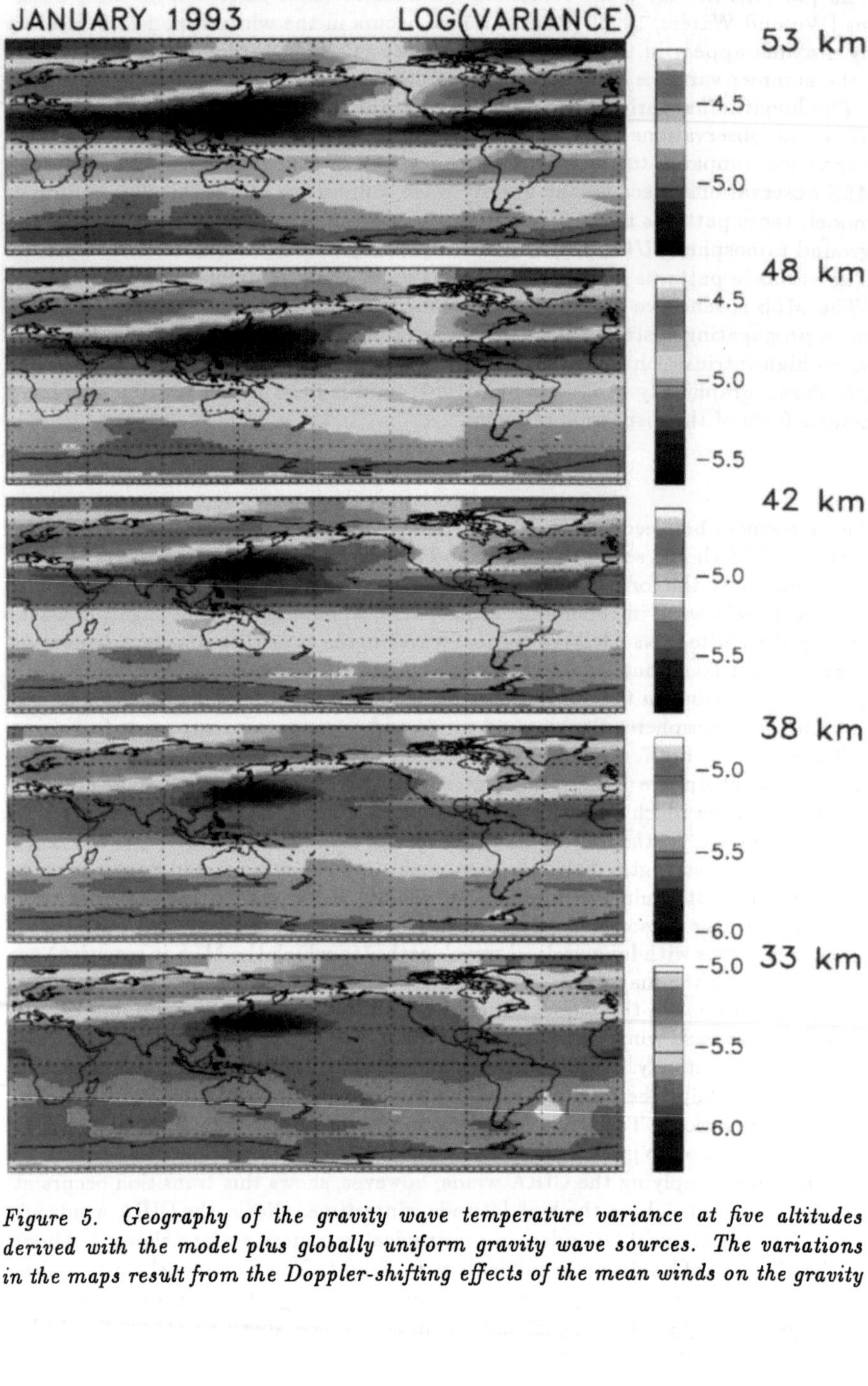


*Figure 5. Geography of the gravity wave temperature variance at five altitudes derived with the model plus globally uniform gravity wave sources. The variations in the maps result from the Doppler-shifting effects of the mean winds on the gravity*

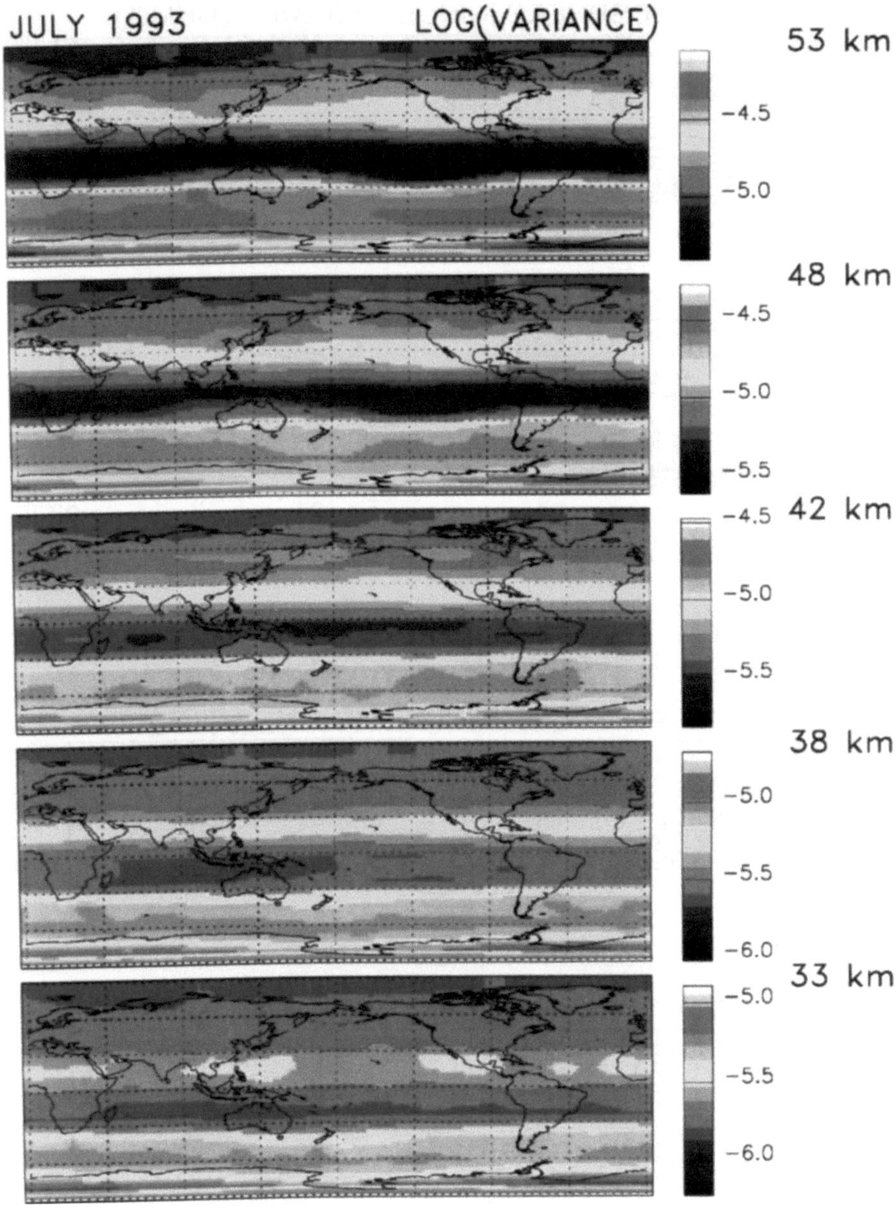


*wave spectrum convolved with the vertical wavelength filtering effects of the MLS observation technique (Fig. 4). (a) From January 1993 UKMO monthly means. (b) From July 1993 UKMO monthly means. (Compare to Wu and Waters [1996].)*

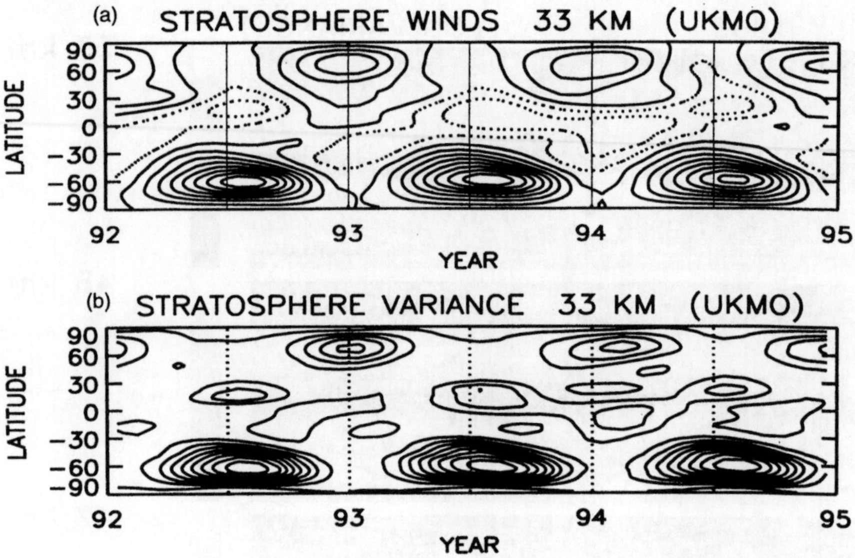


*Figure 6. Latitude-time maps of (a) zonal mean winds and (b) modeled MLS-like temperature variance in the stratosphere at 33 km. In the extratropics, an annual cycle is observed in both. Contour intervals are (a) 10 m $s^{-1}$ wind speed, and (b) $3 \times 10^{-7}$ fractional variance. Dashed contours are negative.*

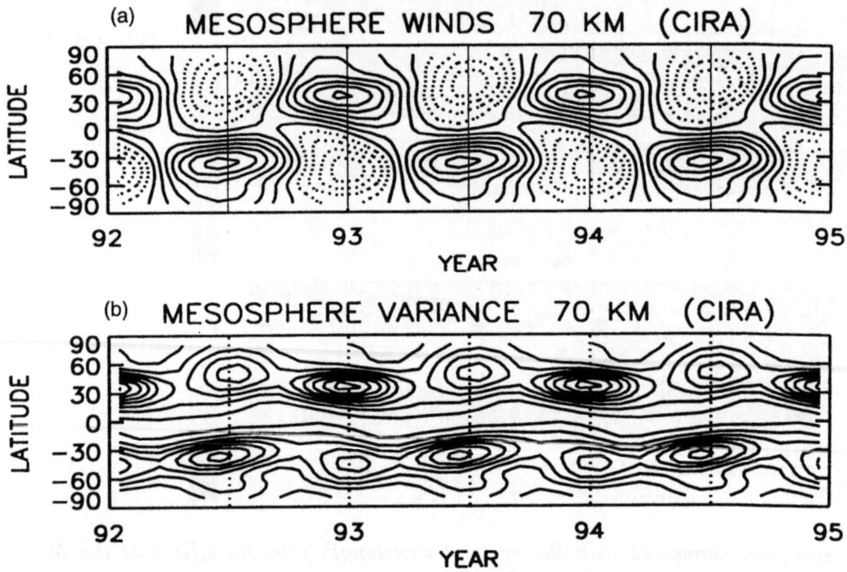


*Figure 7. As in Fig. 6, but in the mesosphere at 70 km. The winds (a) still display an annual cycle, but the variance (b) shows a semiannual cycle. The modeled MLS-like variance is well correlated with the magnitude of the winds. Contour intervals are (a) 10 m $s^{-1}$, and (b) $1.5 \times 10^{-5}$ fractional variance.*

It is important to remember that these profiles of variance are not characteristic of the total gravity wave energy spectrum, but only the very long vertical wavelength end of that spectrum. The traditional concept of gravity wave "saturation", or cessation of energy growth with height, refers instead to the short wavelength end of the spectrum. Profiles of the short vertical wavelength variance would show saturation at much lower altitudes. The wintertime transition to no growth near 55 km in the model marks the altitude where waves with $\lambda_Z$ longer than ~20 km begin to saturate. These are the waves that are propagating upstream, and have zonal phase speeds slower than or opposite in sign to the mean wind. This altitude then marks a transition level where the gravity wave-driven zonal force begins to decrease and quickly changes sign in the upper stratosphere as in Figs. 1 and 2. The MLS data then, with the interpretation provided by the gravity wave model, provides an important confirmation of this property of the observationally derived stratospheric forcing in Fig. 1.

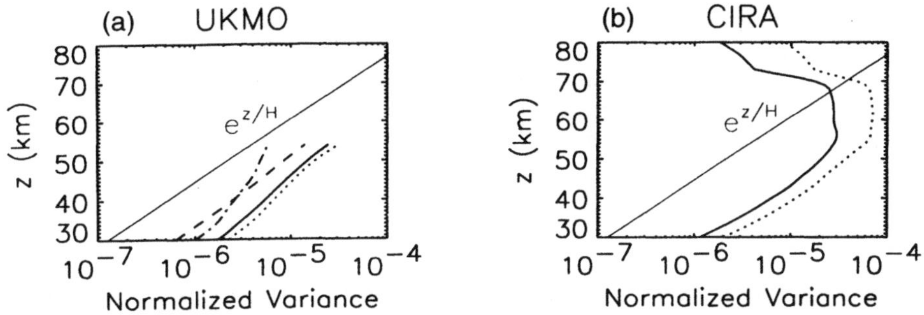


*Figure 8. Vertical profiles of modeled MLS-like variance using (a) UKMO winds for Jan 50-70° N (solid), Jul 50-70° S (dotted), Jan 6-26° S (dashed), and Jul 6-26° N (dash-dotted). CIRA winds at tropical latitudes are unrealistic, so the tropical variances are omitted here.*

## Discussion

The MLS/model comparison dramatically confirms the ability of the MLS to detect gravity wave perturbations by the Wu and Waters technique, but also suggests that the global patterns they observe have little-to-no relationship to gravity wave source variations. Instead, the patterns show the effects of Doppler-shifting by the background winds on the gravity wave spectrum convolved with the vertical wavelength filtering associated with the MLS observation technique itself. Differences between the MLS observations and the model with uniform gravity wave sources are only subtle. The small differences between observations and model are more likely explained by either differences between the actual winds during the MLS observations and the UKMO monthly means used in the model, or by the neglect of meridionally propagating waves in the model. The changing viewing geometry during an MLS orbit may also affect the modeled results.

The seasonal variations in the MLS observed variance are not very meaningful in terms of the effects of gravity waves on the mean flow in the middle atmosphere. The annual cycle in the stratosphere and semiannual cycle in the mesosphere simply follow the magnitude of the zonal mean winds at these levels, showing only the effects of Doppler-shifting on the long vertical wavelength end of the gravity wave spectrum.

The vertical profile of the variance observed by the MLS may be an important piece of information for gravity wave parameterizations. This profile describes the energy growth of the low $m$ portion of the gravity wave energy spectrum, corresponding to $m < m_*$ in the Fritts and Lu [1993] parameterization. The MLS results then provide new constraints for that portion of the spectrum that is difficult to observe by most other methods. The "saturation" of these variance profiles may also provide a constraint for the Hines [1995] parameterization (see also Hines, this book). Cessation of growth of waves with $\lambda_Z$ longer than ~20 km seems to suggest a very large value of $\sigma_T$ that describes the variability in the background wind and controls what portion of the vertical wavelength spectrum is dissipated in his parameterization.

The gravity wave model applied here is very similar to the Lindzen [1981] parameterization, so these results suggest similar limits on wave momentum flux and amplitudes near the sources for applications of this parameterization. The Lindzen parameterization does not treat wave reflection and makes simplifying assumptions that lead to the prediction of lower breaking levels, so the details of the source input parameters must be adjusted from those used here, but similar results can be achieved.

## Conclusions

A model with isotropic, uniform gravity wave sources in the troposphere can capture many key characteristics of the monthly mean distributions of gravity wave effects in the middle atmosphere. Within the currently rather weak constraints on gravity wave-mean flow interactions, this interaction, on monthly mean time scales and above altitudes of ~ 25 km, is dominated by variations in the background winds. Many important aspects of the seasonal, latitudinal and vertical variations in the zonal force distribution can be described without any variability in gravity wave sources. Details of the gravity wave source spectrum are more important in the lower stratosphere, and variations in sources are likely important on shorter time scales.

The recently reported observations of gravity wave temperature variance by the MLS can also be described quite well by a model with uniform and isotropic gravity wave sources. The longitudinal, meridional, and vertical variations in the background winds dominate the patterns observed in these data. Information on source variations, if present, cannot be inferred until the dominant Doppler-shifting effects of the background winds on these observations are first taken into account.

*Acknowledgments.* This research was supported by the Physical Meteorology Program of the National Science Foundation, ATM-9322480, and the National Aeronautics and Space Administration, NASA grants NAG-11803 and NAG5-3188.

**References**

Alexander, M. J., A simulated spectrum of convectively generated gravity waves: Propagation from the tropopause to the mesopause and effects on the middle atmosphere, *J. Geophys. Res., 101*, 1571-1588, 1996.

Alexander, M. J. and K. H. Rosenlof, Non-stationary gravity wave forcing of the stratospheric zonal mean wind, *J. Geophys. Res., (accepted)*, 1996.

Brasseur, G., M. H. Hitchman, S. Walters, M. Dymek, E. Falise and M. Pirre, An interactive chemical dynamical radiative two-dimensional model of the middle atmosphere, *J. Geophys. Res., 95*, 5639-5655, 1990.

Dunkerton, T.J., and N. Butchart, Propagation and selective transmission of internal gravity waves in a sudden warming, *J. Atmos. Sci., 41*, 1443-1460, 1984.

Eckermann, S.D., Effect of background winds on vertical wavenumber spectra of atmospheric gravity waves, *J. Geophys. Res., 100*, 14,097-14,112, 1995.

Fleming, E. L., S. Chandra, J. J. Barnett and M. Corney, Zonal mean temperature, pressure, zonal wind and geopotential height as functions of latitude, *Adv. Space Res., 10*, (12)11-(12)59, 1990.

Fritts, D. C. and W. Lu, Spectral estimates of gravity wave energy and momentum fluxes, II: Parameterization of wave forcing and variability, *J. Atmos. Sci., 50*, 3695-3713, 1993.

Fritts, D. C. and R. A. Vincent, Mesospheric momentum flux studies at Adelaide, Australia: observations and a gravity wave-tidal interaction model, *J. Atmos. Sci., 44*, 605-619, 1987.

Fritts, D. C. and L. Yuan, Measurement of momentum fluxes near the summer mesopause at Poker Flat, Alaska, *J. Atmos. Sci., 46*, 2569-2579, 1989.

Garcia, R. R. and S. Solomon, The effect of breaking gravity waves on the dynamics and chemical composition of the mesosphere and lower thermosphere, *J. Geophys. Res., 90*, 3850-3868, 1985.

Hines, C. O., Parameterization of gravity wave momentum deposition in the middle atmosphere. Part I: Extrowaves, *J. Atmos. Terr. Phys., (submitted)*, 1995.

Holton, J. R., The influence of gravity wave breaking on the general circulation of the middle atmosphere, *J. Atmos. Sci.*, 2497-2507, 1983.

Lighthill, J., *Waves in Fluids*, Cambridge Univ. Press, New York, 504, pp., 1978.

Lindzen, R. S., Turbulence and stress owing to gravity wave and tidal breakdown, *J. Geophys. Res., 86*, 9707-9714, 1981.

Marks, C. J. and S. D. Eckermann, A three-dimensional nonhydrostatic ray-tracing model for gravity waves: formulation and preliminary results for the middle atmosphere, *J. Atmos. Sci., 52*, 1959-1984, 1995.

Rosenlof, K. H., Summer hemisphere differences in temperature and transport in the lower stratosphere, *J. Geophys. Res., (in press)*, 1996.

Saravanan, R., A multiwave model of the quasi-biennial oscillation, *J. Atmos. Sci.*, *47*, 2465-2474, 1990.

Swinbank, R. and A. O'Neill, A stratosphere-troposphere data assimilation system, *Mon. Wea. Rev.*, *122*, 686-702, 1994.

Wu, D. L. and J. W. Waters, Satellite observations of atmospheric gravity waves, *Geophys. Res. Lett.*, *(in press)*, 1996.

# THE PROPAGATION AND BREAKING OF GRAVITY WAVES EXCITED BY FORCING IN THE TROPOSPHERE

Rolando R. Garcia
National Center for Atmospheric Research
P.O. Box 3000
Boulder, CO 80307-3000
U.S.A.

Joseph M. Prusa
Dept. of Mechanical Engineering
Iowa State University
Ames, IA 50011
U.S.A.

**ABSTRACT.** A two-dimensional numerical model is used to study the propagation and breaking at high altitudes of gravity waves forced by tropospheric sources. The model is based on the anelastic approximation to the Navier-Stokes equations with Coriolis forces neglected. Gravity waves are generated by air flow over deflections of the lower boundary. It is shown that dispersion produces a wave field at high altitudes that is dominated by nearly monochromatic wavetrains, even though the boundary forcing excites a localized wavepacket. The morphology of wave breaking depends on the horizontal wavelength: for longer waves, breaking is confined to a relatively small fraction of the wavelength, near the wave crest but, for sufficiently short waves, it involves a large fraction of the wavelength. In common with other studies, it is found that breaking waves undergo "self-acceleration", such that the intrinsic frequency remains approximately constant in spite of large changes in the background wind. It is also shown that many of the features obtained in the calculations can be understood in terms of linear wave theory. In particular, linear theory provides insight into the position of the wave breaking region relative to the source and the minimum and maximum altitudes where breaking occurs. Wave breaking ceases at the altitude where the background dissipation rate (which in our model is a proxy for molecular diffusion) becomes greater than the rate of dissipation due to wave breaking. This altitude – in effect, the model turbopause – is shown to depend on a relatively small number of parameters that characterize the waves and the background state.

## 1. INTRODUCTION

Observational studies of the middle atmosphere (e.g., Theon et al, 1967; Moreels and Herse, 1977; Vincent and Reid, 1983; Balsley et al, 1983; Phillips and Vincent, 1989) indicate that small-scale gravity waves are a common occurrence at mesospheric altitudes.

NATO ASI Series, Vol. I 50
Gravity Wave Processes
Their Parameterization in Global Climate Models
Edited by Kevin Hamilton

The observations yield wave properties that are consistent with those of internal gravity waves, as predicted by linear stability theory (Bretherton, 1966). More sophisticated arguments based upon linear theory indicate that gravity waves could affect profoundly the dynamics of the atmosphere at upper mesospheric altitudes. The work of Lindzen (1981), based upon the WKB method, established that wave breaking could produce a zonal mean force per unit mass of 50-100 m $s^{-1}$ $day^{-1}$ at mesopause altitudes. Lindzen also noted that gravity wave breaking was probably the dominant source of turbulence, and hence mixing, in the extratropical mesosphere and lower thermosphere.

In order to describe approximately the breaking process, Lindzen (1981) used the hypothesis that wave amplitudes are limited by *saturation*. That is, when gravity waves produce a negative vertical gradient of potential temperature, $\partial\theta / \partial z$, they become unstable and turbulent breakdown ensues. This turbulence hypothesis is one of several which have been put forward to explain the dynamics of breaking waves. Additional hypotheses which have been advanced incorporate various elements of dynamical instability (Kelvin-Helmholtz instability), wave transience, and nonlinear wave-wave interactions. These various mechanisms are well summarized in Dunkerton (1989) and Walterscheid and Schubert (1990), among others.

The purpose of the present work is to elucidate some of the physical mechanisms involved in the breaking of gravity waves and their effects upon the basic state of the atmosphere. In particular, the numerical model was designed to test the effects of forcing by a "realistic" tropospheric source, i.e., a source that excites a spectrum of waves in both frequency and wavenumber. Three-dimensional effects, which have recently been studied by Fritts et al (1993), are not addressed in this work. We emphasize instead how forcing and vertical dispersion influence the wave field that propagates to high altitudes, and what processes determine the upper and lower limits of the region of wave breaking.

## 2. Model Formulation

### 2.1 Governing Equations

The model is based on the anelastic approximation to the Navier-Stokes equations in the absence of rotation (Lipps and Hemler, 1982). The relevant fields are decomposed into a basic state $\Phi_o$ which is in hydrostatic equilibrium and a perturbation state $\Phi'$ which represents the departure of the atmosphere from this equilibrium. The decomposed variables $\Phi = \Phi_o + \Phi'$ for $\Phi = p$, $\rho$, $T$, and $\theta$ (the velocity does not have to be decomposed explicitly) are substituted into the equations of continuity, horizontal and vertical momentum, and energy, ignoring rotation and spherical geometry. From these equations the equations for hydrostatic equilibrium are subtracted off, leaving only perturbation equations for the disturbances from equilibrium. The governing equations are solved using an efficient semi-Lagrangian method (Smolarkiewicz and Pudykiewicz, 1992) with monotone interpolation (Smolarkiewicz and Grell, 1992). Due to space limitations, the numerical techniques used in the solution are not discussed here. The reader is referred to Prusa et al (1996) for a complete description of the numerical method.

*2.2 Boundary and Initial Conditions*

The bottom boundary of the computational domain is assumed to be a material surface; at the top of the domain, we set $w=0$. At both boundaries, the potential temperature is specified by its basic state values. The lateral boundary conditions are of open type, with lateral sponge layers. A sponge layer, distributed in the vertical coordinate $z$, is also placed adjacent to the top boundary. Although it acts to produce a radiation boundary condition at the top of the domain, this sponge layer is more than just a computational device, since its has been tuned to mimic the molecular diffusion profile of the atmosphere.

It is possible to choose the damping time to match approximately the effect of molecular diffusion upon waves of a given wavelength by noting that $\tau^{-1} \approx m^2 K_d$, where $m$ is the vertical wavenumber. The results to be discussed in Section 3 indicate that a single wavelength dominates the wave field. It is to this dominant wavelength that the vertical and lateral sponges are tuned. Since $K_d$ increases exponentially as the inverse of the atmospheric density, the damping rate in the vertical sponge layer is specified as follows:

$$\tau^{-1} = \tau_{\min}^{-1} \exp\left(\frac{z - z_{top}}{H_s}\right) \tag{2.1}$$

where $z_{top}$ denotes the top boundary, $\tau_{min}$ is the damping time at $z_{top}$, and $H_s$=5.7 km. This scale height is somewhat small (it should correspond to the density scale height). However, this value was selected in order to match smoothly diffusivity data for the lower thermosphere (Whitten and Popoff, 1971) with corresponding data for the upper mesosphere (U.S. Standard Atmosphere, 1976). The value of $\tau_{\min}$ is set to 172 sec at $z_{top}$=135 km, which provides a good match for $K_d$ at that level.

Calculations are started from a basic state representative of summer conditions at temperate latitudes. The initial wind field is set to a constant value $u_O$=–50 m s$^{-1}$, and the potential temperature, density and pressure distributions are set to background values $\theta_0 = \theta_{00} \exp[(z - z_0) / H_\theta]$ and $(\rho_0, p_0) = (\rho_{00}, p_{00}) \exp[-(z - z_0) / H_\rho]$ where $H_\theta$ and $H_\rho$ are the potential temperature and the density scale heights, respectively, and $\theta_{00}$, $\rho_{00}$, and $p_{00}$ denote values at the lower boundary.

*2.3 Lower Boundary Forcing*

The forcing at the lower boundary is specified as follows:

$$z_s(x,t) \;=\; A \cdot \exp\left\{-\left(\frac{x-x_m}{\sigma_x}\right)^2\right\} \cdot \exp\left\{-\left(\frac{t-t_m}{\sigma_t}\right)^2\right\} \tag{2.2}$$

which represents a Gaussian pulse of amplitude $A$ and half-width $\sigma_t$, centered at time $t_m$. For the calculations discussed in this paper, we have set $A$=200 m, $\sigma_t$=60 min, $t_m$=120 min, and $\sigma_x$ =2 km.

*2.4 Domain Size and Resolution*

Horizontal and vertical resolution of 0.625 km was used in all the calculations presented here. A horizontal domain size of 900 km was chosen in order to capture the full response at high altitudes. The vertical size of the domain is 120 km; the lower boundary is located at 15 km, while the upper boundary is at 135 km. Typically a timestep of between 10 and 20 s was used at the beginning of the computations, while a timestep between 2 and 6 s was required as the wave field approached saturation. In all the computations, the *Courant number* was often greater than unity, but the semi-Lagrangian advection scheme remained stable.

## 3. Results

*3.1 Evolution of the Mesospheric Wave Field*

Fig. 1 shows the development of the potential temperature field when the model is forced by a deflection of the lower boundary with $A$=200 m and $\sigma_x$ =2 km, centered at $x$=0 km; the background wind is –50 m s$^{-1}$ (see Section 3). Results are shown at $t$=100, 160, 220 and 280 minutes. Note that, although the computational domain is 900 km wide, only the central 320 km are shown for clarity.

In the early stages of the simulation ($t$=100, 160 min), a nearly monochromatic wavetrain of short horizontal wavelength is present downstream of the forcing (recall that the background wind is easterly and thus is directed from right to left in the figure). These short waves have horizontal and vertical wavelengths of about $\lambda_x$=25 km and $\lambda_z$=17 km, respectively. They begin breaking shortly after 100 min, and by 160 min the entire region downstream of the forcing is undergoing vigorous overturning. As a consequence of the large aspect ratio of the waves, $\lambda_z / \lambda_x \approx 0.68$, the overturning of potential temperature surfaces that accompanies breaking involves a considerable fraction of the total wavelength.

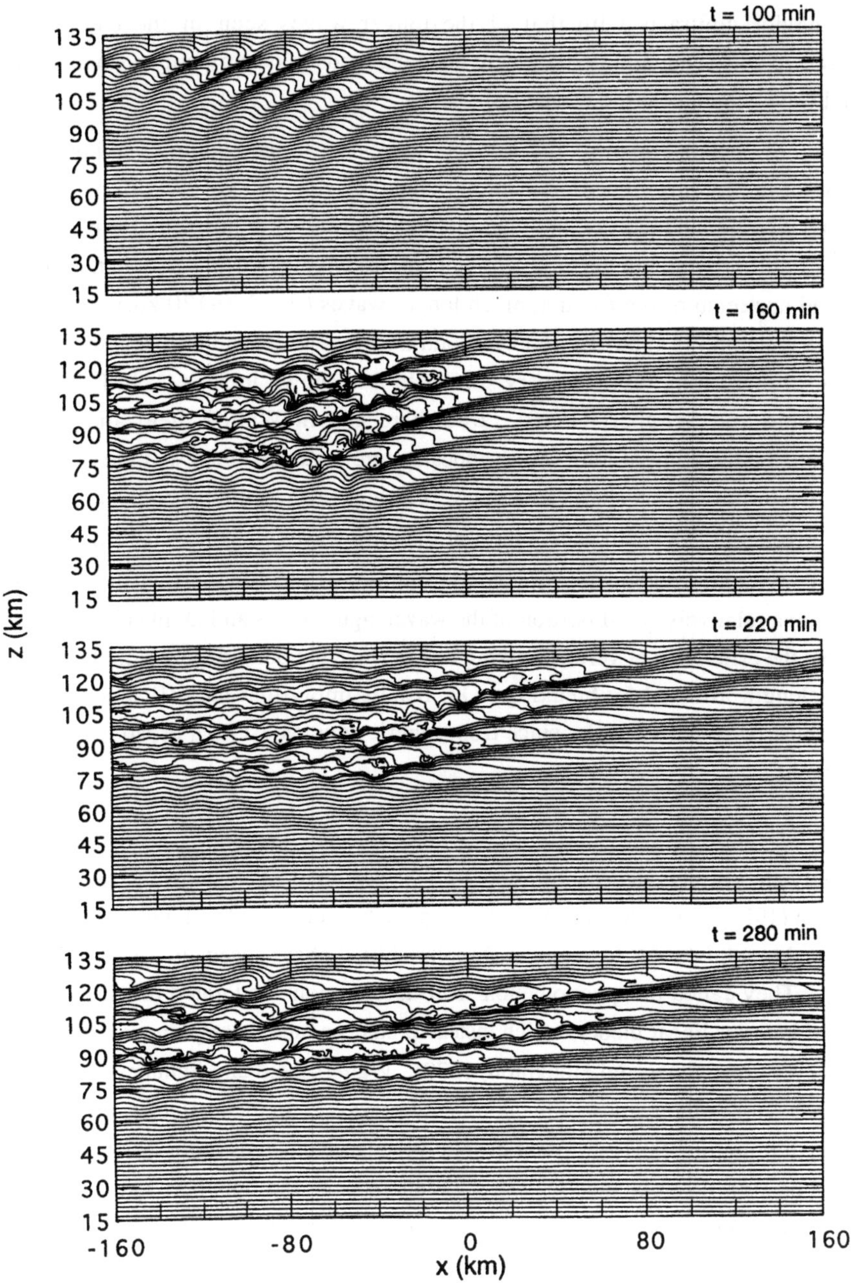


Fig. 1. Potential temperature field at various times during the simulation. The lower boundary forcing is located at $x$=0 km. Waves of short horizontal wavelength appear at high altitudes downstream of the forcing early on in the simulation. At later times, much longer waves are present above and upstream of the forcing.

This behavior contrasts with that of the longer waves seen in the calculations of Walterscheid and Schubert (1990), where breaking tends to occupy only a small fraction of the total wavelength, near the wave crests.

At later times (*t*=220, 280 min), much longer waves appear and begin breaking above and upstream of the forcing. The dominant horizontal wavelength varies with position in the horizontal direction. Downstream of the forcing waves of short horizontal wavelength (25-50 km) are still present, but the wavelength increases with increasing $x$ so that, immediately above and upstream of the forcing, much longer waves ($\lambda_x$=100-120 km) are seen. The vertical wavelength is about 14-15 km. Contrary to the short waves discussed above, the behavior of these long waves is very similar to that shown in Figs. 1-3 of the study by Walterscheid and Schubert (1990). Both here and in Walterscheid and Schubert's calculation, the breaking of long waves appears more organized than was the case for short waves: overturning layers alternate with layers of high static stability, both inclined slightly to the horizontal. The small aspect ratio of the long waves ($\lambda_z / \lambda_x \approx 0.12$ in the present results) is evidently responsible for this morphology, and for the fact that breaking is confined to a relatively small portion of the wavelength. Fritts and Dunkerton (1985) have noted that the efficiency of vertical mixing due to wave breaking depends on the fraction of the wavelength affected by breaking. Our results imply that the efficiency of vertical mixing must be rather variable, since tropospheric sources excite a broad range of zonal wavelengths.

### *3.2 Self-Acceleration*

In a study of gravity wave saturation in a background shear flow, $u_0(z)$, Fritts and Dunkerton (1984) found that, as the waves approached a critical level, their phase velocity, $c$, appeared to adjust so as to keep the Doppler-shifted phase speed, $u_0 - c$, approximately constant. They called this phenomenon "self-acceleration" and noted that it could not be explained in terms linear theory, which instead predicts a rapid change in wavelength near a critical level (e.g., Jones and Houghton, 1972). Walterscheid and Schubert (1990) found similar behavior in their study of breaking gravity waves. In their calculations the vertical wavelength remained fairly constant despite very large changes in the background wind, implying no significant change in $u_0 - c$. Walterscheid and Schubert also determined $c$ from the movement of constant-phase lines, and noted that the result was consistent with the approximate constancy of the vertical wavelength.

We find evidence of self-acceleration in all of our calculations. Examination of Fig. 1 indicates that the vertical wavelength does not change drastically as the waves approach saturation, or indeed at any time during the breaking process, even though the background wind itself undergoes large changes. In common with Walterscheid and Schubert, we find that the approximate constancy in $u_0 - c$ implied by the approximately constant wavelength is consistent with the results of a direct determination of $c$. Obtaining the phase velocity from our results is complicated by the fact that the wavefield is not exactly monochromatic. Accordingly, quadratic regression fits were made to the calculated perturbation potential temperature fields at 60 and 105 km to determine the position of extrema and zeroes. The curve fits included sufficient points to smooth out fine scale variations due to breaking. Phase velocities were then determined by computing the horizontal displacement of particular extrema/zeroes during specified time intervals. Typically, four to eight such points were analyzed. These local values were then averaged to give an estimate of the phase velocity.

Fig. 2 shows the results as a function of time at $z$=105 km for the region above and upstream of the forcing, which is dominated by long waves at times $t \geq 220$ min. The near constancy of $u_0 - c$ is especially striking after 220 min, when the deceleration of the background wind becomes large. The phase velocity changes from about 8 m s$^{-1}$ in the early stages of the simulation to over 33 m s$^{-1}$ at 237 min, and its evolution tracks that of the zonal wind so changes in $u_0 - c$ are always smaller than changes in $u_0$ or $c$ individually.

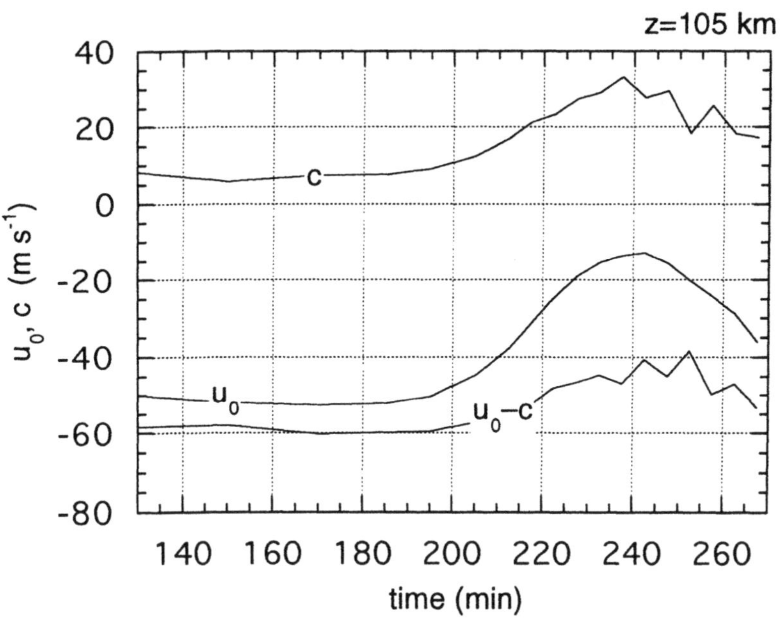


Fig. 2. Evolution of the background wind, $u_0$, and the phase velocity, $c$, of waves for $x \in (0, 160$ km$)$ at $z$=105 km. The waves begin breaking at 200-210 min, causing the background wind to decelerate rapidly, but the phase velocity also changes, keeping $u_0 - c$ approximately constant.

## 4. Interpretation in Terms of Linear Theory

Linear theory for nonhydrostatic gravity waves (Bretherton, 1971) yields the following dispersion relation when the basic state wind is uniform:

$$m^2 = \left(\frac{N}{u_0 - c}\right)^2 - k^2 - M^2 \tag{4.1}$$

where $M = 0.5 / H_\rho$, $m$ is the vertical wavenumber, $k$ is the horizontal wavenumber, $u_0$ is the background wind velocity, and $c$ is the phase velocity. The corresponding vertical and horizontal group velocities are:

$$c_{gz} = \frac{kN}{K^2}\left(1 - \frac{k^2 + M^2}{K^2}\right)^{1/2} = \frac{kNm}{K^3} \tag{4.2a}$$

$$c_{gx} = u_0 - \frac{N}{K}\left(1 - \frac{k^2}{K^2}\right) = u_0 - \frac{N\left(m^2 + M^2\right)}{K^3} \tag{4.2b}$$

where

$$K^2 = k^2 + m^2 + M^2 = \left(\frac{N}{u_0 - c}\right)^2 \tag{4.3}$$

The parameter $K^2$ is the total wavenumber, defined by the middle sum in Eq. (4.3). As $k \to 0$, Eqs. (4.1)-(4.3) collapse to the familiar hydrostatic relations.

To understand the behavior of the mesospheric wave field shown in Fig. 1 we need to consider the dependence of the vertical group velocity on zonal wavenumber, which may be computed from Eq. (4.2). The permitted range of horizontal wavenumbers is $0 \le k \le m_c$, where

$$m_c = \sqrt{\left(\frac{N}{u_0 - c}\right)^2 - M^2} \tag{4.5}$$

The limit $k \to 0$ corresponds to hydrostatic waves while the limit $k \to m_c$ corresponds to evanescent waves. The corresponding range in horizontal wavelength is $\infty > \lambda_x \ge 2\pi / m_c$.

Fig. 3 shows the variation of the components of the group velocity as functions of zonal wavelength, $k$, and frequency, $\omega = kc$, for $u_0 = -50$ m s$^{-1}$. At all frequencies shown, the vertical group velocity has a maximum at zonal wavenumbers of 0.25-0.30 km$^{-1}$ ($\lambda_x \approx$ 21-25 km). At longer wavelengths, $c_{gz}$ decreases gradually to zero as $k \to 0$ ($\lambda_x \to \infty$), while at shorter wavelengths (not shown) it decreases rapidly as $k \to m_c$. The horizontal group velocity depends more strongly on wave frequency. At zero frequency, it varies monotonically from zero as $k \to 0$ to $-c_{gz}$ at $k = m_c / \sqrt{2}$. Its magnitude continues to increase even beyond $k = m_c$, but this is not relevant to the present problem since the corresponding waves do not propagate vertically.

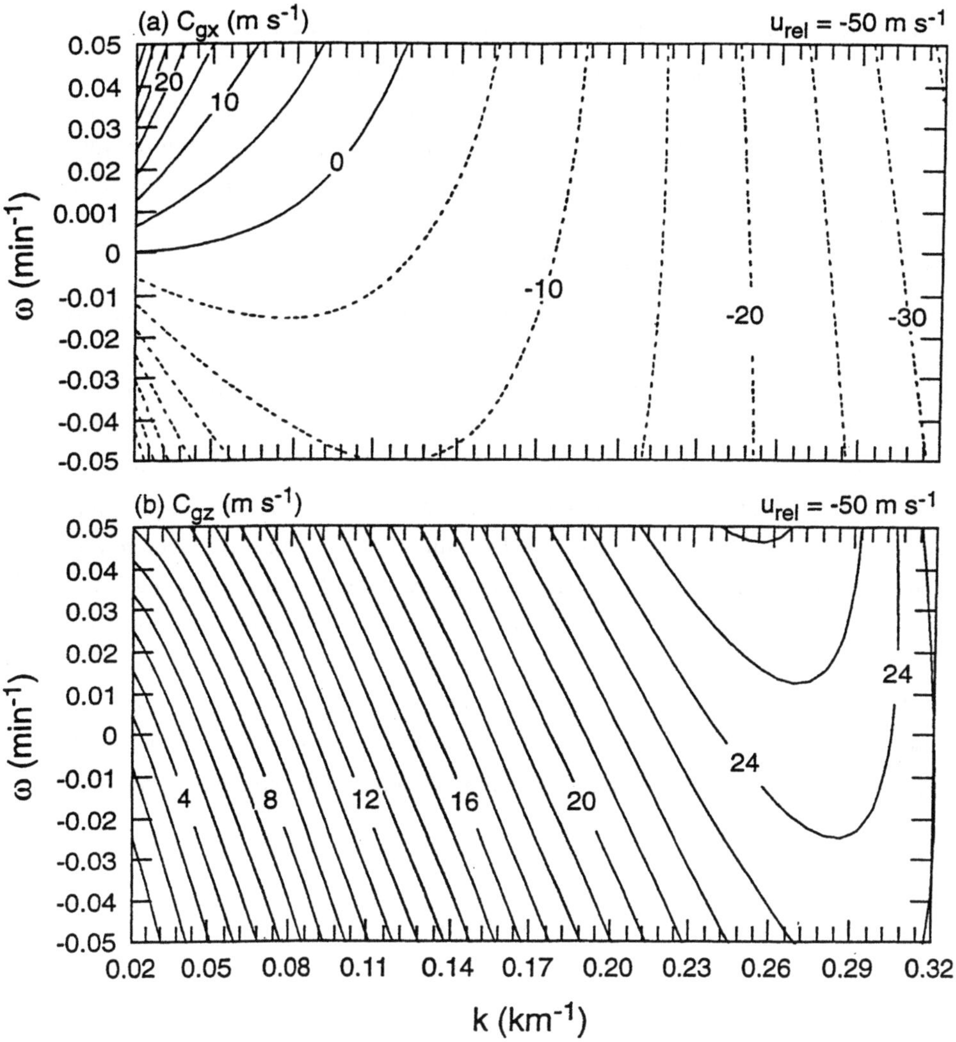


Fig. 3. Horizontal and vertical components of the group velocity as functions of frequency and zonal wavenumber.

Fig. 3 suggests that the early appearance of short waves in the mesosphere (Fig. 1, $t$=100, 160 min) is a result of their large vertical group velocities. In fact, the dominant horizontal wavelength of these waves is about 25 km, which coincides well with the maximum $c_{gz}$ shown in Fig. 3. The magnitude of $c_{gz}$ at this wavelength is about 24 ms$^{-1}$, which gives a travel time of some 63 min between the lower boundary and 105 km. This is consistent with the presence of large amplitude short waves by $t$=100 min in Fig. 1. On the other hand, long waves ($\lambda_x > 50$ km, $k < 0.125$ km$^{-1}$), should reach mesospheric altitudes considerably later than shorter waves because of their smaller vertical group velocity. At wavelengths near 100 km ($k$=0.062 km$^{-1}$), the vertical group velocity is only 6.5 m s$^{-1}$ at zero frequency, and about 9 m s$^{-1}$ at $\omega$=0.025 min$^{-1}$. The corresponding travel times between the lower boundary and 105 km is approximately 231 min for waves of zero frequency and 167 min at $\omega$=0.025 min$^{-1}$. These times bracket the time, $t$=220 min, when long waves appear at high altitudes in Fig. 1.

The position of the mesospheric wavetrains relative to the forcing can also be understood by considering the variation of group velocity with wavenumber. From Fig. 3, the short waves that dominate the response in the early stages of the calculation have horizontal and vertical group velocities in the ratio $c_{gx} / c_{gz} \approx -0.83$ at $\omega$=0, a value that does not vary significantly as a function of frequency. This value is consistent with the appearance of the mesospheric wavetrain downstream of the forcing, at a zenith angle given by $\tan^{-1}(c_{gx} / c_{gz}) = -40°$ (cf. Fig. 1 at $t$=100, 160 min.) For the long waves seen in Fig. 1 at $t$=20, 280 min, the ratio $c_{gx} / c_{gz}$ varies much more strongly with frequency. At $\omega$=0, $\tan^{-1}(c_{gx} / c_{gz}) = -8°$; at $\omega$=0.025 min$^{-1}$, $\tan^{-1}(c_{gx} / c_{gz})$= 33°. These zenith angles bracket the region, upstream of the forcing, where long waves are seen in Fig. 1. Long waves of negative frequency, for which $c_{gx} < 0$, should appear downstream of the forcing; however, these waves have rather small vertical group velocities (e.g., 5 m s$^{-1}$ at $\omega$=-0.025 min$^{-1}$), and would not be expected to arrive at high altitudes until $t$ >300 min, beyond the range of times shown in Fig. 1.

## 5. External Damping and the Altitude of the Turbopause

### 5.1 *Effect of the Vertical Sponge Layer*

To test the effects of damping by the vertical sponge layer we carried out additional calculations wherein the value of the damping time in the vertical sponge layer was varied. In these calculations the damping time, $\tau_{min}$ in Eq. (2.1), was set to 23.3 and 3.15 s, respectively. Recall that the linear damping associated with the vertical sponge is not merely a numerical device; it is in fact a proxy for molecular diffusion, which becomes important in the upper mesosphere. As noted in section 2.2, the nominal value of $\tau_{min}$, 172 s, was chosen to match the effect of molecular diffusion on the dominant waves at high altitudes. In effect, the vertical damping profile (2.1) is displaced downward by 2 scale heights when $\tau_{min}$=23.3 s and by 4 scale heights when $\tau_{min}$=3.15 s

Fig. 4 presents a comparison of the vertical evolution of the breaking region in the three cases mentioned above. It is clear that the maximum height at which wave breaking occurs is set by the vertical damping and is not due to the presence of the top boundary of the domain. This height decreases steadily with decreasing damping time, from 126 km when $\tau_{min}$=172 s to 104 km for $\tau_{min}$=3.15 s. The average vertical displacement between each of the three cases is 11.33 km, which matches precisely the displacement in the damping profiles. (Recall from section 2.2 that the scale height of the vertical damping profile is 5.7 km.) It is also clear from Fig. 4 that the minimum height at which saturated waves are present is the same for the three cases. Evidently, the different values of damping at these lower altitudes have no significant effect on the dynamics: the damping time is so long at these altitudes that the computations are effectively inviscid.

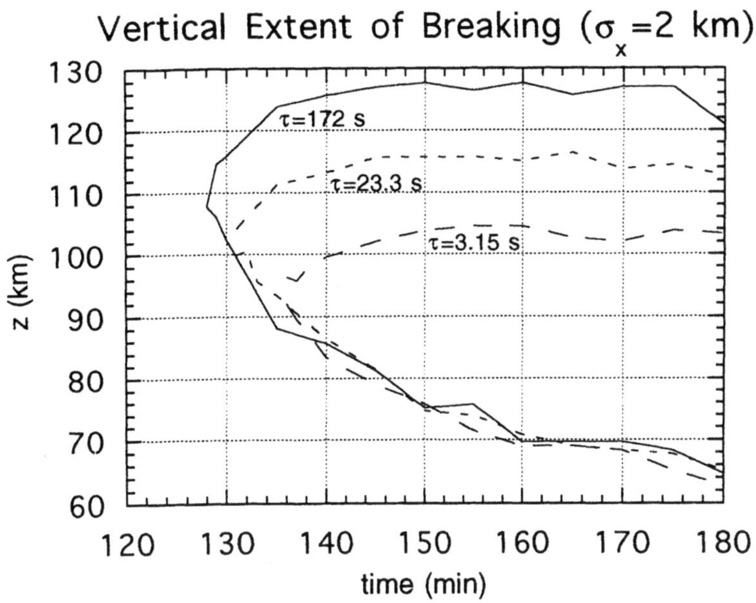


Fig. 4. Change in the vertical extent of the region of wave breaking as the damping time in the vertical sponger layer (2.1) is varied from 172 s to 3.15 s. The lower limit of wave breaking is unaffected by these changes but the upper limit descends as the damping is made faster. See text for details.

To gain some insight into this behavior, we consider the linearized equation for the vertical propagation of gravity wave activity (Andrews et al, 1987):

$$\frac{\partial A}{\partial t} + \frac{\partial (c_{gz} A)}{\partial z} = -2\delta A \tag{5.1}$$

where $c_{gz}$ is the vertical group velocity, $\delta$ is a linear dissipation rate, and $A$ is the wave activity.

If, after the onset of breaking, wave amplitudes cease growing with time, the divergence of the wave activity flux must be balanced by dissipation, i.e.,

$$\delta = -\frac{1}{2A}\frac{\partial (c_{gz} A)}{\partial z} \approx -\frac{c_{gz}}{2}\frac{\partial (\ln A)}{\partial z} \tag{5.2}$$

where we have assumed that the flux divergence is dominated by changes in $A$ itself. If wave breaking prevents wave amplitude growth above the saturation level (Lindzen, 1981), it follows from (5.1)-(5.2) that the corresponding dissipation rate must be given approximately by:

$$\delta = \frac{c_{gz}}{2H_\rho} \tag{5.3}$$

As noted by Garcia (1991), Eq. (5.3) implies that the dissipation rate due to breaking depends on the group velocity of the gravity waves and not on their amplitude. Since the group velocity remains relatively constant while the vertical profile of damping rate $\tau^{-1}$, given by Eq. (2.1), increases exponentially, it may be anticipated that breaking will cease at the altitude where $\delta = \tau^{-1}$. This altitude descends as $\tau^{-1}$ decreases.

We have verified that the calculations are consistent with the assumptions implicit in Eq. (5.2) by computing the vertical profile of wave activity at different times before and during breaking. We used the $u$, $w$, and $\theta$ fields produced by the model to calculate $A(z)$. Fig. 5a displays the evolution of the wave activity profile from 140 to 220 min, at intervals of 20 min. At $t$=140 min, $A$ varies smoothly with altitude, but at later times it decreases very rapidly above 70 km. This is consistent with Fig. 1, which shows that the onset of vigorous wave breaking does not take place before $t$=160 min, and that the breaking region extends down to near 70 km. In Fig. 5b we have plotted $\ln|A|$ for $t$=200-240 min, at 10 min intervals. The logarithmic plot enhances details at the higher altitudes and facilitates evaluation of the vertical e-folding height. Note that $\ln|A|$ is approximately constant with altitude up to 65-70 km, indicating that the waves propagate conservatively below the minimum altitude where breaking occurs. Above this altitude, the wave activity decreases with a scale height very close to the density scale height, $H_\rho$=6.63 km, used in the calculations and indicated by the dashed lines in the figure.

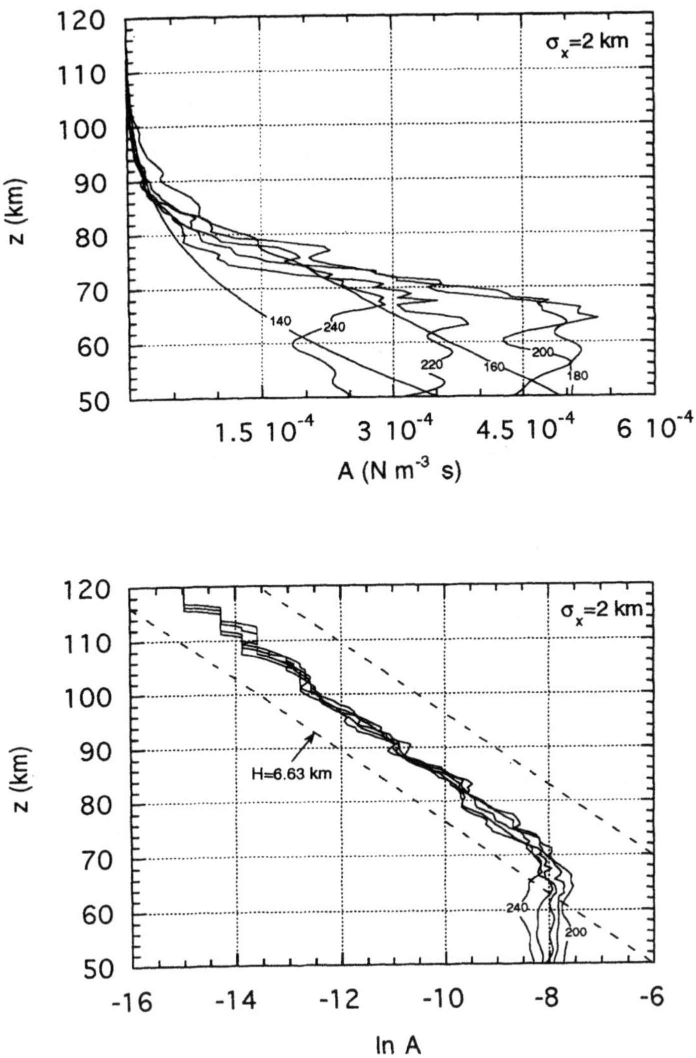


Fig. 5. Wave activity profiles for the region downstream of the forcing (–160, 0 km) as functions of time and altitude. The short waves that break in this region become saturated sometime after $t$=100 min. The saturated profiles decay with height at a rate equal to the density scale height.

*5.2 The Altitude of the Turbopause*

In the Earth's thermosphere dissipation of high-frequency gravity waves is due to molecular diffusion which, like our linear damping profile (2.1), increases exponentialy with altitude. If we express the coefficient of molecular diffusion as

$$\nu = \nu_0 \exp\left(\frac{z - z_0}{H_\rho}\right) \tag{5.4}$$

where $H_\rho$ is the scale height of the atmosphere, and $\nu_0$ is the diffusivity at some reference level $z_0$, then the dissipation rate due to $\nu$ can be written as $\nu(k^2 + m^2 + M^2) = \nu K^2$ for a wave varying as $\exp[i(kx + mz) + Mz]$, where $M$=0.5/$H$, and $k$ and $m$ are the horizontal and vertical wavenumbers, respectively.

Setting $\nu K^2 = \delta$, and using (5.3) and (5.4), we can derive an expression for the height of the turbopause:

$$z_{\text{turbo}} = z_0 + H_\rho \ln\left[\frac{c_{gz}}{2H_\rho} \cdot \frac{1}{\nu_0 K^2}\right] = z_{\text{turbo}} = z_0 + H_\rho \ln\left[\frac{K_{zz}}{\nu_0}\right] \tag{5.5}$$

In the second equality of (5.5) we have set

$$K_{zz} = \frac{kNm}{K^3} \cdot \frac{1}{2H_\rho} \cdot \frac{1}{K^2} = \frac{k(u_o - c)^4}{2H_\rho N^3} \cdot \left(1 + \frac{k^2 + M^2}{m^2}\right)^{-1/2} \tag{5.6}$$

which may be interpreted as a "diffusion coefficient" due to wave breaking. Eq. (5.6) is the counterpart of Lindzen's (1981) expression for $K_{zz}$, and it reduces to Lindzen's formula when $M$, $k$<<$m$..

Eqs. (5.5)-(5.6) imply that the altitude of the Earth's turbopause depends on a relatively small number of parameters characterizing the background atmospheric state and the gravity wave field at high altitudes. If the properties of the wave field (viz., the distribution of power as a function of horizontal wavenumber $k$) are assumed to remain more or less constant with time, then systematic seasonal variations in the altitude of the turbopause may be expected to occur principally as a result of changes in the properties of the background state. Since the magnitude of the zonal mean wind varies significantly between the solstices (when mean zonal winds are strong) and the equinoxes (when they are weak), large changes in the height of the turbopause may be expected as a function of season. Other things being equal, (5.6) predicts that a factor of 2 increase in the magnitude of $u_0$

between equinox and solstice would raise $z_{turbo}$ by 2.77 $H_\rho$, or almost 20 km. It would be of interest to investigate whether this prediction is borne out by observations.

## 5. SUMMARY

We have used a two-dimensional, nonhydrostatic model to study the propagation and breaking at high altitudes of gravity waves forced at tropopause level. The waves are forced by a deflection of the lower boundary, whose shape is a Gaussian with a spatial half-width $\sigma_x$= 2 km and an amplitude $A$=200 m. Our principal findings include the following:

1. Although the lower boundary displacement is Gaussian in shape, the wavefield at mesopause altitudes is nearly monochromatic. The dominance of a single horizontal wavelength at high altitudes is a consequence of the dispersion of the wavepacket forced at the lower boundary. Short horizontal wavelengths ($\lambda_x \approx$25 km) dominate the initial response at high altitudes. At later times, the response at high altitudes is dominated by much longer waves ($\lambda_x \approx$100-120 km). Examination of the linear dispersion relation indicates that the early appearance of short horizontal wavelengths is due to the fact that the vertical group velocity $c_{gz}$ maximizes at $\lambda_x \approx$20-25 km for the parameters used in the present calculations. Horizontal wavelengths near 100 km are associated with vertical group velocities that are two to four times slower than those of the shorter waves, and thus arrive at high altitudes at correspondingly later times. Waves much longer than 100 km have very small vertical group velocities. In our calculations, they radiate out of the side boundaries before they can propagate to high altitudes; in the real atmosphere, they are likely to be subject to much stronger dissipation than shorter, faster waves. These results may help explain why observed wave motions at mesospheric altitudes tend to be monochromatic, with horizontal wavelengths from a few tens to about 100-200 km.

2. The position of the mesospheric wavetrain relative to the lower boundary forcing is strongly dependent on zonal wavelength. For short waves ($k>2\pi/50$ km$^{-1}$), the group velocity is nearly independent of frequency, and the response always appears downwind of the forcing. For longer waves (and especially for wavelengths >100 km) the response is found *upwind* of the forcing. The location of both short and long waves relative to the forcing at the lower boundary can be understood in terms of their group velocity vectors, which point downstream for short waves, and upstream for long waves.

3. For the forcing amplitude employed in the calculations, $A$=200 m in Eq. (2.1), the waves break at altitudes between about 70 and 125 km. The morphology of the wave field in the breaking region is sensitive to the ratio of horizontal to vertical wavelengths. When

the horizontal and vertical scales are comparable, breaking involves a large portion of the wave. On the other hand, when the zonal wavelength is much larger than the vertical wavelength, breaking is confined to a relatively small fraction of the wave, as in earlier calculations, such as that of Walterscheid and Schubert (1990). These differences may be important insofar as the efficiency of mixing due to wave breaking depends on the localization of the ensuing turbulent motion (Fritts and Dunkerton, 1985).

4. Once breaking begins, the phase velocity, $c$, evolves so as to keep the Doppler-shifted phase velocity, $u_o - c$, approximately constant. This behavior, which Fritts and Dunkerton called self-acceleration, has been reported in previous studies (Fritts and Dunkerton, 1984; Walterscheid and Schubert, 1990). Our results suggest that self-acceleration constitutes an actual change in the speed of the waveform and is not due to the presence of a spectrum of frequencies, each with its own critical level.

5. Throughout the breaking region, the wave activity profiles decay with a scale height very close to the atmospheric scale height, $H_\rho$, much as predicted by linear saturation arguments (e.g., Lindzen, 1981). This behavior is present in all of our calculations although the time required to set up a "saturated profile" of wave activity depends on the vertical group velocity of the waves.

6. Wave breaking ceases abruptly at the altitude where external dissipation becomes faster than the dissipation rate due to breaking, $c_{gz}/2H_\rho$, derived from saturation arguments. We find that the altitude where wave breaking ceases –the model turbopause– can be estimated rather accurately as the altitude where the external dissipation rate equals $c_{gz}/2H_\rho$. Since $c_{gz}$ has a strong dependence on the Doppler-shifted phase speed, $u_o - c$, this implies that the altitude of the Earth's turbopause may change significantly as a function of season.

The calculations and analysis presented in this paper have emphasized the behavior of gravity waves as they propagate away from the forcing region, disperse, and finally break at high altitudes. As such, they may be viewed as complementary to the recent work of Alexander et al (1995), who explored in detail forcing mechanisms due to a simulated squall line, and that of Andreassen and co-workers (Andreassen et al, 1994; Fritts et al, 1994; Isler et al, 1994), who addressed the role of wave breakdown in three dimensions. All of these aspects of the gravity wave problem may need to be taken into account to provide a complete understanding of wave breaking in the middle atmosphere and its effect on the atmospheric mean state.

**Acknowledgements:** This work was supported in part by a grant from the National Aeronautics and Space Administration. The National Center for Atmospheric Research is sponsored by the National Science Foundation.

**REFERENCES**


Alexander, M.J., J.R. Holton D.R. Durran (1995) The gravity wave response above deep convection in a squall line simulation. *J. Atmos. Sci., 52*, 2212-2226.

Andreassen, O., C.E. Wasberg, D.C. Fritts and J.R. Isler (1994) Gravity wave breaking in two and three dimensions 1. Model description and comparison of two-dimensional evolutions. *J. Geophys. Res., 99*, 8095-8108.

Andrews, D.G., J.R. Holton and C.B. Leovy (1987) *Middle Atmosphere Dynamics.* Academic Press, Inc., Orlando.

Balsley, B.B., W.L. Eklund and D.C. Fritts (1983) VHF echoes from the high-latitude mesosphere and lower thermosphere: observations and interpretations. *J. Atmos. Sci., 40,* 2451-2466.

Bretherton, F.P. (1971: The general linearised theory of wave propagation. In *Mathematical problems in the geosciences. I. Geophysical fluid mechanics,* ed. W.H. Reid. *Lectures in applied mathematics, 13*, 61-102. American Mathematical Society, Providence, Rhode Island.

Bretherton, F.P. (1966) The propagation of groups of internal gravity waves in a shear flow. *Q. J. R. Meterol. Soc.*, 92, 466-480.

Dunkerton, T.J. (1989) Theory of internal gravity wave saturation. *Pure and Appl. Geophys.,* 130, 373-397.

Fritts D.C. and T.J. Dunkerton (1984) A quasi-linear study of gravity wave saturation and self-acceleration. *J. Atmos. Sci., 41*, 3272-3289.

Fritts, D.C. and T.J. Dunketon (1985) Fluxes of heat and constituents due to convectively unstable gravity waves. *J. Atmos. Sci., 42,* 549-556.

Fritts, D.C., J.R. Isler, and G.E. Thomas (1993) Wave breaking signatures in noctilucent clouds. *Geophys. Res. Lett.* 20, 2039-2042.

Fritts, D.C, J.R. Isler and O. Andreassen (1994) Gravity wave breaking in two and three dimensions 2. Three-dimensional evolution and instability structure. *J. Geophys. Res., 99*, 8109-8123.

Garcia, R.R. (1991) Parameterization of planetary wave breaking in the middle atmosphere. *J. Atmos. Sci.*, 48, 1405-1419.

Jones, W.L. and J.T. Houghton (1972) The self-destructing internal gravity wave. *J. Atmos. Sci., 29*, 844-849.

Isler, J.R., D.C. Fritts, O. Andreassen and C.E. Wasberg (1994) Gravity wave breaking in two and three dimensions. 3. Vortex breakdown and transition to isotropy. *J. Geophys. Res., 99*, 8125-8137.

Lindzen, R.S. (1981) Turbulence and stress owing to gravity wave and tidal breakdown. *J. Geophy. Res.*, 86, 9707-9714.

Lipps, F.B. and R.S. Hemler (1982) A scale analysis of deep moist convection and some related numerical calculations. *J. Atmos. Sci.*, 39, 2192-2210.

Moreels, G. and H. Herse (1977) Photographic evidence of waves around the 85 km level. *Planet. Space Sci.*, 25, 265-273.

Phillips, A. and R.A. Vincent (1989) Radar observation of prevailing winds and waves in the southerm hemisphere mesosphere and lower thermosphere. *Pageoph.*, 130, 303-318.

Prusa, J.M., P.K. Smolarkiewicz, and R.R. Garcia (1996) On the propagation and breaking at high altitudes of gravity waves excited by tropospheric forcing. *J. Atmos. Sci.,* in press.

Smolarkiewicz, P.K. and G.A. Grell (1992) A class of monotone interpolation schemes. *J. Comp. Physics*, 101, 431-440.

Smolarkiewicz, P.K. and J.A. Pudykiewicz (1992) A class of semi-Lagrangian approximations for fluids. *J. Atmos. Sci.*, 49, 2082-2096.

Theon, J.S., W. Nordberg and L.B. Katchen (1967) Some observations on the thermal behavior of the mesosphere. *J. Atmos. Sci.*, 24, 428-438.

*U.S. Standard Atmosphere* (1976) Published by National Oceanic and Atmospheric Administration, National Aeronautics and Space Administration, and United States Air Force. NOAA publication S/T 76-1562. Available from U.S. Government Printing Office, stock no. 003-017-00323-0.

Vincent, R.A. and I.M. Reid (1983) HF Doppler measurements of mesospheric gravity wave momentum fluxes. *J. Atmos. Sci.*, 40, 1321-1333,.

Walterscheid, R.L. and G. Schubert (1990) Nonlinear evolution of an upward propagating gravity wave) Overturning, convection, transience and turbulence. J. Atmos. Sci., 47, 101-125.

Whitten, R.C. and I.G. Poppoff (1971) *Fundamentals of Aeronomy.* John Wiley & Sons, Inc., New York.

# Estimate of Atmospheric Dissipation Derived from UARS/HRDI Measurements

Valery A. Yudin, Marvin A. Geller, and Boris V. Khattatov[1]
Institute for Terrestrial and Planetary Atmospheres
State University of New York at Stony Brook
Stony Brook, New York 11794-5000
USA

**ABSTRACT.** The parameters that appear in gravity wave parameterizations are mostly determined by reasonableness and by being able to derive middle atmosphere mean wind and temperature structures that agree with observations. This puts effective constraints on these parameterizations at middle and high latitudes but does not effectively constrain these parameterizations at low latitudes. The diurnal tide, as derived from UARS HRDI wind observations is used to derive an annual climatology of the implied atmospheric dissipation. This dissipation thus derived is shown to be less than that derived from commonly used gravity wave parameterizations at low latitudes (less than approximately 40 ° latitude). Thus, models using most familiar gravity wave parameterizations will not give structures for the diurnal tide that agree with UARS HRDI observations.

## 1. Introduction

The classic work of Leovy (1964), followed by papers such as those by Schoeberl and Strobel (1978) and Holton and Wehrbein (1980) clearly showed the necessity for the existence of some mechanism for producing a drag acting on the mesospheric mean zonal winds in both the summer and winter hemisphere. Otherwise, as Fels (1985) showed, radiation alone would produce middle to high latitude mesospheric jets whose winds continued to grow with increasing altitude rather than maxima in mesosphere. As Houghton (1978) pointed out, since stationary planetary waves cannot propagate to these levels in summer, it must be the breaking of gravity waves producing this drag on the mean zonal winds, at least in the summer hemisphere. Given the fact that

---

[1] Present affiliation: Atmospheric Chemistry Division; National Center for Atmospheric Research; Boulder, Colorado 80307-3000; USA

NATO ASI Series, Vol. I 50
Gravity Wave Processes
Their Parameterization in Global Climate Models
Edited by Kevin Hamilton

large-scale models of the middle atmosphere are unable to resolve such gravity wave motions, there followed an effort to develop parameterizations for the effects of unresolved gravity waves much in the same manner as is done for cumulus clouds in the lower atmosphere.

Such parameterizations have been developed by Lindzen (1981), Fritts and Lu (1993), and Hines (1996a, 1996b), for example. All of these parameterization schemes have parameters that cannot be determined unambiguously at the present time from observations, so these parameters have been chosen to have plausible values that give modeled mean winds and temperatures that agree with observations. In practice though, this procedure has only given useful constraints on the gravity wave parameterizations at middle and high latitudes. Recently, however, observations have become available that are useful at low latitudes. These are the UARS/HRDI observations of the solar diurnal tide.

## 2. UARS/HRDI Data

The HRDI instrument on UARS observes line-of-sight wind speeds in the mesosphere-lower thermosphere (MLT) region by measuring Doppler shifts in the emission bands of molecular oxygen. By combining measurements along two lines-of-sight and assuming that the vertical wind is negligible and the winds do not change appreciably over the time required to make both line-of-sight measurements, the vector horizontal wind is determined.

An inversion technique is used in the HRDI data processing that uses a sequential estimator technique (Rodgers, 1976, 1990). This has been described in detail by Ortland et al. (1995). Burrage (1993) has described the process of fine tuning the sequential estimator and determining the zero position of the instrument. The process of validating the UARS/HRDI MLT winds has been described by Burrage et al. (1996) and Khattatov et al. (1996a).

## 3. The Diurnal Tide

A clear signature of the solar diurnal tide is seen in the HRDI data. If one plots a single day's meridional wind observations (with most observations being taken near the same local time), one sees oscillations in the sign of the meridional wind where the altitude scale of an oscillation cycle is a bit less than 25 km (for example, see Morton et al., 1993). Furthermore, one sees the amplitude of these oscillating features to be maximum at latitudes of about 25 ° North and South. These are exactly the characteristics expected of the solar diurnal tide (see Chapman and Lindzen, 1970, for example). Morton et al. (1993) and Hays et al. (1994) have used these features in the HRDI observations to determine the observed morphology of the solar diurnal tide.

For extracting the tidal structures from the HRDI wind observations, Khattatov et al. (1996b) have developed a new approach in which a linearized tidal model, including the effects of mean zonal winds, is used together with the HRDI data to determine the climatology of the solar diurnal tidal variations in the meridional wind in the MLT region. Briefly speaking, the tidal model is used to solve for the diurnal tide using climatological estimates for the mean zonal wind and temperature structures. The amplitudes and phases of the modeled diurnal variation in the meridional velocity component are then compared with the inferred amplitudes and phases of the diurnal variation of the meridional wind component from the UARS/HRDI measurements over a month. Quite good agreement is found between the modeled and observed phases with less agreement between the amplitudes. First, the modeled phases are adjusted to agree with the UARS/HRDI observed values, where they are available. This is then followed by an adjustment in the amplitudes. By this procedure then, a climatology for the diurnal variation in the meridional wind component is derived that is in quite good agreement with the UARS/HRDI observations and shows the general global behavior of the model results. These results are shown in figure (1).

At this point, the diurnal variation in the meridional wind component is determined. This is then used together with the governing equations (equations of motion, thermodynamics, continuity, and state) in an iterative scheme to determine the other diurnal tidal variables (zonal wind, vertical wind, temperature, density, pressure). These are then used in the tidal energy equation to derive the implied atmospheric dissipation. More details on this procedure can be found in Khattatov et al. (1996c).

The point stressed in this paper is that the diurnal tide, because of its rather short vertical wavelength and the fact that it has high amplitudes at low latitudes, provides an important constraint on the atmospheric dissipation at low latitudes. Such constraints have not previously been considered to any great extent. In the following, we will illustrate that there are some inconsistencies between the atmospheric dissipation implied by the observed structure of the diurnal tide at middle and low latitudes (between about $\pm 40$ °) and that which is implied by various gravity wave parameterizations that have been used in models of the global middle and upper atmosphere.

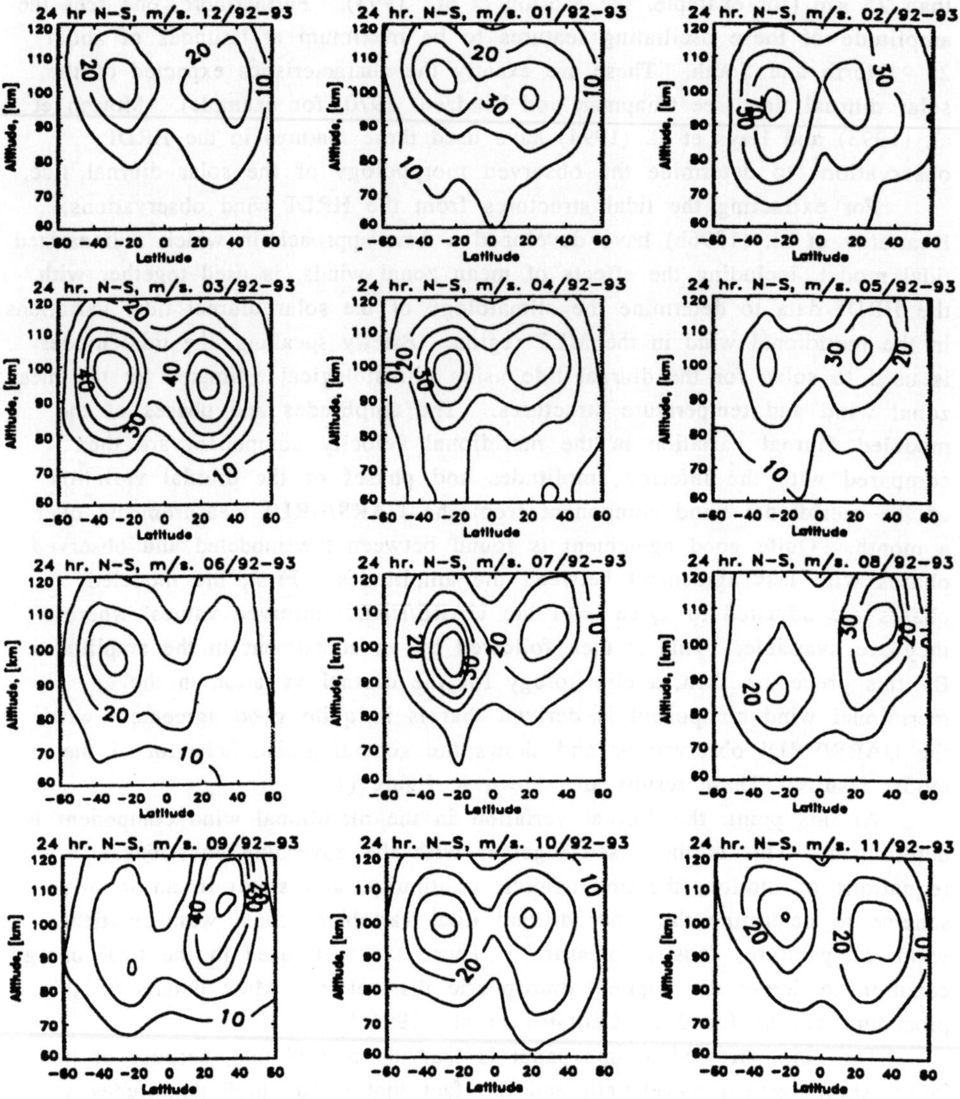


Figure 1. Monthly mean amplitudes of the meridional diurnal tidal winds derived from HRDI data for 1992-93. Contours in all panels except February are at 10 m s$^{-1}$ spacing. The February contours have 20 m s$^{-1}$ spacing. From Khattatov, 1996b.

## 4. Atmospheric Dissipation

Khattatov et al. (1996b) used a tidal model in which atmospheric dissipation was represented by Newtonian cooling and Rayleigh friction for simplicity. Derived monthly mean Rayleigh friction coefficients are shown in figure (2). Two things to keep in mind are that the quality of these results depends critically on the UARS/HRDI data coverage during each month, and these results were obtained using data for the two years 1992-93 to increase the amount of data. Note that, in general, the Rayleigh friction increases with increasing altitude. In general, higher values of Rayleigh friction are seen in the winter hemisphere. We have highlighted the 2 x $10^{-5}$ $sec^{-1}$ contour in each of the panels. June and July are exceptions to this, but these are months with poor data coverage for a number of reasons, the most important being that the UARS spacecraft malfunctioned during these months in 1992. It is to be expected by gravity wave breaking theory (e. g., Lindzen, 1981) that the atmospheric dissipation produced by gravity wave breaking is larger at altitudes below 100 km in the winter hemisphere at middle and high latitudes.

Although Khattatov et al. (1996c) only calculate the implied Rayleigh friction directly, a crude picture of the implied diffusion term, $K_{zz}$, can be obtained for an assumed monochromatic wave by multiplying the Rayleigh friction coefficients by the vertical wavenumber squared. Khattatov et al. (1996c) did this by determining the local vertical wavenumber from the derived diurnal phases. The derived distributions for $K_{zz}$ are shown in figure (3). The distributions of $K_{zz}$ are quite similar to those for the Rayleigh friction in that $K_{zz}$ increases upward and is, in general, larger in the winter MLT regions than in summer. We have highlighted the 200 $m^2$ $s^{-2}$ contour since Forbes and Vincent (1989) have suggested that at the altitude where this value is reached the diurnal tide will cease growing and decay above. The results shown in figure (1) and figure (3) are consistent with this result. Also, by comparing figures (2) and (3), one sees that the Rayleigh friction value of 2 x $10^{-5}$ $sec^{-1}$ corresponds very closely to the $K_{zz}$ value of 200 $m^2$ $s^{-2}$.

It is interesting to compare Khattatov et al.'s (1996c) derived values of $K_{zz}$ from HRDI wind observations with those of Garcia and Solomon (1985). Their derived $K_{zz}$ is shown in figure (4) for solstice conditions. We have highlighted their 2.1 x $10^2$ $m^2$ $s^{-2}$ for comparison with those shown in figure (3). Note that their results show a 2.1 x $10^2$ $m^2$ $s^{-2}$ center that extends down to about 85 km at about 30 °S with a rather flat 2.1 x $10^2$ $m^2$ $s^{-2}$ contour lying at about 106 km elsewhere. For comparison, Khattatov et al.'s 200 $m^2$ $s^{-2}$ contour lies well above 110 km altitude everywhere. Thus, we expect that

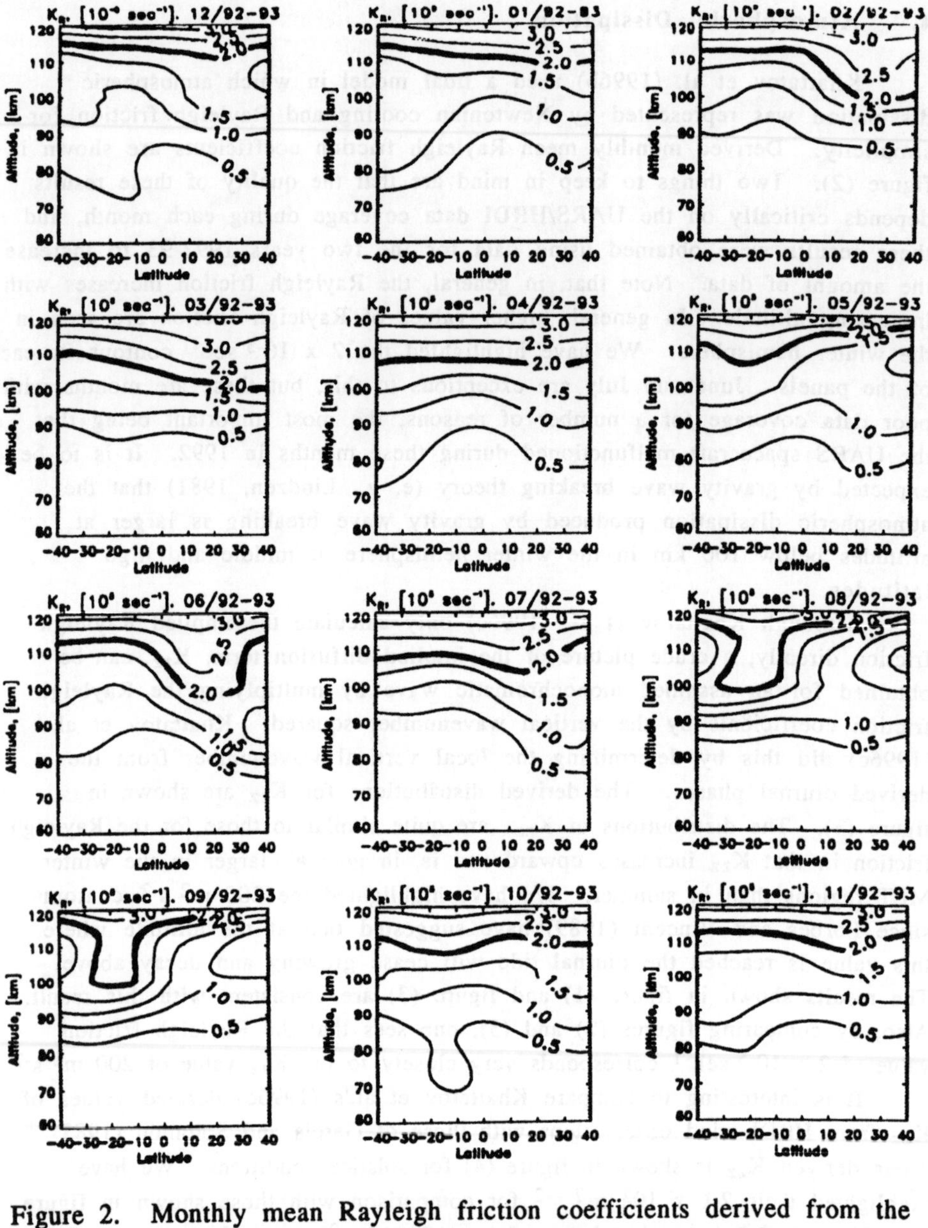


**Figure 2.** Monthly mean Rayleigh friction coefficients derived from the diurnal tide implied by the HRDI data for 1992-93. The 2.0 contour has been highlighted for emphasis. From Khattatov et al., 1996c.

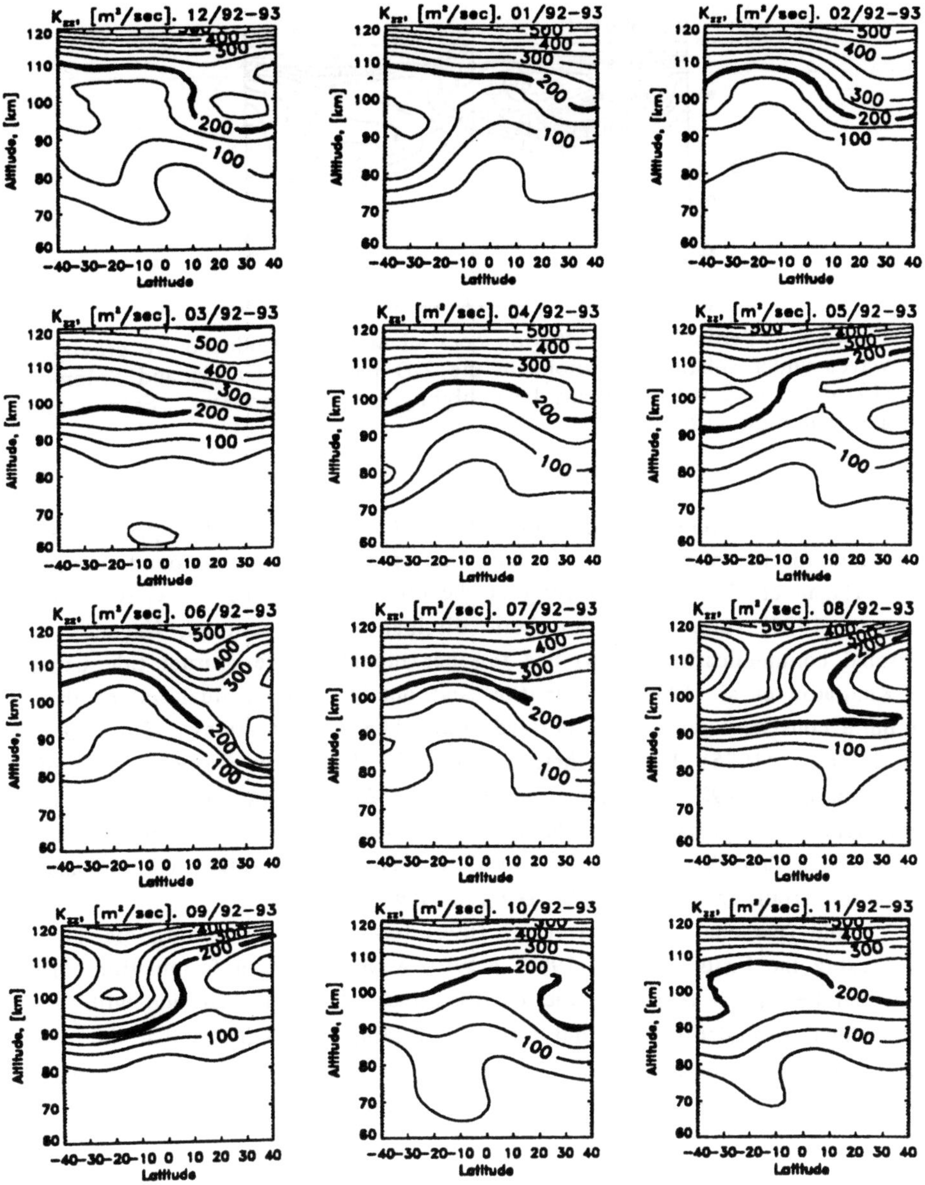


Figure 3. Monthly mean vertical diffusion coefficients derived from the diurnal tide implied by the HRDI data for 1992-93. The 200 contour has been highlighted for emphasis. From Khattatov et al., 1996c.

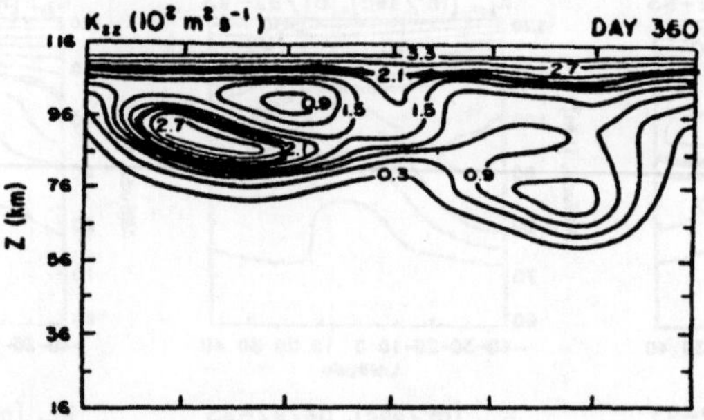


Figure 4. Vertical eddy diffusion coefficient, $K_{zz}$, produced by gravity waves under solstice conditions in the 2-D model of Garcia and Solomon (1985). The 2.1 contour has been highlighted for emphasis. From Garcia and Solomon (1985).

the Garcia and Solomon (1985) atmospheric dissipation would imply excess dissipation for the diurnal tide. In fact, this is exactly what is found to occur in the Hagan et al. (1995) modeling of the diurnal tide using the Garcia and Solomon (1985) values for atmospheric dissipation. In contrast, Hagan (personal communication, 1996) has run her model of the diurnal tide using the Khattatov et al (1996c) dissipation and obtains diurnal tidal values that are quite consistent with those derived from UARS observations. Since Garcia and Solomon (1985) derived their atmospheric dissipation values essentially using Lindzen's (1981) gravity wave breaking formulation, adjusting open parameters to get reasonable looking mean zonal winds and temperatures, this suggests that this formulation does not give proper values for atmospheric dissipation from gravity wave breaking at low latitudes.

Roble (personal communication, 1996) has attempted running his TIMED GCM using the Fritts and Lu (1993) gravity wave saturation formulation. He finds that he cannot reproduce the UARS values of the diurnal tide with conventional formulations of tidal forcing. Again, this suggests that the atmospheric dissipation inferred from the Fritts and Lu (1993) gravity wave saturation formulation is excessive at low latitudes.

## 5. Comparison of UARS and MF Radar Observations

A great deal of current research is ongoing to resolve the differences in MF radar and UARS HRDI and WINDII observations of MLT winds (see Burrage et al., 1996, and Khattatov et al. (1996a) for more details on this. Since MF radar observations seem to give smaller winds than UARS measurements above about 85 km., quite a different structure is observed for the diurnal tide using these two techniques. Figure (5) shows an example of a comparison of the modeled diurnal tide from Hagan et al. (1995) with HRDI and MF radar observations for the month of April at 20 °N. This figure shows that the MF diurnal tidal amplitudes for the northward wind variation maximize at about 85 km while the model maximum occurs at about 105 km, and the HRDI measurements show the maximum at about 95 km. Thus, the atmospheric dissipation would be very different if the MF radar values were correct.

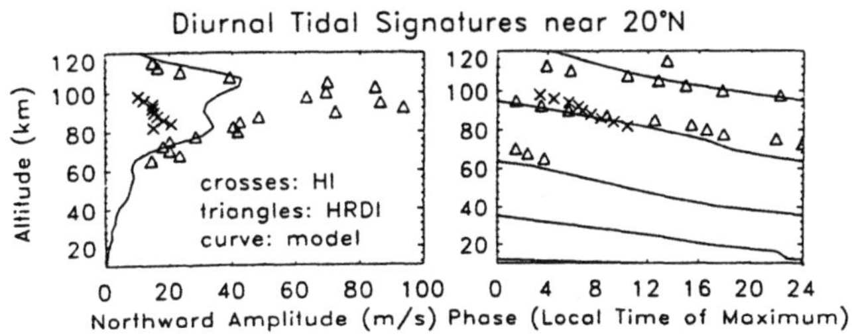


Figure 5. Northward diurnal tide wind amplitudes and phases from the global-scale wind model of Hagan et al. (1995) at 21 °N (curves); from 1992 monthly averaged UARS/HRDI data (triangles); and MF radar measurements over Kauai, Hawaii (20 °N). From Hagan et al. (1995).

## 6. Conclusions

We have illustrated that a certain distribution of atmospheric dissipation in the MLT regions is implied by the diurnal tidal structure that is consistent with UARS measurements at low latitudes. These values for the atmospheric dissipation are less than those from the gravity wave parameterization of Garcia and Solomon (1985) and that inferred from the Fritts and Lu (1993) parameterization. It should be noted that reproducing the observed structure of the diurnal tide in the MLT region is a new constraint that has not yet been applied to gravity wave parameterizations. Furthermore, these are the first stringent constraints to be applied to gravity wave parameterizations in the low latitude MLT region. Previously, only constraints on the middle and high latitude mean zonal flow have been used. It is clear that for a gravity wave parameterization to realistically apply to the actual atmosphere, it must lead to proper modeling of all MLT circulation phenomenon such as the mean zonal flow, planetary waves, tides, etc.

**References**

Burrage, M. D., W. R. Skinner, A. R. Marshall, P. B. Hays, R. S. Lieberman, S. J. Franke, D. A. Gell, D. A. Ortland, Y. T. Morton, F. J. Schmidlin, R. A. Vincent, and D. I. Wu (1993) Comparisons of HRDI wind measurements with radar and rocket observations. *Geophys. Res. Lett.*, *20*, 1,259-1,262.

Burrage, M. D., W. R. Skinner, D. A. Gell, P. B. hays, A. R. Marshall, D. A. Ortland, A. H. Manson, S. J. Franke, D. C. Fritts, P. Hoffman, C. McLandress, R. Niciejewski, F. J. Schmidlin, G. G. Shepherd, W. Singer, T. Tsuda, and R. A. Vincent (1996) Validation of mesosphere and lower thermosphere winds from the high resolution Doppler imager on UARS. *J. Geophys. Res.*, *101*, 10,365-10,392.

Chapman, S., and R. S. Lindzen (1970) *Atmospheric Tides*. Reidel, Dordrecht, Netherlands.

Fels, S. B. (1985) Radiative-dynamical interactions in the middle atmosphere. *Adv. Geophys.*, *28A*, 277-300.

Forbes, J. M., and R. A. Vincent (1989) Effects of the mean winds and dissipation on the diurnal propagating tide: an analytic approach. *Planet. Space. Sci.*, *37*, 197-209.

Fritts, D. C., and W. Lu (1993) Spectral estimates of gravity wave energy and momentum fluxes. Part II: Parameterization of wave forcing and variability. *J. Atmos. Sci.*, *50*, 3,695-3,713.

Garcia, R. R., and S. Solomon (1985) The effect of breaking gravity waves on the dynamics and chemical composition of the mesosphere and lower thermosphere. *J. Geophys. Res.*, *90*, 3,850-3,868.

Hagan, M. E., J. M. Forbes, and F. Vial (1995) An updated model of migrating tides in the middle atmosphere: Initial results and measurement comparisons. In *Proceedings of the workshop on "Wind Observations in the Middle Atmosphere,"* held in Paris 15/18 November, 1994 (available through CNES Headquarters).

Hays, P. B., D. I. Wu, M. D. Burrage, D. A. Gell, H. J. Grassl, R. S. Lieberman, A. R. Marshall, Y. T. Morton, D. A. Ortland, and W. R. Skinner (1994) Observations of the diurnal tide from space. *J. Atmos. Sci.*, *51*, 3,077-3,093.

Hines, C. O. (1996a) Doppler-spread parameterization of gravity-wave momentum deposition in the middle atmosphere. Part 1: Basic Formulation. In press, *J. Atmos. Terr. Phys.*

Hines, C. O. (1996b) Doppler-spread parameterization of gravity wave momentum deposition in the middle atmosphere. Part 2: Broad and quasi monochromatic spectra and implementation. In press, *J. Atmos. Terr. Phys.*

Holton, J. R., and W. M. Wehrbein (1980) A numerical model of the zonal mean circulation of the middle atmosphere. *Pageoph*, *118*, 284-306.

Houghton, J. T. (1978) The stratosphere and mesosphere. *Q. J. R. Meteorol. Soc.*, *104*, 1-29.

Khattatov, B. V., M. A. Geller, V. A. Yudin, P. B. hays, W. R. Skinner, M. D. Burrage, S. J. Franke, D. C. Fritts, J. R. Isler, A. H. Manson, C. E. Meek, R. McMurray, W. Singer, P. Hoffmann, and R. A. Vincent (1996a) Dynamics of the mesosphere and lower thermosphere as seen by MF radars and by the high-resolution Doppler imager/UARS. *J. Geophys. Res.*, *101*, 10,393-10,404.

Khattatov, B. V., M. A. Geller, V. A. Yudin, P. B. Hays, and R. A. Vincent (1996b) Meridional diurnal tidal winds derived from the HRDI/UARS observations. Submitted to *J. Geophys. Res.*

Khattatov, B. V., V. A. Yudin, M. A. Geller, and P. B. Hays (1996c) Global monthly mean climatology of the diurnal tide and dissipation derived from the HRDI/UARS data. Submitted to *J. Geophys. Res.*

Leovy, C. B. (1964) Simple models of thermally driven mesospheric circulation. *J. Atmos. Sci.*, *21*, 327-341.

Lindzen, R. S. (1981) Turbulence and stress owing to gravity wave and tidal breakdown. *J. Geophys. Res.*, *86*, 9707-9714.

Morton, Y. T., R. S. Lieberman, P. B. Hays, D. A. Ortland, A. R. Marshall, D. L. Wu, W. R. Skinner, M. D. Burrage, D. A. Gell, and J.-H. Yee (1993) Global mesospheric tidal winds observed by the High Resolution Doppler Imager on board of the Upper Atmosphere Research Satellite. *Geophys. Res. Lett.*, *20*, 1,263-1,266.

Ortland, D. A., P. B. Hays, W. R. Skinner, V. J. Abreu, J.-H. Yee, M. D. Burrage, A. R. Marshall, and D. A. Gell (1995) A sequential estimation technique for recovering atmospheric data from orbiting satellites. *The Upper Mesosphere and Lower Thermosphere: A Review of Experiment and Theory* . Geophysical Monograph Series, Vol. 87, edited by R. M. Johnson and T. L. Killeen, AGU, Washington, DC, 329-337.

Rodgers, C. D. (1976) Retrieval of atmospheric temperature and composition from remote measurements of thermal radiation. *Rev. Geophys. and Space Phys.*, *14*, 609-624.

Rodgers. C. D. (1990) Characterization and error analysis of profiles retrieved from remote soundings measurements. *J. Geophys. Res.*, 95, 5587-5595.

Schoeberl, M. R., and D. F. Strobel (1978) The zonally averaged circulation of the middle atmosphere. *J. Atmos. Sci.*, *35*, 577-591.

# On the nonlinear interactions between gravity waves in shear flows

J. Vanneste[1] and F. Vial
LMD/CNRS
Ecole Polytechnique
91128 Palaiseau cedex
France

## 1 Introduction

The gravity waves which propagate upward in the middle atmosphere attain very large amplitudes, so large that they break through dynamical or convective instability. When they are strictly isolated, they can be satisfactorily described by the linear theory because of the (almost) incompressibility of the atmosphere. In general, however, several waves coexist, and the nonlinear interactions can significantly influence their propagation far below their breaking level.

The importance of the nonlinear effects is manifest in the mechanism of wave instability: small disturbances to a gravity wave are amplified, extracting energy from the wave at a rate proportional to its amplitude (see e.g. Klostermeyer 1991). This can strongly affect the transport of momentum by the wave, and hence influence the wave–mean flow interaction (Dunkerton 1987).

The interactions between gravity waves have been fruitfully studied using the weakly nonlinear theory, following the approach employed for other types of hydrodynamic waves (e.g., Craik 1985). In this theory, the waves are assumed to keep the spatial structure they have in the absence of nonlinearities; the nonlinear interactions only slowly modulate the wave amplitudes. The dynamical fields can thus be written as

$$\varphi = \sum_a A_a(t) \exp\left[\mathrm{i}(\mathbf{k}_a \cdot \mathbf{r} - \omega_a t)\right] \hat{\varphi}_a,$$

where $\omega_a = \Omega(\mathbf{k}_a)$ satisfies the linear dispersion relation, and $\hat{\varphi}_a$ is the vector of the dynamical fields determined by the polarisation relations. Introducing this form in the equations of motion, one can get (through various techniques) interaction equations, describing the evolution of the amplitudes $A_a$. These equations take the form

$$\frac{\mathrm{d}A_a}{\mathrm{d}t} = \frac{1}{2} \sum_{bc} I_a^{bc} A_b^\star A_c^\star \exp\left[\mathrm{i}(\omega_a + \omega_b + \omega_c)t\right], \quad \mathbf{k}_a + \mathbf{k}_b + \mathbf{k}_c = 0.$$

Owing to the assumption of weak nonlinearity, the study can be further simplified by considering only isolated triads for which $\omega_a + \omega_b + \omega_c \approx 0$. The interactions can then be analysed in two steps:

[1]Present affiliation: Department of Physics, University of Toronto, Toronto, Canada M5S 1A7

NATO ASI Series, Vol. I 50
Gravity Wave Processes
Their Parameterization in Global Climate Models
Edited by Kevin Hamilton

1. The conditions for resonance are investigated, i.e. the system

$$\begin{cases} \mathbf{k}_a + \mathbf{k}_b + \mathbf{k}_c = 0 \\ \omega_a + \omega_b + \omega_c = 0 \end{cases}$$

is solved using the dispersion relation $\omega_i = \Omega(\mathbf{k}_i)$. This indicates which waves can interact significantly.

2. The interaction coefficients $I_a^{bc}$ – given as functions of the three wave structures $\hat{\varphi}_a$, $\hat{\varphi}_b$ and $\hat{\varphi}_c$ – are calculated. Their values govern the energy transfers inside a triad.

A vast amount of literature is devoted to this type of analysis, which has been applied to various models of stratified fluids (e.g., McComas and Bretherton 1977; Yeh and Liu 1981; Dong and Yeh 1991). Interestingly, properties of the interaction coefficients – which can be obtained by manipulating their explicit form – directly result from the conservation laws of the models (Ripa 1981; see also below).

The wave instability can be understood in terms of triad interactions (e.g. Craik 1985). Indeed, an infinitesimal disturbance to a (primary) wave $a$ can consist of two waves $b$ and $c$, with $\mathbf{k}_a+\mathbf{k}_b+\mathbf{k}_c = 0$, and $|A_a| \gg |A_b|, |A_c|$. The evolution of the disturbance amplitude is readily obtained as

$$A_b, A_c = \exp(\sigma_a t), \quad \sigma_a = \sqrt{I_b^{ca} I_c^{ab} |A_a|^2 - \phi^2/4} - i\phi, \tag{1}$$

where $\phi = \omega_a + \omega_b + \omega_c$. It is then easy to see that a wave $a$ is unstable at any amplitude if it can be involved in a resonant triad such that $I_b^{ca} I_c^{ab} > 0$. From the properties of the interaction coefficients, an equivalent condition is found to be $|\omega_a| > |\omega_b|, |\omega_c|$. An infinitesimal wave is thus unstable if it can be the highest-frequency member of a resonant triad. This is Hasselmann's criterion (Hasselman 1967). Finite-amplitude waves are also unstable through non-resonant interactions.

For non-resonant triads, the term $-i\phi$ in (1) modifies the propagation properties of the growing waves compared to those of the corresponding free waves. Consequently, the group velocity of the growing waves is modified; in fact, it is given by

$$\mathbf{v}_b^g = \mathbf{v}_c^g = \left[\nabla_{\mathbf{k}}\Omega(\mathbf{k}_b) + \nabla_{\mathbf{k}}\Omega(\mathbf{k}_c)\right]/2,$$

i.e. the mean of the free wave group velocity (Vanneste 1995*a*). An evaluation of this group velocity in resonance conditions shows that it is generally negative (i.e. oppositely signed to the group velocity of the primary wave $a$); it can have the same sign only for high-frequency primary waves ($|\omega_a| \approx N$, where $N$ is the Brunt–Väisälä frequency). Moreover, the group velocity of the fastest-growing disturbance is negative for low-frequency waves, and tends to zero for high-frequency waves. Growing disturbances to an upward propagating wave can thus be expected to propagate downward or to remain at a constant level.

All of these results have been obtained assuming that the gravity waves propagate in a resting atmosphere. However, the waves are generally superposed to a large-scale shear flow, which can include contributions from planetary waves in addition to a zonal mean component. The presence of a basic shear can be expected to significantly affect the nonlinear wave interactions, as it strongly modifies the linear wave propagation. This has been confirmed by numerical simulations (Dunkerton 1978; Vanneste 1995*a*). The theoretical work dealing with this problem (Grimshaw 1988, 1994; Fritts *et al.* 1992) relies upon a WKB-like approach, and hence is limited to slowly-varying shear flows. (Grimshaw (1994) nevertheless considers interactions with a wave packet approaching its critical level.) The results of the standard approach can thus be maintained, but with wave characteristics $\mathbf{k}_i$, $\omega_i$ and $\hat{\varphi}_i$ which are slowly varying in time and space. Consequently, the resonance condition cannot be satisfied globally, but is only met on a certain space-time surface.

The purpose of this paper is to present a theory for the interactions of gravity in a (possibly) strong shear. This theory, which parallels as much as possible the standard approach for waves in a medium at rest, relies upon the conservation laws relevant to disturbances in shear flows, i.e. the wave-activity conservation laws. It provides a technique to derive the interaction equations in a systematic manner, and allows one to relate the interaction coefficients to the wave activities. As a result, an extension of Hasselmann's criterion for wave stability can be established.

When a basic shear is present, the wave activities are not necessarily sign-definite even if the basic flow is linearly stable. Interactions between waves with wave activities of different sign can lead to a phenomenon called explosive resonant interaction, which results in the simultaneous growth – and finite-time singularity – of three waves forming a resonant triad. This phenomenon is very interesting, since it corresponds to a nonlinear instability of the basic flow; we will discuss its possible existence in stratified fluids. Another consequence of the basic flow is the presence of waves with critical levels. As shows a result of Becker and Grimshaw (1993) recalled below, this is tightly connected to the existence of explosive interaction. Although the theory described in this paper does not strictly apply to waves with critical levels, we briefly discuss its possible extension in this direction.

## 2 Derivation of the interaction equations

We consider a two-dimensional stratified fluid ($x$ and $y$ being the horizontal and vertical directions) and, for simplicity, we use the Boussinesq approximation[2]. The equation governing the evolution of a disturbance to a parallel shear flow $U(y)$ can be written

$$(\partial_t + U\partial_x)\zeta - U''\partial_x\psi + \partial_x\eta + \partial(\psi,\zeta) = 0, \quad (2)$$

$$(\partial_t + U\partial_x)\eta - N^2\partial_x\psi + \partial(\psi,\eta) = 0. \quad (3)$$

[2]The approach is actually very general, since it relies on the Hamiltonian structure of the model equations; it can thus be applied to a variety of models of inviscid stratified fluids.

Here, the vorticity $\zeta$ is related to the streamfunction $\psi$ through $\zeta = \nabla^2\psi$, $\eta$ is the buoyancy, $N$ is the Brunt–Väisälä frequency, and $\partial(f,g) = \partial_x f\, \partial_y g - \partial_y f\, \partial_x g$ is the Jacobian operator. It is convenient to view $\eta$ and $\zeta$ as the dynamical variables and to introduce the vector

$$\varphi = (\zeta, \eta).$$

The system (2–3) has two invariants which are quadratic to the lowest order in $\varphi$; these wave activities are the pseudoenergy $\mathcal{E}$ and the pseudomomentum $\mathcal{P}$. They have been derived by Shepherd (1990), Ripa (1990) and Scinocca and Shepherd (1992), and can be expanded in powers of $\varphi$ according to

$$\begin{aligned} \mathcal{E} &= \mathcal{E}^{(2)} + \mathcal{E}^{(3)} + \ldots, \\ \mathcal{P} &= \mathcal{P}^{(2)} + \mathcal{P}^{(3)} + \ldots \end{aligned}$$

In what follows, we will only use the lowest order parts of this invariants, $\mathcal{E}^{(2)}$ and $\mathcal{P}^{(2)}$, which are the invariants of the linearized equations. The corresponding densities are given by

$$E^{(2)} = \frac{|\nabla\psi|^2}{2} + \frac{\eta^2}{2N^2} + \frac{U\zeta\eta}{N^2} - \frac{UU''\eta^2}{2N^4}, \tag{4}$$

$$P^{(2)} = \frac{\zeta\eta}{N^2} - \frac{U''\eta^2}{2N^4}. \tag{5}$$

Note that the first two terms in $E^{(2)}$ represent the perturbation energy. Note also that in a reference frame moving at a velocity $\alpha$, $P^{(2)}$ is unchanged, whereas $E^{(2)}$ transforms into $E^{(2)} - \alpha P^{(2)}$. In terms of $\varphi$, the two densities can be written in the compact form

$$E^{(2)} = \varphi^{+}\mathsf{E}\varphi, \quad \text{and} \quad P^{(2)} = \varphi^{+}\mathsf{P}\varphi,$$

where E and P are self-adjoint operators easily derived from (4–5).

Consider now the linearized system, obtained from (2–3) by neglecting the Jacobians. Its solution can be sought in the form of modes

$$\varphi = \hat{\varphi}_a(y)\exp\left[\mathrm{i}k_a(x - c_a t)\right].$$

The phase velocity $c_a = \omega_a/k_a$ and the vertical structure $\hat{\varphi}_a$ are found as eigenvalue and eigenfunction of the Taylor–Goldstein equation derived from (2–3). This eigenvalue equation can be written

$$\mathsf{E}_{k_a}\hat{\varphi}_a = c_a \mathsf{P}_{k_a}\hat{\varphi}_a, \tag{6}$$

where $\mathsf{E}_{k_a}$ and $\mathsf{P}_{k_a}$, obtained by making $\partial_x \mapsto \mathrm{i}k_a$ in E and P, are Hermitian[3]. As associated boundary conditions, we take

$$\hat{\varphi}_a(0) = \hat{\varphi}_a(H) = 0,$$

i.e. that we consider a vertically bounded domain of height $H$. Of course, other boundary conditions can be chosen.

For each $k_a$, the complete set of solution to (6) includes (see Banks *et al.* 1978)

[3]This form comes from the fact that the conservation of $\mathcal{E}^{(2)}$ is associated with the symmetry of the system with respect to translation in $t$ (with generator $\partial_t = -\mathrm{i}\omega$), while that of $\mathcal{P}^{(2)}$ is associated with the symmetry in $x$ (with generator $\partial_x = \mathrm{i}k$).

1. a discrete set of regular waves, with $c_a \notin [U_{\mathrm{m}}, U_{\mathrm{M}}]$;

2. a continuous spectrum of singular modes, with $c_a \in [U_{\mathrm{m}}, U_{\mathrm{M}}]$.

Here $U_{\mathrm{m}}$ ($U_{\mathrm{M}}$) is the minimum (maximum) value of the basic velocity. The singular modes have a critical level where their phase velocity is equal to the basic flow velocity; there, the buoyancy and the vorticity are singular. Let us ignore this fact (for now), and consider an expansion in normal modes in the form

$$\varphi = \sum_a A_a(t)\hat{\varphi}_a(y) \exp\left[ik_a(x - c_a t)\right].$$

Introducing this expansion in the equations of motion yields

$$\sum_a \hat{\varphi}_a \left[\frac{\mathrm{d}A_a}{\mathrm{d}t} + \mathrm{i}\omega_a A_a\right] \exp(\mathrm{i}\vartheta_a) + \sum_b \sum_c N(\hat{\varphi}_b, \hat{\varphi}_c) A_b^\star A_c^\star \exp\left[-\mathrm{i}(\vartheta_b + \vartheta_c)\right] = 0, \qquad (7)$$

where $\vartheta_a = k_a x - \omega_a t$, and $N(\hat{\varphi}_b, \hat{\varphi}_c)$ is the vector of the nonlinear terms.

To obtain evolution equations for the amplitudes $A_a(t)$, it is necessary to project this expression onto each mode. This can be done using the projection associated to one of the following orthogonality relations (see Vanneste and Vial 1994):

$$\int_0^H [\hat{\varphi}_a(y)]^+ \mathsf{E}_{k_a}(y)\hat{\varphi}_b(y)\,\mathrm{d}y = E_a\,\delta_{a,b},$$

$$\int_0^H [\hat{\varphi}_a(y)]^+ \mathsf{P}_{k_a}(y)\hat{\varphi}_b(y)\,\mathrm{d}y = P_a\,\delta_{a,b}.$$

In these relations, which are consequences of the Hermicity of $\mathsf{E}_{k_a}$ and $\mathsf{P}_{k_a}$, we consider that $k_a = k_b$, since the orthogonality of modes with different wavenumbers is obvious. Clearly, the orthogonality relations are connected to the conservation of $E^{(2)}$ and $P^{(2)}$ for the linear system (see Held 1985). As a consequence, these quantities are diagonal in the normal mode basis:

$$E^{(2)} = \sum_a E_a \left|A_a(t)\right|^2, \quad P^{(2)} = \sum_a P_a \left|A_a(t)\right|^2, \qquad \text{with } E_a = c_a P_a.$$

Applying

$$\int_{-\infty}^{+\infty} \exp(-\mathrm{i}k_a x)\mathrm{d}x \int_0^H [\hat{\varphi}_a(y)]^+ \mathsf{E}_{k_a}(y)(\cdot)\,\mathrm{d}y$$

to (7) yields the amplitude equations in the form

$$\frac{\mathrm{d}A_a}{\mathrm{d}t} = \frac{\mathrm{i}}{2}\sum_{bc} I_a^{bc} A_b^\star A_c^\star \exp\left[\mathrm{i}(\omega_a + \omega_b + \omega_c)t\right], \quad k_a + k_b + k_c = 0. \qquad (8)$$

with the interaction coefficients

$$I_a^{bc} = E_a^{-1} \int_0^H [\hat{\varphi}_a(y)]^+ \mathsf{E}_{k_a}(y)\,\left[N(\hat{\varphi}_b, \hat{\varphi}_c) + N(\hat{\varphi}_c, \hat{\varphi}_b)\right]\,\mathrm{d}y. \qquad (9)$$

Under the assumption of weak nonlinearity, the system (8) can then be truncated to a single triad. A strong influence of the nonlinear interactions is expected when the triad is resonant, i.e.

$$\omega_a + \omega_b + \omega_c = 0.$$

# 3 Properties of the interaction equations

As in the case of waves propagating in a medium at rest (Ripa 1981), properties of the interaction coefficients $I_a^{bc}$ can be derived from the conservation laws. Obviously, the relevant conservation laws are those of pseudoenergy and pseudomomentum. The conservation of pseudoenergy can be written approximately as

$$\frac{\mathrm{d}}{\mathrm{d}t}\left(E^{(2)} + E^{(3)}\right) = O(A_a^4).$$

The first term, which is associated with the 'slow' amplitude modulations, is $O(A_a^3)$; the second term has a contribution of the same order, associated with the 'fast' oscillations at frequency $\omega_a$. As the two contributions must be equal for any wave amplitudes, it is easily seen that

$$E_a I_a^{bc} + E_b I_b^{ca} + E_c I_c^{ab} \sim (\omega_a + \omega_b + \omega_c) = 0 \tag{10}$$

for resonant triads (Vanneste and Vial 1994). A similar expression relates the interaction coefficients of a resonant triad to the pseudomomentum of the waves. A consequence of these relations is the conservation of $E^{(2)}$ and $P^{(2)}$ for isolated resonant triads. (When $U(y) = 0$, the energy and pseudomomentum are conserved in any isolated triad, as they are exactly quadratic (see Ripa 1981).) Note that (10) can also be derived by a direct manipulation of the explicit form (9) of the interaction coefficients.

Combining (10), its equivalent for the pseudomomentum, and the resonance conditions yields the two equivalent relations

$$\frac{E_a I_a^{bc}}{\omega_a} = \frac{E_b I_b^{ca}}{\omega_b} = \frac{E_c I_c^{ab}}{\omega_c} \quad \text{and} \quad \frac{P_a I_a^{bc}}{k_a} = \frac{P_b I_b^{ca}}{k_b} = \frac{P_c I_c^{ab}}{k_c},$$

or, in a reference frame moving at $\alpha$,

$$\frac{E_a^\alpha I_a^{bc}}{\omega_a^\alpha} = \frac{E_b^\alpha I_b^{ca}}{\omega_b^\alpha} = \frac{E_c^\alpha I_c^{ab}}{\omega_c^\alpha}, \tag{11}$$

where $E_a^\alpha = E_a - \alpha P_a$, and $\omega_a^\alpha = \omega_a - \alpha k_a$ is such that

$$\omega_a^\alpha + \omega_b^\alpha + \omega_c^\alpha = 0. \tag{12}$$

From (11) and (12), the properties of the interactions inside a resonant triad can be discussed in terms of the pseudoenergy and frequency of the waves in a given reference frame. Two types of behaviour are possible:

1. If one of the interaction coefficients is oppositely signed to the other two, the three amplitudes oscillate periodically. This occurs in two situations: (i) if the three $E_a^\alpha$ have the same sign, or (ii) if the wave with a $E_a^\alpha$ oppositely signed to the other two does not have the largest frequency $|\omega_a^\alpha|$.

2. If the three interaction coefficients have the same sign, the three amplitudes grow simultaneously (and have a finite-time singularity); this phenomenon is called *explosive resonant interaction* (ERI) and corresponds to a nonlinear instability of the basic flow. It occurs when the highest-frequency member of the triad has a $E_a^\alpha$ oppositely signed to the other two.

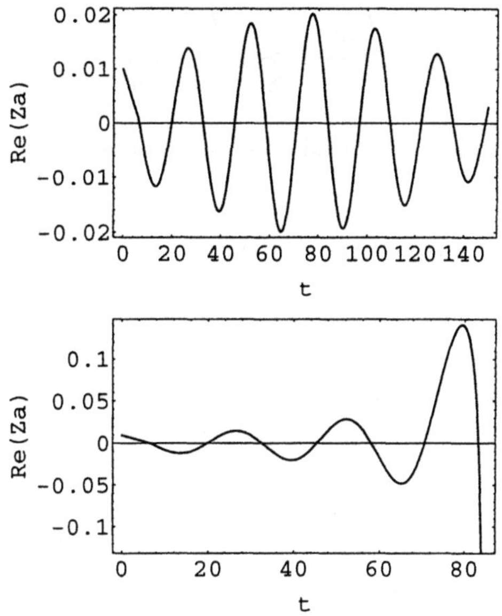


Figure 1: The typical behaviour of a wave amplitude $Z_a = A_a \exp(-\mathrm{i}\omega_a t)$ in a non-explosive triad (top), and in an explosive triad (bottom).

These two types of behaviour in a resonant triad are illustrated in figure 1. Of course, the behaviour of the triad does not depend on the reference frame. To ensure that explosive interaction does not occur, it therefore suffices to find an $\alpha$ such that the three $E_a^\alpha$ have the same sign.

The ERI has been studied in layered models of irrotational (non-stratified) flows (Craik and Adams 1979; Tsutahara 1984), and models of baroclinic instability (Romanova 1987; Meacham 1988; Vanneste 1995*b*). Its importance stems from the fact that it can destabilize flows which are spectrally stable.

For stratified shear flows, Becker and Grimshaw (1993) showed using a Lagrangian approach that ERI requires the presence of modes with critical levels. To see this, let us write the pseudoenergy as

$$E_a^\alpha = (c_a - \alpha) \int_0^H (c_a - U)(\xi_y^2 + k^2 \xi^2)\,\mathrm{d}y,$$

where $\xi = N^{-2}\eta$. For (regular) waves without a critical level (i.e. $c_a \notin [U_\mathrm{m}, U_\mathrm{M}]$) it is clear that the pseudoenergy is always positive when $\alpha \in [U_\mathrm{m}, U_\mathrm{M}]$ (see figure 2). Hence, ERI cannot occur in triads composed of regular waves only. But, for singular modes (with critical level), the previous development are not strictly applicable, as they would involve singular integrals. The possible extension of these developments to take into account the singular modes is discussed in section 5. In the next section, we consider regular modes only, and study the influence of the shear on the instability of waves through interactions with other regular waves.

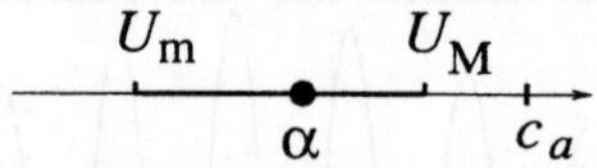


Figure 2: The choice of $\alpha$ showing that explosive resonant interaction requires the presence of critical levels.

# 4 Instability of regular modes

For regular waves, the presence of the shear affects the explicit form of the interaction coefficients, but not the form of the amplitude equations. The standard approach, briefly recalled in the introduction, can thus be applied to study the stability of an infinitesimal wave (see e.g. Craik 1985). A wave $a$ is unstable at any amplitude if it constitutes a resonant triad with two waves $b$ and $c$ such that $I_b^{ca} I_c^{ab} > 0$ (see (1)). To formulate the extension of Hasselmann's criterion to shear flows, it is convenient to choose, as above, $\alpha \in [U_\mathrm{m}, U_\mathrm{M}]$. In such a reference frame, the pseudoenergy of the three regular waves $a$, $b$, and $c$ is positive. From (11) we see that $a$ is unstable if $\omega_b^\alpha \omega_c^\alpha > 0$. Taking (12) into account, we find the equivalent condition

$$|\omega_a^\alpha| > |\omega_b^\alpha|, |\omega_c^\alpha|.$$

An infinitesimal wave is thus unstable if it is involved in a resonant triad in which it has the largest frequency as seen by an observer moving with any point of the flow. This is the extension to shear flows of Hasselmann's criterion.

When the disturbance consisting of $b$ and $c$ is not resonant with $a$, the instability occurs if the amplitude of $a$ exceeds the threshold

$$A_a^{\mathrm{cr}} = \frac{|\phi|}{2\sqrt{I_b^{ca} I_c^{ab}}},$$

where $\phi = \omega_a + \omega_b + \omega_c$, provided that $I_b^{ca} I_c^{ab} > 0$. The growth rate of the instability is given by

$$\sigma_a = \sqrt{I_b^{ca} I_c^{ab} |A_a|^2 - \phi^2/4}.$$

To illustrate the influence of the basic shear on the instability of a gravity wave, we now present explicit calculations of the interaction coefficients and instability characteristics for two examples. The chosen basic shear is

$$U = U_\mathrm{M} \sin\left(\frac{\pi y}{2H}\right), \tag{13}$$

and an uniform (dimensionless) stratification $N = 1$ is considered. In the range of $U_\mathrm{M}$ analysed here, this flow is linearly stable.

When there is no basic flow ($U_\mathrm{M} = 0$), the vertical structure of the waves is simply $\sin(n_a \pi y/H)$. With a shear, $n_a$ can be defined as the number of zeroes of the vertical structure $\hat{\varphi}_a$ in $]0, H[$ minus one; it can be used as a meridional index to fix the vertical mode for a given horizontal wavenumber. As two modes with opposite directions of

| wave | $kH$ | $n$ | $kH$ | $n$ |
|---|---|---|---|---|
| $a$ | −1.3 | 5 | −1.8 | 5 |
| $b$ | −0.5 | −6 | 0.53 | −6 |
| $c$ | 1.8 | 11 | 1.27 | 9 |

Table 1: Horizontal wavenumbers and vertical indices defining the two triads studied with the shear (13).

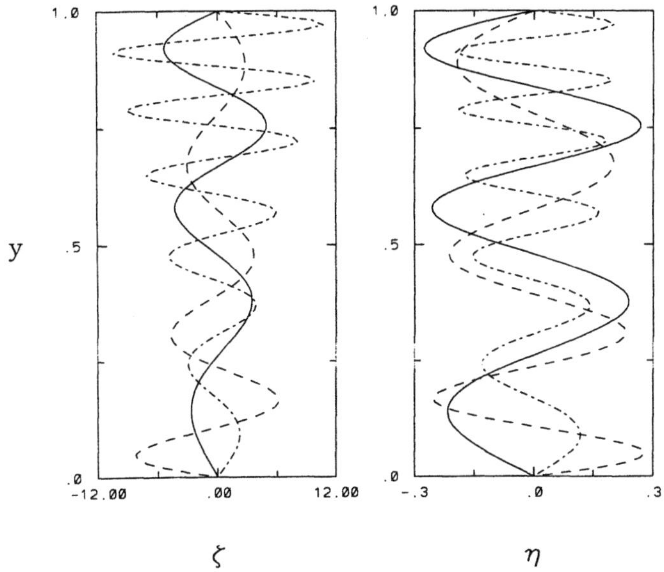


Figure 3: Vertical structure of the vorticity $\zeta$ and buoyancy $\eta$ for the three waves $a$ (solid line), $b$ (dashed line), and $c$ (dash-dotted line) defined in the first column of table 1, in the basic shear (13) with $U_{\mathrm{M}} = 0.04$.

propagation exist for each value of $n_a$, we attribute to $n_a$ a sign corresponding to the direction of propagation.

The characteristics of the first triad considered are given in the first column of table 1. Clearly, this triad satisfies the interaction condition $k_a + k_b + k_c = 0$; it also satisfies

$$n_a \pm n_b \pm n_c = 0, \tag{14}$$

which is a necessary condition for nonlinear interaction to occur when $U_{\mathrm{M}} = 0$. The vertical structure of the three waves is displayed in figure 3 for $U_{\mathrm{M}} = 0.04$. The influence of the shear is apparent: the amplitude (wavelength) of waves $a$ and $c$, which propagate in the direction of the shear, increases (decreases) with the shear amplitude, whereas the reverse is true for wave $b$, which propagates against the shear. Figure 4 shows the frequencies $\omega_a$, $\omega_b$ and $\omega_c$, and (half) their sum as functions of the shear amplitude $U_{\mathrm{M}}$.

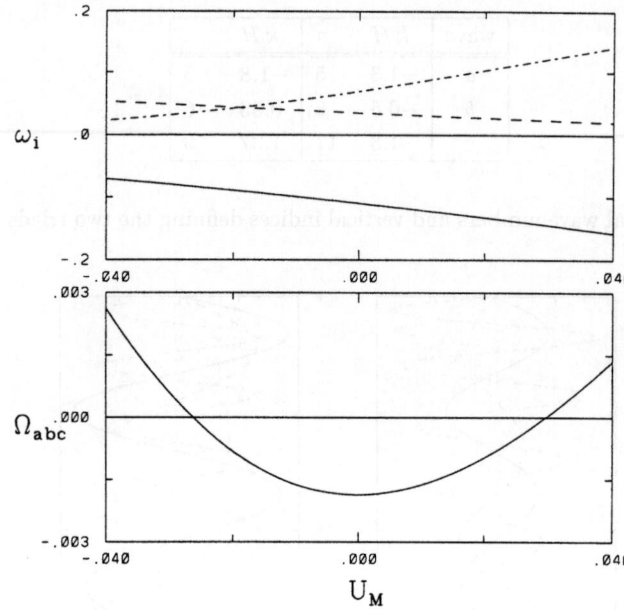


Figure 4: Frequencies and sum of the frequencies ($\Omega_{abc} = \dot{\phi}/2$) of the three waves displayed in figure 3, as functions of the maximum shear.

It can be seen that the triad is resonant for two distinct values of $U_{\rm M}$. Figure 5 displays the three interaction coefficients $I_a^{bc}$, $I_b^{ca}$ and $I_c^{ab}$; it shows that they strongly depend on the shear amplitude. As expected from the theory, the wave with the largest frequency $a$ has an associated interaction coefficient $I_a^{bc}$ which is oppositely signed to the other two at resonance . This wave is thus the 'unstable' member of the triad. Interestingly, $I_a^{bc}$ remains oppositely signed to $I_b^{ca}$ and $I_c^{ab}$ in the entire range of $U_{\rm M}$, although the coefficients vanish several times. This can be explained by the fact that the three waves remain close to resonance – even though their structures change considerably – because the main effect of the shear on the frequencies is a simple Doppler shift (by the average velocity) which does not affect the resonance condition.

Figure 6 shows the critical amplitude for the instability of wave $a$ to a disturbance consisting of waves $b$ and $c$. Since the presence of the shear can render the three waves resonant, it can lead to a vanishing of $A_a^{\rm cr}$, but it can also render wave $a$ unconditionally stable (to this disturbance). The growth rate of the instability is also displayed, assuming an dimensionless amplitude $|A_a| = 0.1$. This figure confirms the strong influence of the shear flow on wave instability.

While condition (14) drastically restricts the possibility of interactions for waves propagating in a resting medium, it has no direct equivalent when a basic shear is present. The possibilities of interactions – and in particular resonant interactions – are thus much more numerous with a shear. To see how the shear can lead to the interaction of waves otherwise isolated, we consider a second triad, whose characteristics are summarized in

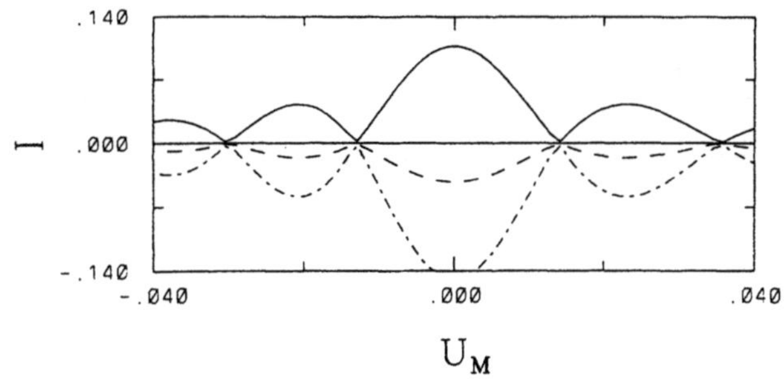


Figure 5: Interaction coefficients for the three waves displayed in figure 3 as functions of the maximum shear (solid line: $I_a^{bc}$; dashed line: $I_b^{ca}$; dash-dotted line: $I_c^{ab}$).

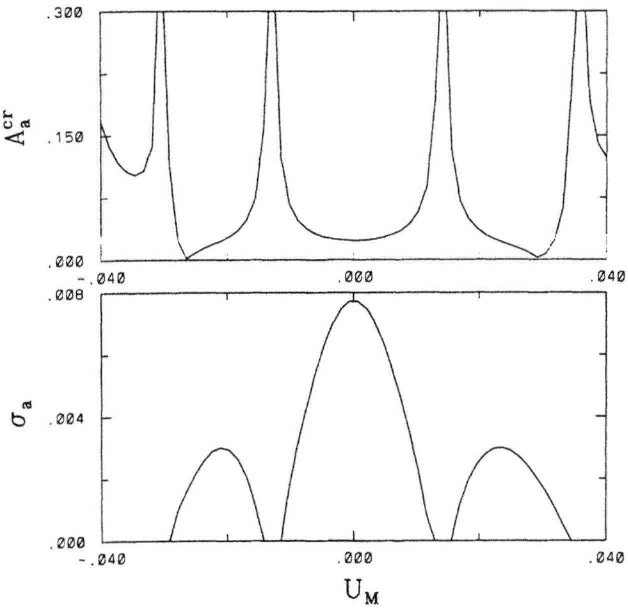


Figure 6: Critical amplitude for wave $a$ and growth rate for waves $b$ and $c$ (with $|A_a| = 0.1$) for the three waves defined in the first column of table 1, as functions of the maximum shear.

the second column of table 1, again with the shear (13). Figure 7 presents the three interaction coefficients. Since (14) is not satisfied, the interaction coefficients vanish when $U_M = 0$, but they attain significant values (comparable to those obtained for the previous triad, which interacts when $U_M = 0$) for sufficiently large values of $U_M$. The triad is resonant for two distinct values of $U_M$; this can be deduced from figure 8, which displays $A_a^{cr}$, and the growth rate of the disturbance for $|A_a| = 0.1$. It can be seen that the instability of wave $a$, which is impossible without shear, is rendered possible by the shear, and that the growth rate attains significant values. Again, this illustrates the strong importance of the shear. (Similar results hold for different types of waves (Vanneste and Vial 1994).)

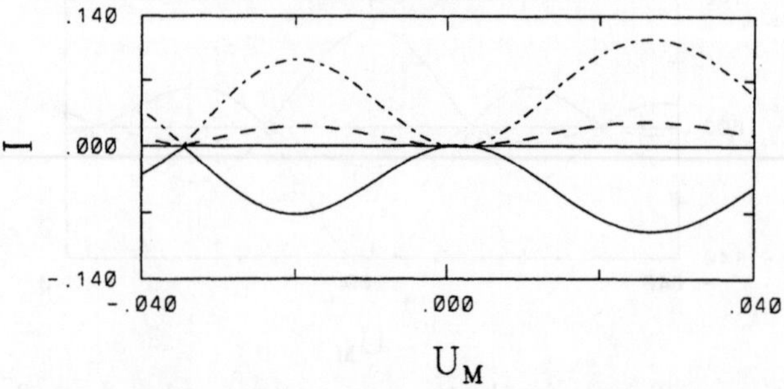


Figure 7: Interaction coefficients for the three waves defined in the second column of table 1, as functions of the maximum shear (solid line: $I_a^{bc}$; dashed line: $I_b^{ca}$; dash-dotted line: $I_c^{ab}$).

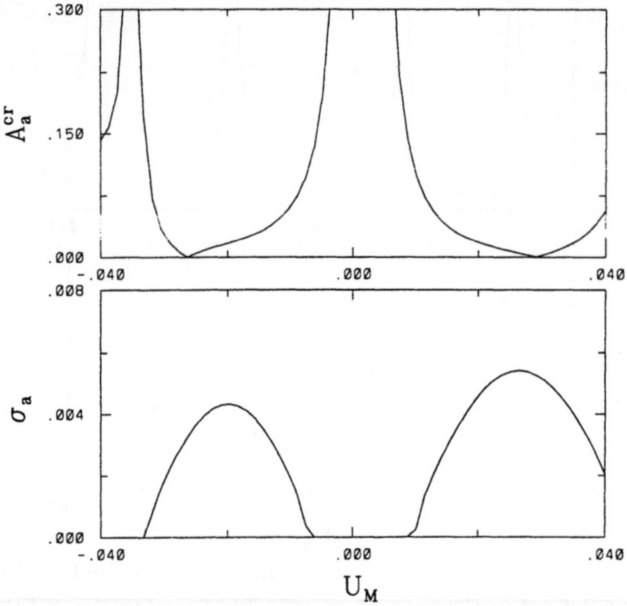


Figure 8: Critical amplitude for wave $a$ and growth rate for waves $b$ and $c$ (with $|A_a| = 0.1$) for the three waves defined in the second column of table 1, as functions of the maximum shear.

# 5 Interactions including singular modes

As indicated at the end of section 3, the approach presented for wave interactions in shear flows breaks down for triads including singular modes. Indeed, if a mode with phase velocity $c_a$ possesses a critical level at $y_{c_a}$ (i.e. $U(y_{c_a}) = c_a$), a Frobenius analysis in the vicinity of $y_{c_a}$ gives

$$\hat{\psi}_a \sim (y - y_{c_a})^{1/2}, \quad \hat{\eta}_a \sim (y - y_{c_a})^{-1/2}, \quad \hat{\zeta}_a \sim (y - y_{c_a})^{-3/2},$$

for the streamfunction, buoyancy, and vorticity, respectively. (We assume the basic flow to be linearly stable, with a Richardson number greater than 1/4.) Clearly, the integrals for $E_a, P_a, I_a^{bc}$ are not defined in the usual sense, and the approach must be reconsidered.

The role of the singular modes in wave interactions is an important issue: as described above, they may be responsible for a nonlinear destabilization through ERI of flows which are linearly stable according to the Miles–Howard criterion. Furthermore, even in cases of stable interactions (i.e. between waves with pseudoenergies of the same sign in some reference frame), the singular waves can be expected to induce interesting effects, since they strongly interact with the basic shear. So far, the issue of incorporating the singular modes in the weakly nonlinear theory remains open. In what follows, we suggest a way of attacking this problem, based on linear studies (Case 1960; Kamp 1991; Balmforth and Morrison 1996) and on recent results for non-stratified fluids (Vanneste 1996). The approach we propose relies on two remarks:

1. Critical layers due to the presence of an isolated singular mode appear in forced systems, or in instability problems, because of the existence of a dominant frequency. In free systems, however, the singular modes are not isolated; rather, a continuous spectrum of modes with $U_{\rm m} < c < U_{\rm M}$ is excited, and the superposition of these modes represents a smooth transient disturbance, so that no extra physical effects (such as viscosity) need to be invoked at a particular location. Critical levels then appear as a mathematical difficulty, induced by the modal decomposition.

2. If a given singular mode is involved in a triad with two other modes, the singular modes in its vicinity (i.e. those with the same horizontal wavenumber but slightly different phase velocity and thus slightly different critical level location) will also interact with the two modes. In other words, there is no physical justification for truncating the continuum of singular modes corresponding to a given horizontal wavenumber.

The singular modes must thus be considered in complete packets, with $U_{\rm m} < c < U_{\rm M}$; this leads to a consistent description of the interactions, and to smooth physical fields without resorting to some artificial regularization process. Consequently, the interacting 'triads' consists of regular waves or complete packets of singular modes, and the non-resonant interactions involving singular modes must be taken into account. This is schematically represented in figure 9, which displays the simple situation of the interaction between two regular modes $b$ and $c$, and a packet of singular modes $a$. One of this mode, $a^r$ is assumed to be resonant, i.e.

$$\omega_a^r + \omega_b + \omega_c = 0,$$

where $\omega_a^r = k_a c_a^r = k_a U(y_a^r)$.

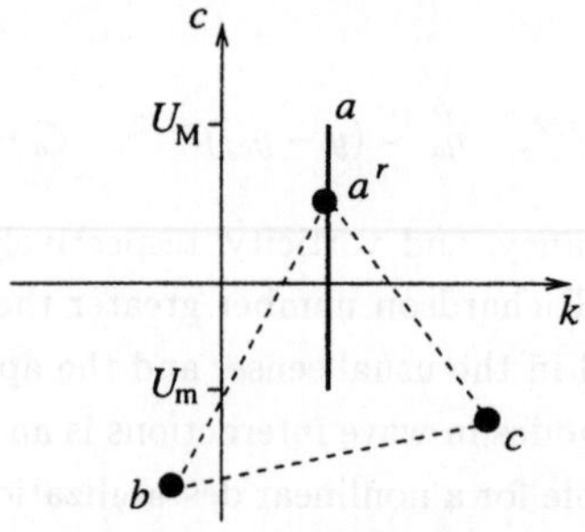


Figure 9: A schematic view of the interaction between a packet of singular modes $a$ and two regular modes $b$ and $c$ in the $(k, c)$-plane. The regular modes constitute a resonant triad (represented by the dashed triangle) with a particular singular mode $a^r$.

Albeit conceptually simple, the approach proposed encounters a number of technical difficulties. First, the complete structure of the singular modes involves Dirac contributions, namely

$$\hat{\zeta}_a \sim (y - y_{c_a})^{-3/2} + \lambda\, \delta(y - y_{c_a}) + \mu\, \delta'(y - y_{c_a}), \tag{15}$$

where $\lambda$ and $\mu$ are constants, and it must be derived from a singular differential equation. Second, if we write the vorticity induced by a packet of singular modes as

$$\zeta = \int A_a(t)\, \hat{\zeta}_a(y)\, \exp[ik_a(x - c_a)t]\, \mathrm{d}c_a,$$

we must be able to calculate this expression in a consistent manner. From Case's comparison between the normal mode approach and the Laplace transform approach (Case 1960), it is clear that the singularities must be integrated by parts; this corresponds to Hadamard's interpretation of singular integration. Finally, we must compute $E_a$ and $P_a$, and derive the interaction equations, by (carefully) manipulating the distributions and finite-part integrals. Results obtained without stratification give us confidence in the possibility of applying this approach to obtain explicit results for gravity waves.

# 6 Summary and discussion

The theory of weakly nonlinear interactions between gravity waves is extended here to incorporate the influence of a strong background shear. Specifically, we describe an approach for deriving the wave interaction equations that is based on the orthogonality relations associated with the wave-activity conservation laws of the system. This allows us to relate the interaction coefficients to the wave activities (pseudoenergy and pseudomomentum), and to discuss the qualitative properties of interacting triads in terms of the properties of the linear waves. The condition found by Becker and Grimshaw (1993) for explosive interaction in a triad – namely, the presence of modes with critical levels – is then recovered, and an extension of Hasselmann's criterion is derived, which determines the 'unstable' member of a non-explosive triad.

The influence of the basic shear on wave instability is illustrated by two examples. Although these examples suggest a strong influence of the shear, they are only indicative: instead of focusing on a unique disturbance (fixed by the wavenumbers and zonal indices), a complete study should consider the entire set of possible (resonant) disturbances in order to find the one with the largest growth rate. A systematic effect of the shear on a given wave (for instance a destabilisation) could only be detected by such a study.

A major consequence of the shear is the existence of a continuum of singular modes possessing a critical level whose role in wave interactions is not fully understood. Here, we propose an approach to study this role, which naturally extends that applied to regular waves. The key-point of this approach is that the singular modes must be considered in packets, with their phase velocity in the entire range covered by the basic flow velocity. The study of weakly nonlinear interactions involving singular modes should bring to light interesting new phenomena, the first of which is the destabilisation of linearly stable flows through explosive resonant interactions. We want to point out that this mechanism likely leads to distributions of singular modes which peak for particular modes (those corresponding to exact resonance), and hence to small-scale vertical structures. This is reminiscent of the conclusion that nonlinear stability cannot be guaranteed in stratified shear flows because of the (possible) existence of small-scale density fluctuations (see Shepherd 1992). As mentioned above, the strong interaction between the singular modes and the basic flow renders the nonlinear interactions including those modes very interesting even in non-explosive situations. An important point is that the linear evolution of a packet of singular modes ultimately leads to the decrease of its energy. Because of the nonlinear interactions, this (non-dissipative Landau) damping may affect the regular waves. The presence of shear, through the existence of singular modes, could thus strongly modify the behaviour of regular waves. These different phenomena are the subject of current work.

# References


Balmforth, N.J. and Morrison, P.J. 1996 Singular eigenfunctions for shearing fluids I. *J. Fluid Mech.*, submitted.

Banks, W.H.H., Drazin, P.G. and Zaturska, M.B. 1978 On the normal modes of parrallel flow of inviscid stratified fluid. *J. Fluid Mech.*, **75**, 149–171.

Becker, J.M. and Grimshaw, R.H.J. 1993 Explosive resonant triads in a continuously stratified shear flows. *J. Fluid Mech.*, **257**, 219–228.

Case, K.M. 1960 Stability of an idealized atmosphere. I. Discussion of the results. *Phys. Fluids*, **3**, 149–154.

Craik, A.D.D. 1985 *Wave interactions and fluid flows.* Cambridge University Press.

Craik, A.D.D. and Adam, J.A. 1979 Explosive resonant interaction in a three-layer fluid flow. *J. Fluid Mech.*, **92**, 15–33.

Dong, B. and Yeh, K.C. 1991 On nonresonant interactions of atmospheric waves in a rotating earth. *Physica Scripta*, **43**, 534–544.

Dunkerton, T.J. 1987 Effect of nonlinear instability on gravity wave momentum transport. *J. Atmos. Sci.*, **44**, 3188–3209.

Fritts, D.C., Sun, S. and Wang, D.Y. 1992 Wave-wave interactions in a compressible atmosphere 1. A general formulation including rotation and wind shear. *J. Geophys. Res.*, **97**, 9975–9988.

Grimshaw, R.H.J. 1988 Resonant wave interaction in a stratified shear flow. *J. Fluid Mech.*, **190**, 357–374.

Grimshaw, R.H.J. 1994 Resonant wave interaction near a critical level in a stratified shear flow. *J. Fluid Mech.*, **269**, 1–22.

Hasselmann, K. 1967 A criterion for nonlinear wave stability. *J. Fluid Mech.*, **30**, 737–739.

Held, I.M. 1985 Pseudomomentum and the orthogonality of modes in shear flows. *J. Atmos. Sci.*, **42**, 2280–2288.

Kamp, L.P.J. 1991 Integral-equation approach to the instability of two-dimensional sheared flow of inviscid fluid in a rotating system with variable Coriolis parameter. *J. Phys. A: Math. Gen.*, **24**, 2029–2052.

Klostermeyer, J. 1991 Two- and three-dimensional parametric instabilities in finite-amplitude internal gravity waves. *Geophys. Astrophys. Fluid Dynam.*, **61**, 1–25.

McComas, C.H. and Bretherton, F.P. 1977 Nonlinear interactions of internal gravity waves. *J. Geophys. Res.* **82**, 1397–1412.

Meacham, S. 1988 Non-modal baroclinic instability. *Dyn. Atmos. Oceans*, **12**, 19–45.

Ripa, P. 1981 On the theory of nonlinear wave-wave interactions among geophysical waves. *J. Fluid Mech.*, **103**, 87–115.

Ripa, P. 1990 Positive, negative and zero wave energy and the flow stability problem in the Eulerian and Lagrangian-Eulerian descriptions. *Pure Appl. Geophys.*, **133**, 713–732.

Romanova, N.N. 1987 Explosive instability in a three-layered rotating liquid. *Izv. Atmos. Ocean. Phys.*, **28**, 343–350.

Scinocca, J.F. and Shepherd, T.G. 1992 Nonlinear wave-activity conservation laws and Hamiltonian structure for two-dimensional anelastic equations. *J. Atmos. Sci.*, **49**, 5-25.

Shepherd, T.G. 1990 Symmetries, conservation laws, and Hamiltonian structure in geophysical fluid dynamics. *Adv. Geophys.*, **32**, 287–338.

Shepherd, T.G. 1992 Arnol'd stability applied to fluid flow: successes and failures. In *Nonlinear phenomena in atmospheric and oceanic sciences* (eds. G.F. Carnevale & R.T. Pierrehumbert), Springer-Verlag, 187–206.

Tsutahara, M. 1984 Resonant interaction of internal waves in a stratified shear flow. *Phys. Fluids*, **29**, 2812–2818.

Vanneste, J. 1995*a* The instability of internal gravity waves to localised disturbances. *Annales Geophysicae*, **13**, 196–210.

Vanneste, J. 1995*b* Explosive resonant interaction of baroclinic Rossby waves and stability of multilayer quasi-geostrophic flow. *J. Fluid Mech.*, **291**, 83–107.

Vanneste, J. 1996 Rossby wave interactions in a shear flow with critical levels. *J. Fluid Mech.*, in press.

Vanneste, J. and Vial, F. 1994 On the nonlinear interactions of geophysical waves in shear flows. *Geophys. Astrophys. Fluid Dyn.*, **78**, 121–144.

Yek, K.C. and Liu, C.H. 1981 The instability of atmospheric gravity waves through wave–wave interactions. *J. Geophys. Res*, **C10**, 9722–9728.

# Gravity wave spectral models and the shapes of gravity wave spectra at low vertical wavenumbers

Christopher D Warner and Michael E McIntyre
Centre for Atmospheric Science
University of Cambridge
Department of Applied Mathematics and Theoretical Physics
Silver Street
Cambridge CB3 9EW
United Kingdom

**ABSTRACT.** Examples of the evolution of gravity wave spectra launched upward from the lower stratosphere are computed at unprecedented spectral resolution, as a function of altitude $z$, in a one-dimensional model. Empirical model energy spectra of the kind proposed by Fritts and VanZandt are used as launch spectra. The dependence of these spectra on vertical wavenumber $m$ have a power-law behavior $m^s$ at small $m$, to which the total wave-energy profile $E_{\mathrm{tot}}(z)$ is sensitive. The effect on $E_{\mathrm{tot}}(z)$ of varying $s$ is investigated for several cases, and the resulting $E_{\mathrm{tot}}(z)$ profiles are compared with an empirical profile $E_{\mathrm{emp}}(z)$, derived by Fritts and Lu from observational studies. To the extent that $E_{\mathrm{emp}}(z)$ is realistic and the power law $m^s$ is realistic, $s$ values around 1.5 are arguably a better choice than the value $s = 1$ that is often assumed *faute de mieux*.

## 1. Introduction

A recently implemented theoretical gravity wave spectral-evolution model (Warner and McIntyre 1996, hereafter WM96), tries to separate clearly the more theoretically justifiable from the more empirical elements of our knowledge of gravity wave propagation and breaking. A properly resolved numerical implementation of this model of spectral evolution is far too expensive to be used as a general circulation model (GCM) parameterization scheme, essentially because spectral elements undergo shearing distortions as in fluid flow. However, we believe this model yields useful insight into the evolution of gravity wave spectra with fewer neglected effects than are necessary for a parameterization scheme.

The gravity wave-energy spectrum at large vertical wavenumber $m$ has a shape $\propto m^{-t}$ where the exponent $t$ is roughly independent of time, place, and altitude and close to 3. This large-$m$ spectral shape seems to be observationally robust provided the shape is averaged over a sufficient number of measurements. But the low-$m$ behavior is not constrained observationally, at least not directly, because the corresponding vertical wavelengths are too big to fit into practical altitude ranges for spectral analysis. There-

NATO ASI Series, Vol. I 50
Gravity Wave Processes
Their Parameterization in Global Climate Models
Edited by Kevin Hamilton

fore, traditionally, model gravity wave-energy spectra (e.g., Fritts and VanZandt 1993) often assume that the small-$m$ shape is a simple power law $m^s$. The most popular choice for the exponent has been $s = 1$. This paper uses the new spectral evolution model to test sensitivity to the choice of $s$, with a view to constraining it better.

## 2. Theoretical gravity wave spectral-evolution model

As explained in more detail in WM96, the spectral evolution model considers a spectrum of gravity waves as a function of $(k_0, \omega_0, \phi)$, where $k_0$ is the magnitude of the horizontal wavenumber $\mathbf{k_0}$, $\omega_0$ is absolute frequency, and $\phi$ is azimuthal direction so that $\mathbf{k_0} = k_0(\cos\phi, \sin\phi)$. We consider modifications to the spectrum as it propagates through changing conditions of background wind $\mathbf{U}(z)$, density $\rho_0(z)$, and buoyancy frequency $N(z)$. The propagation model is linear, locally Boussinesq, and includes both Coriolis and nonhydrostatic effects.

Each spectral element $dk_0\, d\omega_0\, d\phi$ in $(k_0, \omega_0, \phi)$ spectral space is treated as a "generalized wavepacket" that evolves independently of all the other wavepackets. Thus each spectral element propagates conservatively upwards (with $k_0, \omega_0, \phi$, and $dk_0, d\omega_0, d\phi$ invariant), can reach saturation amplitude, can break, and can be back-reflected as $\hat{\omega}$ approaches the buoyancy frequency $N$. The spectral evolution model is quasi-steady and quasi-one-dimensional (like radiation schemes) in that everything is assumed to occur in one (large) GCM-sized grid box on a time scale small compared with the time scale for variation of the background atmosphere.

The dispersion relation is

$$\frac{m^2}{k_0^2} = \frac{N^2 - \hat{\omega}^2}{\hat{\omega}^2 - f^2} \tag{1}$$

where $m$ is the vertical wavenumber and $\hat{\omega}$ is the intrinsic frequency, which is related to $\omega_0$ by

$$\hat{\omega} = \omega_0 - \mathbf{k_0} \cdot \mathbf{U} = \omega_0 - k_0 U \tag{2}$$

where $U$ is the component of the local background wind in the direction of $k_0$. Model spectra are usually expressed as a function of $(m, \hat{\omega}, \phi)$, and relations (1) and (2) can be used to transform between the $(m, \hat{\omega}, \phi)$ spectral space and the $(k_0, \omega_0, \phi)$ spectral space. The Jacobian $J$ of this transformation turns out to be given simply by

$$dm\, d\hat{\omega}\, d\phi = J\, dk_0\, d\omega_0\, d\phi = \frac{m}{k_0}\, dk_0\, d\omega_0\, d\phi\ . \tag{3}$$

Note that $dk_0\, d\omega_0\, d\phi$, is propagation-invariant, but that $dm\, d\hat{\omega}\, d\phi$ is not.

When conservative propagation theory is used to propagate each spectral element through an altitude increment, the pseudomomentum flux can be conveniently expressed as

$$\rho_0 \hat{\mathbf{F}}_{\mathbf{p}}(m,\hat{\omega},\phi)\, dm\, d\hat{\omega}\, d\phi = \rho_0 c_{gz} \hat{E}(m,\hat{\omega},\phi) \frac{\mathbf{k_0}}{\hat{\omega}}\, dm\, d\hat{\omega}\, d\phi \ , \tag{4}$$

where $c_{gz} = -\partial\hat{\omega}/\partial m$ is the vertical component of group velocity and $\hat{E}$ is the spectral density of wave-energy per unit mass. It can be shown that $\hat{\mathbf{F}}_{\mathbf{p}}(m,\hat{\omega},\phi) \propto J^{-1}$; this is because of (3) and the known theoretical result that the pseudomomentum flux density is invariant under conservative propagation when re-expressed as a function of $(k_0,\omega_0,\phi)$.

The empirical wave-breaking criterion used here is to impose as an upper bound an azimuthally isotropic saturation wave-energy $\hat{E}_S(m,\hat{\omega})$ per unit mass, corresponding to a saturation pseudomomentum flux

$$\rho_0 \hat{\mathbf{F}}_{\mathbf{p}S}(m,\hat{\omega},\phi)\, dm\, d\hat{\omega}\, d\phi = \rho_0 c_{gz} \hat{E}_S(m,\hat{\omega})\ (\mathbf{k_0}/\hat{\omega})\, dm\, d\hat{\omega}\, d\phi \ . \tag{5}$$

In other words, a 'chopping function' is used to chop down the pseudomomentum flux, as necessary, to the saturation threshold. This is arguably the simplest possible broad-band wave-breaking criterion. We avoid discussing its theoretical justification, which is controversial. (See, however, the discussion in WM96.) The precise form of $\hat{E}_S(m,\hat{\omega})$ used in the present numerical spectral-evolution model calculations is given in §3, (10), and is proportional to $m^{-3}$.

## 3. Numerical implementation of gravity wave spectral-evolution model

A model calculation is started with a "launch" spectrum that, for convenience (other launch spectra can be used), is here taken to be that of Fritts and VanZandt (1993),

$$\begin{aligned}\hat{E}(m,\hat{\omega},\phi) &= E_0\ A(m)\ B(\hat{\omega})\ \Phi(\phi)\\ &= \frac{\beta N^2}{m_*^2}\ A_0(s,t)\frac{m^s/m{*}^{s+1}}{1+(m/m_*)^{s+t}}\\ &\times B_0(p)\hat{\omega}^{-p}\ \ \Phi_0\ ,\end{aligned} \tag{6}$$

where $E_0$ is the total wave-energy per unit mass; $\beta$ is an empirical coefficient of order $10^{-1}$ (in fact 0.10472 throughout these calculations), $m$ is the vertical wavenumber, $\hat{\omega}$ the intrinsic frequency, and $\phi$ the azimuthal direction of propagation; $A_0$, $B_0$, and $\Phi_0$ are constants chosen so that $A(m)$, $B(\hat{\omega})$, and $\Phi(\phi)$ are each normalized to integrate to unity (see WM96 for full detail); $m_*$ is the "characteristic vertical wavenumber," close to but not equal to the value of $m$ at the peak of $A(m)$. The launch spectra are taken to be of this (azimuthally isotropic) form with $\Phi(\phi) = (2\pi)^{-1} =$ const. The corresponding pseudomomentum flux spectrum is

$$\begin{aligned}\rho_0 \hat{\mathbf{F}}_{\mathbf{p}}(m,\hat{\omega},\phi) &= \rho_0 c_{gz} \hat{E}(m,\hat{\omega},\phi)\frac{\mathbf{k_0}}{\hat{\omega}}\\ &= \rho_0 c_{gz}\frac{\beta N^2}{m_*^2}\ A_0(s,t)\frac{m^s/m{*}^{s+1}}{1+(m/m_*)^{s+t}}\times\\ &\times B_0(p)\hat{\omega}^{-p}\ \ \Phi_0\ \frac{\mathbf{k_0}}{\hat{\omega}}\ .\end{aligned} \tag{7}$$

The launch altitude is typically $\approx 19\,\text{km}$; thus the launch spectrum is conceived of as representing waves escaping upwards from the lower stratosphere, whatever their original source. The exponents $t$ and $p$ will be taken in this paper to be 3 and 5/3 respectively, again following Fritts and VanZandt (1993).

Numerical calculations are done both in $(k_0, \omega_0, \phi)$ space and in $(m, \hat{\omega}, \phi)$ space. Grids of spectral elements $\Delta k_0\, \Delta\omega_0\, \Delta\phi$ and $\Delta m\, \Delta\hat{\omega}\, \Delta\phi$ are defined for typically 4 or 8 azimuthal directions $\phi_j$. In $(m, \hat{\omega}, \phi)$ space, each spectral element in the $j^{\text{th}}$ azimuthal sector $\phi_j \pm \frac{1}{2}\Delta\phi$, centered on $\phi_j$, contributes pseudomomentum flux

$$\rho_0 \hat{\mathbf{F}}_j(m, \hat{\omega}, \phi_j)\, \Delta m\, \Delta\hat{\omega}\, 2\sin(\Delta\phi/2)\,. \tag{8}$$

Equation (8) is obtained by replacing $dm$ with $\Delta m$ and $d\hat{\omega}$ with $\Delta\hat{\omega}$ in (4), and by integrating with respect to $\phi$ for the range of $\phi$ corresponding to the $j^{\text{th}}$ azimuthal sector, that is from $\phi_j - \frac{1}{2}\Delta\phi$ to $\phi_j + \frac{1}{2}\Delta\phi$. The pseudomomentum flux is assumed to be piecewise constant in magnitude over the sector but is summed vectorally, giving rise to the factor $2\sin(\Delta\phi/2)$. The corresponding chopping function $\hat{\mathbf{F}}_{\mathbf{p}\mathrm{S}j}$ is such that $\hat{\mathbf{F}}_j \leq \hat{\mathbf{F}}_{\mathbf{p}\mathrm{S}j}$, with (5) replaced by

$$\rho_0 \hat{\mathbf{F}}_{\mathbf{p}\mathrm{S}j}(m, \hat{\omega}, \phi_j)\, \Delta m\, \Delta\hat{\omega}\, 2\sin(\Delta\phi/2) = \rho_0 c_{\mathbf{gz}} \hat{E}_{\mathrm{S}}(m, \hat{\omega})\, (\mathbf{k_0}/\hat{\omega})\, \Delta m\, \Delta\hat{\omega}\, 2\sin(\Delta\phi/2)\,. \tag{9}$$

The saturation threshold wave-energy is that given by VanZandt and Fritts (1989),

$$\hat{E}_{\mathrm{S}}(m, \hat{\omega}) = \beta N^2\, A_0(s,t)\, B_0(p)\, \Phi_0\, \hat{\omega}^{-p}\, m^{-3}\,, \tag{10}$$

with $p = 5/3$ or $p = 1$, the latter being more consistent with the usual empirical notion of broadband saturation, as described in detail in WM96. The pseudomomentum flux chopping function corresponding to this choice for $\hat{E}_{\mathrm{S}}$ is

$$\begin{aligned} \rho_0 \hat{\mathbf{F}}_{\mathbf{p}\mathrm{S}j}(m, \hat{\omega}, \phi_j)\, \Delta m\, \Delta\hat{\omega}\, 2\sin(\Delta\phi/2) &= \rho_0 c_{\mathbf{gz}} \hat{E}_{\mathrm{S}}(m, \hat{\omega})\, (\mathbf{k_0}/\hat{\omega})\, \Delta m\, \Delta\hat{\omega}\, 2\sin(\Delta\phi/2) \\ &= \rho_0 c_{\mathbf{gz}}\;\; \beta N^2\, A_0(s,t)\, B_0(p)\, \Phi_0\, \hat{\omega}^{-p}\, m^{-3} \times \\ &\times (\mathbf{k_0}/\hat{\omega})\, \Delta m\, \Delta\hat{\omega}\, 2\sin(\Delta\phi/2)\,. \end{aligned} \tag{11}$$

Stretched grids are used to concentrate resolution at the smallest values of $k_0$ or $\hat{\omega}$. The framework can be extended to accomodate other wave-breaking criteria (e.g., Gardner 1994, 1995, 1996; Hines 1991a, 1991b, 1991c, 1993, 1996; Medvedev and Klaassen 1995) while leaving conservative propagation aspects unchanged. It is hoped to test some of these extensions in the near future.

The calculation yields spectral densities of upward and downward propagating pseudomomentum flux. These are appropriately summed to give the gravity wave-induced force $\mathbf{G} = (G_x, G_y)$, as the vertical divergence of the total pseudomentum flux (including back-reflected contributions), and the total wave-energy $E_{\text{tot}}$.

## 4. Results for July, 40°N, CIRA (standard summer) case

Several test cases of sensitivity to the low-$m$ index $s$ in (6) have been looked at so far. The first (standard summer) case, with $s = 1$, is that of gravity wave spectral evolution through the July, 40°N, CIRA model atmosphere (see Figure 1). Here the $m_*$ value of the launch spectrum (6) is taken as $2\pi/2000$m, corresponding to a characteristic wavelength of 2 km. We launch the spectra from the zero of zonal wind at an altitude of 19 km and choose $p = 5/3$, both for the launch spectrum (6) and for the saturation threshold (10). This example of spectral evolution is the standard summer case discussed in detail in WM96.

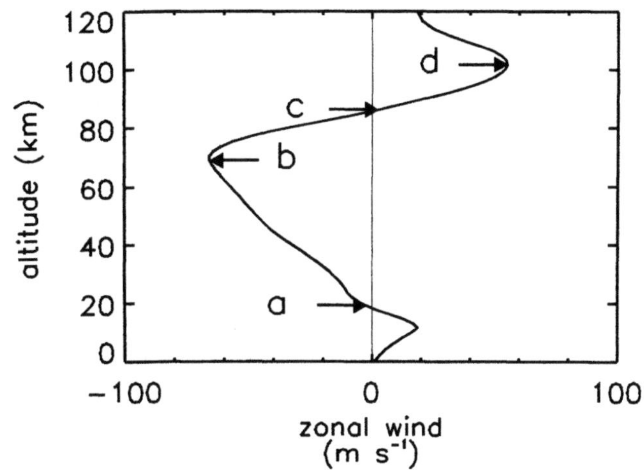


Figure 1. July, 40°N, CIRA 1986 model atmosphere: Mean zonal background wind profile from 0 to 120 km altitude. Arrows show the altitudes for Figure 2.

Figure 2 shows spectra viewed in $(m, \hat{\omega})$ spectral space at the altitudes shown by the arrows in Figure 1, namely 19 km, 69 km, 86 km, and 102 km, for the eastward propagating azimuthal sector of the spectrum ($\phi_j = 0°$). Figure 3 shows the total wave-energy and the zonal wave-induced force profiles. Looking first at Figure 2, we see that from (*a*) (19 km) to (*b*) (69 km), eastward propagating spectral elements are Doppler-shifted to larger $\hat{\omega}$ and smaller $m$, because of the negative wind shear between (*a*) and (*b*) in Figure 1. The Jacobian $J$ decreases, the area of the spectral element decreases correspondingly, and the spectral density increases. Back reflection is likely, and indeed 38% of launched eastward propagating pseudomomentum flux is back-reflected in this case. Between (*b*) (69 km) and (*d*) (102 km), in positive background shear, spectral elements are Doppler-shifted to smaller $\hat{\omega}$ and larger $m$, the Jacobian increases, the area of the spectral element increases correspondingly, and the spectral density decreases, though not nearly fast enough to avoid wave-breaking, because of the $m^{-3}$ factor in (11). Indeed a very considerable cutting down of the high-$m$ end of the spectrum is visible in Figures 2(*c*) and 2(*d*).

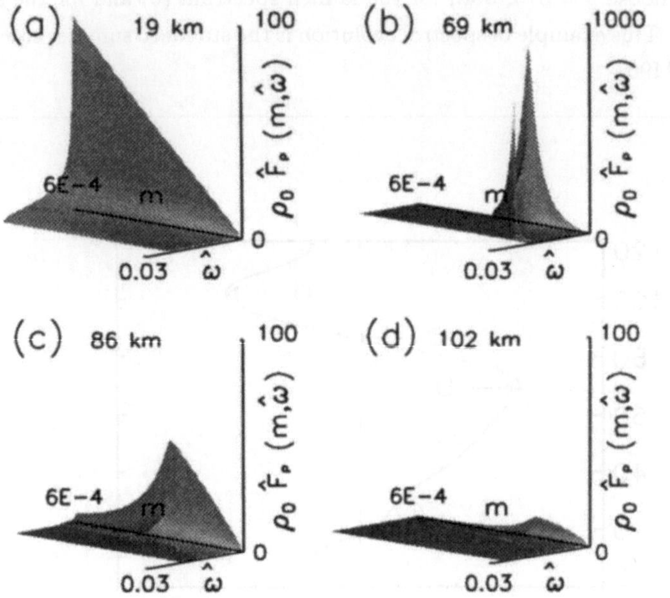


Figure 2. Eastward propagating ($\phi_j = 0°$) pseudomomentum flux spectra showing only a narrow range of $m \ll m_* = 3.142 \times 10^{-3}$rad m$^{-1}$, for the standard summer case. Shaded surface plots in $(m, \hat{\omega})$ spectral space at altitudes of (*a*) 19 km, (*b*) 69 km, (*c*) 86 km, and (*d*) 102 km (arrows in Figure 1). Note that the range of the vertical axis of (*b*) is ten times larger than for (*a*), (*c*), and (*d*). Units are rad m$^{-1}$ for $m$, rad s$^{-1}$ for $\hat{\omega}$, and kg m$^{-1}$ s$^{-2}$ rad$^{-1}$ for $\rho_0|\hat{\mathbf{F}}_\mathbf{p}|$. The edge of the shading shows $N(z)$.

Figure 3(*a*) compares the total wave-energy profile $E_{\text{tot}}(z)$ from the model calculation (solid curve) with the empirical wave-energy profile $E_{\text{emp}}(z)$ derived by Fritts and Lu (1993) from observational studies (dotted curve), all per unit mass as before,

$$E_{\text{emp}}(z) = E_0 \left(\frac{N(z)}{N(0)}\right)^{1/2} \exp(z/H_E)\,, \qquad (12)$$

where $H_E = 2.3H_\rho$ and $H_\rho = 7$ km, a nominal pressure scale height, and $N(z)$ is the buoyancy frequency. The exponential growth of wave-energy per unit mass with altitude is reproduced, but with a greater rate of wave-energy increase than $E_{\text{emp}}(z)$.

Figure 3($b$) also shows the wave-induced force, which is zonally directed because meridional background winds are set to zero. Wave-induced force peak values are $90\,\text{m}\,\text{s}^{-1}\,\text{day}^{-1}$ eastward at 80 km altitude and $350\,\text{m}\,\text{s}^{-1}\,\text{day}^{-1}$ westward at 110 km altitude. A more complete analysis of this example is given in WM96.

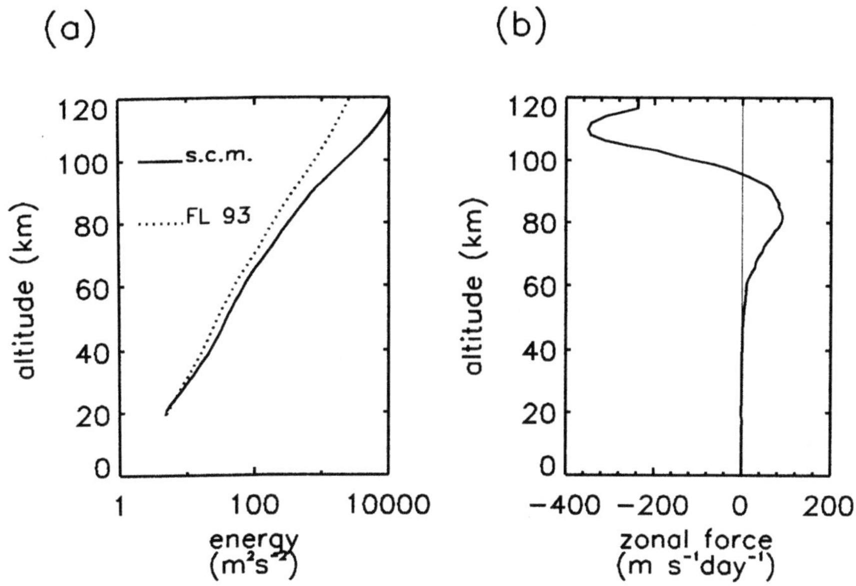


Figure 3. ($a$) Total wave-energy profile $E_{\text{tot}}(z)$ (solid curve marked "s.c.m." for single column model) in standard summer case, which has $s = 1$, as well as $m_* = 2\pi/2\,\text{km}$ and $p = 5/3$. Fritts and Lu's empirical profile $E_{\text{emp}}(z)$, (12), is dotted and marked "FL 93". ($b$) Wave-induced force $\mathbf{G}(z)$ which is zonal because meridional background winds are set to zero.

## 5. Experiments on sensitivity to $s$

We now test the sensitivity of the growth rate of wave-energy to changing the value of $s$ in the standard launch spectrum (6) from the value $s = 1$ used above. For larger values of $s$, the evolving spectrum is expected to contain less energy at high altitudes than for smaller values of $s$, and similarly for the wave-induced forces. Figure 4 shows the standard summer case for a range of values of $s$ and for the same $m_*$ and $p$ values as in Figure 3. Figure 4($a$) compares total wave-energy profiles $E_{\rm tot}(z)$ with the empirical wave-energy profile $E_{\rm emp}(z)$, now shown solid. Figure 4($b$) shows wave-induced force profiles. All these are seen to be sensitive, increasingly with altitude, to the $s$-value hence small-$m$ shape of the launch spectrum.

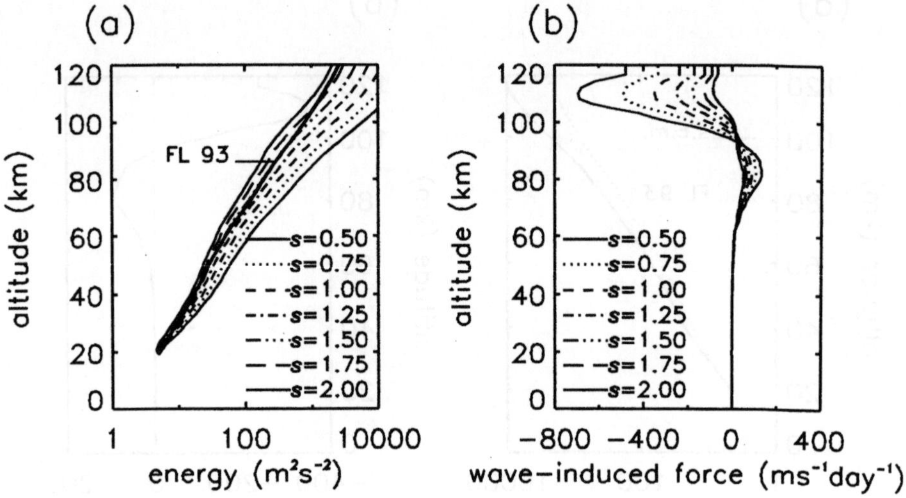


Figure 4. Sensitivity to $s$ in the standard summer case, with $m_* = 2\pi/2\,\text{km}$ and $p = 5/3$. ($a$) Total wave-energy $E_{\rm tot}(z)$ for various values of $s$. Fritts and Lu's empirical profile $E_{\rm emp}(z)$, (12), is shown as a thick solid curve marked "FL 93" (same as the dotted curve in Figure 3($a$)). ($b$) Wave-induced force $\mathbf{G}(z)$, which is zonal because meridional background winds are set to zero, for various values of $s$.

As stated in §3, it is arguable (see WM96) that the same broadband saturation arguments that lead to an $m^{-3}$ dependence in the saturation threshold (10) lead also to an $\hat{\omega}^{-1}$ ($p = 1$) dependence, rather than the $\hat{\omega}^{-5/3}$ ($p = 5/3$). The variation of $E_{\rm tot}(z)$ with $s$ is therefore also investigated for this shallower $p = 1$ saturation threshold (so $p = 1$ in (10) though still 5/3 in (6)). The values of $s$ giving the best fit to $E_{\rm emp}(z)$ in the middle atmosphere for these and other test cases are tabulated in Table 1. The "November 28 1991", ISAMS, 37.5°N, 0°E case is a 24 day average of zonal and meridional winds centered on November 28 1991 (B.N. Lawrence, personal communication, 1995).

Table 1

| Data source | $\hat{\omega}$ dependence of saturation threshold | Approximate best value of $s$ |
|---|---|---|
| Jul, 40°N, CIRA | $\hat{\omega}^{-5/3}$ | 1.5 |
| Jan, 40°N, CIRA | $\hat{\omega}^{-5/3}$ | 1.75 |
| Jul, 40°N, CIRA | $\hat{\omega}^{-1}$ | 1.25 |
| Jan, 40°N, CIRA | $\hat{\omega}^{-1}$ | 1.5 |
| "Nov 28 1991", 37.5°N, 0°E ISAMS | $\hat{\omega}^{-5/3}$ | 1.75 |
| "Nov 28 1991", 37.5°N, 0°E ISAMS | $\hat{\omega}^{-1}$ | 1.5 |

For the cases considered so far, the values of $s$ that fit best are in the range 1.25–1.75, distinctly greater than 1. Both the total wave-energy $E_{\mathrm{tot}}(z)$ and the wave-induced force $\mathbf{G}(z)$ decrease as $s$ increases. This decrease is most pronounced at the highest altitudes. It also appears that the shallower, $\hat{\omega}^{-1}$ dependence in the saturation threshold gives rise to a slightly smaller optimal value of $s$ than the steeper, $\hat{\omega}^{-5/3}$ case.

## 6. Concluding remarks

The experiments on sensitivity to $s$, described in §5, suggest that the value of $s = 1$ assumed by Fritts and VanZandt (1993) may be too low and that a more appropriate choice could be more like $s \approx 1.5$. Further cases need to be considered, especially winter $\{u(z), v(z)\}$ profiles, to assess variability and to see if best $s$ values emerge that are similar to those found here. Other small-$m$ launch spectral shapes besides the simple power law need to be considered; for example, a cutoff at some small $m_c$ might fit observations better.

Future developments to the numerical spectral-evolution model itself include drastically simplifying it for use as a GCM gravity wave parameterization scheme. Several promising ideas are under consideration. It is also intended to use the numerical model as a testbench for existing as well as future gravity wave parameterization schemes. Finally, there will be a continuing need to refine and improve the spectral-evolution model, including the wave-breaking criterion, to take account of new observational data and theoretical ideas as they become available.

**References**

Fritts DC, and W Lu, 1993: Spectral estimates of gravity wave energy and momentum fluxes. Part II: Parameterisation of wave forcing and variability, *J. Atmos. Sci.*, **50**, 3695–3713.

Fritts DC, and TE VanZandt, 1993; Spectral estimates of gravity wave energy and momentum fluxes. Part I: Energy dissipation, acceleration and constraints, *J. Atmos. Sci.*, **50**, 3685–3694.

Gardner CS, 1994: Diffusive filtering theory of gravity wave spectra in the atmosphere, *J. Geophys. Res.*, **99**, 20601–20622.

Gardner CS, 1995: Scale-Independent diffusive filtering theory of gravity wave spectra in the atmosphere, from *The upper mesosphere and lower thermosphere: a review of experiment and theory*, Geophysical monograph 87, AGU, pps 153–175.

Gardner CS, 1996: Testing theories of atmospheric gravity wave saturation and dissipation, *J. Atmos. Terres. Phys.*, NSF CEDAR issue (to appear).

Hines CO, 1991a: The saturation of gravity waves in the middle atmosphere. Part I: Critique of linear-instability theory, *J. Atmos. Sci.*, **48**, 1348–1359.

Hines CO, 1991b: The saturation of gravity waves in the middle atmosphere. Part II: Development of Doppler spread theory, *J. Atmos. Sci.*, **48**, 1360–1379.

Hines CO, 1991c: The saturation of gravity waves in the middle atmosphere. Part III: Formation of the turbopause and of turbulent layers beneath it, *J. Atmos. Sci.*, **48**, 1380–1385.

Hines CO, 1993: The saturation of gravity waves in the middle atmosphere. Part IV: Cutoff of the incident wave spectrum, *J. Atmos. Sci.*, **50**, 3045–3060.

Hines CO, 1996: Doppler-spread parametrization of gravity-wave momentum deposition in the middle atmosphere. Part 1: Basic formulation, *J. Atmos. Terres. Phys.*, submitted.

Medvedev AS, and GP Klaassen, 1995: Vertical evolution of gravity wave spectra and the parametrization of associated wave drag, *J. Geophys. Res.*, **100**, 25841–25853.

VanZandt TE, DC Fritts, 1989: A theory of enhanced saturation of the gravity wave spectrum due to increases in atmospheric stability, *Pure Appl. Geophys.*, **130**, 399–420.

Warner CD, and ME McIntyre, 1996: On the propagation and dissipation of gravity-wave spectra through a realistic middle atmosphere, *J. Atmos. Sci.*, in press.

# Nonlinearity of the saturation spectrum

Colin O. Hines
15 Henry Street
Toronto M5T 1W9
Canada

Abstract: An argument based upon general principles and a minimum of development is presented to establish the role of nonlinearity in producing the observed saturation of gravity-wave spectra in the middle atmosphere. The complementary role of instability of the nonlinear spectrum is identified. The nonlinear models previously developed are reviewed in the context thus provided.

## Motivation

The tail portion of observed gravity-wave spectra is subject to strong nonlinearity in the Eulerian formulation. This nonlinearity acts via the Doppler shift implicit in the advective terms, those containing $\mathbf{v}.\nabla$. Any theory of the tail must be compatible with, if not dependent on, this fact.

In 1959, when it was first presented, the wave interpretation of irregular winds at meteor heights was challenged on the grounds (amongst others): How can the use of linear wave theory be justified when the irregular winds are as large as those observed? To which the answer was: Because the most important potential nonlinearity is that associated with the $\mathbf{v}.\nabla$ term, and its potential for producing nonlinearity is reduced in a single dominant gravity wave by virtue of the near orthogonality of the $\mathbf{v}$ and $\mathbf{k}$ vectors of such a wave; but nonlinearities will certainly be important in wave superpositions.

Thirty years later, if not before, the problem of the tail spectrum should have reopened the question: How can the use of linear wave theory be justified in any interpretation of the tail? The new answer is: It can't!

This answer was almost implicit already in the work of Smith et al (1989), who found (for model calculations based on observations) that their characteristic vertical wavenumber $m_*$, which marked the onset of the tail, was inversely proportional to the square root of the energy density and so to the rms wind fluctuation. It was made explicit in Hines (1991a), where a

NATO ASI Series, Vol. I 50
Gravity Wave Processes
Their Parameterization in Global Climate Models
Edited by Kevin Hamilton

similar model calculation showed (in restated form) that the intrinsic horizontal phase speed of a wave at $m_*$ -- given approximately by $N/m_*$, with N the buoyancy frequency -- was no greater than four times the rms wind speed in the direction of propagation. Since the **k** vector of tail waves had no reason to be orthogonal to the **v** vector of the full spectrum, much of the tail spectrum then had to be subject to severe nonlinearity.

This conclusion can be repeated time and again with the use of observations already available in 1989 or newly available now, though the observations often peter out at the smaller wavenumbers that produce the bulk of the irregular wind. Some reasonable assumptions must be made for those wavenumbers. The most popular current assumption is that the spectrum varies as m at values of m appreciably less than $m_p$, the m at the peak of the spectrum. Observers would do a kindness to their readers, when they publish, if they were to compute an rms horizontal wind speed $\sigma_h$, based on this (or some other, stated) assumption, and if they were to indicate on their illustrations of spectra the corresponding wavenumber $m_h = N/\sigma_h$. Whenever I have done this for myself, using published data, the indicated wavenumber has been found to lie in the tail within a factor of four or so of the apparent peak of the spectrum.

This behavior is illustrated in Fig. 1 for a model spectrum of the horizontal wind fluctuation (assumed to be horizontally isotropic) that I shall call the "general canonical" spectrum. It is given by $AN^2m^{-3}$ in the tail, it reaches a peak value at some $m = m_p$, and it varies as $Bm^s$ at smaller m, with continuity at $m_p$. The permissible values of s lie between 0 and $\infty$, while observational extreme values for A are about 0.5 and 0.1. Within these ranges, the horizontal phase speed at $m_p$ is found to lie within the range 1.6 - 6.3 times the rms wind in the direction of propagation, with a value of 3.5 applying at the currently most popular choice of parameters (s=1, A=1/6). Thus, all tail waves with $m > 3.5\ m_p$ or so have intrinsic horizontal phase speeds that are less than the rms speed of the fluctuating winds through which they are attempting to propagate. Similar results are found for other spectral models, such as the various modified Desaubies spectra.

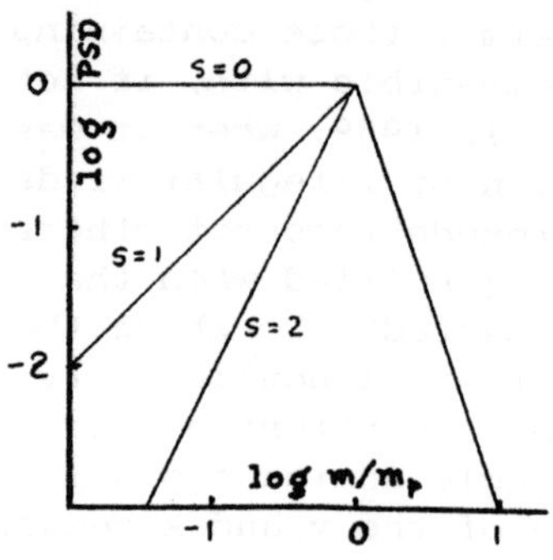

Fig. 1: The general canonical spectrum

It is not sufficient to claim, as some do, that the waves of the tail come in individual wave packets that manage to propagate independently of the other waves and so are free of any serious nonlinear effect. In the first place, virtually all observational evidence is against this view: the tail is almost always well-developed throughout its observed range, with few if any apparent gaps in the spectrum. But more importantly, the rms winds are generated primarily by the small-m portion of the spectrum, and that portion must be present throughout the middle atmosphere in order to provide the tail portion observed at greater heights. Thus, waves of the tail are rendered nonlinear by the non-tail waves, whether or not also by the other tail waves. Any theory of the tail, or any reduction of data with respect to the tail portion of the spectrum (e.g., by the use of linear polarization relations to convert density fluctuations into wind fluctuations), that ignores this basic fact does so at its peril.

## Heuristic approach describing the consequences

Here I attempt to establish in their broadest terms the consequences that follow from the occurrence of strong nonlinearity, leaving for subsequent comment the detailed studies that have attempted to compute those consequences.

Consider the upward propagation of a spectrum of waves through an idealized atmosphere (one that is otherwise wind-free and isothermal, at the equator to remove obfuscation by Coriolis effects, and nondissipating). At some initial height in this model, I constrain the one-dimensional m spectrum only by assuming some physically acceptable form in m, broad and more-or-less smoothly varying, terminating at some upper wavenumber $m_C$, and sufficiently weak that the system is effectively devoid of nonlinearity. (The termination at $m_C$ is introduced to make this last condition possible; were the spectrum not so terminated, there would always be waves with m sufficiently large that they would be subject to nonlinear interaction, no matter how small the rms winds of the whole might be. That is a fact worth noting in its own right.) I will ignore instability to begin with but will return to it before long.

The waves propagate upward and all of them amplify exponentially at the standard rate, with the spectral shape remaining entirely unchanged because there is no mechanism for changing it. Until, that is, the rms winds become large enough that they are comparable to the smallest horizontal phase speed, $N/m_C$. Thereupon, the waves at the upper end of the spectrum must suffer nonlinear interaction with the remainder of the spectrum (and it with them, though with trivial consequences to it). The spectrum must alter in response, and all knowledgeable sentiment would assert that structures with larger m must be produced: the spectrum must extend itself into the range $m > m_C$, and the spectral density just below $m_C$ must be somewhat attenuated. (There may be other redistribution going on as well, at considerably smaller m, but it would be trivial and not of immediate concern.) And the higher the waves propagate, the farther out this tail must extend and the farther down in wavenumber the diminution must extend. The transition between deformed and undeformed portions of the spectrum, at a wavenumber we may designate $m_*$, must move to successively lower m values.

These changes must occur, and they must occur in this way. The only serious questions that arise concerning them are: (1) When the tail is sufficiently well developed, does it adopt an asymptotically stable shape; and, if so, what is its form? (2) How does the transitional $m_*$ vary with height? And (3), how does the intensity of the tail vary with height?

As to (1): Those who accept "similitude" arguments in support of alternative models of the tail spectrum (such as the Dewan/Good and Smith/Fritts/VanZandt linear-instability models) must surely accept that there will be an asymptotic form for the tail and that it must be an $m^{-3}$ form. (I have never seen this similitude argument presented in a fully developed, and therefore challengeable or adaptable, form; I simply quote what it's proponents seem to me to be forced to accept as an undoubted consequence of nonlinearity in the circumstances now being postulated.) The similitude argument seems to be based in part at least on the assertion, employed explicitly by some, that there is no set of underlying parameters (such as N) that combine to yield a characterizing m in the tail region, hence the spectrum must be of a power-law form; and the only form having acceptable dimensions is $N^2m^{-3}$, presumably with a factor of proportionality of order unity.

This argument, if valid at all in other models, should surely be valid in the presently postulated circumstances. It yields precisely the nonlinear wave equivalent of the "inertial subrange" portion of a turbulence spectrum, and I propose calling this portion of the spectrum "wavulence" in consequence -- a point that I will develop shortly.

There is, however, one major distinction between the nonlinear model and all linear models that invoke similitude. It arises from the fact that, with nonlinearity now taken to be important, there is indeed a characterizing wavenumber. This is because the rms horizontal wind speed $\sigma_h$ is added to the list of underlying parameters available for generating a characterizing vertical wavenumber, and it combines with N to provide the wavenumber $m_h = N/\sigma_h$. One must presume that something of interest should happen in the vicinity of this wavenumber, even if not at others. The obvious role for this wavenumber to play, in view of its corresponding horizontal phase speed $\sigma_h$ and the resultant transitional importance of the $v.\nabla$ term in an Eulerian formulation, is to mark the region of transition from relatively unaffected waves to strongly affected waves -- the transition previously recognized and identified as occurring at $m_*$. Thus the picture is painted of a spectrum that is relatively unchanging at $m << N/\sigma_h$ (except for continued growth with height) and that converts in the vicinity of $N/\sigma_h$ to the spectral form $AN^2m^{-3}$. These are, of course, precisely the characteristics that have been attributed to observed spectra in the past, if only we identify $m_*$ as some multiple of $N/\sigma_h$ of order 1.

Prior to the advent of the Doppler-spread theory (Hines, 1991b), the proportionality of the observed $m_*$ to the observed $N/\sigma_h$ was recognized (in a different form) only as a possibly interesting deduction from an assumed model of the observations (Smith et al., 1987). Now, however, the near equivalence of the two is being asserted, and it is being identified as a fundamental aspect of the physics of the problem, one to which the observations would have to conform if the postulated model is anywhere close to reality.

To examine questions (2) and (3), concerned with the height variations of $m_*$ and A, it is necessary to adopt some model spectrum. I adopt for the purpose the general canonical spectrum introduced above, having the form $Bm^s$ for $m < m_p$, $AN^2m^{-3}$ for $m > m_p$, and continuity at $m_p$, and I take B to be growing exponentially with height z as exp (z/H), where H is the atmospheric scale height. The transitional wavenumber $m_*$, which in this model is identically $m_p$, is to be expressed as $qN/\sigma_h$, with q an unknown constant presumed to be of order 1. The imposition of continuity at $m_p$ establishes that $A = BN^{-2}m_p^{s+3}$. Integration of the spectrum over the full (infinite) range of m establishes that $\sigma_h^2 = B(s+3)(2s+2)^{-1}m_p^{s+1}$. The identification $m_p = qN/\sigma_h$ then establishes that $\sigma_h^{s+3} = B(qN)^{s+1}(s+3)(2s+2)^{-1}$. It follows that $\sigma_h$ varies with height as exp (z/[s+3]H), and $m_p$ varies with height as exp -(z/[s+3]H). Finally, after a bit of manipulation, A is found to adopt the value $(2s+2)(s+3)^{-1}q^2$, which is seen to be a constant, independent of height: an appearance of saturation has been produced, automatically.

If the currently most popular choice s = 1 is adopted, then $\sigma_h$ varies as exp (z/4H), $m_p$ varies as exp -(z/4H), and $A = q^2$. If now the currently most popular choice A = 1/6 is adopted, then q = 0.41, which is reasonably of order 1. Alternatively, if the scale height of $\sigma^2$ is taken as 2.3 H (Fritts and VanZandt, 1993), then s = 1.6; the choice A = 1/6 then yields q = 0.38. (Physically, s would be determined by sources, q by the nonlinear processes; A would then follow and be available for observation.)

I submit that all of these results are in good accord with observation and with the principles on which they were derived, and that departures from them with the use of other reasonable spectral models are unlikely to be substantial. (Cf. Hines, 1991b, where a similar calculation is made for a modified Desaubies spectrum.)

The foregoing conclusions are all direct consequences of the claim that the tail spectrum is produced by Doppler-induced nonlinearity in the wave system, extended only by the further use of a standard similitude argument and with the aid of a very general form of spectral model. In particular, it does not depend on detailed calculations such as those of Allen and Joseph (1989) or Hines (1991b, 1993a), to which I will turn shortly.

## The role of instability

The spectrum that was assumed initially is necessarily stable. This is because the integral of the shear spectrum is necessarily substantially less than $m_c^2$ times the integral of the spectrum itself, the latter being $\sigma_h^2$, and the postulated linearity requires that $\sigma_h^2$ is less than $N^2/m_c^2$, from which it follows that the integral of the shear spectrum is necessarily substantially less than $N^2$ -- a circumstance that produces only a trivial probability of instability anywhere in space-time (Hines, 1991a). Even as nonlinearity sets in, the imbalance continues: the spectrum must become substantially nonlinear before it can become significantly unstable. This conclusion stands in marked contrast to the case of a single or dominant wave, for which nonlinearity and instability go hand in hand. A broad spectrum requires the development of a nonlinear tail before it can become unstable.

Once the tail is indeed sufficiently developed, it renders the spectrum as a whole unstable. This becomes apparent mathematically upon the introduction of some shear or lapse-rate or Richardson-number criterion for use with the integral of the associated shear spectrum: the criterion for marginal instability will be met if and only if the tail extends to some wavenumber $m_M$ whose precise value will depend on the criterion employed, a matter that need not be debated here. The natural conclusion is that the tail will then be prevented by instability from extending substantially beyond $m_M$, and that the energy associated with the missing portion of the spectrum will have gone to generate turbulence.

The turbulence will extend the spectrum of structure sizes

to greater values of m, though not necessarily with continuity of spectral density at $m_M$, and will ultimately deposit the energy via molecular dissipation (which is now admitted to the discussion). It is possible to conceive of a continuous transition from wavulence to turbulence, probably with continuity of spectral form and possibly of spectral intensity, with the wavulence being characterized by deterministic physics and the potential for reversibility, the turbulence by chaotic physics and irreversibility.

To be sure, the turbulence once formed may well do some leeching of wave energy from the spectrum at $m < m_M$, possibly computable with the use of a turbulence diffusivity, but I see no reason to suppose that such leeching will appreciably affect the form of that portion of the spectrum except perhaps in the immediate vicinity of $m_M$. (The spatial rate of attenuation imposed by viscosity varies as $m^3$/[intrinsic frequency], which would make the consequences of turbulence an extremely strong function of position in the tail even if the turbulence could act like a uniformly available viscosity, which itself is questionable.) In consequence, I believe it appropriate to consider that the tail extends as already described, out to $m_M$, and that wave energy is transferred to turbulence at or very near $m_M$.

It has been argued to me in private that, since the spectrum as a whole is unstable, wave energy might well be lost directly to the instability at wavenumbers substantially smaller than $m_M$. I disagree, for reasons that emerge from the discussion in Hines (1991a). There I exhibit a phasor chain to illustrate the fluctuation of stability within the wave system, with individual waves contributing individual phasors, each of which rotates at its own pace in time and space; the chain as a whole then thrashes around, with its end occasionally piercing an instability-criterion surface and then retreating again. The resultant instability does not arise uniformly in space or time, but rather in small patches, "here and there, now and then", much like whitecaps on a roughened ocean surface. It simply has no opportunity to extract energy from waves of substantially larger scale but will act rapidly to extract energy from the particular combination of small-scale waves that has established the scale size of each briefly unstable region in turn. This is not to deny that there will be some occasions when a dominant wave acts to destabilize a large region, but it is to deny that such occasions are representative of the normal situation. My view in this appears to be upheld observationally by such measurements as those of Bob Sica (this meeting), whose lidar-determined space-time maps of unstable regions depict exactly the small-scale patchiness just described. The widespread occurrence of turbulence, when found, must then indicate fossil turbulence currently in decay, repeatedly regenerated rather than continuously generated.

The foregoing discussion assumes, of course, that the wave spectrum is permitted to extend nonlinearly in m out to the point required for instability. At sufficiently great heights, with their enhanced effects of molecular viscosity and heat conduction, such a spectral range can be precluded by molecular dissipation. The transition would mark the turbopause, as argued in Hines (1991c). At those heights, the spectrum is known to be substantially narrowed, even to the point that it may, at last, consist of only a few isolated wave packets. In such circumstances, the individual waves may indeed become unstable (while becoming also nonlinear, individually or in small groups), and so may produce associated regions of turbulence, but rapid quenching would render the turbulence itself quite patchy.

Allen and Joseph (1989), hereafter AJ

The analysis of AJ was conducted as a transformation from a Lagrangian formalism to an Eulerian formalism. They made the important point that there are certain physically intrinsic nonlinearities, found in both, but that the additional advective nonlinearity of the **v**.$\nabla$ term is found only in the Eulerian formalism: it is a construct of the choice of coordinate system but is devoid of physical content or intrinsic implications.

They found that the effects of this nonlinearity set in, in the Eulerian formulation, at an earlier stage of amplification than did the effects of the intrinsic nonlinearities in either. (They discussed only oceanic waves and so lacked the exponential growth with height found in atmospheric waves. The term "amplification", as I use it here, refers simply to the consideration of a succession of cases of increasing integrated spectral intensity.) This was their conclusion, at least, with respect to the dispersion relation; whether or not also with respect to the polarization relations, I do not know.

More specifically, when the linear dispersion relation was viewed as a 3-D surface in a 4-D {omega,**k**} space, they found that Lagrangian solutions would remain on that surface even as Eulerian solutions were forced to "diffuse" off it under the influence of the advective nonlinearity: a given **k** vector no longer corresponded to a single omega. Simultaneously, a spectral tail was formed in the Eulerian representation (with the Lagrangian spectrum remaining quite unchanged) and, under specified circumstances, evolved into the $m^{-3}$ form, as has now been anticipated by the similitude argument. This result, arising from detailed evaluation, provides an explicit and calculable confirmation of the conclusions reached above from the more general point of view.

To me, it also implies that waves remain identifiable as waves in the tail, no matter how nonlinear they may be when described in an Eulerian system of coordinates. In a thought experiment, for example, the AJ analysis implies that the wave system could be amplified to produce Eulerian nonlinearity and the development of a tail, and then be attenuated again to reproduce the initial spectrum, provided only that instability had not robbed the spectrum of any energy in the process. Such a thought experiment could be reproduced in real life by the propagation of a suitably chosen spectrum into a region of suitably chosen wind shear, and then reversing the shear to recover the original state: in the absence of instability, the waves would return to their initial spectral distribution, with no exchange of energy amongst modes having occurred despite the transient formation and subsequent disappearance of a spectral tail. It is on this basis that I claimed reversibility at an earlier stage in my remarks and that I distinguished wavulence from turbulence.

In the AJ analysis, a quasi-limiting vertical wavenumber somewhat analogous to $m_C$ was assumed in the Lagrangian formulation, as was a quasi-limiting horizontal wavenumber $h_C$. In the conditions that led to the $m^{-3}$ Eulerian tail, it was found that the rms vertical displacement approximated to $m_C^{-1}$ and that the rms horizontal displacement approximated to $h_C^{-1}$. The first of these conditions is, equivalently, the condition that the rms horizontal wind speed $\sigma_h$ should approximate to $N/m_C$, which is the condition employed above and the obvious one to employ for purposes of Doppler shifting. I have no idea what role, if any, is played by the second of the AJ conditions; it may turn out to be a red herring in the atmospheric context. (In AJ, a form of equipartition of energy amongst modes was assumed, and this led to a specific frequency spectrum. The horizontal and vertical wavenumber spectra were not entirely independently assignable, in consequence, and the two conditions might have been equivalents of one another.) The transition to tail-like behavior occurred, as is to be expected, near to these same wavenumbers.

The AJ analysis provides no explanation for the postulated limits at $m_C$ and $h_C$, other than a passing comment proposing some linkage to instability (whose effects were not otherwise admitted to the analysis). I submit that my own comments above, and others to be found below, give adequate reason for believing that instability does indeed provide the needed mechanism (below the turbopause, in the atmospheric case).

The waves that reach the vicinity of $m_M$ and are obliterated will have arrived there preferentially from the large-m end of the AJ spectrum, near $m_C$. Any substantial spectral component initially at m values appreciably greater than $m_C$ would be quickly attenuated if not obliterated. The assumed limitation of the Lagrangian spectrum at $m_C$ then simply represented the form that would be imposed by instability, in a self-consistent analysis that included instability.

In the AJ analysis the limitation was a stationary property, whereas in the atmospheric case it is progressive as viewed following the propagation of the waves upward in height: the $m_C$ that pertains to any given height will be a function of height, decreasing progressively with increase of height as previously described, and it represents the upper edge of the portion of the initial spectrum that actually survives to that height.

Hines (1991b, 1993a), hereafter H2 and H4 respectively

The analyses of H2 and H4 have been misunderstood in some quarters, and correspondingly misrepresented. It is appropriate to look at them anew, particularly in the light of the foregoing discussion.

H2 was designed to explore the effects of nonlinearity, whose importance in the context had been established in Hines (1991a). There had been many studies of nonlinear effects in atmospheric gravity waves over the years, but virtually all (if not all, save for Weinstock's) had based any quantitative development on the use of small numbers of waves -- usually triads --, this being the most tractable case when accurate conclusions are to be drawn. But the current problem, that of saturation, inherently demanded the use of a broad spectrum.

I was as unable as others had been (with the exception of Allen and Joseph, unknown to me at the time of my analysis) to deal accurately with a broad spectrum. I chose to deal inaccurately with it: to cut the Gordian knot of the $\mathbf{v}.\nabla$ term simply by ignoring the structural coupling that it implied and so by treating the $\mathbf{v}$ part as if it represented a constant, uniform, horizontal background wind, albeit a wind whose statistical characteristics were to be taken as those of the wave-induced, fluctuating wind field. My motivation was simply to see what would emerge -- a leap, not of faith, but of curiosity.

What did emerge I found to be stunning in its apparent compatibility with observations. With a loosely justified assumption as to the appropriate value for $m_C$ (namely, $N/2\sigma_h$), I produced spectral conclusions that provided an approximately

$m^{-3}$ tail, one that had an amplitude essentially independent of height in certain appropriate circumstances. I found this to be compelling evidence, not that the calculation had been done properly -- since it was obviously an improper calculation --, but rather that the basic picture of nonlinear control was correct.

I was further confirmed in this belief when, with the same leap of curiosity, in H4 I sought support for my previously assumed $m_C$. To my own satisfaction, I found support in full measure from back-of-the-envelope calculations; and, prodded by Dave Fritts, who seems to have taken my analysis (if not its conclusions) more seriously than I myself did, I entered upon a full numerical calculation of the resultant spectrum (graciously performed for me by Qihou Zhou of the Arecibo Observatory). This was done for various initial spectra, and each produced much the same tail spectrum in a set of ab initio calculations. One of the resultant curves is shown in Fig. 2 and is compared there with the corresponding modified Desaubis spectrum, which spectrum is often taken to represent the observations best. To me, the comparison is astounding regardless of the shortcomings of the cutting of the Gordian knot on which it was based. The results are, undoubtedly, far better than the analysis that led to them.

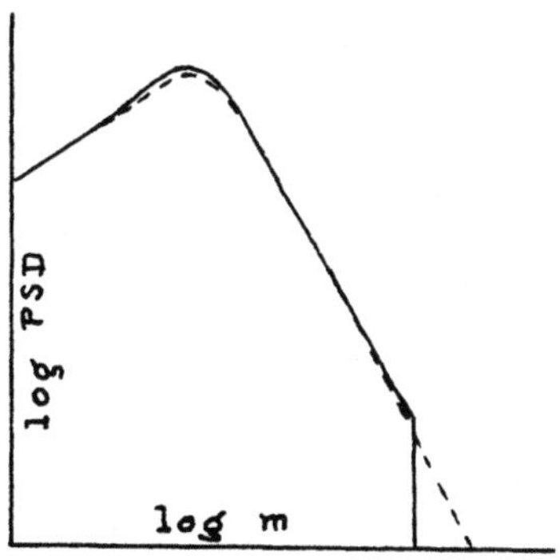


Fig. 2: Spectrum calculated ab initio from Doppler-spread theory (solid curve) and the corresponding modified Desaubies spectrum (dotted curve). The former might be extended to larger m by turbulence.

My cutting of the Gordian knot was never intended to be anything other than a means of exploring what sort of thing might be going on -- what the general nature of the effects of Doppler spreading might be -- and I continue to view it in this light regardless of how remarkably well it has served to provide realistic spectra. I will come shortly to what I think it has revealed that is useful, but I wish first to remove, or at least to deflect, some of the charges that have been made against it.

For example, it has been said that my analysis should have assumed as its v, in application to a given test wave, only that part of the wave-induced wind spectrum that is of substantially longer time and horizontal scales than the test wave, and that something quite different must be done with the rest. The basis of the remark is, of course, that scale separation would then act to minimize the changes of frequency and of horizontal wavenumber that would have been imposed on the waves generated by nonlinear interaction, which changes had been ignored -- quite justifiably, under this restriction. But this particular charge against my analysis attributes to it a degree of sophistication that I have never claimed for it. I did employ, and I would again employ, the **v** of the full spectrum. This is not because I believe the result, but because I believe that the result is as indicative of the true consequences of nonlinearity as any result based upon a partial **v** would be. Put otherwise, anyone who chose to employ only the winds at larger scales for the purpose of a Doppler-shift calculation would be fully justified in doing so but would be left with the residual problem of what to do with the remaining portion of the wind. They are welcome to attempt a solution to that problem; I simply forecast that the result will not be qualitatively different from what my own, possibly cruder, approximation produces, however more reliable it may be within its own limitations. Indeed, I ask what can the effect of the remainder of **v** be on the final spectrum, other than some additional shifting and spreading of a similar type?

As an extension of this scale-separation issue, it has been argued that a wave will experience a "true" Doppler shifting toward critical-layer conditions, and will respond accordingly, only if the wind doing the shifting exists on a time scale (and presumably horizontal scale) much greater than that of the test wave in question. I reject this allegation, if it is intended to imply that the consequences of interactions with smaller-scale waves cannot be like those with larger-scale waves. The test wave will experience, however locally in time and/or in space, a **v** that enters its governing equations as importantly as does its omega, if that **v** is comparable to its own horizontal phase speed. It must respond, however locally, in some fashion; and it must do so repeatedly, in time and/or space. The process produces wave-wave coupling, of course, with the generation of other temporal and horizontal scales; but who is to say that its net effect on the test wave must be substantially different from the net effect that a larger-scale wave would have imposed, insofar as the generation of locally large shears and instability is concerned? By all means, one should have doubts; but one should not prejudge the issue, except perhaps via the similitude argument, and we have seen where that leads.

It has also been remarked, as if importantly, that my H2 and H4 calculations have produced tail slopes of -2.7 or -3.1 or whatever, rather than the wanted -3. To which I reply: So what? I could certainly have juggled some of the governing parameters and so have produced slopes of -3, had my intention been to attempt to delude naive readers into thinking that I had really succeded in matching the data (whose variability and margins of error far exceed those of the computed results in any event). The remarkable fact is that slopes based on my poorly justified analysis come as close to -3 as they do, and I for one am prepared to defer to the AJ analysis, or to the similitude argument, as an appropriate means of going the remaining distance.

Again, the complaint has been made that my analysis does not produce anything like a -3 slope in the asymptotic limit as $m \to \infty$, however well it may do so within a decade of the peak of the spectrum. Instead, the spectrum asymptotes to the form $m^{-1}$ if Coriolis effects are ignored and to a constant if they are included. But this is a totally specious cavil, as I have pointed out in H4: my analysis, of whatever limited validity it may be at m values near the start of the tail, falls apart completely at larger m values and is admitted to do so. This is because it takes into account only the horizontal component of the wave-induced winds, whereas the $\mathbf{v}.\nabla$ term clearly requires in principle the inclusion of the vertical component as well. Omission of that component can be justified for m not too large, but cetainly not for extreme m values (including, in particular, the asymptotic limit); and any attempt to include the vertical wind component for larger m values encounters mathematical problems (see Appendix B of H4) that are simply not worth resolving in view of the heuristic nature of the analysis as a whole. Thus, while I have presented (in all honesty) the unsavory results to which my approximations give rise at large m, I have also shown why those results do nothing to undermine the much more acceptable conclusions at the m values that are of concern observationally. The more acceptable conclusions may indeed be suspect, but certainly not because of misbehavior as $m \to \infty$.

What, then, are the lessons to be learned from H2 and H4?

I submit, first, that these papers give insight into the details of the generation of the tail. H2 provides a prescription, however approximate, for identifying the spread of locations, in part into the tail, of waves that were initially at a specified m value not in the tail; and a prescription, however approximate, for the degree to which their amplitudes would have been suppressed by broadening as the shifting was imposed. I believe that the spreading in m corresponds in its crude fashion to the diffusion off of the dispersion surface that was deduced by AJ: a given initial omega and horizontal wavenumber now produce a spread of vertical wavenumbers, as is illustrated in Fig. 3. Accompanying that spread, there must be a reduction of spectral intensity, increasingly so as one moves to higher initial m values, and it is in this way that the -3 slope is ultimately fashioned.

(Observationally, it would be possible to misinterpret this combined behavior as a saturation effect imposed by the self-destabilization postulated in linear-instability theories. Thus, even when the tail is "patchy" and apparently made up of a few individual wave packets spread over a limited range of m, that structure may in fact result from initially quasi monochromatic waves that have been Doppler spread to their present state, and reduced in amplitude, by the action of the larger-scale, non-tail, waves. It should be noted explicitly that the reduction of amplitude of an individual wave packet has been produced, not by dissipation as in linear saturation theories, but by spreading in wave number -- except only for that part of the spread spectrum that succumbs to instability near $m_M$ and for some minor effects of diffusive dissipation through the bulk of the tail, which effects I have ignored.)

This view of the matter is of course tempered by the imprecision with which the Gordian knot was cut. But I submit that <u>any</u> interpretation of the the tail waves in Eulerian terms must be subject to imprecision, simply because the wavulence of the tail does not consist of linearized Eulerian waves. I submit, further, that the Doppler-spread analysis I have given, or some variant on it, is as close as we are likely to get to describing that wavulence in terms of linearized Eulerian waves.

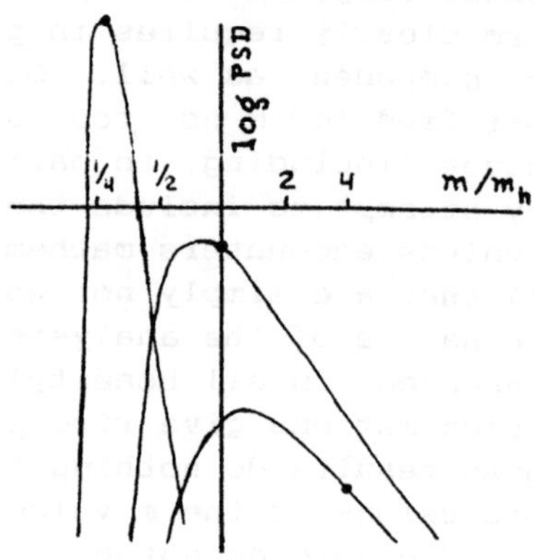

Fig. 3: Doppler spreading of waves initially sited as delta functions at the various dots, upon $m_h$ moving into their proximity. See Hines (1991b) for further explanation and details.

I submit, finally, that the analysis provides a "handle" whereby one can generalize to include the superimposed effects of true background winds. Indeed, that is precisely what is required for purposes of parameterization.

Such winds were excluded from H2 and H4; but, once the Doppler spreading is admitted as a primary feature of the tail and the tail's dissipation, their effects can be incorporated in principle in a fairly straightforward manner. The Doppler shifting imposed by the background winds simply provides a new point of departure from which the Doppler spreading occurs. That spreading depends on the fluctuating winds in the azimuth of propagation, determined for each wave (or each azimuth) in turn, and these depend on the fluctuating winds produced by waves propagating into all non-orthogonal azimuths. Analytic treatment of the problem would be complicated, but numerical treatment is not.

Weinstock (1990), henceforth W90

Weinstock presented a theory of saturation based upon nonlinear waves and a scale-dependent diffusivity that they were hypothesized to produce. It included the deduction of an $m^{-3}$ spectral form at large m. I previously suggested (Hines, 1993b), albeit with some hestation, that his results might well constitute an all-Eulerian equivalent of the asymptotic behavior (at large m) previously obtained by AJ. I must now reverse myself: I do not believe the two to be equivalent. And, since I do believe in the validity of the AJ analysis for wavulence, I must disbelieve the validity of the Weinstock analysis as an analysis of wavulence (though it may well represent the appropriate extension into turbulence; see below).

My previous hesitation was based in part on the fact that the magnitudes assigned to the tail spectrum by AJ and by W90, respectively, appeared to have different parametric structures (e.g., in their dependence on the Coriolis parameter). My present, more conclusive, position is as follows.

With instability and molecular dissipation absent from the AJ analysis (except in peripheral ways), that analysis is devoid of any dissipative mechanism. In these circumstances, the Eulerian response to nonlinearity was found to be a diffusion of {omega,**k**} combinations from the linear dispersion surface, as already stated. The diffusion proceeded in a manner that happened to keep omega and **k** real. In W90, however, the response was hypothesized to be a transition to a single, new dispersion surface, one that required a complex **k** space for its description. But this was only an assumption, one which was forced upon the presumed solution and to which that solution was obliged to adapt. The remainder of the W90 analysis then determined how the adaptation might best take place.

The distinction between the two types of {omega,**k**} space makes it clear that the very structure of the solutions is quite different in the two cases: the two cannot be equivalents of one another. And, since the AJ analysis was purely deductive and straightforward (however difficult), its results can scarcely be challenged for the nondissipative circumstances that it treated.

But, even if the two analyses cannot be large-m equivalents of one another, they can be complements to one another. This is because, as argued above, the AJ spectrum must be unstable if it extends to sufficiently large m values, and some form of turbulence must then set in. I see no argument against a new view: that W90 provides a valid approach to the description of this turbulence, a turbulence wherein the spectral form $m^{-3}$ is also anticipated (e.g., Lumley, 1964).

Such a view, if accepted, would open the way to a new question, that of the "marriage" of the two $m^{-3}$ spectra: does the wavulence spectrum blend smoothly into the turbulence spectrum in the vicinity of some $m_M$, with or without a noticeable shift of spectral intensity; or might the two overlap, with different levels of intensity, in the vicinity of $m_M$; or might there be a gap in the spectrum, with turbulence starting only at m values considerably in excess of $m_M$?

My own expectation is that there would be continuity, with wavulence giving way smoothly to turbulence in the vicinity of some $m_M$, but not necessarily at the same level of intensity. The latter caveat arises because the intensity of wavulence at $m_M$ is determined by the nonlinear wave processes already discussed, combined with the instability criterion that determines $m_M$, whereas the intensity of turbulence is typically determined by the rate of input of energy at its large-scale end, taken here to be $m_M$. I see no reason to suppose that the rate of delivery of energy by wavulence to the vicinity of $m_M$ should be just that which would be required to maintain equality of the two levels of intensity at $m_M$. Indeed, the disparaty of parametric forms for the tail intensities, mentioned above, argues otherwise. However, the question remains for future resolution.

## Consequences for parameterization

When dissipation of the waves occurs in the manner argued here -- specifically, only after they have been spread by nonlinear processes far into the tail -- the details of the dissipation process cease to be of major concern for parameterization purposes. Instead, the details of the nonlinear spreading become the essential feature, with dissipation being understood to occur once adequate spreading has taken place.

This totally altered point of view has been incorporated into a parameterization scheme published elsewhere (Hines, 1997a, b) and has been implemented with considerable success in a large-scale numerical model (Mengel et al., 1995). A brief outline of that scheme is contained in Section 2 of an accompanying paper (Mayr et al., 1997). See also McLandress (1997) in this volume.

Implementation of the parameterization requires the choice of an input spectrum. If this is taken to be of the form $m^s$, then simplified formulae become available if s is chosen to be integral or half-integral. The value s = 1.6, found above, suggests that s = 1.5 be used for parameterization purposes.

## References

Allen KR and Joseph RI (1989) A canonical statistical theory of oceanic internal waves. J. Fluid Mech., **204**, 185-228.

Fritts DC and VanZandt TE (1993) Spectral estimates of gravity wave energy and momentum fluxes. Part I: Energy dissipation, acceleration, and constraints. J. Atmos. Sci., 50, 3685-3694.

Hines CO (1991a) The saturation of gravity waves in the middle atmosphere. Part I: Critique of linear-instability theory. J. Atmos. Sci., **48**, 1348-1359.

Hines CO (1991b) The saturation of gravity waves in the middle atmosphere. Part II: Development of Doppler-spread theory. J. Atmos. Sci., 48, 1360-1379.

Hines CO (1991c) The saturation of gravity waves in the middle atmosphere. Part III: Formation of the turbopause and of turbulent layers beneath it. J. Atmos. Sci., **48**, 1380-1385.

Hines CO (1993a) The saturation of gravity waves in the middle atmosphere. Part IV: Cutoff of the incident wave spectrum. J. Atmos. Sci., 50, 3045-3060.

Hines CO (1993b) A critical comparison of theories of gravity wave saturation. Coupling Processes in the Lower and Middle Atmosphere, Ed. E. V. Thrane et al., Kluwer Academic Publ., 233-239.

Hines CO (1997a) Doppler-spread parameterization of gravity-wave momentum deposition in the middle atmosphere. Part 1: Basic formulation. In press, J. Atmos. Terr. Phys.

Hines CO (1997b) Doppler-spread parameterization of gravity-wave momentum deposition in the middle atmosphere. Part 2: Broad and quasi monochromatic spectra, and implementation. In press, J. Atmos. Terr. Phys.

Lumley JL (1964) The spectrum of nearly inertial turbulence in a stably stratified fluid. J. Atmos. Sci., 21, 99-102.

Mayr HG, Mengel JG, Hines CO, Chan KL and Arnold NF (1997) Modeling the seasonal cycles and equatorial oscillations with a parameterization of the Doppler spread gravity wave theory. This volume.

McLandress C (1997) Sensitivity studies using the Hines and Fritts gravity wave drag parameterizations. This volume.

Mengel JG, Mayr HG, Chan KL, Hines CO, Reddy CA, Arnold NF and Porter HS (1995) Equatorial oscillations in the middle atmosphere generated by small scale gravity waves. Geophys. Res. Lett., 22, 3027-3030.

Smith SA, Fritts DC and VanZandt TE (1987) Evidence for a saturated spectrum of atmospheric gravity waves. J. Atmos. Sci., **44**, 1404-1410.

Weinstock J (1990) Saturated and unsaturated spectra of gravity waves, and scale dependent diffusion. J. Atmos. Sci., **47**, 2211-2225.

# Sensitivity studies using the Hines and Fritts gravity wave drag parameterizations

Charles McLandress
Institute for Space and Terrestrial Science
York University, 4700 Keele Street
North York, Ontario, Canada M3J 1P3

## Abstract

The gravity wave drag parameterizations of *Hines* [1997a,b] and *Fritts and Lu* [1993] are compared using a simple middle atmosphere model capable of simulating the zonal mean circulation and solar tides. Results from the zonal mean version of the model indicate that although both schemes close off the stratospheric jets, the Fritts scheme produces somewhat unrealistic latitudinal structures in the mesospheric zonal winds. The two schemes have a very different impact on the propagating diurnal tide. The Fritts scheme strongly damps the tide in the lower thermosphere and amplifies it in the mesosphere whereas the Hines scheme amplifies it throughout this entire region.

## 1. Introduction

Previous modelling studies of the middle atmosphere have frequently employed gravity wave drag (GWD) parameterization schemes in which the gravity waves are treated in a discrete and linear fashion [*Lindzen*, 1981]. Two recently developed spectral parameterizations [*Fritts and Lu*, 1993 and *Hines*, 1997a,b] (henceforth referred to as FL93 and H97a,b) treat the gravity waves as a continuous spectrum. Their sensitivity and characteristics in middle atmosphere models have neither been fully examined nor the schemes themselves compared.

In this paper the two parameterizations are studied using a linearized primitive equations model that is capable of simulating the coarse features of both the zonal mean circulation and the propagating solar diurnal tide.

The organization of the paper is as follows. In Section 2 the GWD schemes are discussed. In Section 3 are given details about the spectral model. The discussion of the model results follows in Section 4. A brief summary and a discussion of the significance of these results are given in Section 5.

NATO ASI Series, Vol. I 50
Gravity Wave Processes
Their Parameterization in Global Climate Models
Edited by Kevin Hamilton

## 2. Gravity wave drag parameterizations

### 2.1 Hines GWD scheme

The Doppler-spread theory of gravity wave saturation [*Hines* 1991a,b,c and 1993] attributes the $m^{-3}$ slope of the tail portion of the vertical wavenumber $m$ spectrum to the Doppler-spreading that arises in a statistical sense from nonlinear $\vec{v} \cdot \nabla \vec{v}$ wave-wave interactions (where $\vec{v}$ is the horizontal velocity resulting from a broad spectrum of gravity waves). The parameterization scheme that followed from this [H97a,b] has already been used successfully in a 2-D middle atmosphere model [*Mengel et al.*, 1995]. More recent results are also presented in several papers in this volume. To date, however, little or no discussion has been made of the sensitivity of the Hines scheme to changes in the various tunable parameters. In this section, the parameterization will be briefly outlined and the results of several sensitivity experiments presented.

In order for the parameterization scheme to be of practical use in general circulation models, considerable simplification of the original theory is required. First and foremost, is the assumption that waves that have been Doppler spread to sufficiently large vertical wavenumbers are immediately and abruptly removed from the spectrum. Although the details of this dissipation mechanism are not considered, strong critical layer interactions that result are assumed to absorb the waves and cause them to deposit their momentum. The wavenumber at which point obliteration occurs is referred to as the cutoff wavenumber $m_C$. Since it decreases with altitude as a result of decreasing density, the variation of $m_C$ with height provides information about the vertical variation of the momentum fluxes and hence the GWD. The second important simplification is that both the wave-induced and background winds that produce the Doppler spreading and shifting are assumed to provide a time-independent wind field upon which the other waves propagate. This, in combination with the dispersion relation and the principle of conservation of wave action for the unobliterated waves, permits $m_C$ at a given height to be "mapped" onto a single vertical wavenumber in the low end portion of the spectrum at some initial level, generally located in the troposphere (i.e., $m_C(z) \rightarrow m(z = 0)$). The inclusion of the background winds results in selective transmission of gravity waves propagating in different directions - it is this anisotropy in the horizontal momentum field that produces the GWD. Consequently, a cutoff wavenumber is required for each of the $J$ discrete horizontal azimuths.

At the heart of the Doppler-spread parameterization lies the equation for the "mapped" cutoff wavenumber $m_j$

$$m_j = \frac{N_0}{\Phi_1\sigma_j + \Phi_2\sigma_h + V_j - V_{j0}} \tag{1}$$

where $N_0$ is the buoyancy frequency at the initial reference level denoted by the subscript "0", $V_j$ is the background wind component in the $j$th azimuth, and $\Phi_1$ and $\Phi_2$ are constants. The wave-induced rms winds in the $j$th azimuth (a unique feature of the scheme) is denoted by $\sigma_j$, while the total horizontal rms wind speed is given by $\sigma_h$. The calculation of the $\sigma$'s requires evaluating a vertical wavenumber integral which contains the functional form of the prescribed initial $m$-spectrum (i.e., $\propto m^s$) and has limits of integration that range from $m = m_{min} \geq 0$ to $m = m_j$.

The evaluation of $m_j$ is not entirely trivial since the $\sigma$'s are also dependent upon the $m_j$'s. For convenience and practicality, however, an approximate solution to Equation (1) is found by using the $\sigma$'s at the adjecent underlying level $l-1$ to calculate $m_j$ at level $l$. In the course of testing the scheme, it was found that vertical smoothing of the $m_j$'s (using a 1-2-1 stencil) was required to avoid sudden large drag values. It should also be noted that $m_j$ given by Equation (1) is not permitted to increase with height or become negative for this would indicate the reappearance of waves obliterated at lower levels.

In order to examine the sensitivity of the scheme to changes in some of the important parameters it is instructive to use climatological wind and temperature profiles. For this the CIRA-86 reference atmosphere is used and the zonal wind and temperature profiles at 40°N for January are chosen. The corresponding wind profile is shown in Figure 1a.

Since there is considerable uncertainty as to the shape of the low wavenumber end of the $m$-spectrum at tropospheric heights, it is of obvious importance to assess the sensitivity of the drag to this parameter. For computational considerations, however, only certain values of the spectral slope $s$ are useable and Figure 1b shows the resulting drag profiles for three such values ($s = 1$, $s = 3/2$ and $s = 2$). Note that for these and all subsequent results, twelve equally-spaced azimuths are employed along with the specification of an isotropic initial gravity wave source at the ground of $\sigma_{h0} = 1$ m/s. Examination of the results indicates that in all cases the profiles below 85 km are relatively insensitive to changes in the spectral form. However, at the highest levels where the drag arises from the wavenumbers at the very lowest end of the initial $m$-spectrum, the differences are more marked. The short-dashed curve, in fact, exhibits

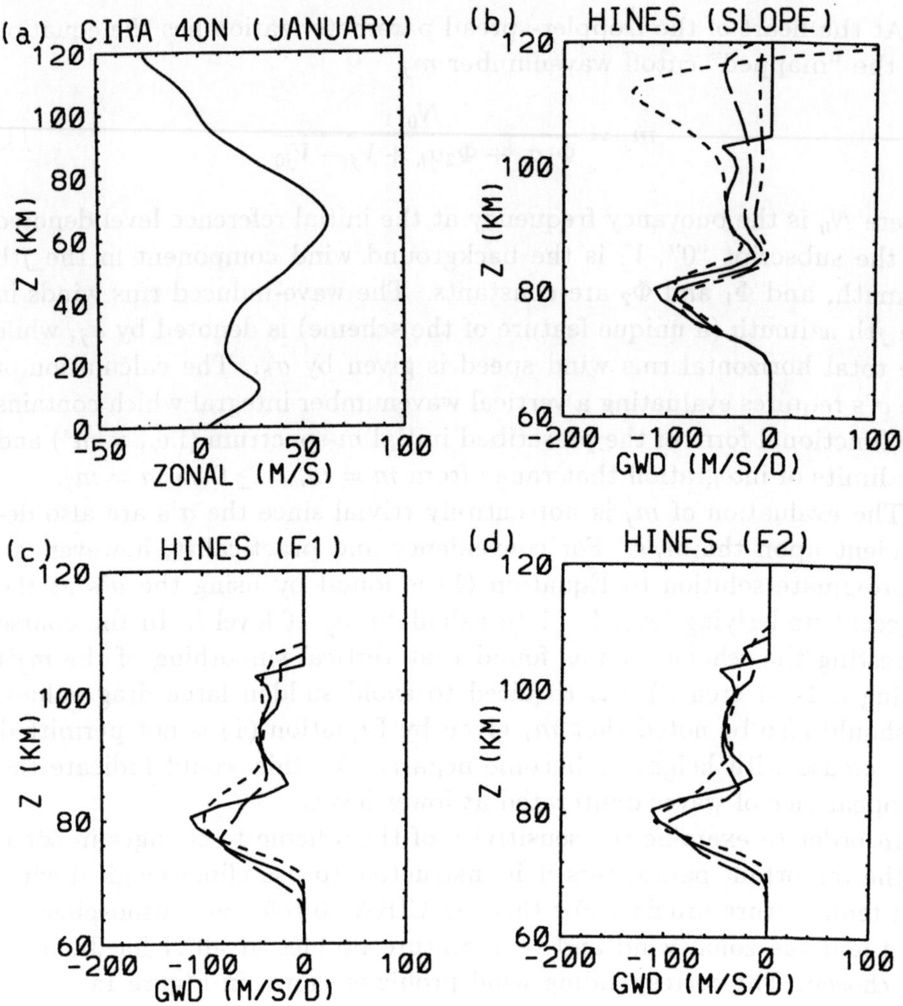


**Figure 1.** Hines GWD using climatological wind and temperature profiles. (a) CIRA-86 zonal mean wind at 40°N for January. (b) GWD for $s = 1$ and $m_{min} = 1/(3\text{km})$ (solid), $s = 1$ and $m_{min} = 0$ (short dashes), $s = 3/2$ and $m_{min} = 0$ (long dashes), and $s = 2$ and $m_{min} = 0$ (medium and two short dashes). (c) GWD for $\Phi_1 = 1.5$ (solid), $\Phi_1 = 1.2$ (short dashes), and $\Phi_1 = 1.9$ (long dashes). (d) GWD for $\Phi_2 = 0.3$ (solid), $\Phi_2 = 0.1$ (short dashes), and $\Phi_2 = 0.5$ (long dashes). Refer to Section 2.1 for more details.

large and rapidly changing drag in the lower thermosphere. Due to both the lack of observational constraints on the very low end of the initial wavenumber spectrum and the inapplicability of the Doppler-spread theory at these very large heights where the spectrum is no longer broad and strongly interacting, it is permissable and in fact advisable to "chop off" the initial spectrum at some minimum value $m_{min}$ as suggested in H97b. For the $s = 1$ case, the solid curve corresponds to $m_{min} = 1/(3\text{km})$. As can clearly be seen, the alteration to the low wavenumber portion of the spectrum has effectively removed the waves responsible for the upper level drag.

Figures 1c and 1d show results for different values of the constants $\Phi_1$ and $\Phi_2$ that appear in Equation (1) for a spectral slope $s = 1$ and $m_{min} = 1/(3\text{km})$. The range of values chosen here represents the expected limiting values suggested in H97a. The sensitivity to changes in these constants is not great and can be easily understood with aid of Equation (1). Increases in $\Phi_1$ and $\Phi_2$, in effect, broaden the fuzzy band around a wave's critical circle, rendering its destruction more probable and causing the wave to "break" lower down.

In the model results that will be described shortly, parameter settings of $s = 1$, $m_{min} = 1/(3\text{km})$, $\Phi_1 = 1.5$, and $\Phi_2 = 0.3$ are used. Another parameter $\Phi_3$, which controls the height at which molecular diffusion begins to dominate, is also needed and is given a value of 1. (It should be noted that in both the calculations of Figure 1 and the subsequent model results, the effects of molecular viscosity are considered in the calculation of the cutoff wavenumber. These effects only become important above 100 km and were included to further reduce the possibility of large drag in the thermosphere.) Another constant $k^*$ is set to $(150\text{ km})^{-1}$, the value used in *Mengel et al.* [1995]. The purpose of $k^*$ is twofold. First, it is required to calculate $m_C$ near and above the turbopause where molecular viscosity determines the maximum permissable value of the cutoff wavenumber. More importantly, it is the value chosen for the integral of the horizontal energy spectrum, denoted $k_j$ in H97a. In the latter role it acts as a multiplicative factor for the momentum fluxes and thus increasing $k_j$ will increase the GWD. Although this is a tunable parameter, it will not be examined here. Since isotropy at the ground has been assumed, a constant value of $k_j = k^*$ has also been chosen.

### 2.2 Fritts GWD scheme

Unlike the Doppler spread scheme which is based solely on theory, the Fritts scheme is constructed from a combination of observed characteristics of the gravity wave spectrum and theoretical notions about the saturation process. Since this scheme has been discussed in detail in FL93, only an overview of the pertinent assumptions will be presented here. The parameterized momentum fluxes are derived from their spectral forms and involve the imposition of several important constraints. First, the total energy at a given height is assumed to be equipartitioned between the zonal and meridional directions. Second, the energy in an individual spectral component (the 4 cardinal azimuths are used) is not permitted to drop to an arbitrarily small value. This constrains the maximum possible amount of azimuthal anisotropy. Third, the vertical profile of the total energy is prescribed: below 85 km FL93 assume a mean scale height for energy growth of $\approx 2.3H$ and a transition region to constant energy above 100 km. A value of 0.2 is used for the anisotropy parameter (denoted by $\alpha$ in FL93). An isotropic energy source $E_0$ of 1 $m^2s^{-2}$ is applied at $z = 0$ and values of 1/18 and 1/22 are chosen for the parameters $C$ and $D$ of FL93, respectively.

## 3. Spectral primitive equations model

A modified version of the linearized spectral primitive equations model of *Vial et al.* [1991] (henceforth VFM91) is employed to examine the two GWD schemes. The log pressure vertical coordinate $z = \ln(p_0/p)$ extends from $z = 0$ to $z = 20$ with $\Delta z = 0.4$. A total of 20 Associated Legendre polynomials are used to represent the horizontal structure. A number of changes to the original VFM91 model have been made. First, the Rayleigh friction approximation of vertical diffusion has been replaced by the actual diffusion which comprises both molecular and eddy terms. Although the Hines scheme calculates an eddy diffusion it is not used. Instead a prescribed height-dependent profile which reaches a value of 200 $m^2/s$ in the upper mesosphere is used. Additional boundary conditions of zero vertical gradients of velocity and temperature are assumed at the model top and bottom. Second, a semi-implicit time formulation has been implemented which enables the time step to be increased to one hour. Third, the transformation from spectral space to real space and back that is required to evaluate the GWD terms is performed using Gaussian quadrature.

The model can be used to simulate either the zonal mean (by setting the zonal wavenumber $m$ in the coefficients of VFM91 to zero) or the diurnal tide (by setting $m$ to one) as was done by those authors. Different heating terms are used depending upon the model configuration. In the tidal case, the diurnal (1,1) Hough mode heating rates for April of *Groves* [1982a,b] are used. In the zonal mean case, the following idealized solsticial ozone heating rate is specified

$$Q = -10^{-4} \exp\left[-\left(\frac{(z_{km} - 50)}{15}\right)^2\right] \sin\theta \qquad (2)$$

where $z_{km}$ denotes height in km. Similar forms of zonal mean heating have been used in other simplified middle atmosphere models [e.g., *Garcia*, 1987]. To simulate long wave cooling in both the zonal mean and tidal versions of the model, a Newtonian cooling formulation is employed, using a coefficient which attains a maximum value of $\approx 0.35$ days$^{-1}$ at 50 km.

The model is integrated forward in time from zero initial conditions until a steady-state is achieved. For this a 50 day integration period was found sufficient.

## 4. Spectral model results

First the zonal mean heating term given in Equation (2) is used in the integration of the zonal mean model. At steady-state and away from regions of strong vertical diffusion, the zonal mean equations are identical to the quasi-geostrophic system which has been used extensively in middle atmosphere studies [e.g., *Garcia*, 1987]. Although the log pressure vertical coordinate is plotted on the figures, approximate altitudes (at least below $z = 14$) can be obtained by multiplying the vertical coordinate by 7 km.

In Figure 2a are shown the zonal winds without GWD. As expected the jets do not close off and the wind speeds reach values of approximately 90 m/s. Figures 2b and 2c show the model simulations using the Hines and Fritts GWD schemes. Although both schemes have closed off the stratospheric jet there are some significant differences. The Hines results exhibit a complete reversal of the mesospheric jet and fairly large vertical wind shears above the stratospheric jet. In contrast, the Fritts scheme produces much weaker vertical shears and at mid-latitudes is unable to reverse the mesospheric jets. Additional experiments (results not shown) employed different values of the lower boundary gravity wave energy input $E_0$ and anistropy parameter $\alpha$ in an attempt to produce a more realistic

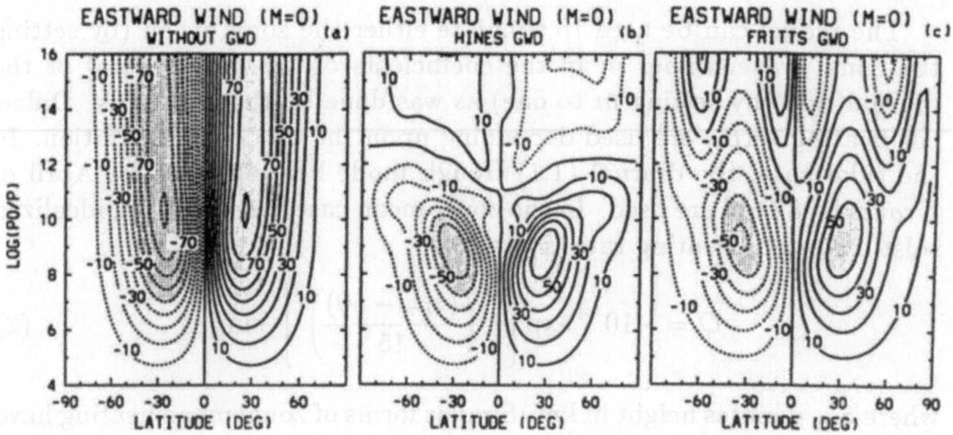


**Figure 2.** Zonal mean zonal wind (m/s) at solstice: (a) without GWD, (b) using Hines GWD, and (c) using Fritts GWD. Wind speeds in excess of 50 m/s are shaded. Refer to Section 2 for specification of parameters.

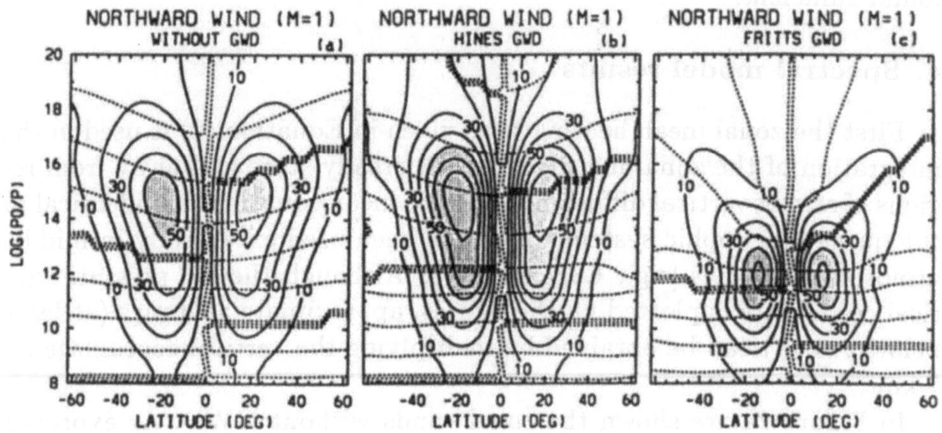


**Figure 3.** Meridional wind component of the diurnal tide at 0h UT: (a) without GWD, (b) using Hines GWD, and (c) using Fritts GWD. Amplitude in m/s (solid) and phase defined as longitude of maximum (dashes). Vertical wavelength given by distance between two sets of the 3 closely-spaced phase lines where the phase jumps from 360° to 0°. Amplitudes in excess of 50 m/s are shaded.

mesospheric zonal wind reversal. In all cases the Fritts scheme produced mesospheric winds having large latitudinal variations with the largest vertical shears (and consequently the strongest reversals) occurring at low latitudes and near 60°. Clearly, to rectify this in this simple model, a latitudinally varying energy source would have to be used.

The final results pertain to the vertically-propagating diurnal tide that is forced in the spectral model by the (1,1) Hough mode heat sources. For practical considerations, the model is integrated without GWD for the first 20 days after which the GWD is slowly turned on and the integration continued for the remaining 30 days. The final solutions are steady. It should be noted that in this case there are no zonal mean winds so that the gravity waves 'see' only the tidal winds.

Figure 3 shows the meridional wind component of the tide for the cases without GWD (a), with the Hines scheme (b), and with the Fritts scheme (c). The differences between Figures 3b and 3c are quite striking - the Hines scheme has amplified the tide at all levels (with respect to the no-GWD case in Figure 3a) while the Fritts scheme has strongly damped it in the thermosphere but amplified it below. Both schemes have advanced the tidal phase and shortened the vertical wavelength. The Fritts scheme however has a significantly shorter wavelength in the upper mesosphere than occurs when the Hines scheme is used. The integrations were repeated using modified parameter values and the overall results were unchanged.

## 5. Summary and conclusions

In this paper the spectral gravity wave drag parameterizations of *Hines* [1997a,b] and *Fritts and Lu* [1993] have been compared. Since the Hines scheme is the more recent of the two, a closer examination of the sensitivity to changes in its tunable parameters was performed. These results showed that the slope of the initial spectrum, one of the more important parameters that must be prescribed in this scheme, was important mainly in the lower thermosphere above 90 km.

A simplified time-dependent model of the middle atmosphere was then used to explore the behaviour of the two schemes. Both schemes were able to close off the stratospheric winter jet, although for all the various parameters settings used the Fritts scheme was unable to produce a realistic zonal wind reversal in the mesosphere. The model simulations of the diurnal tide revealed striking differences between the two GWD schemes.

The Fritts scheme strongly damped the tide in the lower thermosphere while the Hines scheme amplified it.

The question of tidal wind amplification or reduction by GWD is an important one. Recent results from the Upper Atmosphere Research Satellite [e.g., *McLandress et al.*, 1996; *Burrage et al.*, 1995] have revealed that the observed amplitude of the diurnal tide at equinox is considerably larger than the model-simulated one of *Hagan et al.* [1995]. Although inaccuracies in the calculated lower atmosphere heating rates used in the model may account for this, uncertainty in the role of internal gravity waves is the more likely culprit. The diurnal tide model of *Hagan et al.* [1995], in fact, incorporates the effects of gravity wave momentum deposition using an equivalent Rayleigh friction coefficient calculated by *Miyahara and Forbes* [1991]. In this latter paper a monochromatic parameterization like that of *Lindzen's* [1981] was used and found to cause a tidal amplitude reduction. Clearly, the question of gravity wave/tidal interactions must be reexamined.

**Acknowledgements**

I would like to thank Drs. Jeff Forbes, François Vial, Maura Hagan, Rashid Akmaev, Colin Hines and Ray Roble for helpful discussions.

## References

Burrage MD, Hagan ME, Skinner WR, Wu DL, Hays PB (1995) Long-term variability in the solar diurnal tide observed by HRDI and simulated by the GSWM. Geophys Res Lett 22:2641-2644

Fritts DC, Lu W (1993) Spectral estimates of gravity wave energy and momentum fluxes, Part II: Parameterization of wave forcing and variability. J Atmos Sci 50:3695-3713

Garcia RR (1987) On the mean meridional circulation of the middle atmosphere. J Atmos Sci 44:3599-3609

Groves GV (1982a) Hough components of ozone heating. J Atmos Terr Phys 44:111-121

Groves GV (1982b) Hough components of water vapour heating. J Atmos Terr Phys 44:281-290

Hagan ME, Forbes JM, Vial F (1995) On modeling migrating solar tides. Geophys Res Lett 22:893-896

Hines CO (1997a) Doppler-spread parameterization of gravity-wave momentum deposition in the middle atmosphere, Part 1: Basic formulation. J Atmos Terr Phys (in press)

Hines CO (1997b) Doppler-spread parameterization of gravity-wave momentum deposition in the middle atmosphere, Part 2: Broad and quasi monochromatic spectra and implementation. J Atmos Terr Phys (in press)
Hines CO (1993) The saturation of gravity waves in the middle atmosphere, Part IV: Cutoff of the incident wave spectrum. J Atmos Sci 50:3045-3060
Hines CO (1991a) The saturation of gravity waves in the middle atmosphere, Part I: Critique of linear-instability theory. J Atmos Sci 48:1348-1359
Hines CO (1991b) The saturation of gravity waves in the middle atmossphere, Part II: Development of Doppler-spread theory. J Atmos Sci 48:1360-1379
Hines CO (1991c) The saturation of gravity waves in the middle atmosphere, Part III: Formation of the turbopause and of turbulent layers beneath it. J Atmos Sci 48:1380-1385
Lindzen RS (1981) Turbulence and stress owing to gravity wave and tidal breakdown. J Geophys Res 86:9707-9714
McLandress C, Shepherd GG, Solheim BH (1996) Satellite observations of thermospheric tides: Results from the Wind Imaging Interferometer on UARS. J Geophys Res 101:4093-4114
Mengel JG, Mayr HG, Chan KL, Hines CO, Rėddy CA, Arnold NF, Porter HS (1995) Equatorial oscillations in the middle atmosphere generated by small scale gravity waves. Geophys Res Lett 22:3027-3030
Miyahara S, Forbes JM (1991) Interactions between gravity waves and the diurnal tide in the mesosphere and lower thermosphere. J Meteorol Soc Jpn 69:523-531
Vial F, Forbes JM, Miyahara S (1991) Some transient aspects of tidal propagation. J Geophys Res 96:1215-1224

# Modeling the Seasonal Cycles and Equatorial Oscillations with a Parameterization of the Doppler Spread Gravity Wave Theory

Hans G. Mayr
Goddard Space Flight Center
Greenbelt, MD, USA
with J. G. Mengel[1], C. O. Hines[2], K. L. Chan[3], and N. F. Arnold[4]

**Abstract:** The Doppler spread parameterization (DSP) for the deposition of small scale gravity wave (GW) momentum and energy in the middle atmosphere (Hines, 1996a, b) has been incorporated into the 2D version of a numerical spectral model (NSM) of the middle atmosphere. With a globally uniform flux of (quasi) isotropically propagating GW emanating from the troposphere, the NSM was integrated for several model years, and preliminary results were presented by Mengel et al. (1995). The model describes the global temperature and wind fields during solstice and produces significant semi-annual and quasi-biennial oscillations, SAO and QBO, in the zonal winds near the equator. It also shows the QBO extending into the upper mesosphere, in agreement with UARS measurements (Burrage et al., 1996). When the initial GW flux is enhanced, the SAO and QBO amplitudes increase, but the QBO period decreases. When the eddy diffusivity is reduced by a factor of two, the QBO amplitude and period increase significantly, but the SAO amplitude decreases. With time independent solar heating (perpetual equinox) and standard diffusivity, periodicities near 17 months and 8 months are generated, respectively, in the lower and upper stratosphere, the latter being close to the 6 months period of the semi-annual component of solar heating.

## 1. Introduction

Lindzen (1981) suggested that the anomalous reversal in the seasonal variations of the mesospheric temperature, observed to reach a minimum in summer at altitudes near 85 km, is caused by the momentum deposition of breaking gravity waves. He parameterized the process by describing a small number of discrete, independent waves which individually become unstable and deposit their momentum to drive atmospheric winds. This wave breaking process has since been studied in great depth (e.g. Weinstock, 1982; Dunkerton, 1982a, b, 1989; Schoeberl et al., 1983; Schoeberl and Strobel, 1984; Fritts, 1984, 1989; Fritts and Dunkerton, 1985; Schoeberl, 1985, 1987, 1988), and numerical models (e.g., Holton 1982; Gaertner et al., 1983; Geller, 1984; Garcia and Solomon, 1985) have demonstrated that it can account for the observed mesospheric temperature anomaly.

---

[1]Applied Research Corporation, Landover, MD
[2]Hines Consultancy, Toronto, Canada
[3]Hong Kong University of Science and Technology, Hong Kong
[4]Universities Space Research Association, Seabrook, MD

NATO ASI Series, Vol. I 50
Gravity Wave Processes
Their Parameterization in Global Climate Models
Edited by Kevin Hamilton

The theories of Lindzen and Holton (1968) and Holton and Lindzen (1972) have shown how tropical planetary scale waves can accelerate the zonal circulation to produce the Quasi-Biennial Oscillation (QBO) observed at low latitudes. Eastward propagating Kelvin waves and westward propagating mixed Rossby gravity waves would interact with the zonal circulation to deposit their momentum due to critical level interaction and radiative damping. This theory was extended by Dunkerton (1979, 1982c) to explain also the eastward phase of the Semi-Annual Oscillations (SAO) at equatorial latitudes observed in the upper stratosphere and lower mesosphere, which cannot be accounted for by radiative forcing and interhemispheric momentum advection associated with the meridional circulation (Hamilton, 1986). The observed reversal in the SAO of the zonal winds near the mesopause at low latitudes was attributed to breaking gravity waves (Dunkerton, 1982c).

Considering Kelvin waves inferred from satellite observations, Hitchman and Leovy (1988) concluded that these waves produce only 20% to 70% of the required eastward acceleration in the stratospheric SAO and that gravity waves are likely to be important too. Recent results obtained by Hamilton et al. (1995) from a GCM that resolves planetary waves but not small scale gravity waves showed that in the stratosphere the model could not reproduce the eastward phase of the SAO and produced only a relatively weak QBO with amplitudes of less than 2 m/s, which are an order of magnitude smaller than those observed.

While it is accepted that wave mean-flow interactions are of central importance for the zonal circulation of the middle atmosphere in general and for generating the observed equatorial oscillations in particular, the role played by small scale gravity waves in relation to large scale and planetary scale waves is not known.

A parameterization for small scale gravity waves (GW) has recently been developed by Hines (1996a, b) based on his Doppler Spread Theory (Hines, 1991a, b, c, 1993), which accounts for wave-wave and wave-mean flow interactions and provides the momentum source for the circulation and the eddy diffusivity. We have incorporated this scheme into a numerical spectral model (Chan et al., 1994a, b) of the middle atmosphere, and preliminary results have been reported by Mengel et al. (1995). In Section 2 of the present paper, the numerical model and implementation of the gravity wave parameterization are briefly described. In Section 3, numerical results are presented that describe the computed temperature and zonal winds during December solstice. In Section 4, the computed equatorial oscillations (SAO and QBO) are presented for different GW parameters and for perpetual equinox. In Section 5, the results are summarized.

## 2. Model

The Numerical Spectral Model of Chan et al. (1994a, b, 1995), referred to as NSM, has earlier been applied to the mesosphere to describe the mean zonal circulation, the thermally driven diurnal tides and the generation of planetary waves with the use of Rayleigh friction

(Leovy, 1964). This model is time-dependent, 3D, nonlinear, and is formulated in terms of spherical harmonics. In order to limit the nonlinearity to quadratic order so that alias-free transformations can be performed, however, one approximation is made which is of no consequence for the present application: the horizontal density variations are neglected in the nonlinear advection terms (but retained in the linear terms). The computation is performed efficiently with the use of an implicit numerical procedure that avoids severe time step restrictions imposed by the fast traveling waves. The model computes the perturbations from fixed, globally averaged temperature and density variations, which are adopted as functions of height. Normally, the integration is carried out from the surface up to 400 km, assuming that the vertical and horizontal winds vanish at the lower boundary and that, due to molecular dissipation, the vertical gradients of temperature and horizontal winds vanish at the upper boundary.

In the present application, a 2D version of the NSM is employed with emphasis on the stratosphere and mesosphere. The integration is carried out over periods of several model years, and in order to speed up the computation, the upper boundary is chosen at 130 km, which is well above the region of primary interest. With inclusion of the GW parameterization provided by Hines (1996a, b), numerical tests show that the altitude- and related time- integration steps need to be fairly small and they were chosen to be on the order of 0.75 km and 20 minutes, respectively. The maximum meridional wave number for the spherical harmonics retained in the model is $\ell = 12$; thus, the smallest horizontal wave length resolved in the model is about 30°, corresponding to a grid point resolution of about 7.5°. A control run with maximum wave number $\ell = 16$ produced results that were virtually identical, thus justifying the truncation.

The heat source driving the circulation is based on the parameterization developed by Strobel (1978) and contains components associated with the absorption of solar UV radiation by $O_3$ and $O_2$. The heating efficiency is set to one, assuming that the dissociation energy is locally recovered through recombination. The radiative loss is described in terms of Newtonian cooling, and the rates for that are taken from Dickinson (1973) and Wehrbein and Leovy (1982) as shown in Figure 2 of Chan et al. (1994b). While the model is integrated upward from the surface, where natural boundary conditions apply, it does not account for any of the very complex processes that are associated with solar heating, radiative cooling and convection in the troposphere.

Instead of Rayleigh friction as employed in the NSM by Chan et al. (1994b), the momentum deposition by small scale gravity waves was parameterized in a fashion now to be described.

We employ the Doppler-spread parameterization (DSP) developed by Hines (1996a, b). It is designed to take into account nonlinear interactions among the waves of the spectrum, in addition to interactions between the waves and the background winds. These nonlinear interactions derive primarily from the advective nonlinearity of the $\mathbf{V}\cdot\nabla$ term of the Eulerian fluid dynamic equations, with consequences that have been determined in an approximate way by means of a heuristically-inspired mathematical development (Hines, 1991b, c, 1993). The essential aspects of the DSP are as follows.

A set of discrete azimuths of propagation is adopted, proposed as eight or more but taken to be four in this work, the four cardinal directions. A one-dimensional power spectral density, varying over vertical wave number m, is assumed for each azimuth at some initial (low) height. This spectrum is taken to propagate upward, to amplify in response to the diminishing gas density, and to become deformed in consequence of the ever-increasing role played by the resultant nonlinearities. Implicit in the model (and explicit in the underlying analysis) is the development of the characteristic "knee" of middle-atmospheric m-spectra, at which the spectral intensity changes from being an increasing function of m to being a decreasing function of m, the latter usually being described as an $m^{-3}$ tail. The nonlinear waves that constitute this tail have been spread into it from initial m values close to its lower boundary, at the knee, and they tend to spread to ever-larger values of m as a consequence of ever-increasing nonlinear interaction.

Wind and temperature gradients induced by the wave spectrum -- most notably, by the tail portion of the spectrum -- produce instability if the tail extends to sufficiently large m values. It is assumed that the spectrum extends to but not beyond some corresponding m value, denoted $m_M$, that is determined by a requirement for marginal instability. From earlier analysis (Hines, 1991b), this value is taken to be inversely proportional to the rms wave-induced wind; it therefore decreases with increasing height.

It is further assumed that waves having m values less than $m_M$ propagate without dissipation, and that dissipation occurs abruptly at $m_M$ when the height-varying m comes to equal the height-varying $m_M$. This is at best an approximation, one that is based on the view that instability, when it occurs, does so primarily in small patches "here and there, now and then" (as argued by Hines, 1991a, and as may be illustrated by whitecaps on ocean waves) and acts in a highly preferential way to dissipate small-scale, rather than large-scale, waves. The approximation simply carries this view to an extreme: it ignores any dissipation of the larger-scale waves and takes into account only that of the smallest-scale waves, those at $m_M$, which it takes to be obliterated. This approximation would not be valid in a detailed study of dissipation within the tail region, of course, but the details that would be sought in such a study are of little relevance to large-scale parameterizations of the type required here.

In the DSP, the dissipating edge of the spectrum at $m_M$ is mapped back to a corresponding m value at the initial level, denoted $m_j$. This is different for different azimuths (denoted by j=1,2...J in general; E, N, W, S here), if only because of the role of Doppler shifting induced by the background wind. But in the DSP mapping, a Doppler spreading caused by the wave-induced winds is also taken into account, thereby giving the DSP its unique character. Whereas other parameterizations would invoke dissipation if a wave's vertical wave-number becomes too large (according to some criterion) under Doppler shifting induced by the background wind, the DSP takes into account Doppler shifting induced also by the wave spectrum itself: the rms fluctuation is added to the background wind, or subtracted from it if that is more deleterious, the wind fluctuations in effect acting statistically to spread the background wind over a significant range of values and so increasing the propensity for dissipation. (More properly, it is a multiple $\Phi_1$ of the rms wind that is employed, the multiple being 1.5 in the present work.)

The $m_j$ of a given azimuth depends on the rms wind fluctuation in that azimuth, which in turn depends on the spectra at all non-orthogonal azimuths and so on the $m_j$ for all such azimuths, while the $m_M$ that is employed in determining the $m_j$ depends on the rms total wind fluctuation. In principle, all must be determined together in a self consistent manner. The DSP provides means of circumventing this requirement if the height and time steps of the large scale model are sufficiently small, as they are here, but an iteration to self consistency was nevertheless performed in the present work.

Since $m_M$ necessarily decreases with increase of height (in the presence of a broad spectrum), all $m_j$ tend to do so as well. Their decrease implies that a smaller and smaller portion of the initially incident spectrum continues to propagate upward. The lost portion of the spectrum deposits its associated momentum and energy flux, and it does so at a rate that is calculable from a knowledge of the initial spectrum and of the rate at which each $m_j$ is decreasing. (The decrease of $m_j$ can be impeded, for some azimuths, by a sufficient shear in the background winds, but the DSP never permits an $m_j$ to increase: waves may or may not be obliterated, but they are never resurrected.)

It is this calculation of momentum and energy deposition that the DSP provides as its primary output. An associated calculation provides an estimate of turbulent diffusivity on the assumption that, below the turbopause, the deposited intrinsic energy acts as input to an isotropic Kolmogorov spectrum of turbulence having an outer scale comparable to $m_M^{-1}$.

The DSP contains a number of parameters that are adjustable within rather narrow ranges. One such, $\Phi_1$, has already been identified above as taking the value 1.5 in the present work. A second, $\Phi_2$, which enters the specification of $m_M$, probably should lie between 0.1 and 0.4; the value 0.25 was adopted here. The height of the turbopause, which is determined by molecular diffusivity coming to terminate the wave spectrum at an m value less than the $m_M$ that is required for instability, was taken as 105 km in initial simulations; this led to a certain $\Phi_3 = 0.4$. The vertical flux of vertical momentum was ignored, removing the need for assigning a value to the DSP's $\Phi_4$. Heating produced by wave dissipation was likewise ignored, but the energy deposition was employed to determine the diffusivity and for that purpose $\Phi_5$ of the DSP was set to 1; the further $\Phi_6$ was taken as 0.5 to provide a diffusivity of $1.3 \times 10^2$ $m^2 s^{-1}$ at 80 km, in substantial agreement with the value given by Brasseur and Solomon (1984), though we now believe that this value may be too high. For the Prandtl number, the value $Pr = 3$ was adopted. A characteristic horizontal wave number $k_j$, in principle variable with azimuth j, was taken to equal a single value $k^*$ for all. In most cases this was taken to be $(150\ km)^{-1}$, but exceptions occur and are identified.

Finally, the DSP provides for various forms and intensities of input spectrum at the initial height, taken to be the ground level here. Here the spectral density was taken to be proportional to m (for m less than the ground-level value of $m_M$; zero otherwise) for all (4) azimuths and all grid points. The rms wind fluctuation was likewise taken to be the same for waves at all (4) azimuths and all grid points, of a magnitude such that, at 20 km height, the rms total wind fluctuation was 2.5 m/s. This choice is in rough conformity with mean values reported by Allen and Vincent (1995), for example. In reality, these rms values will be variable with azimuth, with latitude (and in 3-D studies, with longitude), and with season and meteorological conditions. It is to be emphasized that, for the initial results presented here, we have made no attempt to improve upon the crude assumptions of (quasi) isotropic propagation and globally and seasonally uniform input spectra.

## 3. Global Scale Seasonal Variations

With the DSP incorporated, the NSM was run in 2D for a number of model years. Accounting for spin-up, the first two years were ignored, and for the remaining six years the average temperature and zonal wind fields are shown for December solstice in Figure 1. Because of the omission of tropospheric processes, the lowest 25 km of the model calculation

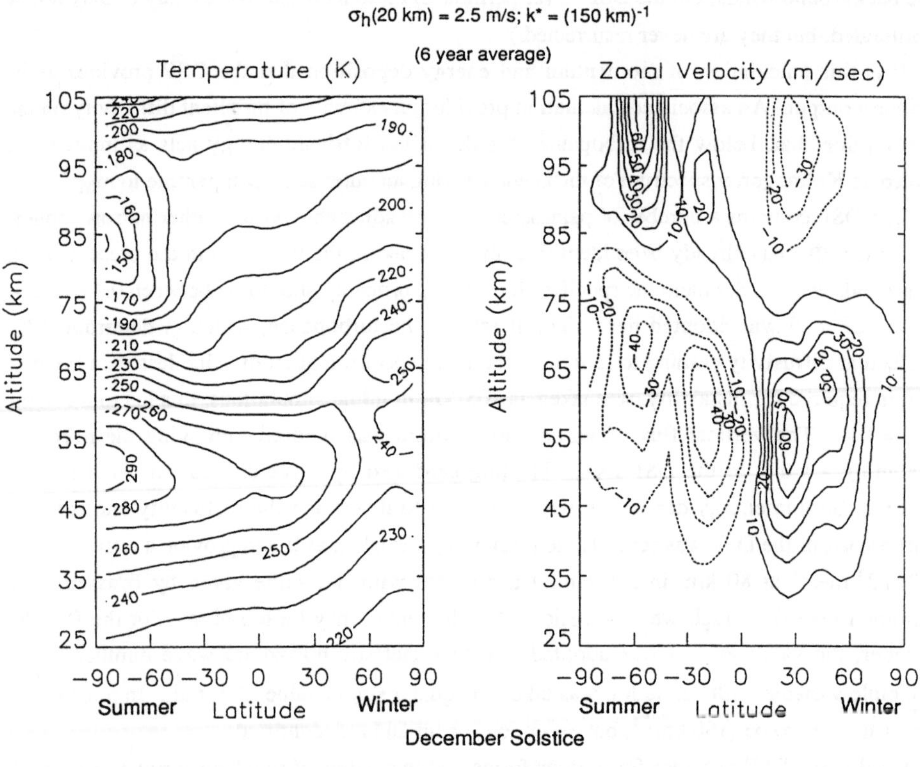


Figure 1: Computed temperatures and zonal winds (positive to the east) during December solstice, with GW momentum deposition and eddy diffusivity (heat conduction and viscosity) provided by the DSP. These results are identical to those shown by Mengel et al. (1995).

are omitted from the figure. Through most of the middle atmosphere, the model reproduces the salient features that characterize the observations. Near 50 km, where the maximum occurs in the absorption of UV radiation by ozone, the temperature peaks in the summer polar region; and due to geostrophic balance, the zonal winds are eastward and westward in the winter and summer hemispheres, respectively. At higher altitudes near 85 km, however, the temperatures are lowest in summer and not in winter, in agreement with observations and contrary to expectations from radiative considerations. Above that altitude, the zonal winds reverse direction, turning eastward in summer and westward in winter, again consistent with observations. As proposed by Lindzen (1981) and demonstrated in earlier models such as that of Garcia and Solomon (1985), these reversals in the latitudinal temperature variations and zonal wind velocities are the direct consequence of wave mean flow interactions.

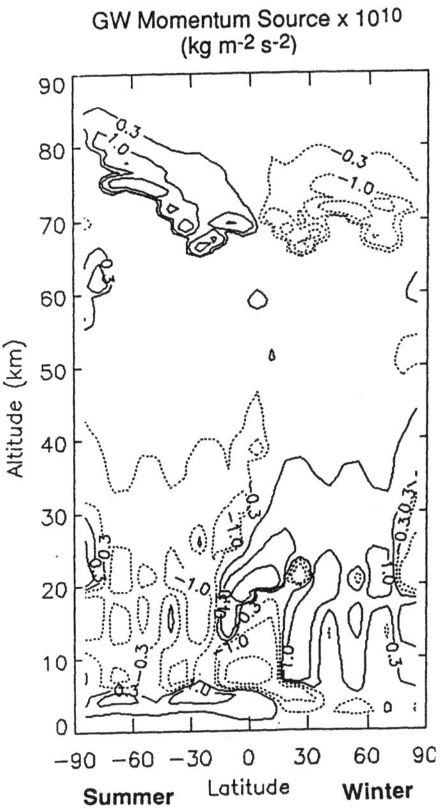


Figure 2: Computed GW momentum source for December solstice (after three model years). In winter, below 40 km, eastward momentum is deposited and removed from the GW flux (generated isotropically); thus, the waves have a net westward momentum in the upper mesosphere, which is deposited causing the winds to reverse there.

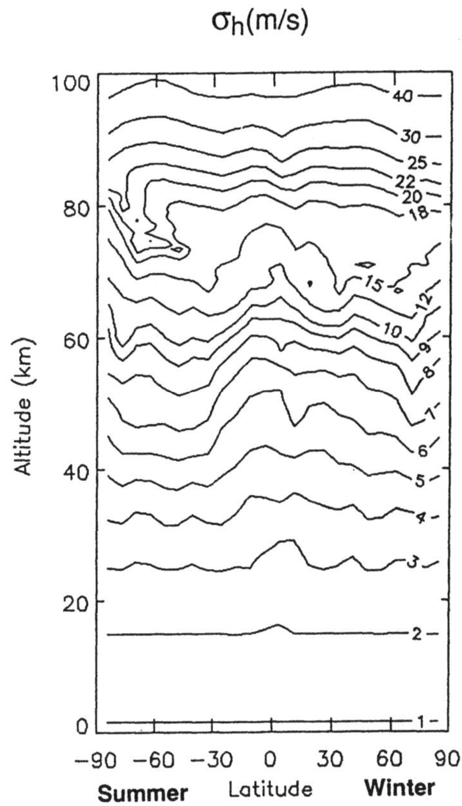


Figure 3: Computed wave induced rms fluctuation of total horizontal wind for December solstice (after three model years). At the surface it is taken to be independent of latitude and season, to produce about $\sigma_h = 2.5$ m/s at 20 km.

The GW momentum source for December solstice after three model years of integration is shown in Figure 2. As the GW propagate through the stratosphere below 50 km where the thermally driven zonal circulation increases with altitude, Doppler shifts, aided by Doppler spreading, cause the waves to break and deposit momentum, which tends to accelerate the winds (though insignificantly due to the large ambient density). Between 55 and 70 km, the zonal winds decrease with height and no longer force the dissipating portion of the GW spectrum to be Doppler shifted to shorter vertical wave numbers; hence the waves are not absorbed there. The GW which reach the upper mesosphere then have a net momentum in the direction opposite to that of the stratospheric circulation. This momentum is ultimately deposited above 70 km due to wave breaking, which causes the zonal circulation to be decelerated and ultimately even to reverse at higher altitudes. At altitudes above 65 km, the GW interaction produces accelerations with peak values close to 100 m/s/day, comparable to the values derived by Brasseur and Hitchman (1986). As shown with earlier models (e.g., Garcia and Solomon, 1985), this GW induced acceleration in the zonal circulation generates a meridional circulation with accompanying vertical winds, which in turn redistribute energy and cause the reversal in the temperature variation. The GW momentum source shown here bears virtually no resemblance to Rayleigh friction, which is formulated as a drag force proportional to the zonal winds.

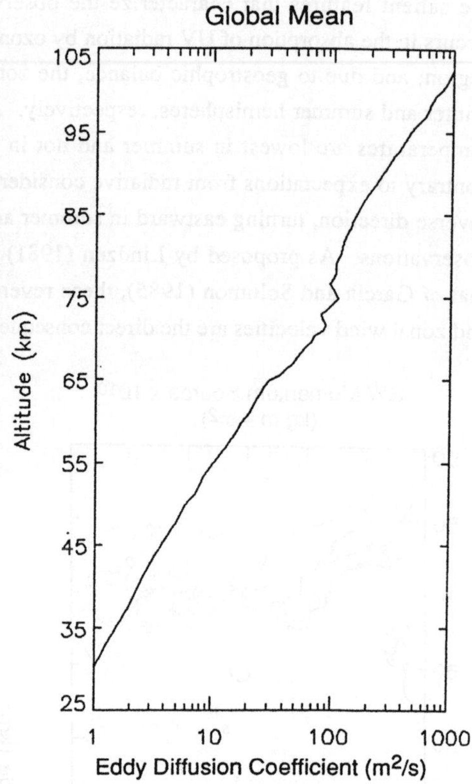


Figure 4: Computed eddy diffusivity for December solstice (after three model years) averaged over latitudes as employed in the model.

The GW spectrum that produces the momentum source and diffusivity derived in the model is characterized by the total wave induced rms horizontal wind, $\sigma_h$, which is shown in Figure 3 for December solstice after three years of integration. As pointed out earlier, the GW are assumed to originate at the surface where $\sigma_h$ is taken to be independent of latitude and season and its value is chosen to produce about $\sigma_h = 2.5$ m/s at 20 km, consistent with observations. This quantity tends to increases with height inversely as the fourth root of the ambient gas density.

Presently, only the global average of the eddy diffusion rate, K, at any given time is considered in the NSM and it is shown in Figure 4 for December solstice after three model years of integration. As pointed out earlier, this quantity is derived by choosing a DSP parameter such that it produces a value of about K(80 km) = $1.3 \times 10^2$ $m^2/s$.

## 4. Equatorial Oscillations, SAO & QBO

For the same GW parameters that produce the seasonal variations in the temperature and wind fields (Figure 1), equatorial oscillations are generated in the zonal winds at equatorial latitudes which are presented in Figure 5 (identical to the results presented by Mengel et al. (1995)). Eliminating the first two years to account for spin-up, the annual and mean components were removed from the remaining six years of integration in order to identify more clearly the residual oscillations

At altitudes above 35 km in Figure 5a, the dominant (residual) periodicity is six months, and the corresponding Fourier component or Semi-Annual Oscillation (SAO) is shown in dashed lines. Near the stratopause at 50 km, the wind velocities in the SAO peak and approach 20 m/s, somewhat smaller than those observed (e.g., Reed, 1966; Hirota, 1980; Reddy and Vijayan, 1993). At 70 km, the variations in the wind velocities abate but higher up increase again and then reverse direction such that near the mesopause at 80 km a SAO is generated with winds of about 10 m/s, which are opposite in phase to those at lower altitudes. This phase reversal in the SAO is observed, but the measured amplitudes are between 15 and 25 m/s (Hirota, 1980; Reddy and Vijayan, 1993 ). We find that these oscillations are essentially confined to latitudes below 20$^{o}$ although the initiated GW flux is taken to be independent of latitude (and season). Since the Coriolis force is unimportant near the equator, the GW induced oscillation in the zonal circulation is not accompanied by a significant oscillation in the meridional winds, which can build up back pressure to dampen the zonal wind oscillation as is the case at middle and high latitudes.

In the stratosphere below 35 km, the computed variations in the zonal circulation shown in Figure 5b have a period of about 21 months, near the lower end of the periods observed in the Quasi-Biennial Oscillation (QBO). The computed downward phase progression of 1.6 km/month is also close to that observed but the wind velocities are about 7 m/s, less than half of those observed (e.g., Andrews et al., 1987). Like the SAO in the model, the computed QBO is confined to low latitudes, in agreement with observations.

Our model results show that with increasing altitude, the computed wind pattern becomes more irregular. This is understandable, considering that the GW that reach higher altitude levels have been filtered and modulated by the winds they traverse and thus effectively transfer information about the zonal winds from lower to higher altitudes. This is seen in Figure 6, where we eliminated the short term variations with a 180 day running average. It reveals that the QBO-like oscillation is not confined to the stratosphere, as was commonly thought,

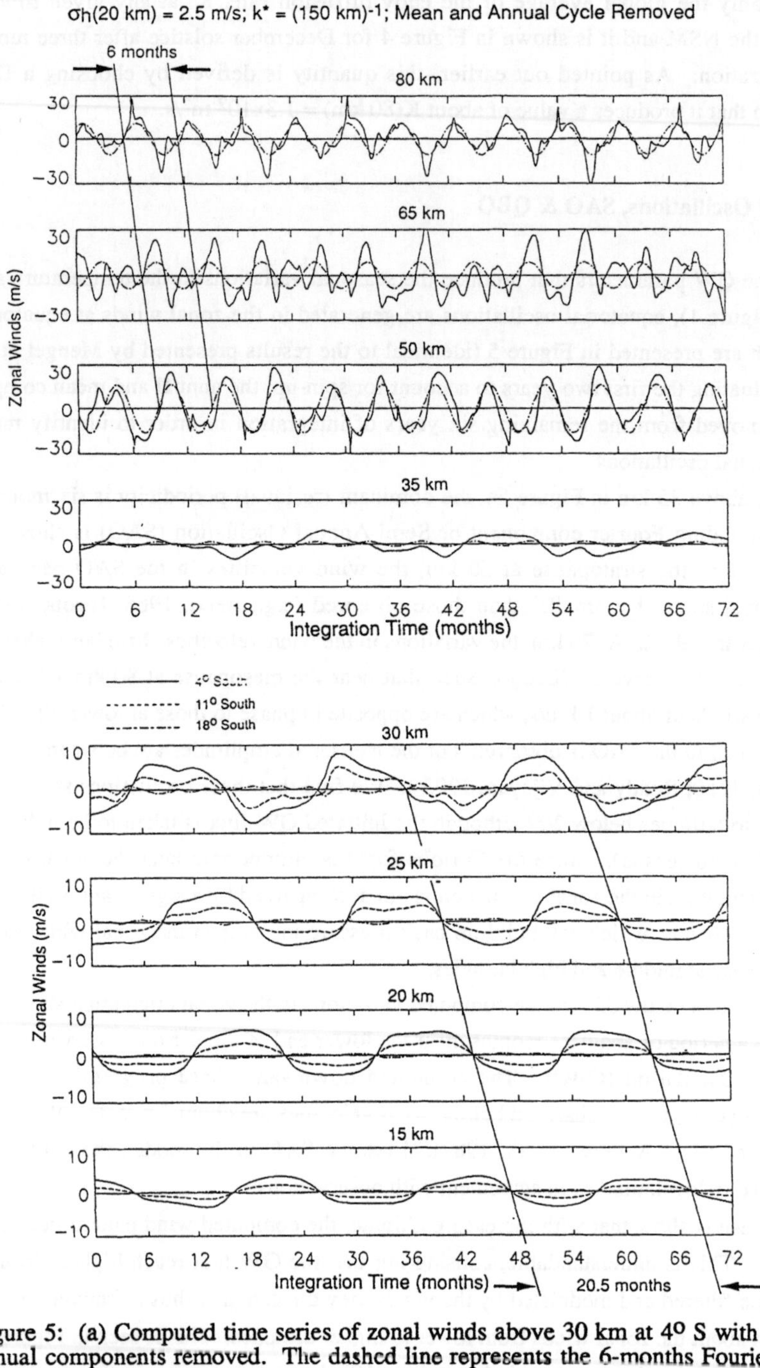


Figure 5: (a) Computed time series of zonal winds above 30 km at 4° S with the mean and annual components removed. The dashed line represents the 6-months Fourier component, the Semi-Annual Oscillation (SAO). The SAO at 80 km is in phase opposite to that at 50 km as observed. (b) Time series of zonal winds below 35 km for different latitudes, showing a periodicity close to 21 months which approaches that of the Quasi-Biennial Oscillation (QBO). These results are identical to those presented by Mengel et al. (1995).

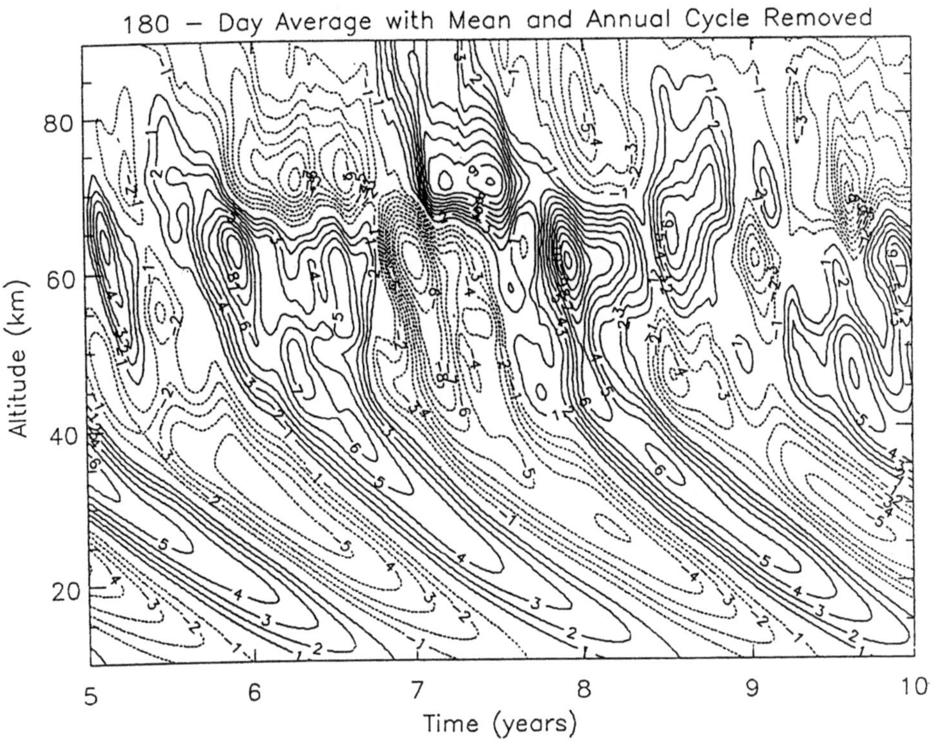


Figure 6: Computed zonal winds (mean and annual components removed) at 4° S, as in Figures 5, but with 180 day running averaging. The QBO like oscillation extends into the upper mesosphere, in agreement with UARS observations (Burrage et al., 1996).

but extends all the way up into the upper mesosphere. Such an oscillation has been observed with the High Resolution Doppler Interferometer (HRDI) measurements onboard the UARS spacecraft (Hays et al., 1993) as reported by Burrage et al. (1996).

The equatorial oscillations presented here are a spin-off from a model that is designed, with the simplest assumptions and minimal tuning of GW parameters, to reproduce the observed temperature reversal near the mesopause during solstice. No attempt has been made to reproduce the observed equatorial oscillations accurately, in part because the model does not yet consider the effects of planetary waves. At this early stage of exploring the effects of small scale gravity waves, however, we have carried out a few computer experiments with the aim of providing some understanding.

In the DSP, the initiated GW momentum flux is proportional to $k^*\sigma_h^3$. When $\sigma_h$(20 km) is reduced in the model from 2.5 to 2 m/s and $k^*$ is increased from (150 km)$^{-1}$ to (77 km)$^{-1}$, the momentum flux remains constant, and our numerical results (not presented here) show that in these two cases the computed quasi-biennial and semi-annual oscillations do not differ significantly.

For $\sigma_h$(20 km) = 3 m/s and k* = (150 km)$^{-1}$, the GW momentum flux is enhanced by 70%, and the computed wind oscillations are different, as shown in Figure 7. Compared to Figure 5, the average SAO amplitude in the stratosphere and mesosphere increased by almost 50% to produce closer agreement with observations. The magnitude of the QBO in the stratosphere increased nearly as much, but its period decreased from 21 to 17 months, moving it away from the observations.

Thus far, we had retained a constant, height dependent eddy diffusivity, which was accomplished by choosing the appropriate DSP parameter to produce the "standard diffusivity", $K_S$, in our model as shown in Figure 4. Here we present a solution with a different value, $\Phi_6$ = 0.25, resulting in a diffusivity that is reduced by a factor of two. In Figure 8, the computed time series in the altitude range from 20 to 80 km are shown for K = $K_S$/2. By comparison with the earlier results in Figure 5 for the standard (and larger) eddy diffusivity, the computed QBO in the lower stratosphere now has a much larger amplitude, about 13 m/s, and its period is about 30 months, close to the upper end of the observed values. At altitudes above 40 km, however, the SAO (in dashed lines) is significantly reduced in magnitude, and the variations are much less coherent. Conversely, when K is increased from $K_S$ to $K_S$x2, the results (not presented) show that, at altitudes below 40 km, the induced period is only 13 months, much shorter than the 21 months seen in Figure 5 and its amplitude is only about half as large; however, above 40 km, the SAO amplitudes are somewhat larger than those in Figure 5, and the variations are more coherent.

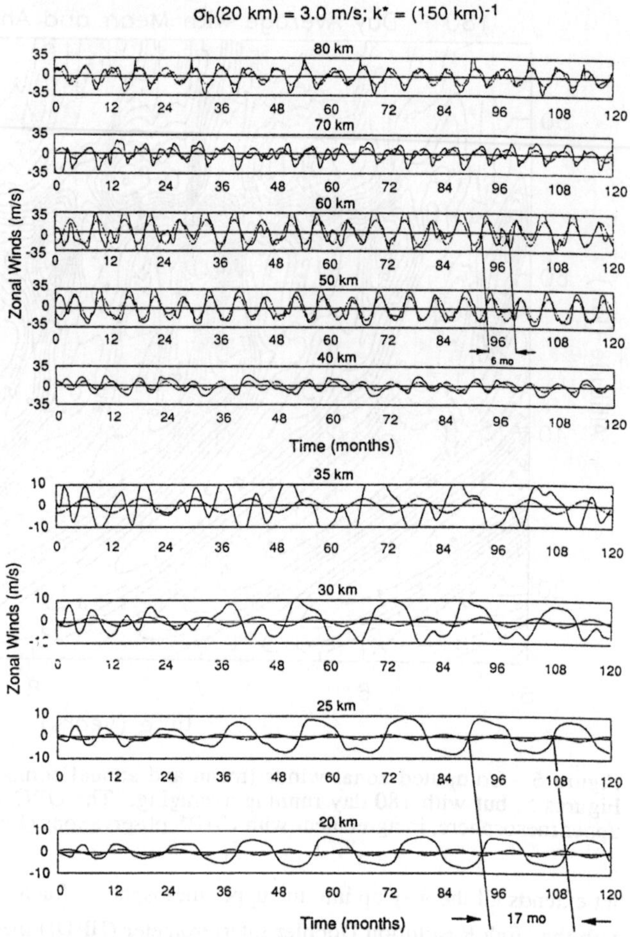


Figure 7: Computed zonal winds at 4° S with mean and annual components removed for $\sigma_h$ (20 km) = 3 m/s, k* = (150 km)$^{-1}$, with GW momentum flux about 70% larger than employed earlier. The SAO and QBO amplitudes are enhanced compared to the results in Figure 5 but the QBO period is shorter.

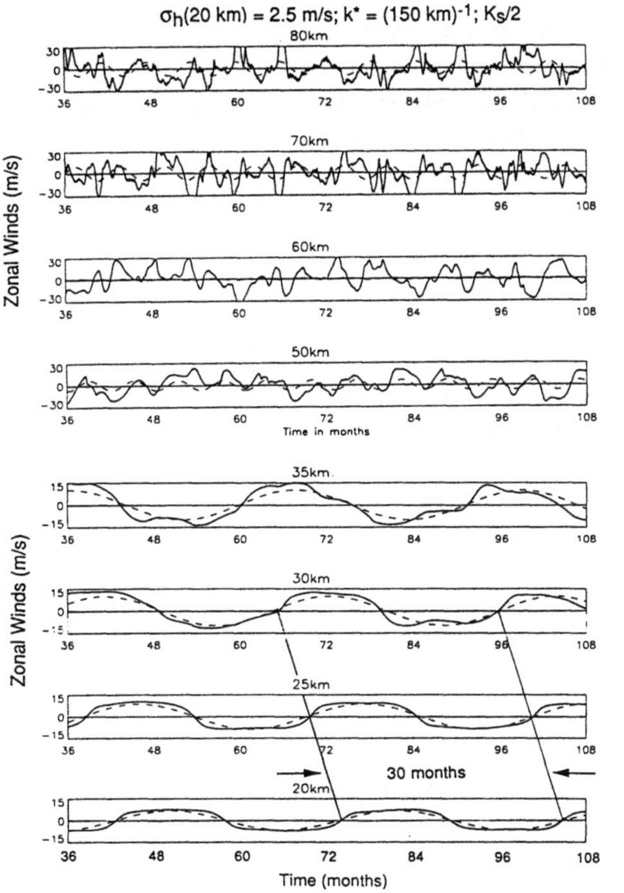


Figure 8: Computed zonal winds at 4° S with a factor of 2 smaller diffusivity, $K = K_S/2$. The Fourier harmonics for SAO and QBO (dashed lines) are shown above and below 40 km, respectively. Compared to Figure 5, the QBO amplitude and period are enhanced but the SAO amplitude is reduced.

In the present model, as in other models, the QBO is solely excited by the GW interaction without an external source of such periodicity. This is not true for the SAO. Such an oscillation can in principle also be excited by the semi-annual component in the solar heating, except that it does not come close to reproducing the observations (Hamilton, 1986). If a periodicity close to two years (as seen in the QBO) can be excited without such a periodicity in the solar forcing, we have asked, what other oscillation periods could be excited if the solar heating were kept constant? For the standard GW parameters of $\sigma_h(20\ km) = 2.5$ m/s, $k^* = (150\ km)^{-1}$, and the standard diffusivity, $K_S$, we show in Figure 9 the computed time series in the zonal winds at 4° south, covering a period of 6 model years. After a period of about two years (needed for spin-up), a long period oscillation resembling the QBO is excited in the lower stratosphere below 40 km, as expected. The amplitudes, however, are smaller than those shown in Figure 5. The period is also considerably shorter, about 17 months compared to almost 21 months in Figure 5. At higher altitudes, between 40 and 60 km, however, very different periods around 8 months are generated, which happen to be close to the 6 months of the semi-annual component in the solar heating. With such a wave induced periodicity, one may understand then that the SAO amplitudes are as large as seen in Figure 5, where the seasonal variations in the solar heating are also feeding the oscillation. In contrast to the earlier results with seasonal heating, the weaker oscillations shown in Figure 9 do not extend to altitudes above 70 km.

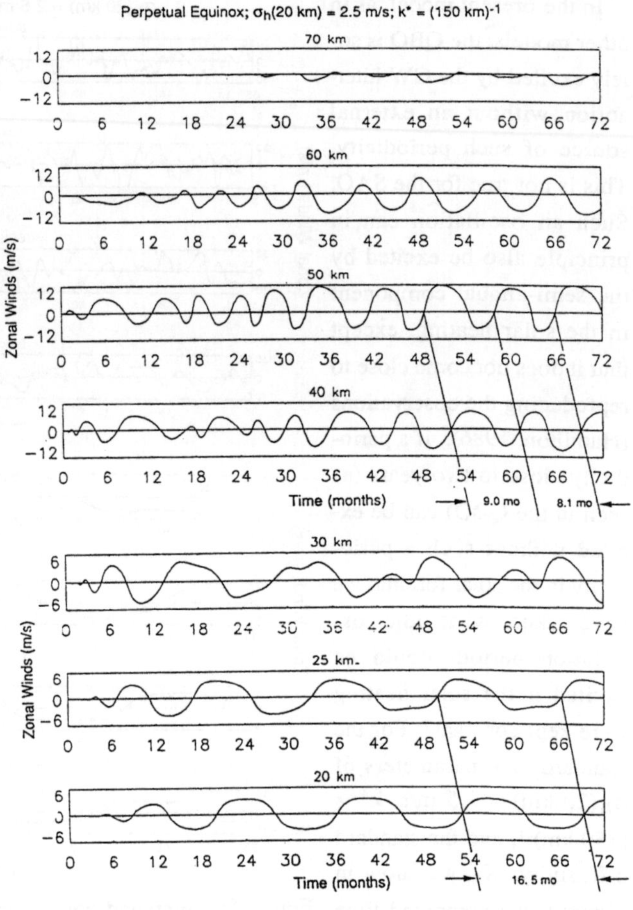


Figure 9: Computed zonal winds at 4° S for perpetual equinox. Above 30 km, oscillations with periods around 8 months are excited; while below, the QBO-like oscillations have a period near 17 months. Compared to earlier results with seasonally varying solar heating, the amplitudes are significantly smaller.

It was shown earlier that with decreasing eddy diffusivity, K, the QBO period significantly increased in the lower stratosphere (see Figures 5 and 8 for comparison). The same might also be expected for the shorter period excited by the GW at higher altitudes as seen in Figure 9. A decrease in K would then cause this period to increase above 8 months, moving it further away from the semi-annual (6 months) period in the solar heating, which could explain the reduced amplitude of the SAO shown in Figure 8.

We have seen in Figure 9 that under perpetual equinox the computed QBO differed significantly from that under seasonally varying solar heating presented in Figure 5, suggesting significant coupling between the SAO and QBO. This notion is confirmed with a computer experiment, in which the GW momentum source was artificially suppressed above 40 km. The numerical results (not presented here) show that this enhances significantly the amplitude and period of the QBO-like oscillation in the lower stratosphere.

## 5. Summary

With minimal tuning of the GW Doppler Spread Parameterization of Hines (1996a, b) and with the simplest assumptions about the initiated GW flux, the model reproduces in the mesosphere the observed anomalous reversals in the temperature and zonal wind fields during solstice (and equinox, not shown here).

With the same GW parameters, the model produces significant equatorial oscillations in the zonal circulation on semi-annual and quasi-biennial time scales. The average computed SAO amplitudes are 17 and 11 m/s (20 - 30 observed) at 50 and 80 km, respectively, and the oscillations at these levels are in opposite phase, consistent with observations. The computed QBO in the lower stratosphere has a period of about 21 months (20 - 32 observed), an amplitude of about 7 m/s (less than half of that observed) and a downward phase speed of 1.6 km/month (1.3 observed). The computed QBO extends into the upper mesosphere, consistent with UARS measurements (Burrage et al., 1996).

With enhanced GW momentum flux, the amplitudes of the SAO and QBO increase, but the QBO period decreases. When the eddy diffusivity is reduced by a factor of two, the QBO amplitude and period increase significantly, but the SAO amplitude decreases. Conversely, when the diffusivity is enhanced by a factor of two, the excited oscillations (not shown here) have small amplitudes and a period of only 13 months in the lower stratosphere; but the SAO amplitude is enhanced.

For perpetual equinox, with constant solar heating, the model generates periodicities of about 17 and 8 months (the latter near the 6 months period in the seasonal variations of the solar heating) in the lower and upper stratosphere, respectively; but the amplitudes are smaller than those obtained with seasonal heating. None of these oscillations extend to altitudes above 70 km. When the GW momentum source is turned off above 40 km, the model results (not presented here) show the amplitude and period of the QBO in the stratosphere to be significantly enhanced. Computer experiments (not presented here) show that the equatorial oscillations completely cease below a certain threshold in the GW momentum flux, which is only about 30% below the standard value adopted here, indicative of the nonlinear nature of the wave interaction.

In closing, we emphasize that for the model presented here only the simplest assumptions were made about the GW flux emanating from the troposphere, i.e., being isotropic and independent of latitude and season; and, with the exception of k*, the parameters for the DSP that describe the GW momentum deposition were simply chosen from the middle of the proposed ranges. Although the model is integrated from the Earth's surface up, it does not describe, even phenomenologically, the circulation of the troposphere. Moreover, the model does not describe the effects of planetary waves, which can affect the mean zonal circulation significantly.

Acknowledgment: The authors are greatly indebted to Dr. Ronald C. Taylor, National Science Foundation, for he stimulated their interest in the QBO and thus led them to discover this oscillation in the numerical model.

**References**

Allen, S. J., and S. A. Vincent, Gravity wave activity in the lower atmosphere: Seasonal and latitudinal variations, *J. Geophys. Res.*, **100**, 1327, 1995

Andrews, D. G., J. R. Holton, and C. B. Leovy, *Middle Atmosphere Dynamics*, pp. 489, Academic Press, 1987

Brasseur, G. and S. Solomon, *Aeronomy of the Middle Atmosphere: Chemistry and Physics in the Stratosphere and Mesosphere*, D. Reidel, Dordrecht, 1984

Brasseur, G., and M. Hitchman, The effect of breaking gravity waves on the distribution of trace species in the middle atmosphere, *Transport Processes in the Middle Atmosphere*, Edts. G. Visconti and R. Garcia, NATO Advanced Science Institutes Series, Vol. 213, 1986

Burrage, M. D., R. A, Vincent, H. G. Mayr, W. R. Skinner, N. F. Arnold, and P. B. Hays, Long-term variability in the equatorial middle atmosphere zonal wind, *J. Geophys. Res.*, **101**, No. D8, 12847, 1996

Chan, K. L., H. G. Mayr, J. G. Mengel, and I. Harris, A 'stratified' spectral model for stable and convective atmospheres, *J. Comp. Phys.*, **113**, 165, 1994a

Chan, K. L., H. G. Mayr, J. G. Mengel, and I. Harris, A spectral approach for studying middle and upper atmospheric phenomena, *J. Atmos. Terr. Phys.*, **56**, 1399, 1994b

Chan, K. L., H. G. Mayr, J. G. Mengel, and I. Harris, A numerical spectral model of the mean zonal circulation and the tides in the middle and upper atmosphere, in *The Upper Mesosphere and Lower Thermosphere: A Review of Experiment and Theory*, Edits. R. M. Johnson and T. L. Killeen, Geophysical Monograph 87, 1995

Dickinson, R. E., Method of parameterization for infrared cooling between altitudes of 30 and 70 kilometers, *J. Geophys. Res.*, **78,** 4451, 1973

Dunkerton, T. J., On the role of the Kelvin wave in the westerly phase of the semiannual zonal wind oscillation, *J. Atmos. Sci.*, **36**, 32-41, 1979

Dunkerton, T. J., Wave transience in a compressible atmosphere, 3 The saturation of internal gravity waves in the mesosphere, *J. Atmos. Sci.*, **39**, 1042, 1982a

Dunkerton, T. J., Stochastic parameterization of gravity wave stresses, *J. Atmos. Sci.*, **39**, 1711, 1982b

Dunkerton, T. J., Theory of the mesopause semiannual oscillation, *J. Atmos. Sci.*, **39**, 2681-2690, 1982c

Dunkerton, T. J., Theory of internal gravity wave saturation, *Pure Appl. Geophys.*, **130**, 373, 1989

Fritts, D. C., Gravity wave saturation in the middle atmosphere: A review of theory and observations, *Rev. Geophys.*, **22**, 275, 1984

Fritts, D. C., and T. J. Dunkerton, Fluxes of heat and constituents due to convectively unstable gravity waves, *J. Atmos. Sci.*, **42** ,549-556, 1985

Fritts, D. C., A review of gravity wave saturation processes, effects and variability in the middle atmosphere, *Pure Appl. Geophys.*, **130**, 343, 1989

Gaertner, V., M. Memmesheimer, and P. W. Blum, A zonally averaged dynamical model for the middle atmosphere including gravity wave mean flow interaction, 1, Solstice condition, *Planet. Space Sci.*, **31**, 1465, 1983

Garcia, R. and S. Solomon, The effect of breaking gravity waves on the dynamical and chemical composition of the mesosphere and thermosphere, *J. Geophys. Res.*, **90**, 3850, 1985

Geller, M. A., Modeling the middle atmosphere circulation, *Dynamics of the Middle Atmosphere*, Ed. J. R. Holton and D. Matsuno, 467, Terra pub. Tokyo, 1984

Hamilton, K., Dynamics of the stratospheric semi-annual oscillation, *J. Meteorol. Soc. Jpn.*, **64**, 227, 1986

Hamilton, K., R. J. Wilson, J. D. Mahlman, and L. J. Umscheid, Climatology of the SKYHI troposphere-stratosphere-mesosphere general circulation model, *J. Atmos. Sci.*, **52**, 5, 1995

Hays, P. B., V. J. Abreu, M. E. Dobbs, D. A. Gell, H. J. Grassl, and W. R. Skinner, The high-resolution Doppler imager on the Upper Atmospheric Research Satellite, *J. Geophys. Res.*, **98**, 10713, 1993

Hines, C. O., The saturation of gravity waves in the middle atmosphere. Part I: Critique of linear instability theory, *J. Atmos. Sci.*, **48**, 1348, 1991a

Hines, C. O., The saturation of gravity waves in the middle atmosphere. Part II: Development of Doppler spread theory, J. *Atmos. Sci.*, **48**, 1360, 1991b

Hines, C. O., The saturation of gravity waves in the middle atmosphere. Part III: Formation of the turbopause and of turbulent layers beneath it, *J. Atmos. Sci.*, **48**, 1380, 1991c

Hines, C. O., The saturation of gravity waves in the middle atmosphere. Part IV: Cut-off of the incident wave spectrum, *J. Atmos. Sci.*, **50**, 3045, 1993

Hines, C. O., Doppler-spread parameterization of gravity-wave momentum deposition in the middle atmosphere. Part 1: Basic Formulation, in press, *J. Atmos. Terr. Phys.*, 1996a

Hines, C. O., Doppler-spread parameterization of gravity-wave momentum deposition in the middle atmosphere. Part 2: Broad and quasi monochromatic spectra, and implementation, in press, *J. Atmos. Terr. Phys.*, 1996b

Hirota, I. Equatorial waves in the upper stratosphere and mesosphere in relation to the semiannual oscillation of the zonal wind *J. Atmos. Sci.*, **35**, 714, 1978

Hirota, I. Observational evidence of the semiannual oscillation in the tropical middle atmosphere - a review , *Pure Appl. Geophys.*, **118**, 217, 1980

Hitchman, M. H., and C. B. Leovy, Evolution of the zonal mean state in the equatorial middle atmosphere during October 1978 - May 1979, *J. Atmos. Sci.*, **43**, 3159, 1986

Hodges, R. R., Generation of turbulence in the upper atmosphere by internal gravity waves, *J. Geophys. Res.*, **72**, 3455, 1967

Hodges, R. R., Eddy diffusion coefficients due to instabilities in internal gravity waves, J. *Geophys. Res.*, **74**, 4087, 1969

Holton, J. R., and R. S. Lindzen, An updated theory for the quasi-biennial cycle of the tropical stratosphere, *J Atmos. Sci.*. **29**, 1076-1080, 1972

Holton, J. R., The role of gravity wave induced drag and diffusion in the momentum budget of the mesosphere, *J. Atmos. Sci.*, **39**, 791, 1982

Leovy, C. B., Simple models of thermally driven mesospheric circulation, *J. Atmos. Sci.*, **21**, 327, 1964

Lindzen, R. S., and J. R. Holton, A theory of the quasi-biennial oscillation, *J. Atmos. Sci.*, **25**, 1095, 1968

Lindzen R. S., Turbulence and stress due to gravity wave and tidal breakdown, *J. Geophys., Res.*, **86**, 9707-9714, 1981

Mahlman, J. D., and L. J. Umscheid, Dynamics of the middle atmosphere: successes and problems of the GFDL "SKYHI" general circulation model, in *Dynamics of the middle atmosphere,* J. R. Holton and T. Matsuno, eds., 501-525, Terrapub. Tokyo, 1981

Mengel, J. G., H. G. Mayr, K. L. Chan, C. O. Hines, C. A. Reddy, N. F. Arnold, and H. S. Porter, Equatorial oscillations in the middle atmosphere generated by small scale gravity waves, *Geophys. Res. Lett.*, **22** (22), 3027, 1995

Reddy, C. A., and L. Vijayan, Annual and semiannual oscillations in the equatorial middle atmospheric winds, *Adv. Space Res.*, **13,** (1)1373, 1993.

Reddy, C. A., and L. Vijayan, Large variability of zonal wind semi-annual oscillation in the equatorial stratosphere, presented at the IAMAS Symposium on "Middle Atmosphere Sciences" IUGG Scientific Assembly, Boulder, USA, July 2-14, 1995

Schoeberl, M. R., D. F. Strobel, and A. J. Apruzese, A numerical model of gravity wave breaking and stress in the mesosphere, *J. Geophys. Res.*, **88**, 5249, 1983

Schoeberl, M. R., and D. F. Strobel, Nonzonal gravity wave breaking in the winter mesosphere, *Dynamics of the Middle Atmosphere*, Ed. J. R. Holton and D. Matsuno, 45, Terra pub. Tokyo, 1984

Schoeberl, M. R, A ray tracing model of gravity wave propagation and breakdown in the middle atmosphere, *J. Geophys. Res.*, **90**, 7999, 1985

Schoeberl, M. R, Dynamics of the middle atmosphere, *Rev. Geophys.*, **25**, 501, 1987

Schoeberl, M. R., A model of stationary gravity wave break down with convective adjustment *J. Atmos. Sci.*, **45**, 980, 1988

Strobel, D. F., Parameterization of atmospheric heating rate from 15 to 120 km due to $O_2$ and $O_3$ absorption of solar radiation, *J. Geophys. Res.*, **83**, 6225, 1978

Vijayan, L. and C. A. Reddy, Radiative damping of equatorial waves in the middle atmosphere, *Q. J. Roy. Met. Soc. (London)*,, 20, 1323, 1994

Wehrbein, W. M., and C. B. Leovy, An accurate radiative heating and cooling algorithm for use in a dynamical model of the middle atmosphere, *J. Atmos. Sci.*, **39,** 1532, 1982

Weinstock, J., Non-linear theory of gravity waves: Momentum deposition, generated Rayleigh friction and diffusion, *J. Atmos. Sci.*, **39**, 1698, 1982

# The representation of sub-grid scale orography in GCMs

François LOTT

LMD-ENS, 24 rue Lhomond, 75231 Paris Cedex 05, France

Martin MILLER

ECMWF, Shinfield Park, Reading, Berkshire RG2 9AX, England

April 1-5, 1996

**Abstract**

The problem of representing Subgrid Scale Orography (SSO) in General Circulation Models (GCM) and Numerical Weather Prediction Models (NWPM) is discussed using a very accurate dataset (185mx185m) of the Alps. The linear approach shows that for horizontal scales varying between few Kms and 100Km mountains excite gravity waves which transport significant angular momentum. Nevertheless, this linear view is not sufficient for real mountains, and upstream blocking, low level wave breaking and flow splitting around the mountains often occur. At smaller scales, the orography strongly affects the Boundary Layer turbulence. These nonlinear effects lead to drag enhancement which are significant compared to the gravity wave drag. Efforts in the literature to represent these different mechanisms (gravity wave drag, increased roughness length, envelope orography) are also presented.

Recent progress in representing these nonlinear effects in a more consistent manner than in the past is also summarized. This includes development and testing against data of a new SSO scheme at ECMWF. An important feature of the scheme proposed is that it deals explicitly with a "blocked" low level flow which does not go over the mountain under certain dynamical configurations and which is slowed down. The upper part (non-blocked) of the low level flow goes over the mountains and forces gravity waves. By comparison to the PYREX data, it appears that the new scheme improves the realism of the ECMWF model. It also has a beneficial impact on the forecast performances. In the LMD-GCM, it improves the occurrence of synoptic blockings.

NATO ASI Series, Vol. I 50
Gravity Wave Processes
Their Parameterization in Global Climate Models
Edited by Kevin Hamilton

# 1 Introduction

Numerical models of the atmospheric general circulation use a combination of three different ways of representing the SSO. The first representation of the SSO consists in increasing the roughness length over mountains (Tibaldi and Geleyn, 1981; Mason, 1985) , to represent the form drag exerted by a hill on a turbulent flow (Taylor, 1977) . The second is the parameterization of the mountain gravity waves (Boer et al., 1984; Palmer et al., 1986) . It has led to improve the simulation of the tropospheric winter's jet in the north-hemisphere at high altitudes. The third representation of the SSO is the envelope orography. It consists in enhancing the model mean orography by a factor which is geographically dependent on the standard deviation of the SSO (Wallace et al., 1983) . The development of these different parameterizations was made relatively independently one from the other, which implicitly assumes that some separation can be made between the different scales of the earth orography. In reality, the atmosphere is dominated by nonlinear effects, and this scale separation does not seem justified. For instance the mesoscale dynamics over mountain certainly depends on the low level boundary layer (Richard et al., 1989) ; the envelope orography which increases the mountain drag in models can also be viewed as a form of mountain drag parameterization.

Section 2 discusses and criticizes these different approaches using a very accurate dataset of the Alps. In Section 2.1, the linear wave theory is applied and the importance of the gravity waves for the General Circulation of the Atmosphere is discussed. In Section 2.2 this linear approach is criticized and the nonlinear processes that are presumed to occur are presented in two limit configurations. The first one is that of stable or slow flow, which in the literature was studied in the context of nonlinear mountain waves. The second one is that of neutral or fast flow, in the literature it was studied in the context of the interaction between the boundary layer and the orography. To a certain extent, the envelope orography is an artefact to parameterize these nonlinear processes.

The importance of the nonlinearities and some important limitations of the concept of envelope orography have led Lott and Miller (1996) to propose a new representation of the SSO in GCMs. This work was also motivated by the results of numerous recent studies (Hunt and Snyder, 1980; Smolarkiewicz and Rotunno, 1989; Schär and Smith, 1993) on nonlinear flow past orography, to which SSO parameterization schemes needed to be adapted. The general principle of the scheme are summarized in Section 3.1. Tests and calibrations with the ECMWF model during Periods of Intense Observation of the PYREX campaign are presented in Section 3.2. Impact on the model forecast performances are also summarized. Some results with the same scheme implemented in the LMD-GCM are presented in Section 3.3.

# 2 General considerations

One of the fundamental problem in representing the impact of mountains in GCMs and NWPMs is that the earth orography has no preferential horizontal wavenumber. Figure 1 for instance, shows a section of the Alps orography, h(x), along a zonal transect of length L=800Km, located at the latitude $45.9^o N$, and according to three different datasets. The first one is a very accurate dataset (185mx185m) close to that presented by Volkert (1990)(hereafter referenced as the IPA [1] dataset). The second one is the USN (10'x10') dataset upon which most of the orographic climate fields are deduced in GCMs. The third is the model orography, explicitly used as a lower boundary condition for the Geopotential height in the T213 ECMWF model with mean orography. Figure 1 illustrates that the real mountains are poorly represented in both the ECMWF model and in the USN dataset. According to the ECMWF model, a typical horizontal extension of a mountain ridge is 100km, according to the USN dataset it is 25Km, while in reality, the IPA dataset shows that it does not exceed few kilometers. In the following, it will be shown that those differences are important for the General Circulation of the Atmosphere, and that the development of parameterization schemes is necessary.

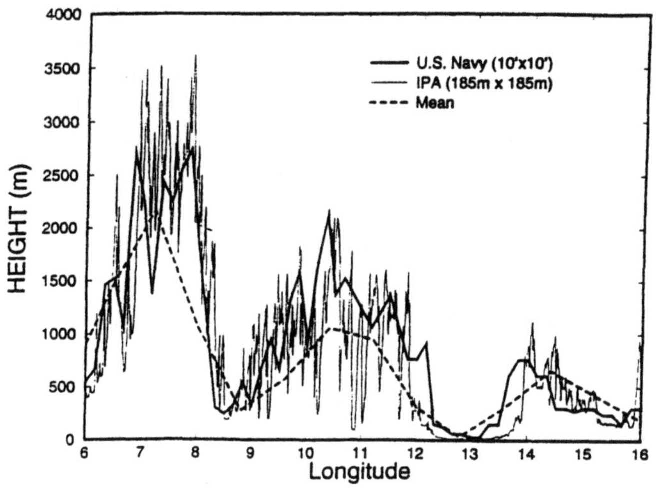


Figure 1: *East West Sections of the Alps at 45.9 N High resolution (185mx185m) IPA dataset (normal); USN dataset (bold); ECMWF model T213 (Bold-dotted)*

---

[1]Institute fur Physiks der Atmosphare.

## 2.1 A linear point of view

In this part, the linear response to the orographic forcing is described. Since the discussion is only qualitative, it will be assumed that the orography is periodic and the linear response can be expanded in series of harmonics. It is also assumed that the response along a given East-West transect of the Alps is two-dimensional. Then, to specify a lower boundary condition, the orography displayed on Figure 1 is written:

$$h(x)=\sum_{K=0}^{M}H_K\cos(K\frac{2\pi}{L}+\phi_K)\,,\; H_K=\frac{2}{L}\int_0^L h(x)\cos(K\frac{2\pi}{L}+\phi_K)dx \qquad (1)$$

where each harmonic, $K$, is related to a zonal wavenumber, $k=K\frac{2\pi}{L}$. $H_K$ and $\phi_K$, are the amplitude and the phase associated to this harmonic respectively. $L$ is the length of the Alps section displayed on Figure 1 and $M$ is the highest order harmonic number. Typically, $M=\frac{N}{2}-1$, where $N$ is the number of points available along the transect. From the IPA dataset, more than $M=2000$ harmonics can be extracted for the Alps, while $M=29$ harmonics can be extracted from the USN dataset and only $M=6$ harmonics can be extracted from the T213 model.

To determine at which scales mountain waves, induce a transfer of angular momentum between the earth and the atmosphere, the governing equations for the linear motion are written on a $f$-plane, and it is assumed that the zonal wind $U$ and the buoyancy frequency $N^2$ are independent of height. In the Boussinesq approximation, the equation for the vertical structure of an harmonic, $W_K(z)$, of the vertical velocity, $w(x,z)$ is:

$$\frac{d^2W_K(z)}{dz^2}+k^2\frac{N^2-k^2U^2}{k^2U^2-f^2}W_K(z)\;=\;0 \text{ with } W_K(0)=kUH_K\,. \qquad (2)$$

Solving equation (2) and coming back to the physical space, gives:

$$w(x,z)=-\sum_{K=1}^{K_f-1}kUH_K\,e^{-m_Kz}\,\sin(kx+\phi_K)$$

$$-\sum_{K=K_f}^{K_N}kUH_K\sin(m_Kz+kx+\phi_K)\;-\sum_{K=K_N+1}^{M}kUH_K\,e^{-m_Kz}\,\sin(kx+\phi_K) \qquad (3)$$

$$\text{where}\quad m_K=k\sqrt{\left|\frac{N^2-k^2U^2}{k^2U^2-f^2}\right|}\,,\; K_f=\frac{fL}{2\pi U}\;\text{ and }\; K_N=\frac{NL}{2\pi U}\,.$$

Among the different harmonics, only those in the second summation terms of Equation (3) correspond to propagating gravity waves, the others are evanescent in the vertical direction. Among those different modes, it is known that only

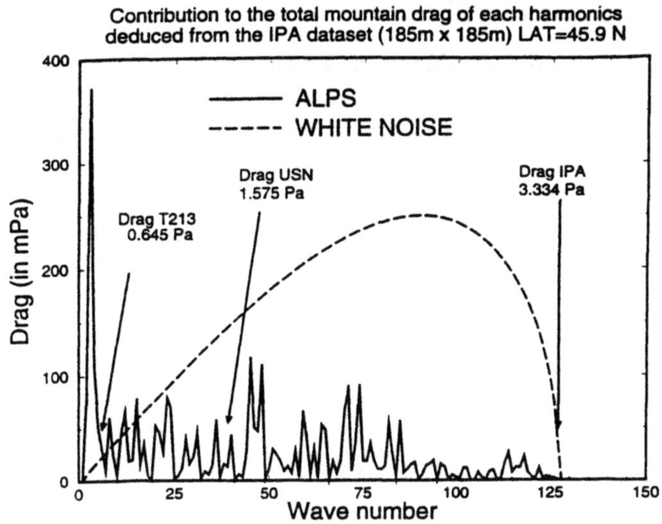


Figure 2: *Amplitude of the drag harmonics by harmonics (bold). The values displayed with arrows correspond to truncated and untruncated summations.*

the propagating waves allow a flux of angular momentum from the earth to the atmosphere (see for instance Gill (1982)). At the ground, this flux is balanced by the gravity wave drag:

$$D = \frac{1}{L}\int_0^L -p\frac{dh}{dx}dx = \frac{1}{2}\sum_{K_f}^{K_N} \rho_0\sqrt{|(N^2 - k^2U^2)(k^2U^2 - f^2)|}{H_K}^2 . \tag{4}$$

Figure 2 shows the contribution of each wave forced by the mountain to the gravity wave drag (4) when $U = 10ms^{-1}$ and $N = 0.01s^{-1}$. Figure 2 also shows a 'white noise' curve which is the function, $\sqrt{|(N^2 - k^2U^2)(k^2U^2 - f^2)|}$ , that modulates the amplitude of the harmonics of the orography $H_K$ into the wave drag expression (4). It shows that beyond the two cut-off wavenumber $K_f = 2$ and $K_N = 128$, the harmonics forced by the orography do not contribute to the drag. In the flow configuration considered here, the wave drag equals, $D = 3.334Pa$, when the summation (4) is done over all the harmonics provided by the IPA dataset. It can be easily shown that such a drag have a significant impact on the General Circulation of the Atmosphere (Lott, 1994) . In a very accurate GCMs, like the ECMWF model at truncature T213, the Alps are roughly represented until the wavenumber 6.The contribution of these modes to the drag (4) gives 0.645Pa. It means that the model explicitly represents a part of the gravity wave spectrum but a lot of waves still need to be parameterized. Furthermore, when the evaluation of D in (4) is truncated at the wavenumber 29 to simulate the wave drag one can estimate from the USN dataset, the drag D=1.575Pa. It shows that the USN dataset, upon which most parameterizations of the SSO are based, is not accurate enough to describe the complete gravity waves spectrum.

## 2.2 Few nonlinear effects

The calculations presented in subsection (2.1) have shown that the shortest modes (i.e., with wavelength smaller than 6.28km in the numerical application presented on Figure 2), do not contribute to the mountain drag. Nevertheless, this is only true in the linear context and can be criticized. Indeed, Figure 1 shows that the individual peaks of the mountains have an heights varying between 200m and 1500m, and can not be considered as infinitesimal. Then, to evaluate if the flow response to an individual peak is linear, it is conventional to compare a typical vertical scale of the disturbance to the height of the mountain. For one single isolated obstacle with horizontal length $l$, equation (3) shows that the flow response is dominated by a vertical wavenumber (or an e-folding decay length),

$$m \approx \frac{U}{N}\sqrt{\left|\frac{1-Ro^{-2}}{1-Fr^2}\right|} \tag{5}$$

where $Ro = \frac{U}{fl}$ and $Fr = \frac{U}{Nl}$ are a Rossby number and a Froude number respectively. Then, the linear approximation assumes that the vertical scale of the disturbance is large compared to the mountain height:

$$\lambda >> H, \text{ i.e., } H_N{}^{-1}\sqrt{\left|\frac{1-Ro^{-2}}{1-Fr^2}\right|} >> 1, \text{ where, } H_N = \frac{HN}{U}. \tag{6}$$

the non-dimensional parameter, $H_N$, is often referred as a non-dimensional mountain height. For an individual mountain peak it is natural to neglect the Rossby number. Then the nature of the flow response only depends on $Fr$ and $H_N$. In the incident flow configuration considered, $H_N$ is of order 1 and $Fr$ is of order 0.2, the horizontal length of the individual peaks varying between $3Km$ and $9Km$. In this context, the linearity condition (6) is not satisfied. Then, to describe the nature of the nonlinear effects that can occur, it is conventional to use the two asymptotic configurations which are presented below.

### Slow flow ($Fr << 1$), linear if $H_N << 1$

The first configuration is the stable (or slow flow) configuration for which, $Fr << 1$. In this case, the linear response of the flow to a single peak is essentially made of radiating gravity waves and the linearity criteria reduces to $H_N << 1$. As shown before, for a typical peak of the Alps, this parameter is of order 1 or even larger. In this case, the nonlinear effects can increase the wave drag by a factor two and even more. Furthermore the related increase of the drag is related to low level effects, like upstream blocking (Pierrehumbert and Wyman, 1985) , flow splitting around the obstacle (Miranda and James, 1992; Schär and Smith, 1993) and low level breaking gravity waves(Durran, 1987) . It is noteworthy that all these effects tend to affect the flow at levels located close and below the mountain top. The new SSO parameterization scheme developed in Lott and Miller (1996) and which is summarized in Section 3, account for these effects.

**Fast flow($Fr >> 1$), linear if $\frac{H}{L} << 1$**

The second configuration is the neutral configuration (or fast flow) for which the advective time scale, $\frac{l}{U}$, is small compared to the buoyancy time scale $N^{-1}$. Then, the linearity condition (6) reduces to the small slope condition: $\frac{H}{L} << 1$. In the literature the neutral flow configuration was mainly studied in the context of the interaction between the boundary layer and the orography (Grant and Mason, 1990) . In this case, the orography increases the Boundary Layer drag, to values that are comparable to the gravity waves drags. For the Alps, this slope parameter has a typical amplitude of 0.4, which is large according to Wood and Mason (1993). In this case, the large drags are related to the flow separation that occurs close to the mountain peaks. In some models, this effect is parameterized by increasing the roughness length over mountain ranges (Tibaldi and Geleyn, 1981) .

**The concept of envelope orography**

It has sometimes been assumed that the low level processes essentially stop the flow beyond the mountain peaks, inducing sheltered zones where the wind is zero. It was further considered that those zones remain isolated from the large scale flow, assuming that the diffusion processes at the altitude of the mountain peaks are small. Under this hypothesis, the drags described in the last two chapters no longer needs to be parameterized, if the flow located inside these sheltered areas is excluded from the model. These arguments have justified the introduction of an envelope orography in the ECMWF model (Wallace et al., 1983) . Nevertheless, the envelope orography has some important defects: the onset of sheltered zones certainly depends on the flow characteristics, while the envelope orography is invariant; the hypothesis that the diffusion processes are small near the top of the peaks is in contradiction with airplane measurements which often show strong turbulence at these levels; the envelope orography leads to poor precipitation forecasts over mountains and has a negative impact on the radiation balance over high plateaus.

# 3 General principles of the new schemes, validation and impacts

This discussion has shown that the parameterization of the SSO in atmospheric models need to include the nonlinear effects, and has to provide a replacement for the envelope orography. These are the premises of the new SSO representation developed by Lott and Miller (1996) at the ECMWF. This scheme assumes that for the large scale flow, the model mean orography is the optimal representation. This choice is motivated dynamically: the largest force exerted by mountains on the atmosphere is a lift force, which is essentially perpendicular to the flow and varies with the mountains' volume (Smith, 1979a) . This lift force accompanies the anticyclonic circulation that occurs above large scale massifs in midlatitudes. Consequently, only forces due to the intersection of sub-grid scale mountains with model levels and those due to the generation of vertically propagating gravity waves have to be parameterized. It this context, the new scheme assumes that the Froude number of the SSO is small. This excludes from the analysis all the features related to the interaction between the boundary layer and the orography described before. This limitation of the scheme is nevertheless consistent with the fact that USN dataset has a poor resolution, and only shows mountain ranges with horizontal scales that varies between 10Km and 100Km (i.e., for which the Froude number is generally small).

## 3.1 General principles and impacts in ECMWF forecasts

The new SSO Scheme assumes that the mesoscale flow dynamics can be described by two conceptual models, the relevance of which depends on the non-dimensional height of the mountain $H_N$, defined by (6). At small $H_N$, all the flow goes over the mountain and gravity waves are forced by the vertical displacement of the fluid. Assuming that the mountain has the shape of one single elliptic mountain whose height variation is determined by the parameter $b$ in the along ridge direction and by the parameter $a$ in the normal ridge direction ($\gamma = a/b < 1$), the geometry of the mountain can be written as:

$$h(x,y) = \frac{H}{1 + \frac{x^2}{a^2} + \frac{y^2}{b^2}} \quad . \tag{7}$$

In simple configurations (i.e., when the incident flow is perpendicular to the ridge), the gravity wave surface stress has the magnitude,

$$\tau_w = \rho_0 b G B(\gamma) N U H^2 \tag{8}$$

assuming that the Boussinesq and the hydrostatic approximations are valid. In equation (3), G is a function of the mountain sharpness (Phillips, 1985) , and $G \approx 1.23$, for the mountain given by (7). The constant $B(\gamma)$ is a function of the mountain anisotropy, $\gamma$ , varying from B(0)=1 for a two dimensional ridge to $B(1) = \pi/4$, for a circular mountain. At large $H_N$, the stratification limits the vertical motion of the fluid and part of the low level flow goes around the mountain. The depth of this blocked layer when $U$ and $N$ are independent is given by,

$$Z_b = H \max\left(0., \frac{H_N - H_{NC}}{H_N}\right) \tag{9}$$

where $H_{NC}$ is a critical mountain height of order 1. The depth $Z_b$, can be viewed as the upstream elevation of the lower isentropic surface, which is raised exactly to the mountain top (see Figure 3). In each layer below $Z_b$, the flow streamlines

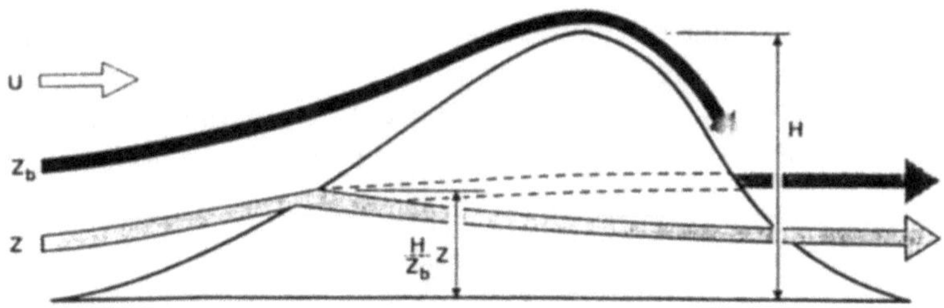


Figure 3: *Schematic representation of the low level flow behaviour parametrized in the new-scheme*

split around the obstacle, and it is assumed that flow separation occurs on the obstacle flank. Then, the drag exerted by the obstacle on the flow at these levels is given by,

$$D_b(z) = -\rho_0 C_d l(z) \frac{U\,|U|}{2} \quad . \tag{10}$$

Here $l(z)$ characterizes the length of the obstacle as seen by the flow with upstream altitude $z$ and $C_d$ is a constant of order 1 according to the theory of jet in ideal fluids (Kirchoff (1876); Gurevich (1965)). According to observations, $C_d$ can be of order 2 when suction effects occur on the rear of the obstacle (Batchelor (1967)). In the parameterization scheme, this drag is applied to the flow level by level. It was shown in Lott and Miller (1996) that the addition of the blocked flow drag to the gravity wave drag allows the scheme to be a rather good predictor of high drag states measured by Stein (1992) and Miranda and James (1992) in nonlinear simulations of mountain flows.

These ideas have been utilized to represent the effect of the subgrid scale orography in the ECMWF model. Following Baines and Palmer (1990), the subgrid scale orography over one gridpoint region (GPR) is represented by four parameters, $\mu$, $\gamma$, $\sigma$ and $\theta$, which represent the standard deviation, the anisotropy, the slope along the principal axis and the orientation of the orography. Then the drag amplitude was estimated combining formulas valid for elliptic mountains with the USN data. Considerable simplifications are then implied and the calculations are virtually scale analyses relating the different amplitudes to the SSO parameters. Hence the calibration and validation of the new scheme with field data is essential.

The local behaviour of the new SSO scheme was examined in the T213 ECMWF model using analysis and short range forecasts covering the PYREX campaign. During this campaign, important quantities like the pressure drag and the momentum fluxes induced by the Pyrénées (a mountain ridge located at the border between France and Spain) were measured [2]. In the following, attention is limited to the PYREX Period of Intense Observations (PIO), which cover a quarter of the two monthes of the PYREX campaign, and during which the wind component perpendicular to the mountain range was often large. This limit the calibration to 60 cases. Figure 4 shows, the T213 total model drag (SSO drag + Boundary Layer Drag + explicit pressure drag) due to the Pyrénées, compared to that measured during the PYREX campaign. In this context, the parameter-

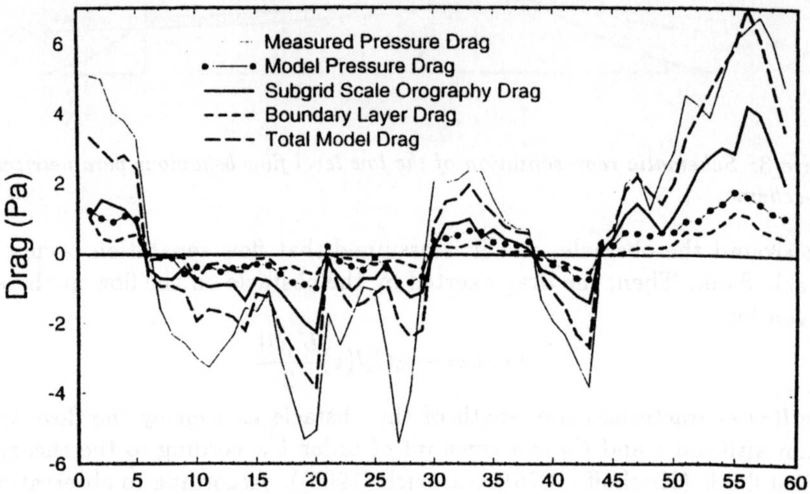


Figure 4: T213 forecasts, ECMWF model with mean orography and the new SSO scheme. Explicit pressure drag and parameterized mountain drags during PYREX

ized drags can be viewed as a way to make up the difference between the model pressure drag and the measured pressure drag. The Figure (4) shows that with the new orography representation, the parameterised drag is large and make up in a very consistent way the difference between the model pressure drag and the measured pressure drag.

In the new scheme, it is assumed that when the flow component perpendicular to the ridge is large, the mountain drag is mainly related to a slow down of the low level flow and to some upward propagating waves. The amplitude of the wave stress significantly above the mountain top is often small. Consequently, the scheme certainly has an impact on the low level flow dynamics at the sub-synoptic scale. Accordingly, the realism of the new scheme has also been evaluated, making

[2]For further details about the PYREX campaign and the ECMWF reanalysis see Bougeault et al. (1993) and Lott (1995) respectively.

diagnostics of the low level flow in the model and comparing it to more accurate simulation of the flow near the Pyrénées. This has shown that the new scheme produced a wake downstream of the ridge, and that this wake is realistic. The fact that this wake is essentially due to the new scheme was proved, by diagnosing Bernoulli function deficit across the Pyrénées as well as the related fluxes of isentropic vorticity Haynes and McIntyre (1987); Schär (1993).

Prior to operational implementation, the new scheme was also tested in forecasts configurations. Results from those on skill scores, precipitation amount and distribution indicate that the combination of mean orography together with the new scheme, performed better than that using envelope orography and the old gravity wave drag scheme. The new scheme is also beneficial to the forecast skill of the ECMWF model and is in operational use since April 1995.

## 3.2 Impact on winter climate (LMD-GCM)

The new scheme is also implemented in the gridpoint climate GCM of the LMD. Since orography is known to be one of the forcings of the steady planetary waves in winter (Valdes and Hoskins, 1989) , the associated storm tracks (Hoskins and Valdes, 1990) , and is also known to influence the onset of blockings (Mullen, 1994) , the sensitivity of these different features to different orographic representation has also been evaluated. The methodology consists in performing series of five winter runs (DJF 85-90) with the LMD-GCM at resolution $96 \times 72 \times 19$, forced by observed SST. The performance of the model was then evaluated by comparison with the operational ECMWF analyses. At first, runs with no orography and with no SSO scheme have been done. They show that in absence of orography the steady planetary wave is very weak, the storm tracks are badly located, and the percentage of blocking events, according to the Tibaldi and Molteni (1990) index is very small.

Then, as indicative experiments to determine which of these features can be affected in GCMs using different types of orography representation, experiment with no orography but with forces acting on the flow at low level have been done. In a first series of experiments the local impact of mountains is parameterized by a lift force (i.e., that is essentially perpendicular to the flow) whith amplitude given by Smith (1979b) in his study of quasi-geostrophic flow over three dimensional ridges. Another set of experiments was done by providing very large drags (i.e., that virtually stop the flow) at the model levels which should intersect the mountains. Those runs show that the lift forces allow the model to produce a realistic steady planetary wave, but does not improve the location of the storm tracks and does not increase the number of blocking events (when compared to the no orography case described before). The impact of the large drag is quite opposite, as it locates properly the storm tracks, but leads the model to overestimate the onset of blockings. Although very idealized, these experiments somehow support the rather conventional view that the steady planetary wave is

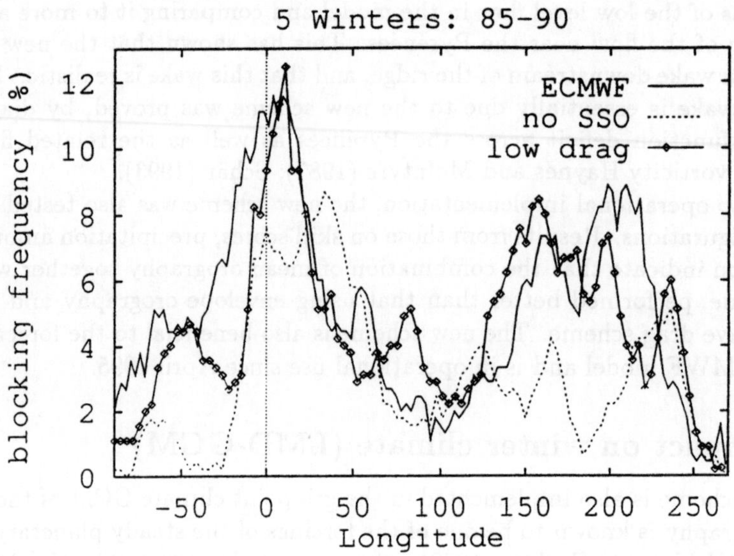


Figure 5: Blocking frequency as a function of the Longitude. The blocking index is calculated following Tibaldi and Molteni (1990). From daily 500Hpa ECMWF data: continuous line; LMD GCM control run: dotted line; LMD GCM with low level drag: continuous line and diamonds.

essentially related to the large scale response of the atmosphere to the process of quasi-geostrophic Taylor column compression. The response of the model to the large drags support the conventional view that synoptic blockings are linked to topographic drag (Charney and Strauss, 1980) .

A more physical manner to represent the lift force in GCMs, is nevertheless to use a mean orography as a lower boundary condition. In this context the last series of run presented compare cases with mean orography and no SSO parameterization to cases with mean orography and SSO parameterization. Regarding the parameterization scheme the result obtained goes in the direction described before. The scheme improves the blocking frequency (Figure 5) and has a week negative influence on the steady planetary wave structure. This last result, together with the fact that the steady planetary wave seems well defined by lift forces, indicate this as an open area for future research.

# 4 Conclusion

This study has illustrated that the orographic features which scales vary from 100m to 100km can play a significant role on the general circulation of the atmosphere. Among those scales, the largest (5Km-100Km) covers the gravity wave spectrum. In a GCM, for which the grid spacing is larger than 100Km, all those waves need to be parameterized. In NWPM which can have smaller gridspacing, part of this spectrum can be resolved explicitly by the model, but a large part remains to be parameterized. Furthermore, for realistic mountains, the linear hypothesis is not valid and nonlinear effects also need to be considered. Conventional gravity waves drag schemes have to be adapted to represent low level wave breaking, upstream blocking and flow splitting. In this nonlinear context, even smaller orographic features can slow down the large scale flow, through interaction between the boundary layer and the irregular bottom. These last effects are sometimes parameterized, by increasing the roughness length over mountains. The concept of envelope orography, which stands that below the mountain peaks remain sheltered zones, where the atmosphere is at rest, can be viewed as an artefact to simulate low level nonlinear effects.

In the present models the three different parametrization of the SSO (i.e., gravity wave drag, enhanced roughness length and envelope orography) have been developed rather independently one from the other. As remarked in the introduction, this distinction is quite arbitrary. For instance, the increase of the roughness length above hilly regions results from study of neutral flow above orography. It is quite clear that in most of the cases of interest (i.e., winter over land), the stratification will modify significantly this theory. Similarly, theories of mountain waves (linear or non-linear) mostly assume a small slope, although this is generally not observed in reality. The reality seems to be at a transition, in the nonlinear range (i.e., at large $H_N$), between small slope configuration (theory of mountain waves with low level breaking or/and flow separation on the mountain flank) and large slope configuration (flow separation in the vertical direction and near the mountain peak) which is not really known.

This shows that there are plenty of theoretical reasons to parameterize the SSO in GCM and to account for the nonlinearities. Aware of this, a new parameterization of Subgrid Scale Orographic Drag has been developed at the ECMWF by Lott and Miller (1996). It encompasses recent developments in the nonlinear theory of stratified flows around obstacles. In this scheme, particular attention has been paid to the drag in the layers of the atmosphere below the mountain peaks. This low level part of the scheme replaces the envelope enhancement of the model orography. The upper part of the scheme still represents gravity waves. It has been revised to allow a better representation of the mountain ridges orientation, and of its anisotropy. After calibration, the new scheme reproduces well the mountain drag measured over the Pyrénées during the PYREX field campaign. It was further shown that the new scheme has a realistic dynamical impact on the model dynamics in the lee of the mountain. These results reinforce the basic hypothesis of the new scheme that the mountain drag in nature, essentially slows down the low level flow. Finally, the scheme was also found to have a beneficial impact on forecast performance. Its implementation in the LMD-GCM, was rather easy. It appears that the strong low level drag implied by the scheme, has beneficial impacts on several aspects of the northern hemisphere winter climatology. In particular, it influences the onset and location of large scale blockings over the Pacific and the Atlantic. It seems to have a beneficial impact on the occurrence of those events. It nevertheless seems to slightly degrade the structure of the steady planetary waves.

# 5 References

Baines, P. G., and T. N. Palmer, Rationale for a new physically-based parametrization of sub-grid scale orographic effects, Technical Memorandum 169, ECMWF, Shinfield Park, Reading, RG1 5AX Berkshire, England, 1990.

Batchelor, G. K., *An introduction to Fluid Dynamics*, Cambridge University Press, 1967.

Boer, G. J., N. A. McFarlane, R. Laprise, J. D. Henderson, and J. P. Blanchet, 1984, The canadian climate centre spectral atmospheric general circulation model, *Atmos.-Ocean*, *22*, 397–429, 1984.

Bougeault, P., A. Jansa, J. Attie, I. Beau, B. Benech, B. Benoit, P. Bessemoulin, J. Caccia, J. Campins, and B. a. Carrissimo, 1993, The atmospheric momentum budget over a major mountain range: first results of the PYREX field program, *Annales Geophysicae*, *11*, 395–418, 1993.

Charney, J. G., and D. M. Strauss, 1980, Form-drag instability, multiple equilibria and propagating gravity waves in baroclinic, orographically forced, planetary wave systems, *Journal of the Atmospheric Sciences*, *37*, 1157–1176, 1980.

Durran, D. R., 1987, Another look at downslope winds. Part 2: Nonlinear amplification beneath wave-overturning layers, *Journal of the Atmospheric Sciences*, *44*, 3402–3412, 1987.

Gill, A. E., *Atmosphere-ocean dynamics*, Academic Press, 1982.

Grant, A. L. M., and P. J. Mason, 1990, Observation of boundary layer structure over complex terrain, *Quarterly Journal of the Royal Meteorological Society*, *116*, 395–418, 1990.

Gurevich, M., *Theory of jets in ideal fluids*, Academic Press, 1965.

Haynes, P. H., and M. E. McIntyre, 1987, On the evolution of vorticity and potential vorticity in the presence of diabatic heating and frictional or other forces, *Journal of the Atmospheric Sciences*, *44*, 828–841, 1987.

Hoskins, B. J., and P. J. Valdes, 1990, On the existence of storm tracks, *Journal of the Atmospheric Sciences*, *47*, 1855–1864, 1990.

Hunt, J., and W. H. Snyder, 1980, Experiments on stably stratified flow over a model three dimensional hill, *Journal of Fluid Mechanics*, *7*, 671–704, 1980.

Kirchoff, G., *Vorlesungen über mathematische physik*, Leipzig, 1876.

Lott, F., and M. Miller, 1996, A new subgrid scale orographic drag parameterization; its testing in the ECMWF model, *Quarterly Journal of the Royal Meteorological Society*, In press, 1996.

Lott, F., The significance of subgrid scale orography and problems in their representation in GCM, in *Parametrization of sub-grid scale physical processes*, Seminar

Proceedings, 277–303, ECMWF, 1994.

Lott, F., 1995, Comparison between the orographic response of the ECMWF model and the PYREX 1990 data, *Quarterly Journal of the Royal Meteorological Society, 121*, 1323–1348, 1995.

Mason, P. J., On the parametrization of orographic drag, in *Seminar on Physical parameterization for numerical models of the atmosphere*, 139–165, ECMWF, 1985.

Miranda, P. M. A., and I. A. James, 1992, Non-linear three dimensional effects on gravity-wave drag: splitting flow and breaking waves, *Quarterly Journal of the Royal Meteorological Society, 118*, 1057–1081, 1992.

Mullen, S. L., 1994, The impact of an envelope orography on low frequency variability and blocking in a low resolution general circulation model, *Journal of Climate, 7*, 1815–1826, 1994.

Palmer, T. N., G. J. Shutts, and R. Swinbank, 1986, Alleviation of systematic westerly bias in general circulation and numerical weather prediction models through an orographic gravity wave drag parametrization, *Quarterly Journal of the Royal Meteorological Society, 112*, 2056–2066, 1986.

Phillips, D. S., 1985, Analytical surface pressure and drag for linear hydrostatic flow on three dimensional elliptical mountains, *Journal of the Atmospheric Sciences, 41*, 1073–1084, 1985.

Pierrehumbert, R. T., and B. Wyman, 1985, Upstream effect of mesoscale mountains, *Journal of the Atmospheric Sciences, 42*, 977–1003, 1985.

Richard, E., P. Mascart, and E. Nickerson, 1989, The role of surface friction in downslope windstorms, *Journal of Applied Meteorology, 28*, 241–251, 1989.

Schär, C., and R. B. Smith, 1993, Shallow water flow past isolated topography. Part 1: Vorticity production and wake formation, *Journal of the Atmospheric Sciences, 50*, 1373–1400, 1993.

Schär, C., 1993, A generalization of bernoulli's theorem, *Journal of the Atmospheric Sciences, 50*, 1437–1440, 1993.

Smith, R. B., 1979*a*, The influence of mountains on the atmosphere, *Advances in geophysics*, 87–230, 1979*a*.

Smith, R. B., 1979*b*, Some aspects of the quasi-geostrophic flow over mountain, *Journal of the Atmospheric Sciences, 36*, 2385–2393, 1979*b*.

Smolarkiewicz, P., and R. Rotunno, 1989, Low Froude number flow past three-dimensional obstacles. Part 1: baroclinically generated lee vortices, *Journal of the Atmospheric Sciences, 46*, 1154–1164, 1989.

Stein, J., 1992, Investigation of the regime diagram of hydrostatic flow over a mountain with a primitive equation model, *Monthly Weather Review, 120*, 2962–2976, 1992.

Taylor, P. A., 1977, Model predictions of neutrally stratified planetary boundary layer flow over ridges, *Boundary-Layer Meteorology, 107*, 111–120, 1977.

Tibaldi, S., and J. Geleyn, 1981, The production of a new orography, land-sea mask and associated climatological surface fields for operational purposes, *ECMWF Research Department, Technical Memorandum, 40*, 1981.

Tibaldi, S., and F. Molteni, 1990, On the operational predictability of blocking, *Tellus, 42A*, 343–365, 1990.

Valdes, P. J., and B. J. Hoskins, 1989, Linear stationary wave simulations of the time-mean climatological flow, *Journal of the Atmospheric Sciences*, *46*, 2509–2527, 1989.

Volkert, H., 1990, An alpine orography resolving major valleys an massifs, *Meteorology and Atmospheric Physics*, *43*, 231–234, 1990.

Wallace, J. M., S. Tibaldi, and A. Simmons, 1983, Reduction of systematic forecast errors in the ECMWF model through the introduction of an envelope orography, *Quarterly Journal of the Royal Meteorological Society*, *109*, 683–717, 1983.

Wood, N., and P. Mason, 1993, The pressure force induced neutral, turbulent flow over hills, *Quarterly Journal of the Royal Meteorological Society*, *119*, 1233–1267, 1993.

# The Effect of Parameterized Gravity Wave Drag on Simulations of the Middle Atmosphere during Northern Winter 1991/1992 - General Evolution.

Bryan N. Lawrence[1]
Rutherford Appleton Laboratory
Chilton, Didcot Oxfordshire, OX11 0QX
United Kingdom
(also at the Department of Physics, Oxford University)

**Abstract**

In this paper we present a comparison between observed data and a number of simulations of early winter 1991/1992, carried out with a mechanistic model of the middle atmosphere (based upon the U.K.M.O. Stratosphere-Mesosphere Model). The model was integrated from early November 1991 through to mid January 1992, a period when the real atmosphere was in the throes of a very significant stratospheric warming.

The simulations differed mainly in that they were performed with a variety of parameterizations of the gravity wave drag. It is seen that the various gravity wave parameterizations have significant effects on the modelled evolution leading up to the observed mid-January warming. The simulated sequence of events is shown to be sensitive to the scheme used and the strength of the wave breaking resulting. It is found that the impact of the mesospheric wave drag is seen even low in the stratosphere. The simulations are compared with observations obtained by the Improved Stratospheric and Mesospheric Sounder (ISAMS) which flew aboard the Upper Atmosphere Research Satellite (UARS) and collected data during late 1991 and some of 1992.

## 1 Introduction

It has long been known that some sort of mesospheric body force is required to close off the stratospheric polar night and mesospheric summer jets (e.g. Leovy 1964). Houghton (1978) seems to have been the first to suggest that gravity waves might be responsible for that force, and the first parameterization to encapsulate the physics of the process was that of Lindzen (1981), who calculated the effective force on the zonal flow due to the breaking of a monochromatic gravity wave.

Many authors have extended Lindzen's work, of particular relevance to this work was the series of papers (Holton 1982, 1983, 1984, and Holton and Zhu 1984) which discussed the effect of monochromatic waves on the mesosphere. It was soon found that the influence of gravity waves was beneficial even in the troposphere (eg Palmer et al. 1988 and McFarlane 1987) and it was not long before efforts at parameterizing sources were attempted (e.g. Rind et al. 1988).

However, even as modellers were attempting to use the new parameterizations in their models, it became obvious that simple monochromatic parameterizations were probably going to be inadequate on their own as observations from all over the world showed the ubiquity of a saturated spectrum of gravity waves in the middle atmosphere (e.g. Fritts 1984).

[1] Current Address: Department of Physics and Astronomy, University of Canterbury, Christchurch, New Zealand. (email: b.lawrence@phys.canterbury.ac.nz)

NATO ASI Series, Vol. I 50
Gravity Wave Processes
Their Parameterization in Global Climate Models
Edited by Kevin Hamilton

Recently efforts at producing parameterizations of such spectra have come to fruition, and in this paper we discuss the effect of two such parameterizations upon the modelled evolution of early winter 1991/1992. These parameterizations are compared to the simplest of all parameterizations - Rayleigh Friction - and to detailed observations of the period. The period chosen has been well described in a series of papers[2] summarising early results from experiments carried aboard the Upper Atmosphere Research Satellite (UARS). In this work we concentrate in particular on the observations obtained by the Improved Stratospheric and Mesospheric Sounder (ISAMS).

Mechanistic models such as the stratosphere-mesosphere model (SMM) used here suffer a major deficiency when compared with General Circulation Models (GCMs) which do include a troposphere: there is no source of realistic short period and small-scale wave activity propagating from below. GCMs do explicitly include some of the dynamical phenomena expected to cause gravity waves (such as mountains, fronts and baroclinic storm systems) and these must cause resolved gravity waves. In addition, they do include parameterizations of processes which must also generate model gravity waves such as convection and boundary layer phenomena. Nonetheless, even GCMs need to include explicit parameterizations of gravity waves to alleviate bias which arise from the fact that they don't generate enough small scale wave activity (or resolve small enough scales to generate the whole spectrum of wave activity).

In either sort of model, it is a moot point whether the presence or absence of such waves has a significant effect on the atmosphere as shown in the model evolution. In the SMM, we must parameterize their effect with simple parameterizations, and we can examine the effect of changing the distribution and/or strength of wave activity. Of course one must also consider the efficacy of the parameterization itself. It is the purpose of this paper, and subsequent work, to examine the effect of such waves as they break in the mesosphere. In this first study we are concentrating only on the simplest sensitivity tests, so we use global mean sources, further work will examine the effect of a distribution of sources. Despite being carried out in a simple mechanistic model, this work has relevance to GCMs as the sensitivity shown to the parameterizations will be similar to that seen in GCMs (both in terms of the GCM sensitivity to parameterizations and their sensitivity to breaking resolved waves). We begin with a description of the model used and of the gravity wave parameterizations. The general evolution of some chosen simulations is presented and compared to the ISAMS data.

## 2 Model Description

The model used is a version of the U.K.M.O. Stratosphere-Mesosphere Model first introduced by Butchart et al. (1982). The model has subsequently been heavily modified by the author to improve the representation and subsequent diagnosis of the mesosphere. It is a primitive equation grid-point model which is integrated from an initial condition (to be described), and is constrained by 100 hPa geopotential height fields linearly interpolated between daily observations valid for the simulation time.

The original model used Rayleigh Friction to mimic the effect of gravity waves in the mesosphere: the most obvious improvement was to try and introduce a more sophisticated scheme. Initially, this was completely unsuccessful, as the model had no vertical diffusion or convection scheme and much of the mesosphere became catastrophically unstable. The introduction of a dry convection scheme has alleviated that problem. A further significant modification has been in the handling of the vertical wind. In order to minimise the effect of model resolved gravity

[2]of particular relevance to this work are the papers in the special issue of the *Journal of the Atmospheric Sciences*, volume **51**, 1994.

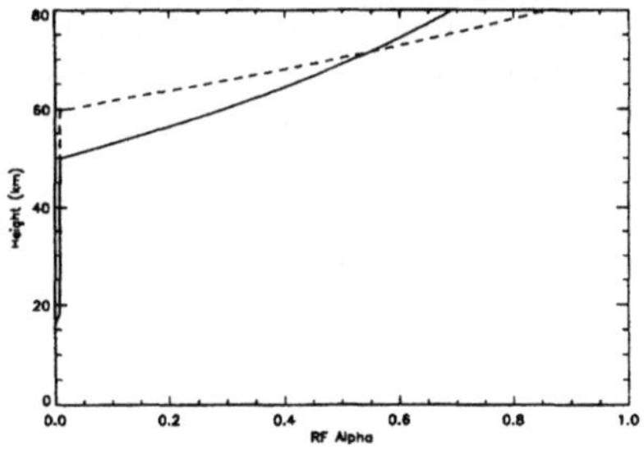


Figure 1: Rayleigh Friction Profiles Used (days$^{-1}$) : Standard coefficients fall on the solid line, the altered coefficients described in section 2.1 fall on the dashed line.

waves on diagnostics and transports, suitable averaged winds have been used in the diagnostics.

The radiation scheme used is an updated version of MIDRAD (Shine 1987), in these simulations we use daily average heating: that is, despite carrying out simulations of the mesosphere, there are no directly generated solar tides in this model (and none propagating from the lower atmosphere). The runs described here use three very different types of gravity wave parameterizations: Rayleigh Friction, the "Fritts" scheme (Fritts and Lu 1993), and the "Hines" scheme (Hines 1996a,b).

The latter two schemes are intended to parameterize the propagation of a (possibly saturated) spectrum of gravity waves, as opposed to previous monochromatic schemes such as those referenced in the introduction. The two schemes are based on very different theoretical grounds, but both purport to track the "characteristic vertical wavenumber", $m^*$, of spectra at each grid point as they propagate up through the model domain. The characteristic vertical wavenumber is defined as the wavenumber above which saturation limits wave amplitudes. The essential feature of both these schemes is that it is the change in this parameter as the spectra propagate vertically which leads to a force on the background flow.

## 2.1 Rayleigh Friction

The standard version of the model (as used by the UKMO and others) uses a Rayleigh Friction scheme, that is, the effect of gravity waves enters the momentum equation through a term of the form

$$\delta u_{\text{gwaves}} = -\alpha(z)\, u(\lambda, \phi, z)\delta t \tag{1}$$

where of course $u$ is the model wind at any grid point $\lambda, \phi, z$ (and the scheme is applied to both the zonal and meridional winds). The standard profile essentially turns on gravity wave *drag* at 50 km and (for a constant wind profile) increases it slowly to a maximum at 80km which, for a 50 m/s wind, would result in a retarding force per unit mass of 35 m/s/day. In order to mimic the effect of the other parameterizations, another profile is also used in this study, termed the "raised" profile. The two profiles are shown in Fig. 1: it can be seen that the raised profile starts higher and increases faster.

## 2.2 Fritts

As explained above, the Fritts scheme is a spectral parameterization: The background to the parameterization can be found in Fritts and VanZandt (1993) and full details of the parameterization in Fritts and Lu (1993). Here, we simply summarise the major features of the scheme needed to understand the application of the scheme in the SMM.

The scheme begins by dividing the total gravity wave energy, $E_0$, at the bottom of a vertical column into components, $E_i$, propagating in the direction of the four quadrants (north, east, south and west). Initial anisotropies in the energy distribution can be specified: however, the energy is constrained to be equally split between the meridional and zonal axes - this constraint is maintained at all levels. In addition to ensure that no component becomes too small, there is a limit on the anisotropy between the two components on any one axis, achieved by ensuring that

$$-0.5 + \alpha_F \leq b_{x,y} \leq 0.5 - \alpha_F. \tag{2}$$

Here, $b_{x,y}$ denotes the amount of anisotropy on that axis: for example, for the east-west axis

$$b_x = (E_2 - E_4)/E_0,$$

and $\alpha_F$ is an arbitrary tunable parameter (chosen to be 0.1 for the simulations described below).

The method uses empirical relationships to associate with each of these component spectra a characteristic horizontal and vertical wave number. Each spectrum is assumed to propagate vertically, and the scheme proceeds by considering (for each column) how the total energy of each component is affected by the vertical gradients in stability, density, and background wind. As the parameterization works up the height profile, the first two of these are expressed as multiplicative factors to the energy below, while the effect of the background wind comes via a Doppler shift of the characteristic vertical wave number (which then, via the empirical relationship, affects the energy). The Doppler shifting factor potentially admits problems with shifting of the characteristic wave number through a critical line, but constraints based on the observations and the conservation of momentum flux are used to ensure that the scheme is stable.

The relationship between the energy content of each component spectrum and the momentum flux ($P_i$) associated with it is also empirically based:

$$P_i \simeq DE_i \tag{3}$$

where $D \simeq 1/22$ is a factor obtained by Fritts and VanZandt who integrated a representative spectral shape for suitable parameters. The net momentum flux along each axis is then the difference between the momentum flux in the appropriate components and the resulting force per unit mass on the flow is simply the density weighted gradient of the momentum flux which is expressed in terms of the energies of the component spectra, e.g.

$$F_x = -\frac{1}{\rho}\frac{\partial}{\partial z}(\rho D[E_2 - E_4]) \tag{4}$$

## 2.3 Hines

Although based on very different theoretical grounds (Hines 1991a,b, 1993), the Doppler spreading theory of Hines also yields a conceptually simple parameterization of the effect of an upwardly propagating spectrum of gravity waves upon the background flow (Hines 1996a).

The Doppler spread theory begins with the assumption that at any given height the vertical wavenumber ($m$) spectrum can be divided into three distinct regions: A conservatively propagating region, where $0 < m \leq m_*$; a region where dissipation is taking place ($m_* < m \leq m_C$) because Doppler spreading is moving some vertical wave numbers into the third region where $m > m_C$ and all wave energy and momentum are totally dissipated. Hines(1996a) and references therein then show that for an azimuthally isotropic spectrum and no background winds $m_C$ is approximately $N/(\psi_1 \sigma_T)$, where $\psi_1$ is a factor justified on both empirical and theoretical grounds. $N$ is the Brunt-Vaisala frequency and $\sigma_T$ is the rms gravity wave induced horizontal wind.

In the parameterization, the total spectrum is divided into $J$ azimuths, and at any height there is a cutoff wavenumber $m_j$ for each direction which depends on $N_i$ (the value of $N$ in the source region), $v_j$ (the background wind in the $j$th direction), $V_j$ (the background wind in the $j$th direction in the source region), and $\sigma_j$ the rms horizontal wind fluctuation in the $j$th direction (which will depend on contributions from waves propagating at all azimuths):

$$m_j = N_i/(\psi_1\sigma_j + \psi_2\sigma_T + v_j) \qquad (5)$$

In this work, $V_j$ is assumed to be zero and the spectra are assumed to be isotropic one level below the model bottom (which, using the Hines parlance, allows both intro-waves and extro-waves). The source strength is introduced via $\sigma_{10\text{km}}$ which is the rms velocity of the total spectrum at 10 km (which is scaled up to the model bottom using density weighting). The two factors, $\psi_1$ and $\psi_2$ are constants which the user has some discretion to modify - Hines suggests ranges of $1.2 \le \psi_1 \le 1.9$ and $0.1 \le \psi_2 \le 0.4$. Initial tests with the parameterization led to two choices of these parameters: in both cases, $\psi_2 = 0.4$, in the "standard" version, $\psi_1 = 1.5$ (chosen as a midrange value), and in the "altered" version $\psi_1 = 1.9$ (chosen to lower the breaking altitude and highlight the effect of the anisotropies). In the work that follows the choice of $\psi_1 = 1.5$ and $\psi_2 = 0.4$ are termed the standard choice and $\psi_2 = 1.9$ and $\psi_2 = 0.4$ the altered choice.

Each of these component spectra propagates vertically: As any given spectral element propagates up, the wavenumber increases due to internal non-linearities in the spectrum (initially, primarily those due to the effect of a given wave number being modified by the Doppler effect of the mean flow, eventually the effect of the other waves becomes important due to the density effect). Provided the spectral element remains within the first region, the vertical flux of momentum and wave action carried by that element is conserved.

However, eventually the element may enter the second region, and there will be some reduction in the momentum carried by that element (and of that component spectrum as a whole, as it is assumed that this is a non-conservative process). The vectorial addition across all azimuths of these momentum flux reductions results in the net gravity wave driven mean flow stresses. The relationship between the cutoff vertical wavenumber and the horizontal momentum flux introduces another constant related to the characteristic horizontal wave number of the portion of the spectrum carrying most of the momentum - here we have chosen horizontal wavelengths of 100 km and 200 km in the standard and altered versions respectively.

## 2.4 Initial Conditions and Bottom Boundary

As this work has progressed, the data used to initialize the runs evolved (with improving retrieval techniques), and an unexpected sensitivity of the gross features of the simulations to the initial condition was found. (Mechanistic models such as the one used here are usually thought of as being less dependent on the initial condition than free-running GCMs.) Accordingly, simulations were carried out with two different initial conditions. The two initial conditions used were from V7 of the ISAMS retrieval software (the first to carry out a reasonable mesospheric retrieval) and from V10 (the best available observations of the mesosphere at the time the first of the simulations reported here was carried out). The major difference between the two sets of data on this day is that the V10 data has greater vertical resolution (insofar as it resolves greater vertical gradients as can be seen from Fig. 2), resulting in a warmer stratopause and a colder polar lower stratosphere.

The observed temperatures were stacked on top of bottom boundary geopotential heights from the so called " assimilation" data provided to the UARS project by the UKMO (Swinbank and O'Neill 1994). The geopotential fields were then used to calculate balanced winds (Randel

a) ISAMS V7 b) ISAMS V10

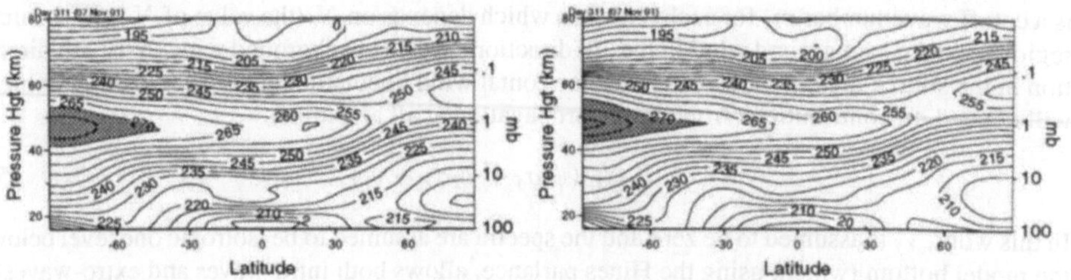


Figure 2: Zonal Mean Initial Temperature Fields

1987).[3] The major differences between the initial conditions are encapsulated in the zonal wind fields shown in Fig. 3. Here we see that there is considerably different structure in the equatorial region, that the polar night jet is stronger in the V7 sub-tropical mesosphere (but further from the pole and broader in the lower stratosphere).

a) ISAMS V7 b) ISAMS V10

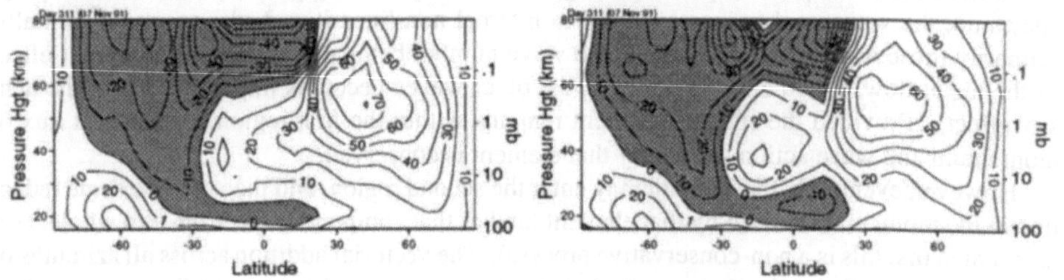


Figure 3: Zonal Mean Initial Zonal Wind Fields

The effect of these and other variants of the initial conditions will be examined in detail in another study, here we concentrate only on the effect of the gravity wave parameterizations upon the different evolutions. Nonetheless, as will be seen the sensitivity to initial conditions means that we must treat the results to be presented with caution: later studies will present results from an ensemble of runs.

## 2.5 Model Runs

Clearly in the real atmosphere gravity waves will be generated by a wide variety of source mechanisms, and indeed most will probably be generated in the troposphere or perhaps at the tropopause. One might expect such source mechanisms to feed into the gravity wave parameterizations as source strengths and/or anisotropies. However, apart from the possible exception of orographic gravity wave drag, no source mechanism has been well characterised for any of the parameterizations used here. Hence, based on this lack of knowledge, and with the benefit of simplifying the factors involved for interpretation, the model runs using the Hines scheme and the Fritts scheme have all used global mean sources. That is, the input to the parameterization at the model bottom (100 hPa) has been the same at each grid-point. Since the source amplitudes are both poorly known and we have been trying to mimic the global response with global mean sources, it has

[3] For reasons unconnected with this study there were slight numerical differences in the wind algorithm used.

| Param. Used | Init. Cond. | Label |
|---|---|---|
| RF $^{A}$ | IS V10 | RFI |
| RF $^{A}$ | IS V7 | RFI7 |
| RF $^{B}$ | IS V10 | RFH |
| Hines $^{C}$ | IS V10 | H1 |
| Hines $^{D}$ | IS V10 | H2 |
| Hines $^{D}$ | IS V7 | H2I7 |
| Fritts $^{E}$ | IS V10 | FTC |
| Fritts $^{E}$ | IS V7 | FTCI7 |

$A$ Standard
$B$ Altered (Raised)
$C$ $\sigma_{10\text{km}} = 2.0$, $k = 100\text{km}^{-1}$, standard constants
$D$ $\sigma_{10\text{km}} = 4.0$, $k = 200\text{km}^{-1}$, altered constants
$E$ $E_0 = 7.5$

Table 1: A summary of the Model Simulations: The label column denotes the shorthand for each simulation as used in the text and following figures. Note that ISV10 is also used as the label for the actual observations

been necessary to try a large number of possible source strengths. The runs reported here are those which gave suitable amplitudes of the polar night jet.

The full set of simulations described are summarized in table 1, and have been chosen to span the range of the parameterizations used and to highlight some aspects of the sensitivity of the simulations.

# 3 Results

The major features of the evolution of the middle atmosphere during this period have been presented by Rosier et al. (1994); O'Neill et al. (1994); Ruth et al. (1994) among others. Here we attempt to focus on the effect of the gravity wave parameterizations upon the evolution of some model simulations of that period. We attempt to qualitatively assess their fidelity by direct comparison with ISAMS V10 data. We begin by displaying time series of the zonal mean winds at 66°N (Figs. 4-7). While the zonal mean is not a very discriminating measure of the actual synoptic structure, we can see immediately that there are gross differences between the simulations and the observed evolution.

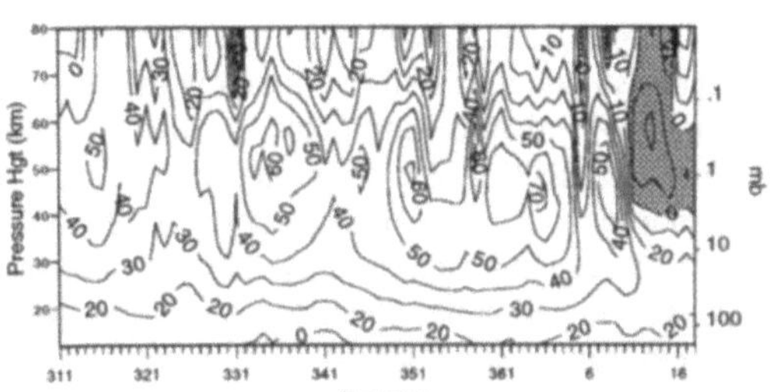


Figure 4: ISV10: Zonal Mean Zonal Wind at 66°N. Balanced winds calculated from the ISAMS V10 data.

The major features evident in the ISAMS winds (Fig. 4) are: the very significant reversal of the winds in the middle and upper stratosphere and mesosphere in mid-January (hereafter called the "significant" warming); the massive deceleration and just as rapid recovery of the jet just prior to the significant warming; and the apparent pulsating nature of the jet maximum prior during the preceeding year, with periods of acceleration and deceleration (these pulsations are associated with a series of minor warmings, as described in Rosier et al. 1994).

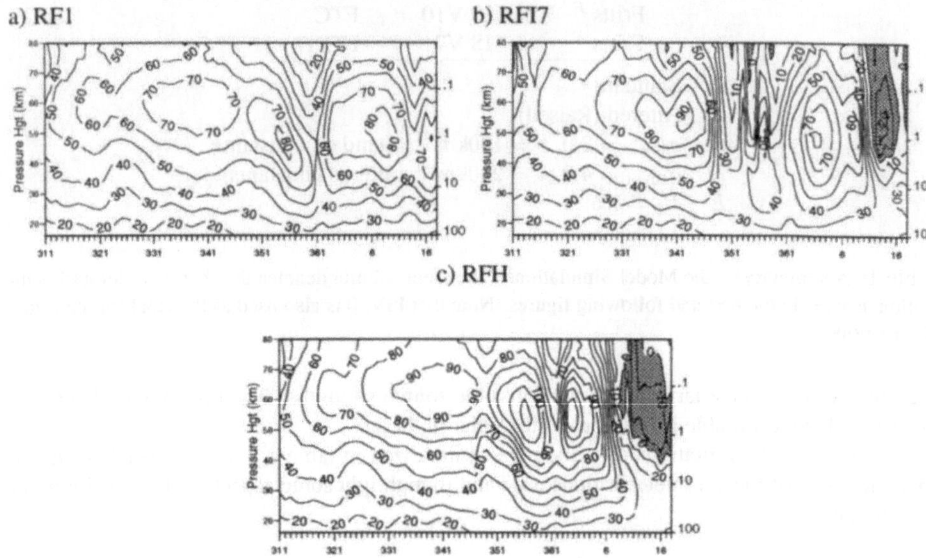


Figure 5: Zonal Mean Wind at 66°N. Rayleigh Friction Runs.

When one considers how well simulations with this model and Rayleigh Friction have done historically, it is obvious by comparing the time series of zonal mean zonal wind at 66°N (Fig. 5) that there is a surprising spread of simulations. The standard configuration of the model (using the "normal" Rayleigh Friction, and initialised with what one might expect to be the most accurate initial condition (ISAMS V10 data) does the poorest (Fig. 5a) in that it simply fails to reproduce the mid-January warming (although if the simulation is continued, it would appear to eventually produce a warming). The evolution does indicate a massive disturbance in late December 1991, in common with the observations, but in general the simulation does not do well. The simulation initialised with ISAMS V7 data does significantly better, and in fact reproduces the January warming (including the defining vortex merger, (Rosier (1996))) with surprising fidelity. The reasons why these simulations differ so much is the subject of a further study (in preparation) but seems in part to be due to wind fields calculated from the observed temperature data. The third RF simulation (RFH, Fig. 5c) shows that the quality of the simulation is not only related to the initial condition - a change in the parameterized gravity wave drag has led to a markedly better simulation (at least insofar as portrayed by the zonal mean wind evolution). It will be seen that although this simulation has obtained a significant warming, it has characteristics which are quite different from those observed. It is included simply to show that even in the regime of a very simple gravity wave drag scheme (Rayleigh Friction), the parameterization has considerable impact on the simulation.

The next set of simulations were those carried out with the Fritts scheme (Fig. 6). Although only the results of two simulations are presented here, a large number of simulations were car-

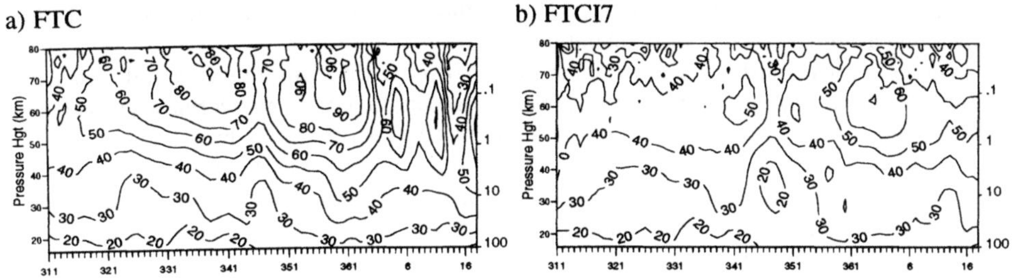


Figure 6: Zonal Mean Wind at 66°N. Fritts Parameterization Runs

ried out, and none of them developed a large warming in January. It can be seen by comparison of Figs. 6a and 6b that this failure is not related to the initial condition. These two runs have been chosen for presentation because the value of $E_0$ used for these simulations gave the best jet magnitude. No more results from the Fritts scheme are presented as it was not possible to find a parameter regime that resulted in a simulation that had anything in common with the observed atmosphere.

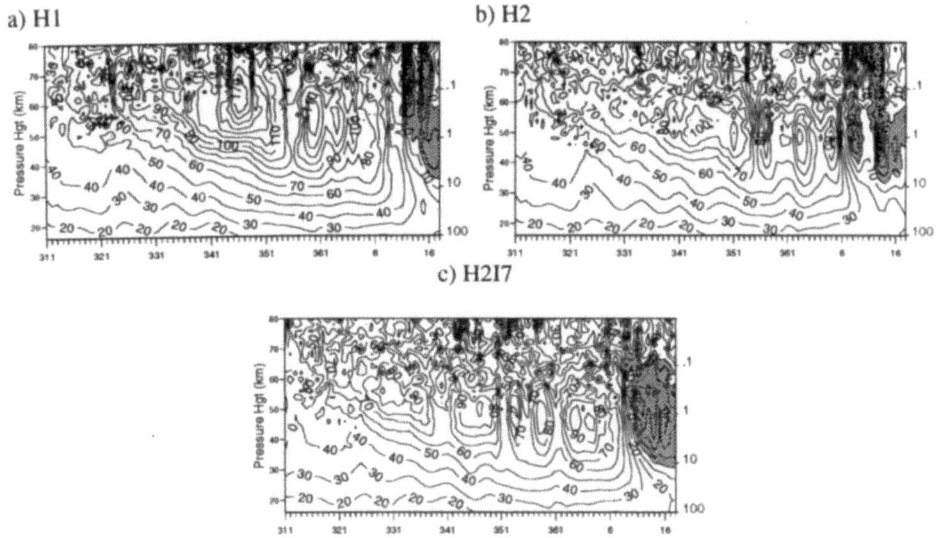


Figure 7: Zonal Mean Wind at 66°N. Hines Parameterization Runs

The final set of simulations used the Hines parameterizations. Once again a large number of simulations were carried out, but only the results of three will be presented here (Fig. 7). It can be seen that all three produced significant warmings, and in fact that run H2I7 was over enthusiastic, insofar as the warming produced could be classified as a major warming. Another obvious characteristic is the pulsating character of the jet which is especially obvious in simulation H2I7. However, what is most obvious about the timeseries is the excessive "noisiness" of the winds in the mesosphere. None of the other parameterizations produced simulations with this feature, however, as will be asserted, at least some of this variability is probably more in keeping with reality! Run H1, which was the first run with the Hines parameterization (the parameters used were chosen from column tests), produced a polar night jet which was too strong, the parameters chosen for run H2 were selected to drop the altitude at which wave breaking occurred while maintaining a similar strength of acceleration.

In order to truely understand the mechanisms involved in the warmings seen in the runs, one should consider the synoptic structures. However, in order to present an overview we use diagnostics of large scale eddy amplitudes. By examining zonal wave one we have a good indicator of when the polar night vortex is perturbed: When it is not centered on the pole the amplitude of wave one increases. We restrict our interest to zonal wave one, as none of the simulations resulted in a wave two warming, and although the amplitude of zonal wave two increased in the incipient stages of the vortex mergers, it is not as useful a diagnostic.

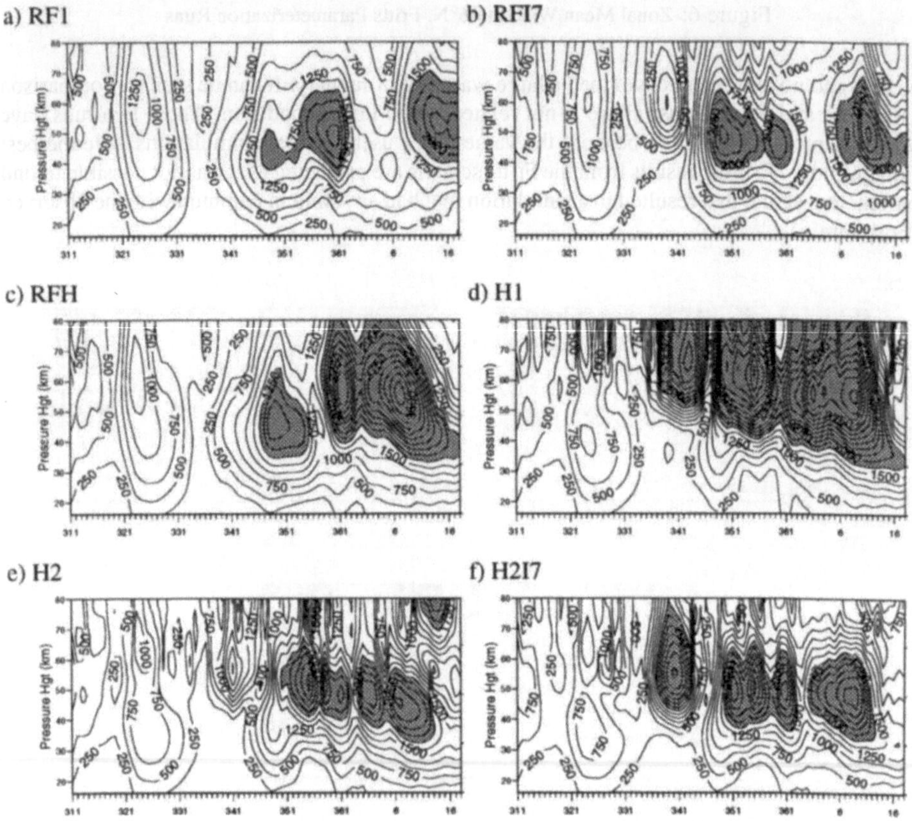


Figure 8: Amplitude of zonal wave one at 66°N in the simulations

Fig. 8 summarizes for both Rayleigh Friction and the Hines scheme the difference between runs discussed here. As one examines Figs. 8a and 8b, the major discernable difference appears to be in December, where we see that there is a vortex disturbance just prior to day 341 in RFI7 in the mesosphere, which doesn't occur in RF1. This disturbance is also obvious in all the Hines runs (Figs. 8d-f), and is the signature of the beginning of the succession of minor warmings which culminate in the mid-January warming (as described in Rosier et al. (1994)). By contrast, there is no sign of this disturbance in RF1, which doesn't have a mid-January warming. It is also not present in RFH, which does have a mid-January warming. The explanation for this is that although RFH does simulate a mid-January warming it is of a very different nature: Fig. 9 shows that in run RFH, the vortex in the lower mesosphere is gradually eroded throughout the season,

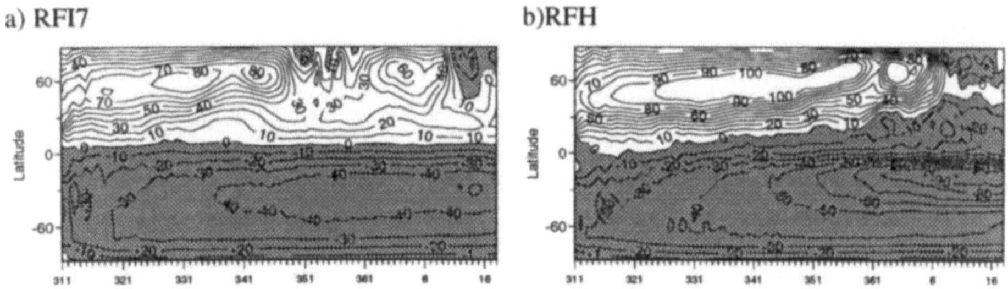


Figure 9: Time series of zonal mean wind on the 60 km pressure surface: contrast between simulations RFI7 and RFH.

and is not subject to a series of pulsations (as seen in the observations and the other simulations which produce warmings).

In order to summarize the evolution of the lower stratosphere, we show the zonal mean temperature at 22km for all latitudes in Fig. 10. In the observations (which are limited to latitudes north of 32.5°S, we see that all the simulations give reasonable simulations south of 60 °N. However, north of 60°N, we see that there are significant differences associated with the interplay between descent induced heating, (large scale) wave-driven heat fluxes and radiative cooling in the polar night.

The observations (Fig. 10a) suggest that there are a few short periods when the zonal mean temperature falls below 195 K at 22 km (about 42 hPa). These periods occur prior to the significant warming, when the jet is strongest. These periods are associated with very cold localized regions, in at least one of these cases, polar stratospheric clouds were observed (Taylor et al. 1994). By contrast, all the simulations produce long periods where the zonal mean temperature falls below 195K, and in the RFH,H2 and H2I7 simulations, below 190K. These do not occur in nature, and if we had a chemistry scheme in these simulations, would have been associated with unrealistic chemical depletion of ozone.

In simulation RF1, the extra cold periods are associated with the fact that the vortex is considerably less disturbed, and radiative cooling has had longer to decrease the temperature in the polar night. In RFI7, which has a more realistic sequence of warmings, we see that the temperature at 22km is more realistic. The three simulations which are colder also have more realistic warming structures, thus it is not as easy to attribute reasons as to why the north polar region is so cold in these runs, although it is clear that in the simulations with the coldest poles (the Hines simulations), the polar night jet, although suffering pulsations associated with minor warmings, is still too strong before the significant warming. It is well known that overly strong jets result in poles which are too cold (via thermal wind, eg Andrews et al. (1987)).

# 4 Discussion

Most of the simulations have captured the essential defining characteristic of the period: a series of oscillations of the zonal mean wind field leading to a significant warming in mid-January 1992. However, three of the simulations described here had not resulted in a warming by the 18th of January (when the simulations were terminated to conserve computer disc space). The three failures are the Fritts simulations and the RF1 simulation.

Not surprisingly, all the model runs show some sensitivity to the initial conditions used to begin the integrations. However, received wisdom has it that the evolution of the stratosphere is dominated by the upward propagating planetary waves from the troposphere, and in particular,

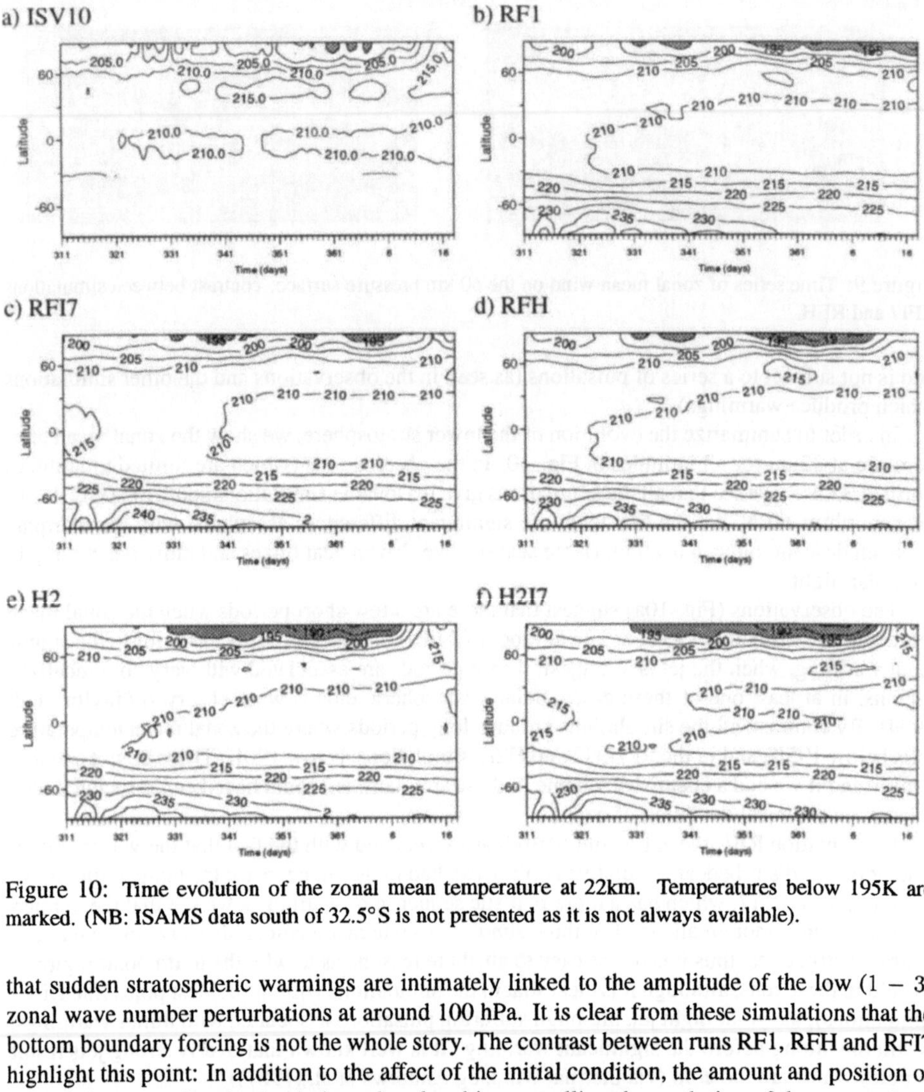


Figure 10: Time evolution of the zonal mean temperature at 22km. Temperatures below 195K are marked. (NB: ISAMS data south of 32.5°S is not presented as it is not always available).

that sudden stratospheric warmings are intimately linked to the amplitude of the low (1 – 3) zonal wave number perturbations at around 100 hPa. It is clear from these simulations that the bottom boundary forcing is not the whole story. The contrast between runs RF1, RFH and RFI7 highlight this point: In addition to the affect of the initial condition, the amount and position of the gravity wave drag is also intimately related in controlling the evolution of the stratosphere below. The effect of the gravity wave driving in the mesosphere penetrates down throughout the model domain, as can be seen in figure 10 which shows the temperature evolution three model levels from the boundary (actually, the effect penetrates right down to, and through - via vertical advection of temperature - the model boundary).

These results are quite in accordance with the ideas of downward control first introduced by Haynes et al. (1991). The effect of gravity wave drag on lower stratospheric temperatures was first highlighted by Garcia and Boville (1994), although they were not convinced as to the importance of the direct effect of gravity waves on lower stratospheric temperatures in the northern hemisphere. The results presented here suggest that in addition to any direct effects (such as those that might occur in an axisymmetric atmosphere), one must consider the affect of the drag upon the evolution of the stratospheric warmings. This seconday effect may in fact be of even greater importance upon both the real and simulated climate.

Early model simulations, such as those of Peltier et al. (1986), found that the quality of model simulations of warmings was very dependant on the form of the Rayleigh Friction used - a result in accordance with those presented here. However, Smith (1992) found that the occurence of warmings in her model depended only on the lower stratospheric winds and was independent of the upper winds. Part of the explanation for this may be in her methodology: She initialized her model runs only nine days earlier than the warming, it is certainly not yet clear from the simulations described here over what timescale the mesospheric evolution was affecting the lower stratosphere in these time-evolving simulations.

Perhaps of greatest relevance to understanding these results is the view that much of the difference between runs which do capture the observations well seem to be related to the presence of absence of the vortex interactions in the upper stratosphere and lower mesosphere. These are the interactions which lead to the pulsations seen in the zonal mean wind. The effect of the gravity waves can be thought of as a damping on the nonlinear planetary-scale wave train (eg O'Neill and Pope 1988) arriving in the mesosphere. This then critically affects the evolution of the warmings, which (when of the wave one type seen here) are observed in both the model and the atmosphere to occur in the lower-mesosphere and progress downwards. This nonlinear wave train is certainly sensitive to the regime through which it begins propagating (at least while it behaves linearly), and so some further part of the explanation of the difference between these results and those of Smith (1992) may be related to the effective damping of the tropospheric disturbances in the lower stratosphere (in accordance with the results of Schoeberl and Strobel 1980).

Although not shown here, tests of the Fritts scheme with various columns of winds showed it to be rather insensitive to variations in the background wind strength (although somewhat more to the windshear) — it seems to result in a given body force on the middle atmosphere for a given source strength regardless of the wind profile it is used upon. With a source strong enough to get the zonal mean winds in the observed ballpark the perturbations from the zonal mean are also strongly damped (for example, the amplitude of zonal wave one in geopotential doesn't exceed 1500 m at 66°N at any time during run FTCI7), which is why neither run FTC nor FTCI7 reproduced the observed mid winter warming.

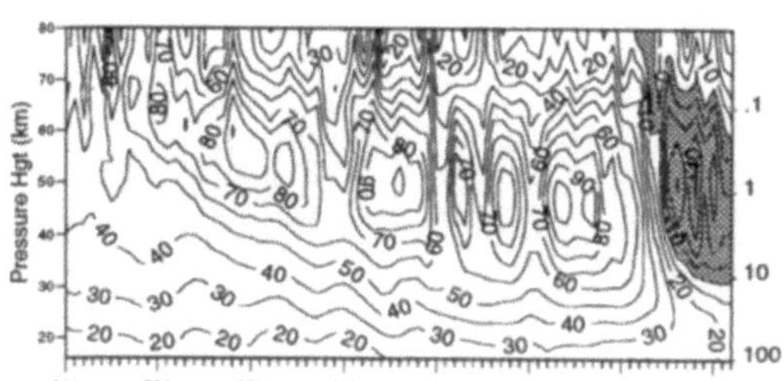


Figure 11: The timeseries of zonal wind at 66N from the H2I7 run, calculated from the model temperature fields, the bottom boundary, and the balanced wind algorithm

The Hines runs display a very noisy mesosphere when the model zonal winds are dumped every day as here. However, much of that noise is in very small scales, probably model resolved gravity waves, as can be seen when the winds are calculated from the model temperature fields using the same balanced wind scheme as used to calculate the winds from the ISAMS data (Fig. 11). It is comforting also to see the balanced wind scheme does well at recovering the model winds right up to the stratopause (compare Fig. 11 with Fig. 7c). Of course it is to be expected that the balanced wind (as opposed to the true wind) should be more consistent in time, as it is represents the winds due to synoptic and greater scales.

Even allowing for the contribution from numerical noise (including the possible influence of both the model lid and spurious model resolved gravity waves) to the noisiness seen in the Hines runs, there is still more variability in the balanced winds in Fig. 11 than there is in the RF runs. However it is believed that this too is more real: Not only is the amount of variability similar to that seen in the observations (Fig. 4), the transition from low temporal variability below the stratopause to greater temporal variability above is consistent with that seen in rocket winds (eg Leovy and Ackerman 1973). This is a clear improvement over the Rayleigh Friction simulations which produce an unrealistically smooth and undisturbed mesosphere.

H1 H2

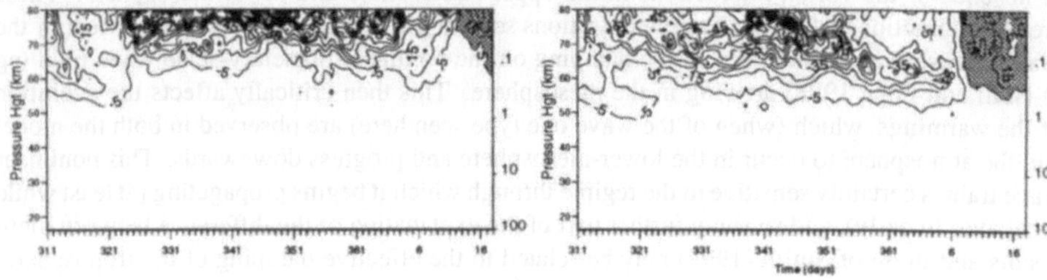


Figure 12: Zonal Mean Tendency of the Zonal Wind due to the Hines Parameterization at 66°N (m/s/day)

The improvement in the simulation between ISAMS H1 and H2 is apparent, not only does the jet maximum drop between the runs, the penetration of the disturbances to the vortex is considerably damped in run H2. This has come about from the lower altitude of wave breaking which can be seen by a comparison between the strength of the wave breaking (Fig. 12). Run H2 has considerably lower maximum values than Run H1, and the altitude is much more like the middle of the mesosphere. These runs show the importance of choosing appropriate parameters for the scheme. It can be seen that the values of the tendencies in run H2 are of the same order as those diagnosed by Marks (1989).

a) RFI7 b) H2I7

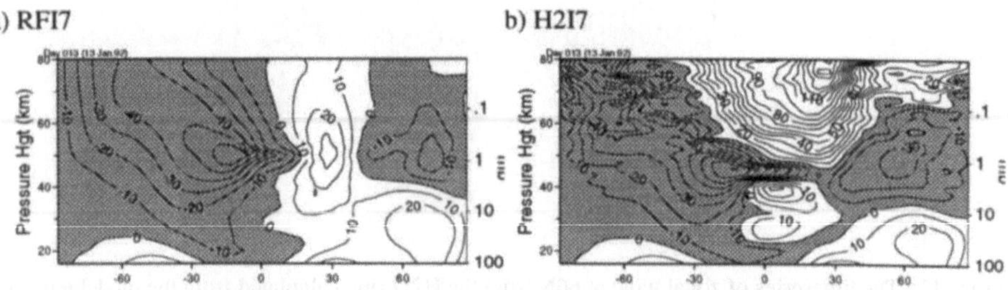


Figure 13: Zonal Mean Zonal wind on the 13th of January 1992

Although this paper has concentrated mainly on summarizing polar phenomena, run H2I7 has produced a more interesting simulation than run RFI7 throughout the atmosphere as can be seen from Fig. 13. The figure shows the winds during the maximum of the warming in both simulations: the major feature of interest is that H2I7 has produced the northern hemisphere mesospheric subtropical jet which is consistent with observations (eg Fig. 1. in Rosier et al. 1994). The RF runs always fail to produce this phenomenon. It can be seen that the tropical structures are quite different - of interest here is the nose of easterlies which push accross from the south into the north in the stratosphere - in H2I7 this nose has moved steadily downward during the simulation. There is obviously much more to be learnt by continuing this run. The strength of the shear zone seen in the summer hemisphere is consistent with other parameterizations (Warwick Norton, *pers com*) and will also be the subject of further work.

# 5 Conclusions

In this paper, we have compared a number of different simulations of a short period of the Northern Winter 1991/1992 with the ISAMS observations. The simulations differed in the type of gravity wave parameterizations used and/or the initial condition and the main aim of this paper was simply to identify the differences introduced by the different gravity wave schemes and qualitatively estimate their impact on the quality of the simulations (and hence upon our ability to model the middle atmosphere).

The major feature observed during the period was the very significant warming in January, preceded by a number of minor warmings as discussed in Rosier et al. (1994). Most of the simulations produced a similar warming in early January, although the nature of the warmings differed. However, some of the simulations did not, and although it is clear that the initial condition has some influence on the subsequent evolution of the simulations, with an appropriate gravity wave scheme the stratospheric dynamics are dominated by the effect of the troposphere below, the nonlinear stratospheric response to that effective forcing, and the mesospheric evolution above (exerting it's effects via downward control). Subsequent work (Lawrence, 1996, in preparation) evaluates the quantitative effect of these schemes on calculated descent rates in the polar night, and on the specific details of the simulated warmings. Further work will concentrate on identifying and quantifying the effects of more realistic estimates of source distributions.

Of the gravity wave parameterization schemes examined here, and given caveats about using global mean sources, it would appear that the Hines scheme gives the best simulations.

# Acknowledgements

Much work on the middle atmosphere would not be possible without the advent of UARS data. Of particular relevance to me was the ISAMS data: the ISAMS team are to be congratulated for producing an excellent observational data set. I am also indebted to David Andrews, Dave Fritts, Steve Eckermann, Colin Hines, Richard Kennaugh, Alan O'Neill, Warwick Norton and Suzanne Rosier for useful discussions and/or correspondence. Finally, I am exceedingly grateful for the inspiration of Lee Germon and Chris Harris, and to the U.K. Natural Environment Research Council who funded this work.

# References

Andrews, D. G., Holton, J. R., and Leovy, C. B. (1987). *Middle Atmosphere Dynamics*. Academic Press, Inc., San Diego, Calif.

Butchart, N., Clough, S., Palmer, T., and Trevelyan, P. (1982). Simulations of an observed stratospheric warming with quasi-geostrophic refractive index used as a model diagnostic. *Q. J. R. Meteorol. Soc.*, 108:475–502.

Fritts, D. C. (1984). Gravity wave saturation in the middle atmosphere: A review of theory and observations. *Rev. Geophys. Sp. Phys.*, 22:275–308.

Fritts, D. C. and Lu, W. (1993). Spectral estimates of gravity wave energy and momentum fluxes. Part II: Parameterization of wave forcing and variability. *J. Atmos. Sci.*, pages 3695–3713.

Fritts, D. C. and VanZandt, T. E. (1993). Spectral estimates of gravity wave energy and momentum fluxes. Part I. Energy dissipation, acceleration, and constraints. *J. Atmos. Sci.*, 50:3685–3694.

Garcia, R. R. and Boville, B. A. (1994). "Downward Control" of the mean meridional circulation and temperature distribution of the polar winter stratosphere. *J. Atmos. Sci.*, 51:2238–2245.

Haynes, P., Marks, C., McIntyre, M., Shepherd, T., and Shine, K. (1991). On the "Downward Control" of extratropical diabatic circulations by eddy-induced mean zonal forces. *J. Atmos. Sci.*, 48:651–678.

Hines, C. O. (1991a). The saturation of gravity waves in the middle atmosphere. Part I: Critique of linear-instability theory. *J. Atmos. Sci.*, 48:1348–1359.

Hines, C. O. (1991b). The saturation of gravity waves in the middle atmosphere. Part II: Development of Doppler-spread theory. *J. Atmos. Sci.*, 48:1360–1379.

Hines, C. O. (1993). The saturation of gravity waves in the middle atmosphere. Part IV: Cutoff of the incident wave spectrum. *J. Atmos. Sci.*, 50:3045–3060.

Hines, C. O. (1996a). Doppler-spread parameterization of gravity-wave momentum deposition in the middle atmosphere. Part 1: Basic formulation. Submitted to J. Atm. Terr. Phys.

Hines, C. O. (1996b). Doppler-spread parameterization of gravity-wave momentum deposition in the middle atmosphere. Part 2: Broad spectra and quasi monochromatic waves, including orowaves. Submitted to J. Atm. Terr. Phys.

Holton, J. and Zhu, Z. (1984). A further study of gravity wave induced drag and diffusion in the mesosphere. *J. Atmos. Sci.*, 41:2653–2662.

Holton, J. R. (1982). The role of gravity wave induced drag and diffusion in the momentum budget of the mesosphere. *J. Atmos. Sci.*, 39:791–799.

Holton, J. R. (1983). The influence of gravity wave breaking on the general circulation of the middle atmosphere. *J. Atmos. Sci.*, 40:2497–2507.

Holton, J. R. (1984). The generation of mesospheric planetary waves by zonally asymmetric gravity wave breaking. *J. Atmos. Sci.*, 41:3427–3430.

Houghton, J. T. (1978). The stratosphere and mesosphere. *Q. J. R. Meteorol. Soc.*, 104:1–29.

Leovy, C. (1964). Simple models of thermally driven mesospheric circulation. *J. Atmos. Sci.*, 21:327–341.

Leovy, C. and Ackerman, T. (1973). Evidence for high-frequency synoptic disturbances near the stratopause. *J. Atmos. Sci.*, 30:930–942.

Lindzen, R. S. (1981). Turbulence and stress owing to gravity wave and tidal breakdown. *J. Geophys. Res.*, 86:9707–9714.

Marks, C. J. (1989). Some features of the climatology of the middle atmosphere revealed by Nimbus 5 and 6. *J. Atmos. Sci.*, 46:2485–2508.

McFarlane, N. A. (1987). The effect of orogaphically excited gravity wave drag on the general circulation of the lower stratosphere and troposphere. *J. Atmos. Sci.*, 44:1775–1800.

O'Neill, A., Grose, W. L., Pope, V. D., MacLean, H., and Swinbank, R. (1994). Evolution of the stratosphere during northern winter 1991/1992 as diagnosed from U.K. Meteorological Office analyses. *J. Atmos. Sci.*, 51:2800–2817.

O'Neill, A. and Pope, V. (1988). Simulations of linear and nonlinear disturbances in the stratosphere. *Q. J. R. Meteorol. Soc.*, 114:1063–1110.

Palmer, T. N., Shutts, G., and Swinbank, R. (1988). Alleviation of a systematic westerly bias in general circulation and numerical weather prediction models through an orographic gravity wave drag parameterization. *Q. J. R. Meteorol. Soc.*, 112:1001–1039.

Peltier, W., Higuchi, K., and Bloxam, R. (1986). Quasi-geostrophic simulations of the 1979 stratospheric major warming: The importance of middle-atmosphere drag. *Pure Appl. Geophys.*, 124:1051–1085.

Randel, W. J. (1987). The evaluation of winds from geopotential height data in the stratosphere. *J. Atmos. Sci.*, 44:3097–3120.

Rind, D., Suozzo, R., Balachandran, N., Lacis, A., and Russel, G. (1988). The GISS global climate - middle atmosphere model. Part I: Model structure and climatology. *J. Atmos. Sci.*, 45:329–370.

Rosier, S., Lawrence, B., Andrews, D., and Taylor, F. (1994). Dynamical evolution of the northern stratosphere in early winter 1991/92, as observed by the improved stratospheric and mesospheric sounder. *J. Atmos. Sci.*, 51:2783–2799.

Rosier, S. M. (1996). *Dynamical Evolution of the Northern Stratosphere in Early Winter, 1991/92: Observational and Modelling Studies.* PhD thesis, University of Oxford.

Ruth, S., Remedios, J., Lawrence, B., and Taylor, F. (1994). Meaurements of $N_2O$ by the UARS Improved Stratospheric and Mesospheric Sounder during early northern winter 1991/92. *J. Atmos. Sci.*, 51:2818–2833.

Schoeberl, M. R. and Strobel, D. F. (1980). Numerical simulations of sudden stratospheric warmings. *J. Atmos. Sci.*, 37:214–236.

Shine, K. (1987). The middle atmosphere in the absence of dynamical heat fluxes. *Q. J. R. Meteorol. Soc.*, 113:603–633.

Smith, A. K. (1992). Preconditioning of stratospheric sudden warmings: Sensitivity studies with a numerical model. *J. Atmos. Sci.*, 49:1003–1019.

Swinbank, R. and O'Neill, A. (1994). A stratosphere-troposphere data assimilation system. *Mon. Wea. Rev.*, 122:686–702.

Taylor, F. W., Lambert, A., Grainger, R. G., Rodgers, C. D., and Remedios, J. J. (1994). Properties of northern hemisphere polar stratospheric clouds and volcanic aerosol in 1991/92 from UARS/ISAMS satellite measurements. *J. Atmos. Sci.*, 51:3019–3026.

# The Parameterization of Gravity Wave Drag Based on the Nonlinear Diffusion of Wave Spectra

Alexander S. Medvedev and Gary P. Klaassen
Department of Earth and Atmospheric Science
York University
North York, Ontario, M3J 1P3
Canada

Byron A. Boville
National Center for Atmospheric Research
Climate Modeling Section
P.O. Box 3000, Boulder, CO 80307
USA

**Abstract** We present a new parameterization of gravity wave drag based on the *Medvedev and Klaassen* [1995] theory of gravity wave spectral evolution and saturation. The scheme has been tested in a column model as well as with a mechanistic version of three-dimensional Middle Atmosphere CCM2 model. For the range of anticipated variability of source spectra in the troposphere the scheme produces plausible results consistent with observations and with theoretical estimates. Results also indicate that the specification of seasonal and latitudinal variability of wave sources is required for three-dimensional modeling.

## 1 Introduction

Gravity wave (GW) drag deposited by waves propagating from below substantially affects the circulation of the middle atmosphere. Parameterization of these processes in general circulation models requires the specification of an absorption mechanism which causes damping of waves, and therefore, a deposition of their momentum to the mean flow.

A manifestation of systematic wave damping processes in the atmosphere is the apparent "universality" of observed gravity wave spectra. Perhaps the most significant feature of "universal" spectra is the power-law behavior of the high vertical wavenumber tails of their power spectral densities (PSD), $S$. The latter can be approximated by $S(m) = A_0N^2/m^3$, where $N$ is the Brunt-Väisälä frequency, $m$ is the vertical wavenumber, and $A_0$ is the spectral amplitude. It is this "saturation" of

NATO ASI Series, Vol. I 50
Gravity Wave Processes
Their Parameterization in Global Climate Models
Edited by Kevin Hamilton

wave amplitudes with height, despite the diminishing gas density, which implies some kind of wave damping mechanism.

Many theories have been devised to explain the observed spectral structure of the variance field. For example, *Zhu* [1994] invokes off-resonant wave-wave interactions as well as radiative damping, while *Gardner* [1994] attributes damping to molecular and/or turbulent diffusion. Some theories assume that the saturated behavior is due to the convective and/or dynamical instability of wave harmonics or wave packets while they propagate upward independently [*Dewan and Good*, 1986; *Smith et al.*, 1987]. Other theories attribute saturation to nonlinear interactions between harmonics within the broad spectra of gravity waves. In *Weinstock*'s [1990] approach harmonics with smaller $m$ are considered as a source of enhanced scale-dependent diffusion which can affect the growth of lower $m$ harmonics with height. The Doppler spread theory [*Hines*, 1991b; 1993a] treats the irregular fluctuations of the wave field as an additional background which force harmonics to spread over the spectrum. *Hines* [1993a] argued that the principle effect of this spreading is to shift wave harmonics from the lower $m$ part of the spectrum to the high-$m$ portion where they dissipate due to instabilities and/or diffusion. Further critical assessment of these theories can be found in [*Hines*, 1991a; 1993b].

In [*Medvedev and Klaassen*, 1995] (hereafter referred to as MK95) we employed *Weinstock*'s [1982] technique and Hines' physical insight regarding nonlinear Doppler effects to develop a more general approach for describing the evolution of GW spectra in the atmosphere. There we assumed that only longer period waves in the spectrum constitute an additional background wind for a given wave component. This low-frequency, spectrum-induced mean wind produces Doppler shifting as well as modified critical levels for higher-frequency harmonics. Interactions near these spectrum induced critical lines result in wave damping and, eventually, in wave obliteration. We have shown in MK95 that these spectral interactions can alternatively be treated in terms of nonlinear enhanced diffusion in the spirit of *Weinstock* [1990]. The MK95 theory distinguishes between wave breaking and wave saturation, and gives the criteria for both processes in terms of wave spectra. It also predicts the development of a "universal" saturated shape for an arbitrary source spectrum of GW. The predictions of the MK95 theory for a single wave under saturation conditions coincide with those of *Lindzen* [1981].

The purpose of this paper is to present a practical gravity wave drag parameterization based on the MK95 theory, which can be used in large-scale circulation models. Parameterizations usually imply further simplifications of the theories they are based on as well as an introduction of some tunable parameters. The scheme we present will retain all physically important features of MK95, and relies on relatively few tunable parameters apart from the source spectrum. To date we are aware of three other spectral GW drag schemes which have been developed and tested in atmospheric models. One approach is to consider harmonics in the spectrum as propagating independently, with each harmonic attaining saturation according to *Lindzen*'s [1981] scheme. Although this approach has been successfully used in a number of numerical models (e.g., [*Garcia and Solomon*, 1985]), it tends to generate step-function GW drag profiles where the momentum deposition is concentrated at and above the breaking levels. The spectral parameterization developed by *Fritts and Lu* [1993] combines theoretical aspects of the assumed instability mechanism of wave saturation, and the semi-empirical *a priori* form of GW spectra. Since observed instantaneous spectra usually deviate from the averaged "universal" shape, one may expect that this approach will be successful in reproducing wave drag profiles when GW spectra are close to the average condition, and will deviate when wave spectra differ from the "universal" one. The parameterization based on Doppler spreading theory [*Hines,* 1996], rather than tracking individual harmonics, calculates the evolution with height of average spectral parameters such as RMS velocity.

In this paper we present a gravity wave drag parameterization based on the MK95 theory. In Section 2 we review the relevant formulae from MK95, while in Section 3 we discuss numerical aspects of the scheme. The range of variability of input source spectra are discussed in Section 4. In Section 5, we discuss the results of column-model tests based on representative distributions of mean fields from CIRA-86, and in Section 6 we present tests with a three-dimensional mechanistic version of the Middle Atmosphere CCM2 model (MA CCM2).

## 2 Theory

According to MK95, the evolution of gravity wave spectra with height is given by the following equation

$$\frac{dS(m_R)}{dz} = \Big(-\frac{\rho_{0z}}{\rho_0} + \frac{m_{Rz}}{m_R} - \beta\Big)S(m_R), \tag{1}$$

where $S$ is the power spectral density (PSD) of horizontal wind associated with gravity waves, $\rho_0$ is the mean density, $m_R$ is the real part of the vertical wavenumber, and $\beta$ is the coefficient of nonlinear damping due to interactions of the harmonic $m_R$ with other harmonics in the spectrum. The vertical wavenumber $m_R$ accounts for an additional nonlinear Doppler shift produced by lower frequency waves in the spectrum and is related to the "linear" vertical wavenumber $m_0$ by the expression

$$m_R = m_0 \sqrt{\pi} \alpha \exp(-\alpha^2) \mathrm{erfi}(\alpha). \tag{2}$$

In (2) the following notation for the error function of an imaginary argument is used

$$\mathrm{erfi}(\alpha) = \frac{2}{\sqrt{\pi}} \int_0^\alpha \exp(x^2) dx,$$

and the dimensionless parameter $\alpha$ is defined by

$$\alpha = \frac{N}{\sqrt{2} m_R \sigma} = \frac{c - \bar{u}}{\sqrt{2}\sigma}, \qquad \sigma^2 = \int_{m_R}^\infty S(m') dm', \tag{3}$$

where $N$ is the buoyancy frequency, and $\sigma$ is the horizontal RMS wind created by all wave harmonics in the spectrum with vertical wavenumbers larger than that of a given harmonic. Note here that if the spectrum consists of a single monochromatic wave, the RMS velocity $\sigma$ coincides with the wave amplitude $u'$, and the parameter $\alpha$ is proportional an inverse Froude number $F_r = mu'/N$ for this wave. The convective instability threshold for monochromatic waves is $u' \to \mid c - \bar{u} \mid$, where $c$ is the horizontal phase velocity of the wave, and $\bar{u}$ is the mean wind. In the case of a broad spectrum our theory gives the instability threshold $\sigma \to \mid c - \bar{u} \mid$, or $\alpha \to 1/\sqrt{2}$. Therefore, it is convenient to consider the parameter $\alpha$ as a straightforward generalization of the inverse Froude number (or equivalently as the square root of the Richardson number) for the case of a broad wave spectrum.

The expression for the nonlinear damping coefficient has the form

$$\beta = \sqrt{2\pi} N \sigma^{-1} \exp(-\alpha^2). \tag{4}$$

The set of equations (1)-(4) for calculating the evolution of the PSD $S$ with height becomes closed if we add the expression for the dependence of the "linear" vertical wavenumber $m_0$ on background parameters in the form

$$\frac{N^{(2)}}{m_0^{(2)}} = \frac{N^{(1)}}{m_0^{(1)}} - \Delta\bar{u}, \tag{5}$$

where superscripts *1* and *2* denote quantities at two height levels, and $\Delta\bar{u}$ is the mean wind shear between these two levels. (5) follows from the dispersion relation for gravity waves and represents the variations of the apparent vertical wavenumber of a harmonic with the constant phase velocity due to vertically inhomogeneous mean wind and temperature.

Once the amplitude of the harmonic $S(m)$ and the coefficient of nonlinear damping $\beta(m)$ are known, the momentum deposition associated with this wave can be found from

$$a_x(m) = \frac{S(m)\beta(m)k_h}{m}, \tag{6}$$

where $k_h$ is the horizontal wavenumber. Note here, that the evolution of spectra is entirely described by (1)-(5) only in terms of vertical wavenumber, while the intensity of wave drag $a_x(m)$ is proportional to $k_h$. In our scheme, $k_h$ enters only as a scaling factor for the momentum deposition, and is consequently the only tunable parameter in the scheme apart the initial spectrum itself. The total wave drag created by the entire spectrum is calculated by adding up the contributions from all harmonics.

MK95 have shown that gravity wave saturation in the atmosphere typically occurs at $\alpha_{\text{sat}} = 1$ to 3. Together with (2) this implies $m_R \approx 1.1$ to 1.3. Therefore for purposes of wave drag parameterization, the systematic nonlinear Doppler shift created by other harmonics in the spectrum can be neglected, and $m_R$ can be replaced by the "linear" vertical wavenumber $m_0$ everywhere in (1)-(6). This result is contrary to Doppler spread theory [*Hines*, 1991b; 1993; 1996] in which all momentum deposition ultimately derives from Doppler shifting of harmonics to higher $m$. However it should be noted that in Hines' theory the entire spectrum accounts for the Doppler spreading of a particular harmonic, whereas we have excluded lower-$m$ waves from our RMS wind $\sigma$ on the basis that for a given $k_h$, they correspond to waves that oscillate more rapidly.

## 3 Numerics

For numerical purposes, we convert the continuous input source PSD $S$ to a discrete spectrum consisting of $M$ harmonics with horizontal velocity fluctuation amplitudes $U_j^2 = S_j \Delta m$, where $\Delta m$ defines the discretization of the spectrum, and $j$ is the index of a given harmonic. The RMS velocity $\sigma$ and the parameter $\alpha$ in (3) are calculated as follows

$$\sigma_j^2 = \sum_{n=j}^{M} U_n^2, \qquad \alpha_j = \frac{N}{\sqrt{2} m_j \sigma_j}, \tag{7}$$

where $j = 1$ and $j = M$ are harmonics with the lowest and highest vertical wavenumbers in the spectrum, respectively.

The solution of the discretized form of (1) has the form (cf. (27) in MK95)

$$U_j^2(z) = U_j^2(z_s)\frac{\rho_{0s} m_j}{\rho_0 m_{js}} \exp\Big[-\int_{z_s}^{z} \beta_j(z')dz'\Big], \tag{8}$$

where subscript $s$ denotes the values at the reference height level $z_s$. If the vertical coordinate is discretized, and if the source spectrum $U_j^2$ is known at some height, (8) can be integrated upward step-by-step as follows

$$U_j^2(z_i) = U_j^2(z_{i-1})\frac{\rho_0(z_{i-1})m_j(z_i)}{\rho_0(z_i)m_j(z_{i-1})} \exp\Big[-\beta_j(z_{i-\frac{1}{2}})\Delta z_i\Big], \tag{9}$$

where $\Delta z_i$ is the grid interval between the higher level $z_i$ and lower level $z_{i-1}$, and $z_{i-1/2}$ denotes the half level between $z_{i-1}$ and $z_i$. In (9) the apparent "linear" vertical wavenumber $m_j$ of the harmonic with a given $k_h$ and phase velocity $c_j$ varies with height according to (5) following variations of the background wind $\bar{u}$ and Brunt-Väisälä frequency $N$

$$m_j(z_i) = \frac{N(z_i)m_j(z_{i-1})}{N(z_{i-1}) - m_j(z_{i-1})\Delta\bar{u}}. \tag{10}$$

The "nonlinear" spectrum-induced Doppler shift of the vertical wavenumber can be calculated from (2) (although in practice it may be neglected).

The discretized form of (6)

$$a_x = \sum_{j=1}^{M} \frac{U_j^2 \beta_j k_h}{m_j}, \tag{11}$$

together with (7), (9), (10) and (4) represent a closed set of equations for calculating the evolution of spectra with height and the corresponding wave momentum deposition. The procedure starts from the lowest level where the source spectral distribution $U_j^2(z_s)$ is specified. At each vertical step RMS velocities $\sigma_j$, parameters $\alpha_j$ and nonlinear damping rates $\beta_j$ are calculated using (7) and (4) for every harmonic beginning with the highest $m_j$, i.e. with $j = M$. Then the vertical wavenumbers $m_j$, squared wave amplitudes $U_j^2$ and wave drag $a_x$ can be found at the next vertical level $z_i$ from (10), (9) and (11), respectively. It is shown in MK95 that the instability threshold for a wave harmonic in terms of the entire spectrum is $\alpha_j^2 \leq 1/2$. Therefore, harmonics meeting this criterion are removed entirely from the spectrum, and their momentum is deposited at the breaking level. This effectively removes harmonics slightly before they reach the "linear" critical levels $c_j - \bar{u} = 0$ of Lindzen-type schemes.

It is worth noting here, that the GW drag parameterization described above coincides with that of *Lindzen* [1981] if the broad spectrum collapses to a single harmonic, as demonstrated by MK95. Unlike Lindzen's approach where the wave breaking occurs suddenly, our scheme produces gradual wave damping and associated wave drag below breaking levels. The similarity of the MK95 scheme to convectively unstable saturation becomes more apparent if we formulate it in terms of phase velocities of gravity waves using the dispersion relation $m^2 = N^2/(c - \bar{u})^2$. Then the spectral distribution of harmonics in $m$ corresponds to a distribution in $c$ with higher vertical wavenumbers corresponding to lower phase velocities, and *vice versa.* This also means that harmonics with $c > \bar{u}$ ("extrowaves") have positive $m$, and "intrawaves" ($c < \bar{u}$) have $m < 0$. This formulation is conceptually similar to the multiple gravity wave drag parameterization employed in many previous studies ([*Holton*, 1983; *Garcia and Solomon*, 1985]). In these schemes, momentum deposition is assumed to be created by a set of wave harmonics with different phase velocities which propagate and saturate independently. However, it should be emphasized that the parameterization we have described systematically accounts for nonlinear interactions between harmonics within the spectrum. Since our saturation criterion $\alpha_j$ depends on RMS velocity $\sigma_j$ of the higher $m$/lower $c$ portion of the spectrum rather than on the amplitude of the wave itself, the altitudes at which saturation occurs are generally lower than in the case of independently propagating waves, and wave drag profiles grow more smoothly with height. Even a small amplitude harmonic can experience significant damping in the presence of strong waves in the spectrum contributing to RMS wind $\sigma_j$, and therefore can produce stronger momentum deposition than when it is considered to propagate independently.

## 4 Input source spectra

To apply the parameterization procedure described above, vertical wavenumber "source" spectra must be specified at some reference altitude. Since the troposphere is believed to be a region of strong gravity wave generation, source spectra are most conveniently specified at some level above the tropopause. Observational studies have provided a number of estimates of amplitudes and spectra shapes as well as their variability. However, these studies were mainly concerned with the behavior of the high-$m$ tail of spectra which is affected by saturation processes, whereas harmonics growing with height in the unsaturated portion of spectrum are most important for GW momentum deposition at higher altitudes.

The problem of estimating vertical wavenumber source spectra from observations is nontrivial because these spectra characterize wave fields over a range of altitudes, and hence they cannot be attributed to a particular height $z_s$. In the presence of vertically inhomogeneous gravity wave sources in the lower atmosphere instantaneous $m$-spectra are expected to be highly variable in time, space and wavenumber. A model spectrum which approximates a typical shape and amplitudes of observed PSD has the form (see [*Fritts and VanZandt*, 1993] for further discussion)

$$S(m) = A_0 \frac{(m/m_*)^s}{1+(m/m_*)^{s+t}}, \tag{12}$$

where $m_*$ is the characteristic wavenumber which describes the transition between spectral forms at large and small $m$, and the parameter $t$ is close to −3. The lower $m$ portion of the spectrum is believed to vary generally from $s = 0$ ("Desaubies" spectrum) to $s = 1$ ("Modified Desaubies"), and it should be noted that the uncertainty in this portion of the spectrum is significantly affected by the imposed vertical analysis window and the vertical variation of background fields, among other things. The characteristic wavenumber $m_*$ increases in altitude, reflecting the entrance of the lower-$m$ portion of the spectrum into the process of saturation as wave amplitudes grow with height.

One constraint for source spectra can be obtained by equating the saturated amplitude in (12) to the observed range $S(m) = N^2/(1 \text{ to } 9)m^3$ for $m \gg m_*$:

$$A_0 = \frac{N^2}{(1 \text{ to } 9){m_*}^3}. \tag{13}$$

The observed values of $m_*$ in the troposphere typically vary from $m_* = 2\pi/2.5\text{km}$ to $m_* = 2\pi/900\text{m}$ [*Allen and Vincent*, 1995; *Fritts and Chou*, 1987; *Tsuda et al.*, 1991]. An example of two such model spectra with $m_* = 2\pi/1\text{km}$ and $s = 0, 1$ are shown in Figure 1. However, we must acknowledge that individual spectra observed in the lower atmosphere have a more complicated form. There are also indications that the lower-$m$ portion consists of several peaks representing a systematic "global" wave source in the lower atmosphere [*Weinstock*, 1996]. Although such features are undeniably important for the purposes of wave drag parameterization, the details of source spectra dependencies are not well known. However the model distributions presented in Fig. 1 likely cover the range of variations for observed spectra.

## 5 Calculations for typical wind profiles

In this section we examine the response of the proposed parameterization scheme to different mean wind profiles and variations of the source

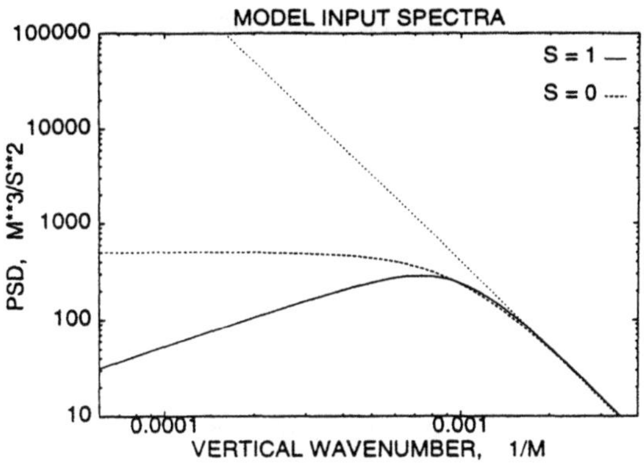


Figure 1: Desaubies ($s = 0$) and Modified Desaubies ($s = 1$) model vertical wavenumber spectra.

spectrum. We have chosen two mean wind and mean temperature profiles from the CIRA-86 model (50°N and 50°S in January), which may be considered as a representative for winter and summer solstices. The temperature is used to calculate the density in (9) and the buoyancy frequency $N$. The distribution of zonal mean wind $\bar{u}$ is shown in Fig. 2. The winter profile exhibits two eastward jets in the troposphere (with maximum $\sim$ 20 m s$^{-1}$) and in the stratosphere ($\sim$ 45 m s$^{-1}$), and a reversal in the mesosphere. The summer profile shows the eastward tropospheric jet ($\sim$ 20 m s$^{-1}$), the westward stratospheric jet with peak value $\sim$ 68 m s$^{-1}$ and a further reversal near 90 km.

All calculations in the one-dimensional model were carried out using a vertical gridstep $\Delta z = 2$ km. This vertical resolution is typical for the middle atmosphere in many contemporary general circulations models. Wave source spectra of the form described above were specified in the troposphere at $z = 10$ km. All results presented below were obtained using the characteristic horizontal wavenumber $k_h = 2\pi/300$ km and assuming that vertical wavenumber source spectra covered the interval from $m = 2\pi/800$ m to $2\pi/18$ km, which is typical of observations. In the calculations we consider waves propagating in both directions with respect to the local wind. This means that at the height of the source spectrum "extrowaves" with $c - \bar{u} > 0$ have $m > 0$, and "intrawaves" with $c - \bar{u} < 0$ are formally described by $m < 0$. The resulting wave drag is calculated as a sum of two profiles (generally it is a difference), while the RMS ve-

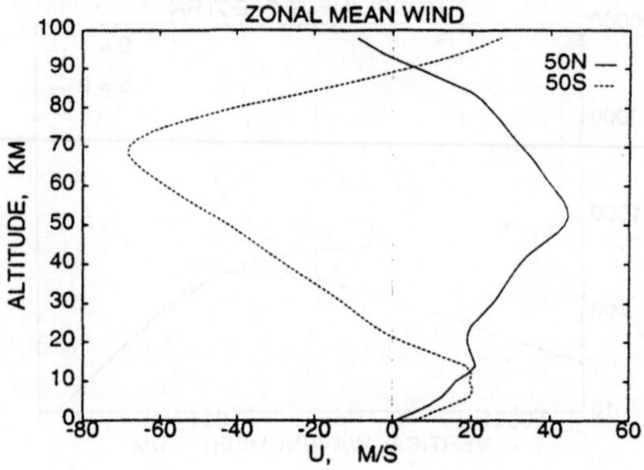


Figure 2: Mean zonal wind for January at 50°N (solid line) and 50°S (dashed line) latitudes from the CIRA-86 model used in the parameterization tests.

locity is created by waves going in both directions. We now consider the sensitivity of the proposed scheme to spectral resolution by discretizing the same source spectrum with varying numbers of harmonics. Fig. 3 exhibits profiles obtained for the "modified Desaubies" source spectrum (12) with $s = 1$, $m_* = 0.006$ rad m$^{-1}$, and the amplitude $A_0 = 100$ m$^3$ s$^{-2}$, employing $M = 200$ (which we regard as an "exact" solution), 30, 15, and 5 harmonics. It is seen from the figure that below 90 km, the resulting wave drag is almost indistinguishable for $M = 200$, 30, and 15 harmonics, while the $M = 5$ case still provides reasonable accuracy. The higher sensitivity of the results to variations of $M$ above 90 km are related to an insufficient resolution of lower-$m$/higher-$c$ portion of the spectrum which contains the most of wave energy at these heights. The accuracy of the parameterization above the mesopause could be increased if one were to employ variable $\Delta m$-grids with resolution concentrated toward lower vertical wavenumbers.

We next consider behaviour of the parameterization scheme when various source spectra are specified. Fig. 4a presents profiles of gravity wave drag for a "winter" distribution of mean wind employing the "modified Desaubies" source spectra (12) with $s = 1$ and $m_* = 0.006$ rad m$^{-1}$ approximated by $M = 15$ harmonics. Profiles are shown for $A_0 = 50$, 100, and 500 m$^3$ s$^{-2}$. It is seen from the figure that the factor of ten increase in the overall spectral amplitude increases the wave drag peak value from $\sim -70$ to $\sim -180$ m s$^{-1}$ day$^{-1}$, although we note that the response is not

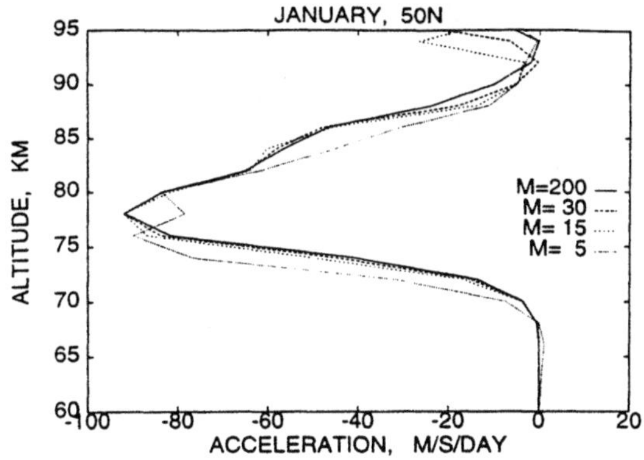


Figure 3: GW drag profiles calculated for "winter" conditions (50°N January) utilizing $M = 200$, 30, 15 and 5 harmonics in the model spectrum.

linear. The heights of drag maxima decrease from $\sim$ 80 km for $A_0 = 50$ to $\sim$ 70 km for $A_0 = 500$ m$^3$ s$^{-2}$ reflecting the obvious fact that waves with weaker amplitudes generally saturate at higher altitude.

The RMS horizontal wind predicted by the parameterization can also be compared qualitatively to that deduced from observations. In the Fig. 4b we have plotted $\sigma$ calculated over the entire spectrum as a function of altitude. At the lower boundary (10 km) the RMS velocity associated with the input spectra varies between 1 and 2 m s$^{-1}$, which is in a good accordance with observations (e.g., [*Sato*, 1994]), and with values recommended for use in gravity wave drag models (e.g., [*Fritts and Lu*, 1993]). In the middle atmosphere the calculated profiles exhibit RMS velocity fluctuations up to a few tens m s$^{-1}$, which are also consistent with observations [*Tsuda et al.*, 1994]. However, the RMS wind reaches a maximum near $z \sim 60 - 70$ km (slightly below the peaks of momentum deposition) and diminishes at higher altitudes. Observations, on the other hand, indicate the RMS wind should increase slowly above 60–70 km. The explanation for this difference may be found in the limited range of our spectrum. To produce stronger parameterized RMS winds at higher altitudes, one must decrease $m_{min}$ of the input spectrum. It is natural to anticipate that upon doing so, the peaks in RMS velocity profiles will flatten, and RMS profiles will vary more slowly with height. However, we have chosen not to do this owing to the large uncertainty in the shape of gravity wave spectra at lower $m$.

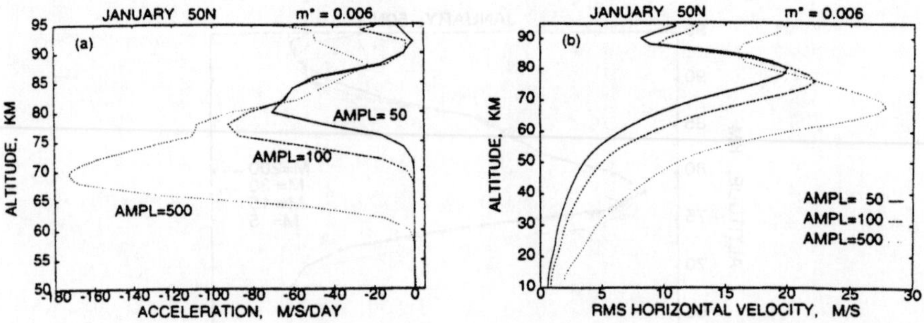


Figure 4: Profiles of (*a*) wave drag and (*b*) horizontal RMS wind calculated for "winter" conditions (50°N January) using the Modified Desaubies input spectrum ($s = 1$) with $A_0 = 50$, 100, and 500 m$^3$ s$^{-2}$.

In the summer hemisphere, strong stratospheric easterlies filter out westward travelling gravity waves almost entirely, while the tropospheric jet filters out all eastward directed harmonics with phase velocities below $\sim 20 - 30$ m s$^{-1}$. As a result, only fast harmonics (with low $m$) can participate in the momentum deposition. In our calculations the saturation of these waves generally occurs above 90 km. Figure 5 presents profiles of RMS wave velocity for the "summer" wind profile calculated with the input spectra described above. It is seen from the figure, that RMS wind fluctuations associated with the spectrum of propagating gravity waves does not exceed $\sim 4$ m s$^{-1}$ up to $\sim 75$ km due to the filtering effect of the mean wind. Only above $\sim 90$ km $\sigma$ becomes comparable with $c - \bar{u}$ for the remaining large $c$ portion of the spectrum. Therefore, wave saturation and momentum deposition in the summer hemisphere occurs somewhat higher than in the winter hemisphere.

Figures 6a and 6b demonstrate sensitivity of the gravity wave drag to variations in the source spectrum slope and in the characteristic vertical wavenumber $m_*$. Both calculations have been done for "winter" conditions using $M = 15$ spectral harmonics and the source amplitude $A_0 = 100$ m$^3$ s$^{-2}$. Fig. 6a presents wave drag profiles obtained for the slopes $s = 0$ and $s = 1$ with $m_* = 0.006$ rad m$^{-1}$. As anticipated, the spectrum containing more energy at lower $m$ (i.e. $s = 0$) saturates lower and produces more momentum deposition at the same height than the

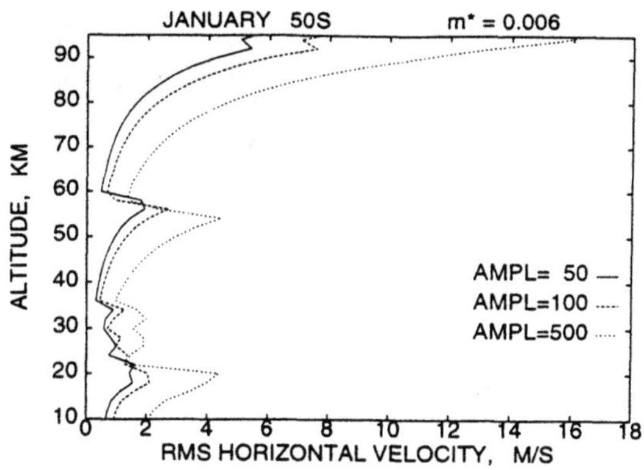


Figure 5: Horizontal RMS velocity calculated for "summer" conditions 50°S using the Modified Desaubies input spectrum ($s = 1$) with $A_0 = 50$, 100, and 500 m$^3$ s$^{-2}$.

spectrum with more steeply decreasing in $m$ amplitudes ($s = 1$). Fig. 6b displays wave drag profiles calculated for the $s = 1$ source spectra with $m_* = 0.006$ and $m_* = 0.0025$ rad m$^{-1}$. The wavenumber $m_*$ characterizes the location of spectrum "knee", or the lowest vertical wavenumber affected by saturation. Fig. 6b clearly demonstrates that spectra with higher $m_*$ saturate lower, and therefore produce maxima of wave drag at lower altitude.

## 6 Testing with a mechanistic version of CCM2

Results presented in Section 5 suggest that our parameterization scheme produces profiles consistent with observations and theoretical expectations within the range of natural variability of input source spectra. In this section we present the results of tests using the three-dimensional mechanistic version of the Middle Atmosphere Community Climate Model (MA CCM2).

The formulation of MA CCM2 as well as aspects of simulated annual cycle have been described by *Boville* [1995]. In the mechanistic version of the model, the troposphere has been replaced by given distributions of the geopotential and the prognostic variables on a pressure surface near the tropopause. The model was driven from below by boundary conditions taken from MA CCM2 runs with the "full" tropospheric physics. In the middle atmosphere, the model contains the physical parameter-

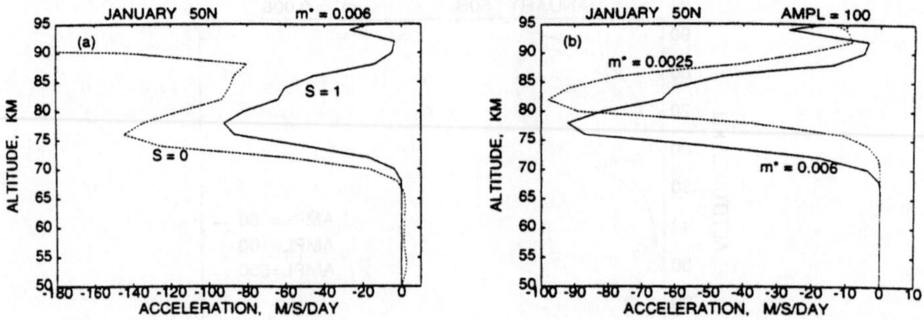


Figure 6: Wave drag profiles calculated for $(a)$ the input source spectra with $(s = 1)$ and $(s = 0)$ with $m_* = 0.006$ rad m$^{-1}$, and $(b)$ for the Modified Desaubies spectrum $(s = 1)$ with $m_* = 0.006$ (solid line) and $m_* = 0.0025$ (dashed line).

izations discussed in [*Boville*, 1995]. To estimate the net effect of the parameterization scheme, in our simulations we have removed the sponge layer at the top as well as the parameterization of orographically generated gravity wave drag. The version used in our calculations had 47 vertical levels covering altitudes up to $\sim$ 0.004 mbar, and T21 horizontal resolution.

In our calculations we considered waves with $k_h = 2\pi/300$km propagating in four directions, meridional and zonal. A GW source spectrum of the form (12) was specified at the lower boundary (234 mbar), and assumed to be isotropic and constant with time. The parameterization has been tested for horizontally uniform source spectra as well as for several spectra with latitudinal dependencies which approximated possible distributions of wave sources in the troposphere.

Figure 7 presents monthly averaged mean zonal wind and the zonal component of the GW induced acceleration for "Modified Desaubies" $(s = 1)$ source spectrum with latitudinally varying $A_0$ of the form $A_0 = 50(1 + 100\sin^2 2\theta)$, where $\theta$ is the latitude. We assumed that this form roughly approximates an enhancement of GW sources in the midlatitude tropospheric jets. It is seen from the figure that the imposed wave drag closes easterlies in the summer hemisphere, and reproduces wind reversal in the mesosphere of the winter hemisphere. The magnitude of the jet in the Southern hemisphere is near 60 m s$^{-1}$, which is somewhat lower

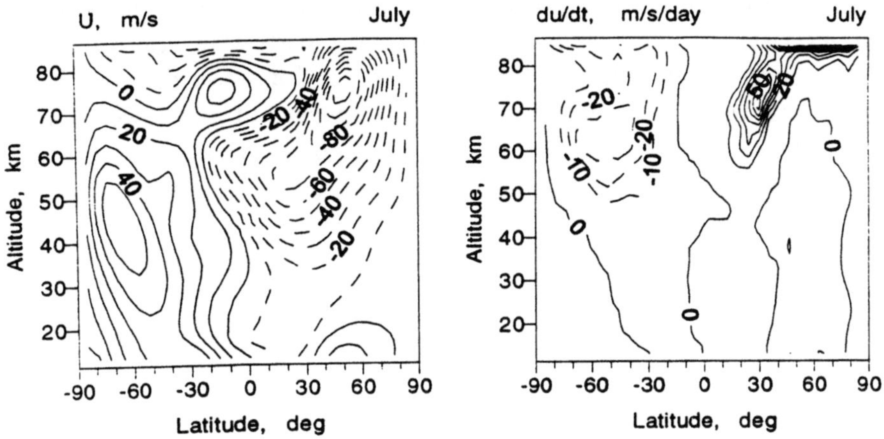


Figure 7: Mean zonal wind ($a$) and mean zonal gravity wave drag ($b$) simulated with the mechanistic MACCM2 for July

than that of CIRA-86 (90 m $s^{-1}$). However the results clearly show the tendency of wave drag to split the jet. The lower magnitude of westerlies (compared to observations) is associated with the lower height where the wind reversal occurs (again compared to CIRA-86). This suggests that we have overestimated the strength of the wave sources in the winter hemisphere. As well, the summer hemisphere jet is stronger ($\sim$ 110 m $s^{-1}$) than in CIRA-86 (70 m $s^{-1}$), and the location of the maximum ($\sim$ 75 km) is somewhat higher than in CIRA-86 ($\sim$ 90 km). Fig 7b shows the corresponding distribution of GW drag. It is seen from the figure, that there is a lack of zonal flow deceleration in mid- and high latitudes of the winter hemisphere. Further tuning of the scheme will require a more realistic specification of wave source spectra in the lower atmosphere.

## 7 Summary and Conclusions

We have presented a spectral parameterization scheme for calculating gravity wave momentum deposition in the middle atmosphere. The scheme is based on the MK95 theory of the vertical evolution and saturation of GW spectra. The parameterization has been tested using a column model for representative distributions of mean wind and temperature, as well as with a three-dimensional mechanistic version of the Middle Atmosphere CCM2 general circulation model.

The parameterization has only one "physical" tunable parameter, the characteristic horizontal wavenumber of subgrid-scale gravity waves $k_h$, which serves as an "efficiency" scaling factor for wave drag. All further tuning can be done by specifying source spectra at the lower boundary. This implicitly characterizes the degree of physical and theoretical completeness of the scheme, and emphasizes the need for a knowledge of GW sources in the atmosphere.

For the range of anticipated variability of source spectra parameters in the troposphere, the scheme produces wave drag profiles as well as horizontal RMS velocities consistent with observations and with theoretical estimates. Results of sensitivity experiments and of the one year run with mechanistic version of MACCM2 suggest that the parameterization successfully reproduces typical patterns of the middle atmosphere circulation both for solstice and equinox conditions, with remarkably little tuning effort. However, as with any other successful GW parameterization, precise tuning of the scheme will require a more detailed specification of source spectra in the lower atmosphere as well as their seasonal and latitudinal variability.

It is worth noting that these results were obtained by including only the effects of nonlinear diffusive damping and Doppler shifting by the background wind. For purposes of GW drag parameterization, the effects of nonlinear spectrum-induced Doppler shifting derived by MK95 can be neglected, Furthermore, the number of harmonics representing the spectrum can be reduced to $\sim$ 10 without a noticeable effect on the resulting wave drag. This indicates that the scheme is as efficient as a multiple-wave Lindzen scheme, and suitable for use in GCMs.

## References


Allen SJ and Vincent RA (1995) Gravity wave activity in the lower atmosphere: Seasonal and latitudinal variations. J Geophys Res 100:1327-1350

Boville BA (1995) Middle atmosphere version of CCM2 (MACCM2): Annual cycle and interannual variability. J Geophys Res 100:9017-9039

Dewan EM and Good RE (1986) Saturation of the "universal" spectrum for vertical profiles of horizontal scalar winds in the atmosphere. J Geophys Res 91:2742-2748

Fritts DC and Chou H-G (1987) An investigation of the vertical wavenumber and frequency spectra of gravity wave motions in the lower stratosphere. J Atmos Sci 44:3610-3624

Fritts DC and Lu W (1993) Spectral estimates of gravity wave energy and momentum fluxes. Part II: Parameterization of wave forcing and variability. J Atmos Sci 50:3695-3713

Fritts DC and VanZandt TE (1993) Spectral estimates of gravity wave energy and momentum fluxes. Part I: Energy dissipation, acceleration, and constraints. J Atmos Sci 50:3685-3694

Garcia RR and Solomon S (1985) The effect of breaking gravity waves on the dynamical and chemical composition of the mesosphere and lower thermosphere. J Geophys Res 90:3850-3868

Gardner CS (1994) Diffusive filtering theory of gravity wave spectra in the atmosphere. J Geophys Res 99:20,601-20,622

Hines CO (1991a) The saturation of gravity waves in the middle atmosphere. Part I: Critique of linear-instability theory. J Atmos Sci 48:1348-1359

Hines CO (1991b) The saturation of gravity waves in the middle atmosphere. Part II: Development of Doppler-spread theory. J Atmos Sci 48:1360-1379

Hines CO (1993a) The saturation of gravity waves in the middle atmosphere. Part IV: Cutoff of the incident wave spectrum. J Atmos Sci 50:3045-3060

Hines CO (1993b) A critical comparison of theories of gravity wave saturation. Coupling Processes in the Lower and Middle Atmosphere Kluwer Academic Publishers, 233-239

Hines CO (1996) Doppler-spread parameterization of gravity-wave momentum deposition in the middle atmosphere. Submitted to J. Atmos. Terr. Phys.

Holton JR (1983) The influence of gravity wave breaking on the general circulation of the middle atmosphere. J Atmos Sci 40:2497-2507

Lindzen RS (1981) Turbulence and stress owing to gravity wave and tidal breakdown. J Geophys Res 86:9707-9714

Medvedev AS and Klaassen GP (1995) Vertical evolution of gravity wave spectra and the parameterization of associated wave drag. J Geophys Res 100:25,841-25,853

Sato K (1994) A statistical study of the structure, saturation and sources of inertio-gravity waves in the lower stratosphere observed with the MU radar. J Atmos Terr Phys 56:755-774

Smith SA, Fritts DC, and VanZandt TE (1987) Evidence for a saturated spectrum of atmospheric gravity waves. J Atmos Sci 44:1404-1410

Tsuda T, VanZandt TE, Mizumoto M, Kato S and Fukao S (1991) Spectral analysis of temperature and Brunt-Väisälä frequency fluctuations observed by radiosondes. J Geophys Res 96:17,265-17,278

Tsuda T, Murayama Y, Nakamura T, Vincent RA, Manson AH, Meek CE, and Wilson RL (1994) Variations of the gravity wave characteristics with height, season and latitude revealed by comparative observations. J Atmos Terr Phys 56:555-568

Weinstock J (1982) Nonlinear theory of gravity waves: Momentum deposition, generalized Rayleigh friction, and diffusion. J Atmos Sci 39:1698-1710

Weinstock J (1990) Saturated and unsaturated spectra of gravity waves and scale-dependent diffusion. J Atmos Sci 47:2211-2225

Weinstock J (1996) Spectra and a global source of gravity waves for the middle atmosphere. Adv Space Res 17:(11)67-(11)76

Zhu X (1994) A new theory of the saturated gravity wave spectrum for the middle atmosphere. J Atmos Sci 51:3615-3626

# Effects of Gravity Wave Drag in the Berlin Troposphere-Stratosphere-Mesosphere GCM

Steven Pawson
Institut für Meteorologie,
Freie Universität Berlin,
Carl-Heinrich-Becker-Weg 6-10,
12165 Berlin,
Germany

ABSTRACT. The effects of including various, generally quite simple, representations of gravity wave drag on the evolution of the circulation of the middle atmosphere of the troposphere-stratosphere-mesosphere general circulation model developed at the Free University of Berlin is investigated. Specifically, the importance of gravity wave drag for the development of stratospheric sudden warmings in the model simulations are discussed. It is shown that changing the form of the drag in the mesosphere of the model is unimportant for the development of the circulation below the upper stratosphere over periods of about 20 days. This is despite the important effects of the mesospheric drag on the climatological state of the model. Topographic gravity wave drag acting in the lower stratosphere can significantly affect the development of sudden warmings via direct effects on the local flow and the planetary wave stuctures in the stratosphere and troposphere.

## 1. Introduction

A series of integrations, which have been made with the troposphere-stratosphere-mesosphere (TSM) general circulation model (GCM) developed at the 'Freie Universität Berlin', are presented. These integrations address the problem of how gravity wave drag (GWD) affects the development of the stratospheric flow over periods of up to 25 days. These can thus be regarded as 'extended range forecasts'. Two different ensembles of experiments are discussed, starting on different days of a reference integration of the model.

Much interest has centered on the importance of gravity waves to the circulation of the middle

NATO ASI Series, Vol. I 50
Gravity Wave Processes
Their Parameterization in Global Climate Models
Edited by Kevin Hamilton

atmosphere. There are two specific aspects. First, in the lower stratosphere, it was shown (Palmer et al., 1986; McFarlane, 1987) that the inclusion of a topographic GWD parametrization can lead to significant improvements in the performance of numerical weather forecast and climate models. Specifically, the force **D** exerted by the resolved waves does not balance the frictional forces and mountain torque, so that anomalous features occur in the zonal-mean (the lower polar stratosphere is too cold and the subtropical jet extends too high) and in the eddy fields in the lower stratosphere and troposphere. Inclusion of additional forcing **G** due to gravity waves can rectify this problem. Second, the problem of closing the polar night jet (PNJ) and changing the meridional and vertical structure of the temperature field in the mesosphere are believed to be directly linked to the GWD. This was demonstrated quite convincingly in a GCM by Rind et al. (1988), who showed the importance of including gravity waves with non-zero phase speeds, forced by convective activity and shear instabilities, as well as the topographic waves.

Most integrations made with the Berlin TSM GCM (Pawson et al., 1991; Langematz and Pawson, 1996) have included a very crude representation of the mesospheric GWD (Rayleigh friction, referred to as the force **X**) and have not considered the topographic GWD in the lower stratosphere. Indeed, Miller et al. (1989) showed how **G** should only be required at higher model resolutions, since at low resolution (e.g. T21) both the northward eddy momentum flux and the mountain torque are too weak and the errors cancel, so that the systematic errors found at higher resolution when the eddy fluxes are better represented is not present.

Pawson et al. (1996) showed how the modeled climatology responded to changes in the mesospheric drag; specifically, weakening the drag, which is applied at the uppermost levels, caused a direct change in the mesosphere (stronger wind) and an indirect change extending down into the middle stratosphere, which resembled the first mode of variability of the zonal-mean state of the model (Pawson et al., 1995), the model favoring a state with a stronger polar night jet (PNJ). These changes are consistent with the diagnostic concept of 'downward control' (Haynes et al., 1991). However, increasing the depth over which the drag was active led to a different type of change. In a second experiment, the drag profile was extended downwards, so that it acted in the entire mesosphere. Then, the differences in **X** led to a response in the planetary wave structures, which changed $D$ (the zonal-mean zonal component of **D**); the consequence was a significantly different structure of the zonal-mean zonal wind ($\overline{u}$) which extended into the troposphere.

The current study complements that of Pawson et al. (1996) in that the short-term (up to 25 day) response of the model to changing the representations of GWD is examined. In section 2 the model is described, the emphasis being on the GWD parametrizations used. The experimental design is presented in section 3. The results are shown in section 4. Finally, the results are discussed and summarized.

Table 1. The parameters used in each of the integrations for the Rayleigh friction coefficients, as defined by Eq. (1).

| Run | $\gamma_1$ (days$^{-1}$) | $\gamma_2$ (days$^{-1}$) | $z_1$ (km) | $z_2$ (km) | PSS | Comments |
|---|---|---|---|---|---|---|
| 1 | 0.0 | 0.250 | 78.0 | 6.3 | No | Reference run |
| 2 | 0.0 | 0.125 | 63.0 | 7.5 | No | |
| 3 | 0.0 | 0.250 | 78.0 | 6.3 | Yes | PSS at all levels |
| 4 | 0.0 | 0.250 | 78.0 | 6.3 | Yes | PSS for $p < 5$ hPa |
| 5 | 0.0 | 0.250 | 63.0 | 7.5 | No | Drag on waves only |

## 2. Model description

The Berlin TSM GCM is based on the global, primitive equations using a spectral representation of the horizontal structure. In the current study it was integrated at T21 resolution, which has a spatial resolution of $\Delta\lambda \times \Delta\phi \simeq 5.6^\circ \times 5.6^\circ$ on the equivalent Gaussian grid. Physical tendencies are determined on this grid, using a comprehensive package of physical parametrizations, based on the troposphere of the Hamburg climate models (Roeckner et al., 1992), with the radiation schemes of Fouquart and Bonnel (1980), Shine and Rickaby (1989), Strobel (1978) and Morcrette (1991), implemented as appropriate in the troposphere, stratosphere and mesosphere as described in Pawson et al. (1996).

The TSM GCM has 34 levels and uses the hybrid coordinate system described by Simmons and Strüfing (1983). The transition from sigma to pressure coordinates takes place in the troposphere and above 50 hPa all levels are based on pressure; the vertical resolution in the stratosphere is about 3.5 km, decreasing in the mesosphere (Pawson et al., 1996). Subgrid-scale drag is represented at the highest levels using a Rayleigh-friction parametrization as in Holton and Wehrbein (1980), with horizontal drag $\mathbf{X} = (X, Y) = -\gamma(z) \times (u, v)$, where $u$ and $v$ represent the eastward and northward velocity, and the linear coefficient is given by:

$$\gamma(z) = \gamma_1 + \gamma_2 \times \tanh\left(\frac{z - z_1}{z_2}\right). \tag{1}$$

The strength of the drag can be adjusted by varying the parameters $\gamma_1$ and $\gamma_2$, while the level and the depth over which it acts is determined by $z_1$ and $z_2$.

No other representation of sub-gridscale drag has previously been used in the TSM GCM, although various GWD schemes have been implemented and tested for various reasons. The Palmer

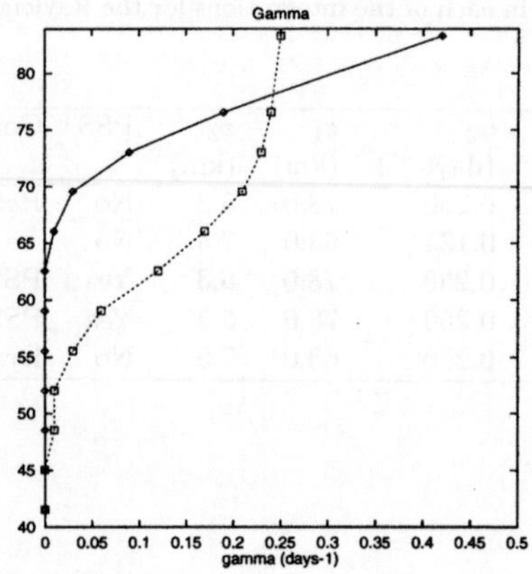


Figure 1. Vertical structure of the Rayleigh friction coefficient $\gamma$, shown on the model levels, for experiments one (solid line) and two (dashed line).

et al. (1986) scheme for topographic GWD was provided with the model, although following the results of Miller et al. (1989) it was not included at the low horizontal resolution usually used (e.g. Langematz and Pawson, 1996). A similar scheme was used by Rind et al. (1988) in their model, where traveling gravity waves were also included.

### 3. Experiment design

In order to obtain some degree of confidence about the results, an ensemble of short experiments was started using different structures of GWD. It proved adequate to perform just two 25-day sets of 'extended prediction' experiments, which were spawned from the control integration described in Pawson et al. (1996) several days before sudden warmings occurred. The intention was to investigate how changes in the GWD affect the development of the flow sufficiently to hasten or delay the occurrence of stratospheric warmings. This is then a complementary study to Pawson et al. (1996), where the effects of changing the mesospheric drag on the long-term mean state of the TSM GCM was investigated. The seasonal cycle was not included in this study; the experiments are short enough for this to be of secondary importance. (Further, the reference integration was made under perpetual January conditions.)

The form of the drag used in the five runs of both experiments is summarized in Table 1. Experiment 1 is the control integration, where **X** is restricted to the top few model levels. In experiment 2 the drag is deeper, extending down to the stratopause. The two profiles of $\gamma$ are shown in Fig. 1.

In experiments 3 and 4 the Palmer et al. (1986) topographic GWD parametrization was used, in conjunction with the weaker Rayleigh friction at the highest model levels; these two experiments differ in that the drag **G** is set to zero for $p > 5$ hPa in experiment 4, so that no forcing occurs below the upper stratosphere (note that this results in a slight loss of total energy from the model). Finally, experiment 5 uses the deeper profile of $\gamma(z)$ but only applied to the zonally-asymmetric part of the field. The two initial fields for the experiments were taken from days (A) 603 and (B) 783 of the reference integration. Note that this long perpetual January integration underwent several large oscillations, similar to those discussed in Pawson et al. (1995), only that the current version of the TSM GCM is capable of simulating major stratospheric warmings (Erlebach et al., 1996). In experiment set A a major warming was simulated, while in set B only minor warmings occurred.

## 4. Results

The evolution of $\overline{u}$ at 10 hPa (Fig. 2a) illustrates the occurrence of a major stratospheric warming on day 12 of experiment A1. Experiment B1 (Fig. 2b) commenced near the peak of a minor warming, after which the PNJ partly recovered (day 6), followed by a further minor warming which peaked on day 14.

The four perturbed integrations used the deeper $\gamma$ profile applied to the entire velocity field (A2, B2) or just the eddy component (A5, B5). The shallow $\gamma$ was used with the Palmer et al. (1986) parametrization of topographic GWD at all levels (A3, B3) and at pressures lower than 5 hPa (A4, B4). In all cases when $\gamma$ was changed in the mesosphere there was no significant effect on the evolution of the stratospheric flow, especially the timing of the warmings (Table 2). This is because $D$ dominates the forcing and the planetary waves originate in the troposphere. The changes in the mesospheric drag require a finite time to propagate downwards into the stratosphere.

However, in experiments A3 and B3 there were significant changes in the stratospheric flow

Table 2. Days of occurrence of the major (series A) and minor (series B) warmings.

| | 1 | 2 | 3 | 4 | 5 |
|---|---|---|---|---|---|
| A | 13 | 12 | 23 | 12 | 13 |
| B | 2,15 | 2,15 | 5 | 2,16 | 2,15 |

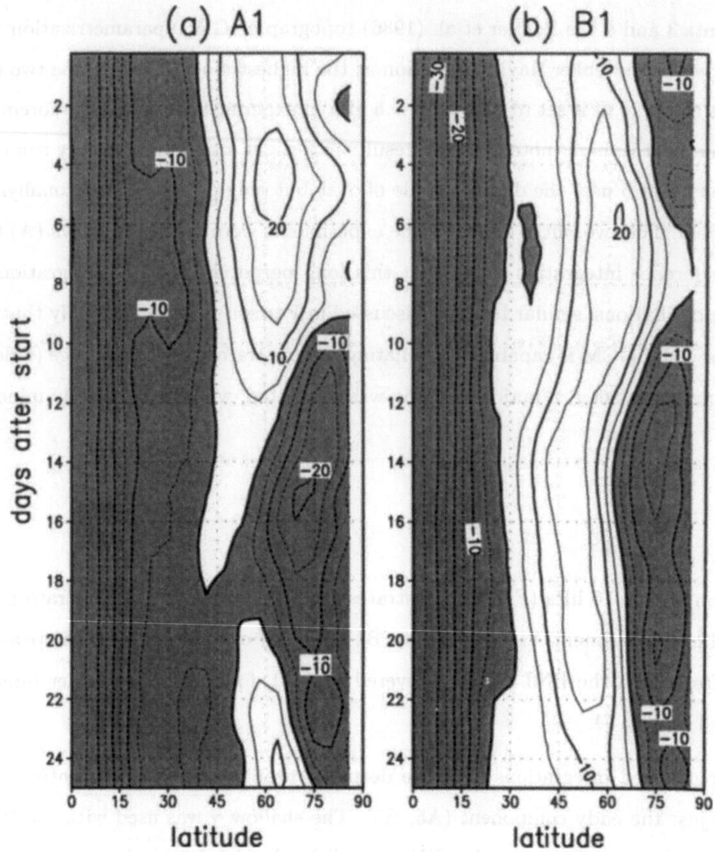


Figure 2. Time series (days 1–25) of $\overline{u}$ at 10 hPa in the northern hemisphere for experiments (a) A1 and (b) B1. The contour interval is 10 $ms^{-1}$, with easterlies shaded light gray. Note that the time axis starts at 06UT on the first day and ends after 24 complete days, at 00UT on the 25th day. Data were saved four times daily and the 'running diurnal mean' is shown.

(Table 2). The major warming is examined more detail. Experiments A1 and A3 commenced from the $\overline{u}$ in (Fig. 3a), with a strong westerly PNJ. In A1, this was been replaced by strong easterlies on day 12 (Fig. 3b). These reach 70 $ms^{-1}$ at 60 km and 60°N. In A3 $\overline{u}$ is quite different (Fig. 3c) as the strong easterly winds have not developed at high latitudes. The PNJ is weaker but it takes a further 10 days for the wind reversal at 10 hPa, 60°N to occur (Table 2), by which time the westerly jet has already started to reform at higher levels (Fig. 3d).

Averaged over days 1–12 of A1, $D$ reaches $-45$ $ms^{-1}$ in the lower mesosphere near 60°N (Fig. 4a). This is almost twice as strong as in A3 (Fig. 4b), revealing that the planetary waves have

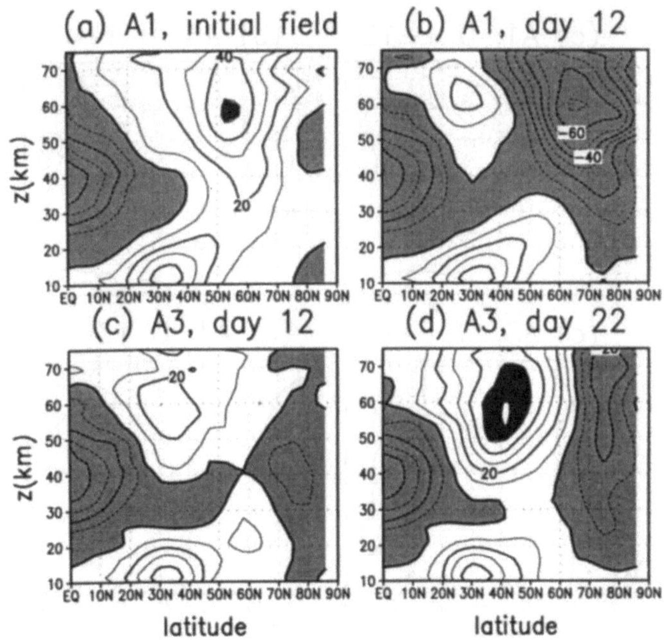


Figure 3. Zonal-mean cross sections of $\overline{u}$ in experiment A, for (a) the starting day; (b) day 12, experiment A1; (c) day 12, experiment A3; (d) day 22, experiment A3. The contour interval is 10 $ms^{-1}$, with easterlies shaded light gray and values of $\overline{u}$ between 40 and 50 $ms^{-1}$ are shaded dark gray.

been affected by including the topographic GWD, whose direct contribution of ($G$) to the forcing averaged over days 1–12 of A3 (Fig. 4c) reaches 20 $ms^{-1}day^{-1}$ in the mesosphere but is quite small in the stratosphere. The mean value of $D$ over days 13–22 of A3 (Fig. 4d) shows easterly forcing, which is only strong in the high latitude mesosphere.

Whereas the planetary wave forcing in A1 forces strong easterlies after 12 days, the topographic GWD in the lower stratosphere leads to a decrease of this eddy forcing and a delay of the major warming, which takes a different form in A3.

Similar results were found in the B-experiments. Changing $\gamma$ had almost no effect on the evolution of the lower stratospheric zonal-mean flow over the 25 days of the experiment. Use of topographic GWD at all levels led to strong changes (e.g. Table 2).

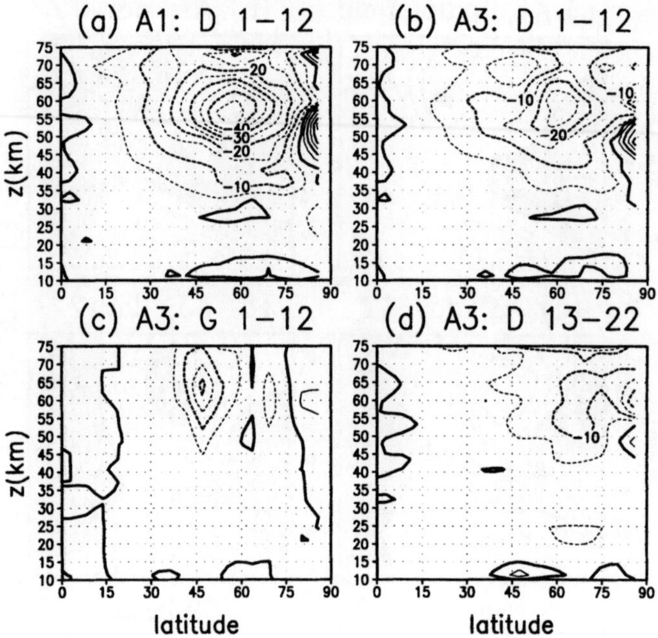


Figure 4. Zonal-mean cross sections of the forcing of $\overline{u}$ in experiment A, for (a) $D$, experiment A1, the mean for days 1–12; (b) $D$, experiment A3, the mean for days 1–12; (c) $G$, experiment A3, the mean for days 1–12; (d) $D$, experiment A3, the mean for days 13–22. The contour interval is 5 $\mathrm{ms^{-1}day^{-1}}$, with negative (easterly) values represented by dashed lines.

## 5. Discussion and conclusions

In this study the effects of changing the GWD on the evolution of the zonal-mean flow over periods of up to 25 days has been examined. This was presented in terms of the influence of the GWD on the forecasting of minor and major stratospheric warmings which occurred in the reference integration of the TSM GCM. The results can be summarized as follows. Changing the representation of the drag at high levels ($p = 5$ hPa and lower) had only a small influence on the development of the flow in the lower and middle stratosphere ($p \geq 10$ hPa). This contrasts with the effects on the climatological structure discussed by Pawson et al. (1996), where changing the form of the mesospheric drag affected the entire atmosphere through direct and indirect effects. The direct effect involves a change in the forcing of $\overline{u}$, a corresponding change in the mean meridional circulation, and a consequent downward communication into the stratosphere; it is consistent with the idea of downward control (Haynes et al. 1991) when the long-term mean is considered. The indirect influence arose because $\gamma$ also acts on the zonally-asymmetric component of the velocity, so that changes in $\gamma$ lead to a different structure in $D$.

Only when the drag acted in the lower stratosphere were any significant changes in the mesospheric flow detected. Then, in experiment series A, the timing and nature of the major warming was affected. In series B the minor warming was delayed. The lower stratospheric GWD has a direct influence on the planetary waves, consequently modifying their forcing $D$ of the zonal-mean flow, leading to large changes in the temporal development in the model.

The current study provides a preliminary look of the effects of gravity waves in the Berlin TSM GCM. Future work using more contemporary parametrizations of gravity wave drag is planned. Both the transient and climatological influences will be studied.

**Acknowledgements.** I am grateful to my colleagues in the stratospheric research group of the meteorological institute in Berlin, particularly to Ulrike Langematz for her work with the model and Gerd Radek for his help with the processing of results. The model development was supported by the climate research program of the German federal government, grant 07VKV01/1-Nr. 15.

## References

Erlebach P, Langematz U, Pawson S (1996) Simulations of stratospheric sudden warmings in the Berlin troposphere-stratosphere-mesosphere GCM. Ann. Geophysicae 14: 443–463.

Fouquart Y, Bonnel B (1980) Computations of solar heating of the Earth's atmosphere: a new parametrization. Beitr. Phys. Atmos. 53: 35–62

Haynes PH, Marks CJ, McIntyre ME, Shepherd TG, Shine KP (1991) On the "downward control" of extratropical diabatic circulations by eddy-induced mean zonal forces. J. Atmos. Sci. 48: 651–678

Holton JR, Wehrbein WM (1980) A numerical model of the zonal-mean circulation of the middle atmosphere. Pure Appl. Geophys. 118: 284–306

Langematz U, Pawson S (1996) The Berlin troposphere-stratosphere-mesosphere GCM: climatology and forcing mecahnisms. Q. J. R. Meteorol. Soc. submitted

McFarlane NA (1987) The effect of orographically excited gravity wave drag on the general circulation of the lower stratosphere and troposphere. J. Atmos. Sci. 44: 1775–1800

Miller MJ, Palmer TN, Swinbank R (1989) Parametrization and influence of subgridscale orography in general circulation and numerical weather prediction models. Meteorol. Atmos. Phys. 40: 84–109

Morcrette JJ (1991) Radiation and cloud radiative properties in the European Centre for Medium Range Weather Forecasts Forecasting System. J. Geophys. Res. 96: 9121–9132

Palmer TN, Shutts GJ, Swinbank R (1986) Alleviation of a systematic westerly bias in general circulation and numerical weather prediction models through an orographic gravity wave drag parametrisation. Q. J. R. Meteorol. Soc. 112: 1001–1039

Pawson S, Langematz U, Meyer A, Strauch P, Leder S, Rose K (1991) A comparison of the climatology of a troposphere-stratosphere-mesosphere model with observations. Clim. Dyn. 5: 161–174

Pawson S, Meyer A, Leder S (1995) Internal Variability in a Perpetual January Integration of a Troposphere-Stratosphere-Mesosphere GCM. Q. J. R. Meteorol. Soc. 121: 369–397

Pawson S, Langematz U, Radek G, Schlese U, Strauch P (1996) The Berlin troposphere-stratosphere-mesosphere GCM: sensitivity to radiation and gravity wave drag. Q. J. R. Meteorol. Soc. submitted

Rind D, Suozzo R, Balachandran NK, Lacis A, Russell G (1988) The GISS Global Climate-Middle atmosphere model. Part 1: Model structure and climatology. J. Atmos. Sci. 45: 329–370

Shine KP, Rickaby AJ (1989) Solar radiative heating due to absorption by ozone. Ozone in the Atmosphere, Bojkov RD and Fabian P (eds), Deepack, Hampton VA, 597–600

Simmons AJ, Strüfing R (1983) Numerical forecasts of stratospheric warming events using a model with a hybrid vertical coordinate. Q. J. R. Meteorol. Soc. 109: 81–111

Strobel D (1978) Parametrisation of the atmospheric heating rate from 15–120km due to O2 and O3 absorption of solar radiation. J. Geophys. Res. 83: 6225–6230

# The Role of Parameterized Drag in a Troposphere-Stratosphere-Mesosphere General Circulation Model

Kevin Hamilton
*Geophysical Fluid Dynamics Laboratory/NOAA*
*Princeton University*
*P.O. Box 308*
*Princeton, New Jersey 08542*
*USA*

ABSTRACT. This paper discusses simulations of the global circulation up to 0.01 mb obtained using a general circulation model with an imposed zonal wind relaxation that forces the zonal-mean state close to observations. This enables estimates to be made of the zonal drag required to bring general circulation model simulations of the middle atmospheric circulation into line with observations. The diagnosed drag is then compared with results from the Lindzen (1981) parameterization of wave drag from unresolved gravity waves. With reasonable values of the wave input parameters, the parameterization can capture many of the features of the diagnosed drag distribution, but only if some modification of the Lindzen procedure is adopted to allow the region of parameterized wave breaking to extend to lower heights.

## 1. Introduction

General circulation models (GCMs) have typically simulated middle atmospheric climates that are closer to radiative equilibrium than observed. This has been a concern from the earliest studies that were made using comprehensive models that extended significantly above the tropopause (*e.g.*, Manabe and Mahlman, 1976). It seems reasonable to ascribe these difficulties to the absence of unresolvably small gravity waves (GWs) in the model, and thus to suppose that models will need parameterizations of the effects of unresolved waves. A truly first-principles approach to this problem is not remotely possible at this time, however, and any GW parametrization will require considerable tuning to make it function so that it leads to a realistic model simulation. This paper reports on an attempt to estimate the detailed requirements for a GW parameterization in a particular troposphere-stratosphere-mesosphere GCM. Then in Section 5 the adequacy of one currently proposed GW parameterization in reproducing the required effects will be considered.

## 2. The Model

The GFDL SKYHI model has been described extensively in a number of earlier publications (e.g., Fels *et al.*, 1980; Hamilton *et al.*, 1995; Hamilton, 1995a,b). The version considered here has $3^o$x$3.6^o$ horizontal resolution and 40 levels in the vertical from near the ground to 0.0096 mb (~80 km). Ozone amounts, cloudiness and sea surface temperatures are all prescribed to reasonable climatological values. The standard version of the model has no parameterization of any subgrid-scale gravity wave effects. This version of the model has

NATO ASI Series, Vol. I 50
Gravity Wave Processes
Their Parameterization in Global Climate Models
Edited by Kevin Hamilton

been integrated for more than 25 years, and the resulting climatology has been documented in Hamilton *et al.* (1995) and Hamilton (1995a,b). These studies showed that the model has serious systematic biases in the middle atmospheric circulation. The problems are particularly severe in the SH in the Austral winter and spring seasons. The June-August (JJA) mean temperatures near the South Pole in the model are much lower than observed, by as much as 70°C near 1 mb. The simulated polar night jet is much too strong (peak zonal-mean JJA-mean winds ~170 $m\text{-}s^{-1}$, *vs.* ~80 $m\text{-}s^{-1}$ for observations). In the Austral spring the SH vortex breakdown is very significantly delayed in the simulation (Mahlman *et al.*, 1994). The NH winter simulation has a similar problem, but it is much less severe. The simulated North Pole December-February (DJF) temperatures appear to be as much as 20° lower than observations, and the climatological NH winter vortex in the model is confined too tightly near the pole.

The present study began with a set of calculations designed to determine the zonal drag that must be added to the model to enable it to accurately reproduce the observed seasonal cycle of wind and temperature over the entire domain. Only zonal drag effects are considered, so both drag on the meridional flow and any possible contributions from a GW parameterization to small-scale turbulent mixing are ignored.

### 3. Design of the Drag Experiments

The model was run with an extra term in the zonal momentum equation, a simple linear relaxation of the zonal-mean wind to a prescribed distribution $\bar{u}_o$.

$$\frac{\partial u}{\partial t} = \; \ldots . -\alpha\,(\bar{u} - \bar{u}_o)$$

The overbar denotes the zonal average. The relaxation rate $\alpha$ is a specified function of pressure and latitude, while the $\bar{u}_o$ field depends on pressure, latitude and time of year. $\bar{u}_o$ was determined by fitting six harmonics of the annual cycle to the 12 monthly mean observed $\bar{u}$ climatologies given by Fleming *et al.* (1988). The application of this procedure is straightforward in the present grid-point model which solves the horizontal momentum equations explicitly - a similar approach might be more difficult to implement in a model solving for the vorticity and divergence. Three cases have been considered. The distribution of $\alpha$ as a function of pressure is shown in Fig. 1 for both Case 1 and Case 2. The drag in Case 1 covers the entire region above 320 mb, but is quite weak at low levels and increases with height. At the top few levels the relaxation timescale is 8 hours. The relaxation rate in Cases 1 and 2 does not depend on latitude. Case 2 differs from Case 1 in having the drag limited to the top six levels (at and above 0.035 mb). The relaxation rate in Case 3 is the same as in Case 1 poleward of 22.5° latitude, but is very much reduced at lower latitudes (in fact it is zero equatorward of 13.5°).

The choice of the vertical modulation of the relaxation rate was motivated by a desire to produce a close agreement between the simulated and observed wind and temperature field, but allowing as much freedom as possible for the model flow to develop fully. Unfortunately the requirement of close agreement requires a rather short relaxation timescale, at least in the mesosphere.

Integrations for each case were started on January 1 using an initial field taken from the middle of the long control run. The model integration continued through March of the third year for Case 1 and through April of the second year for Cases 2 and 3. The first four months were discarded from each integration and most of the analysis reported here is based on the succeeding 12 months.

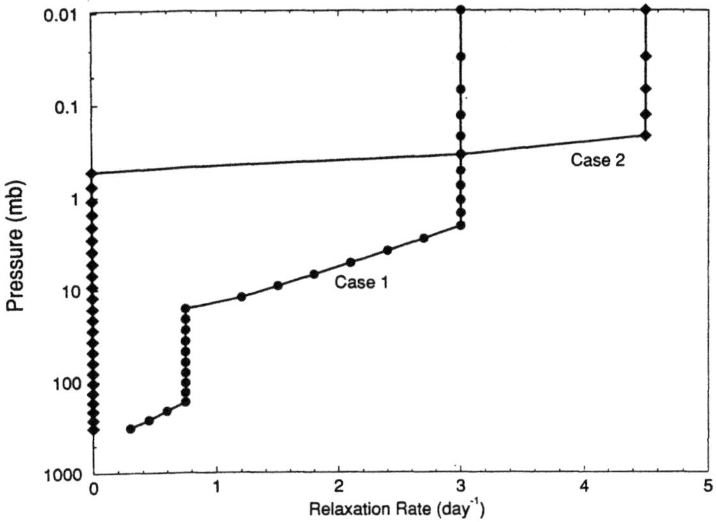


Fig. 1. The distribution of the relaxation rate $\alpha$ used in the zonal momentum equation in Case 1 (circles) and Case 2 (diamonds).

## 4. Results

The left panel of Fig 2. shows the January-mean $\bar{u}$ field for Case 1. The imposition of the relaxation has obviously been quite successful in driving the simulation close to the observations. In particular the equatorward slope of the polar night jet axis with height is well represented in the model with drag, a feature that is not well captured in the control integration (Hamilton *et al.*, 1995). Fig. 3 shows the monthly-mean, zonal-mean temperature for the months of January, April, July and October for Case 1, expressed as differences from the climatological observations of Fleming *et al.* (1988). In January and April the model with the Case 1 drag comes within 10° of the observed temperature except at the top level right near the poles. Note that there are global-mean differences between the simulation and observations at almost all levels. In particular, except in the lower stratosphere, the model simulation is colder than observed. Quite similar global-mean differences were seen between the control run and the temperature climatology of Fleming *et al.* (see Hamilton *et al.*, 1995). In the middle atmosphere such biases likely reflect some combination of inadequacies in the prescribed ozone field and/or the observed temperature climatology (see Hamilton *et al.*, 1995).

In July the temperature differences between the Case 1 simulation and the observations near the high latitude SH stratopause get somewhat larger (up to 15°). This is the remnant of the very severe cold pole problem that afflicts the control run in SH winter. The temperature bias in October is not quite so large as in July, but still exceeds 10° at high latitude near 40 km.

Fig. 4 shows the time series of daily North Pole temperature at 9 mb for the two NH winters in the Case 1 integration. The overall evolution through the winter is in quite good agreement with long-term mean climatology of the annual march of temperature (Naujokat *et al.*, 1988).

Of course, given the strong imposed relaxation of the mean flow, the circulation in the model does not develop a realistic degree of day-to-day or interannual variability, and nothing

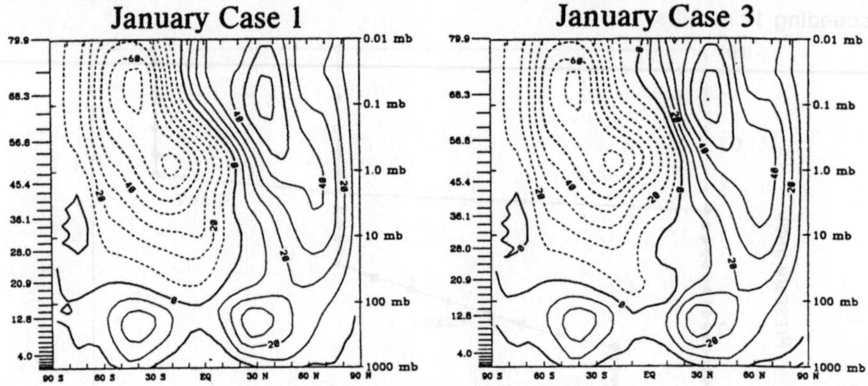


Fig. 2. The zonal-mean zonal wind distribution averaged over January for the model in Case 1 (left) and Case 3 (right). Westerly (easterly) winds are denoted by solid (dashed) contours and the contour interval is 10 m-s$^{-1}$. On these and subsequent panels the tick marks on the left hand side give the locations of the levels. The numbers on the left are heights in km for a standard midlatitude temperature profile.

remotely like a sudden warming appears in the simulation. The NH stationary wave structure is shown in Fig. 5, which compares DJF mean 10 mb heights for observations, the SKYHI control run, and the Case 1 integration. The control model results clearly show the bias of the mean flow towards a unrealistically tight vortex, but the phase and overall amplitude of the stationary wave seem reasonably well reproduced. The Case 1 integration shows a somewhat improved overall structure of the vortex. The stationary wave pattern in Case 1 is also reasonably realistic, although the location of the Aleutian high is displaced considerably westward of its observed location.

Fig. 6 shows the actual drag imposed in the Case 1 simulation averaged for each of January, April, July and October. In the summer hemisphere in both January and July the strong westerly forcing in the mesosphere is apparent. This is required to reduce the strength of the summertime easterly jet and to cool the summer polar upper mesosphere. Interestingly, there is a rather significant region of easterly forcing below the stratopause in both the northern and southern summer hemispheres. This may turn out to be difficult for GW drag parameterizations to reproduce (see next section).

Fig. 7 shows the zonal-mean temperature deviations from observed climatology for the Case 2 simulation. The restriction of the drag to the region above about 60 km leads to some significant deterioration of the simulation at lower levels. This becomes a particular problem in the high latitude SH in July and October where the Case 2 upper stratospheric temperatures are more than 20$^{o}$ too cold. Of course, the Case 2 results are still a vast improvement over the control simulation in the SH winter and spring. A drag restricted to upper levels can close off the polar night jet in the mesosphere quite realistically, but leaves the upper stratospheric flow too strong.

Fig. 8 displays the diagnosed drag in January, April, July and October of the Case 2 integration. In January, April and July, the Case 2 drag looks very roughly like the Case 1 drag, but truncated at 0.35 mb. One interesting exception is the region of easterly drag in the summer hemispheres. The easterly drag regions that were mainly below 1 mb in Case 1 are

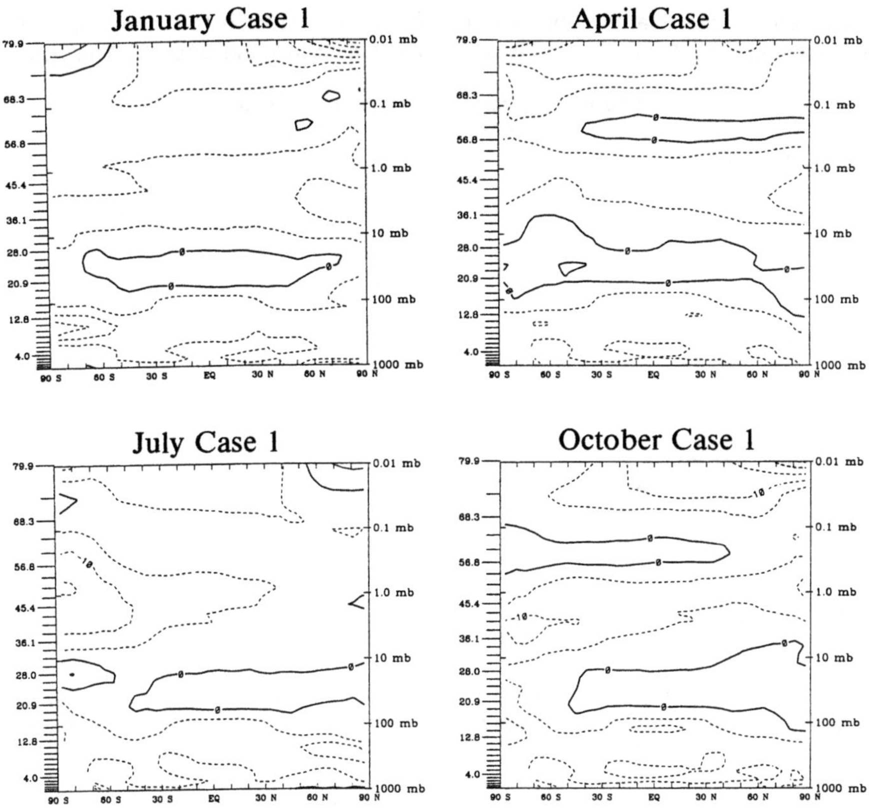


Fig. 3. The zonally-averaged temperatures in the Case 1 experiment minus the climatological observed values from Fleming et al. (1988). The contour interval is 5°C and solid (dashed) contours denote positive (negative) values. Results for January, April, July, and October.

concentrated in Case 2 at the bottom of the region of nonzero α (*i.e.* near 0.3 mb). The diagnosed drag in the SH in October is very different in Case 1 and Case 2. The very strong

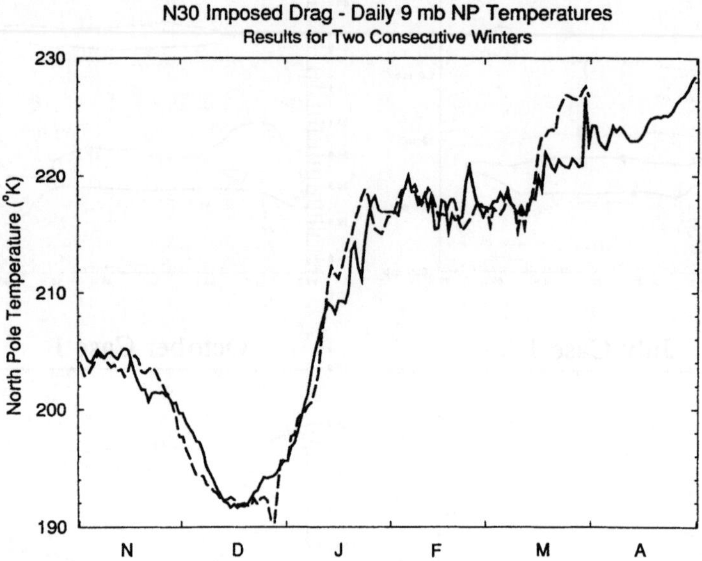


Fig. 4. The daily temperatures near 9 mb in the Case 1 integration averaged over the 88.5°N grid row. Results from November through April during the first full winter (solid line) and during the second winter (dashed line).

regions of westerly drag in the high latitude middle stratosphere (Case 1) or the midlatitude lower mesosphere (Case 2) may be particularly hard to reproduce in a GW drag parameterization.

The right panel in Fig. 2 shows the zonal-mean zonal wind for January in the Case 3 drag experiment (*i.e.* with no imposed drag near the equator). Outside of the equatorial region the results are almost identical to those for Case 1, and even close to the equator the results Case 3 results are not greatly altered from those of Case 1 (the main difference being the structure of the easterly tongue of extending from the summer hemisphere which is somewhat more realistic in Case 1). The implication would appear to be that the GCM can produce reasonable tropical simulations without any imposed drag near the equator and that the results are surprisingly insensitive to the tropical drag. These conclusions appear to be applicable to all months of the year.

The approach described here to diagnose the required subgrid scale drag in a GCM has some obvious limitations. The strong constraint needed on the mean flow means that the actual evolution of the flow through the seasonal cycle will be somewhat unrealistic (for example, there are no sudden warmings in the NH winter, as noted earlier). This could be somewhat alleviated by using a nonlinear constraint on the mean flow in place of the simple linear relaxation adopted here. The restriction of the drag to the zonal component of the flow and the neglect of the possible three-dimensional modulation of the drag (*e.g.* by planetary waves in the NH winter stratosphere) are both significant approximations. The present procedure also obviously fails to find a unique distribution of the drag, since the results depend on the assumed α. This appears to be a particular problem in SH spring. Despite all these

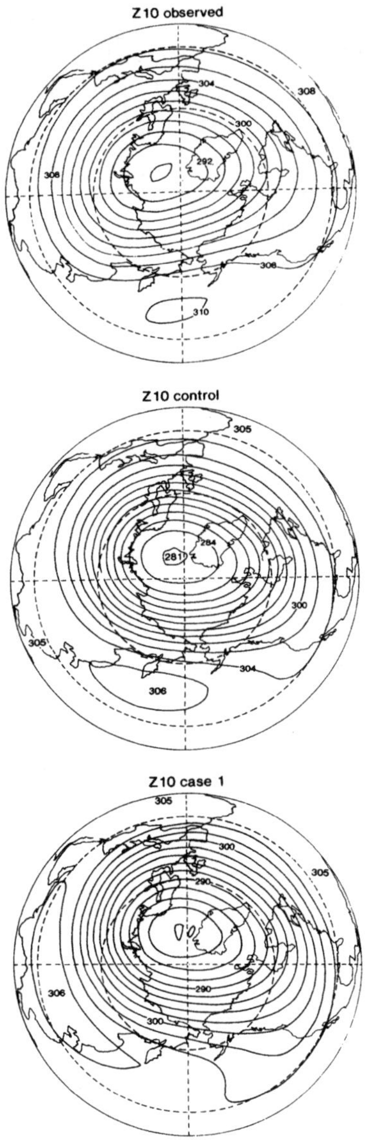


Fig. 5. The December-February mean 10 mb geopotential heights from observations (top), the SKYHI control run (middle) and the Case 1 integration (bottom). The contours are labeled in 100's of m and the contour interval is 200 m. Observations are from 17 years of daily Free University of Berlin analyses. The control result is a mean of 25 years of integration, while the Case 1 results are an average over two December-February periods.

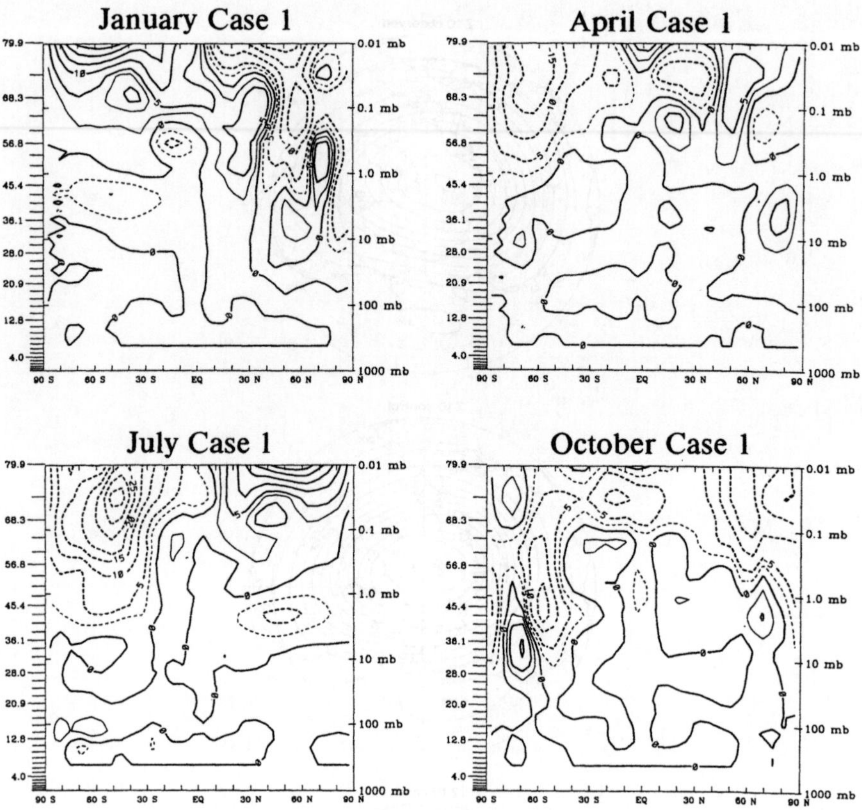


Fig. 6. The imposed zonal drag in Case 1. Results for January, April, July and October. The interval for the heavy contours is 5 m-s$^{-1}$-day$^{-1}$ and solid (dashed) contours represent westerly (easterly) mean flow forcing. Also shown at are the 2.5 m-s$^{-1}$-day$^{-1}$ contours.

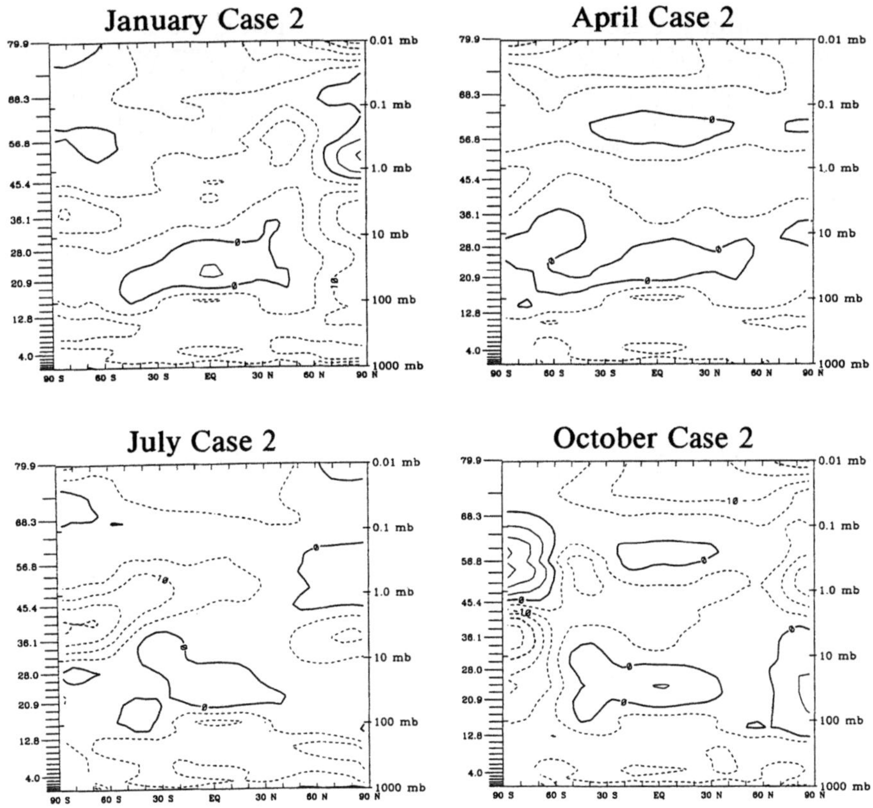


Fig. 7. As in Fig. 3 but for Case 2.

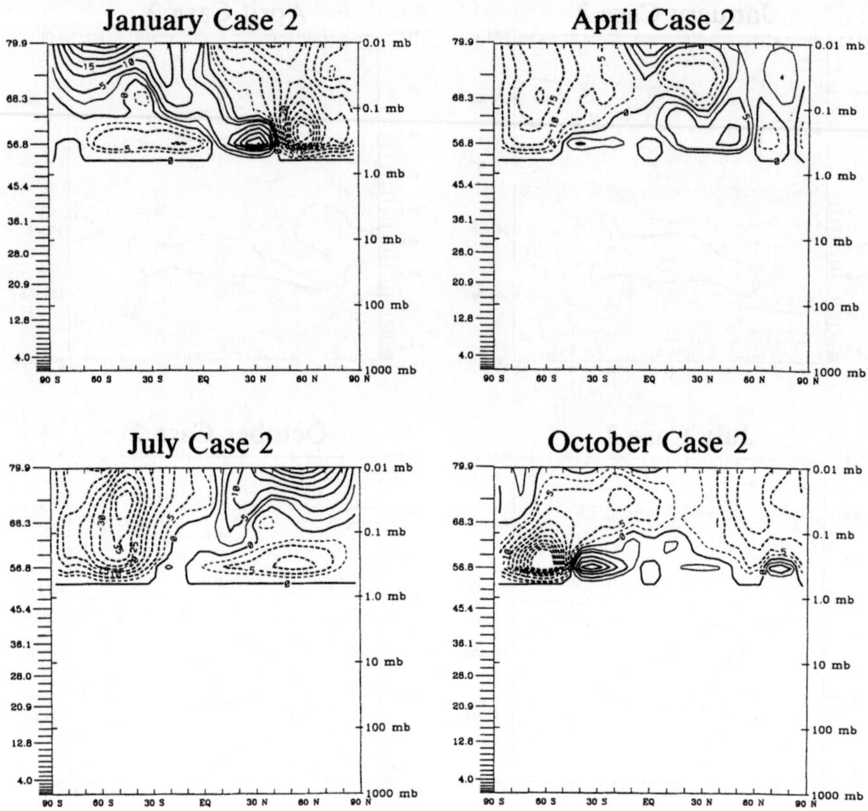


Fig. 8. As in Fig. 6, but for Case 2.

difficulties, however, the diagnosed drag can serve as a useful indication of the requirements of a GW drag parameterization, and can play a role in the tuning of GW parameterizations.

## 5. Comparison with a Simple Gravity Wave Drag Parameterization Scheme

There are a number of parameterization schemes that have been proposed to account for the effects of unresolved gravity waves on the resolved flow. The simplest conceptually is that advanced by Lindzen (1981) and used extensively in various two-dimensional models (*e.g.*, Holton, 1982, Garcia and Boville, 1994). In its basic form the Lindzen parameterization (LP) treats the gravity waves as a set of discrete, non-interacting, upward-propagating, steady, monochromatic waves. The wave fluxes are specified at some lower boundary (typically taken near the tropopause) and the fluxes are then determined using linear WKB theory, with the exception that the amplitudes are reduced at levels where the wave would produce unstable lapse rates. In terms of the vertical eddy momentum flux, the consequences of these assumptions are that for each wave the momentum flux $\rho_o\overline{u'w'}$ is conserved except for the imposition of a saturation condition (written here assuming zonal alignment of both mean flow and wave phase propagation)

$$\overline{u'w'} < \frac{k\,(c-\bar{u})^3}{2N}$$

In these expressions *u* is the zonal wind, *w* is the vertical wind, *k* is the zonal wavenumber of the wave, *c* is the zonal phase speed, N is the Brunt-Vaisala frequency, the overbar denotes the mean flow, while the prime denotes contributions from the wave. In addition one can impose a constraint that $\rho_o\overline{u'w'}$ be zero at all heights above a critical level for the wave. This simple form allows a ready computation of the momentum flux at the discrete model levels and then a differencing of $\rho_o\overline{u'w'}$ between levels produces the forcing to be inserted in the zonal momentum equation of the model:

$$\text{Total forcing} = \sum_j (1/\rho_o)\frac{\partial}{\partial z}\left(\rho_o(\overline{u'w'})_j\right)$$

where the index *j* labels the discrete waves considered. Given the mean flow (*i.e.* resolved scale wind and stratification in a GCM column) and the wavenumber, phase speed and imposed lower boundary flux for each wave, the mean flow forcing can be computed simply.

It seems likely that the straightforward application of the LP (*i.e.* treating each wave independently and neglecting any transience in the wave source) will result in an overestimate of wave breaking heights. Lindzen (1981) proposed an *ad hoc* modification of LP to extend the region of wave breaking somewhat lower. Another approach is to introduce what could be termed an "intermittancy factor" ε, so that the imposed lower boundary fluxes for all the waves are multiplied by 1/ε while the total forcing is multiplied by ε.

$$\text{Total forcing} = \sum_j \varepsilon\,(1/\rho_o)\frac{\partial}{\partial z}\left(\rho_o(\overline{u'w'})_j\right)$$

This would be a consistent LP treatment of a situation in which the waves are present for a fraction ε of the time and completely absent the remainder of the time. Varying ε will not change the effective total imposed flux, but will change the level at which the nonlinear saturation becomes important.

As part of this study a calculation was performed of the drag obtained by applying the LP to the mean state generated in the Case 1 integration. The parameterization has 10 waves with phase speeds of -50, -40, -30, -20, -10, 10, 20, 30, 40, 50 m-s$^{-1}$ and each wave has $|\overline{u'w'}|$=0.0001 m$^2$-s$^{-2}$ at 320 mb. Each wave is associated with a zonal wavelength of 200 km. For practical application to the model level structure a provision was made to absorb the flux from all the waves at the top level (*i.e.* not to allow any flux to escape through the model lid). The calculated drag using the zonal average conditions for the first half of July as the mean state are shown in Figs. 9 and 10. In particular, Fig. 9 displays results for a single column in SH midlatitudes for five different values of $\varepsilon$. In each case the model shows strong effects of mean flow filtering, with weak westerly drag appearing at lower stratospheric levels where the Doppler-shifted phase speed of the westerly waves is small, while very strong easterly drag appears in the upper mesosphere where the easterly waves either saturate or are absorbed at the top level. The drag profile is a strong function of $\varepsilon$. With $\varepsilon$=1 or $\varepsilon$=0.1 the entire easterly drag region is confined to the top level. With smaller values of $\varepsilon$ the drag region extends further downward. It is striking just how small $\varepsilon$ has to be to spread the mesospheric drag over a height range comparable to that diagnosed from the imposed drag experiments. Of course, changing the $k$ for the imposed waves has the same effect as changing $\varepsilon$. Thus including very long horizontal wavelength waves in the input spectrum could allow a somewhat larger value of $\varepsilon$ to be used. Of course this is a problematic issue for GCMs which, in principle at least, can explicitly resolve sufficiently long GWs.

Fig. 10 shows the height-latitude distribution of the drag in the case $\varepsilon$=0.001. The overall distribution has some resemblance to the diagnosed July drag in both the Case 1 and Case 2 experiments (Figs. 6 and 8). In particular, both the LP result and the diagnosed Case 1 and Case 2 drag distributions in the SH mesosphere display a strong easterly drag peaking at values ~30-40 m-s$^{-1}$-day$^{-1}$ in midlatitudes at the second highest model level. The diagnosed westerly drag peaking at the top level in the NH mesosphere is also roughly reproduced in the LP result. Missing from the LP drag is the region of diagnosed easterly drag in the NH upper stratosphere (Case 1) or lower mesosphere (Case 2). The LP predicts a region of much smaller (in acceleration units) easterly drag in the NH middle stratosphere.

These results are based on an assumption of a geographically-invariant input spectrum. In general there is no reason why the input parameters should not vary substantially with location and time of year. In practice with a sufficiently small $\varepsilon$ and freedom to adjust the input spectrum, inclusion of the LP in a model should allow a very realistic middle atmospheric zonal mean temperature distribution to be obtained, at least in the solstitial seasons.

## 6. Conclusion

The calculations of the zonal-mean zonal drag needed to force the model into agreement with observed climatology produced results that in many respects are in accord with the simplest notions of how GWs propagating upward from the troposphere ought to interact with the mean flow. The easterly (westerly) drag in the extratropical winter (summer) mesosphere is what would be expected to result from the filtering of the wave spectrum during propagation through the polar night (summertime easterly) jet and subsequent wave breaking at high levels. These basic results for the summer and winter mesosphere seem to be independent of the choice of relaxation rate (at least within the range of possibilities considered here). While this is encouraging, there are some more problematic features apparent when the diagnosed drag results are examined in detail and actually compared to the predictions of a

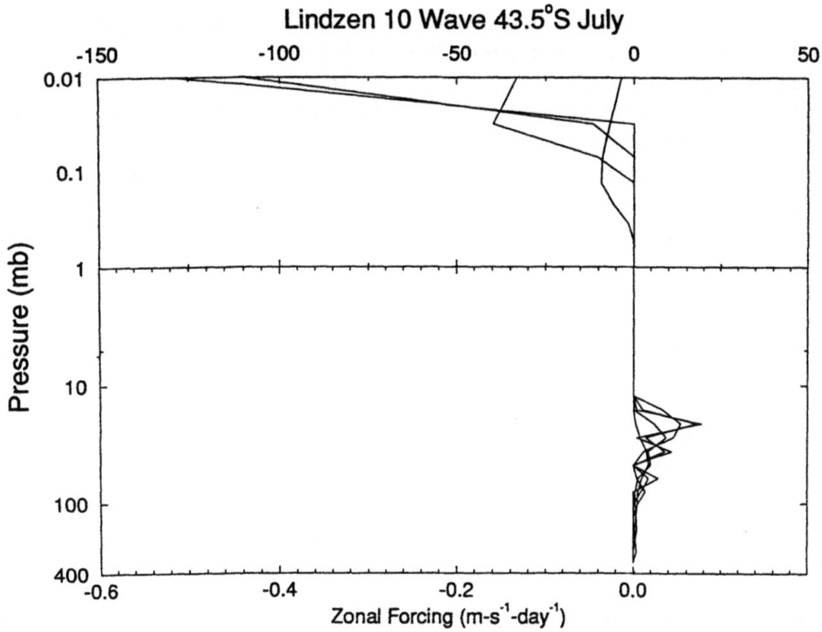


Fig. 9. The drag provided by the 10 wave Lindzen scheme described in the text. Results for 43.5°S using a mean state from the Case 1 run averaged over the first half of July. Results are presented for ε of 1., 0.1, 0.01, 0.001 and 0.0001. Note the change in the scale for the drag between the upper and lower panels. The curves for ε=1 and ε=0.1 are identical in the upper panel.

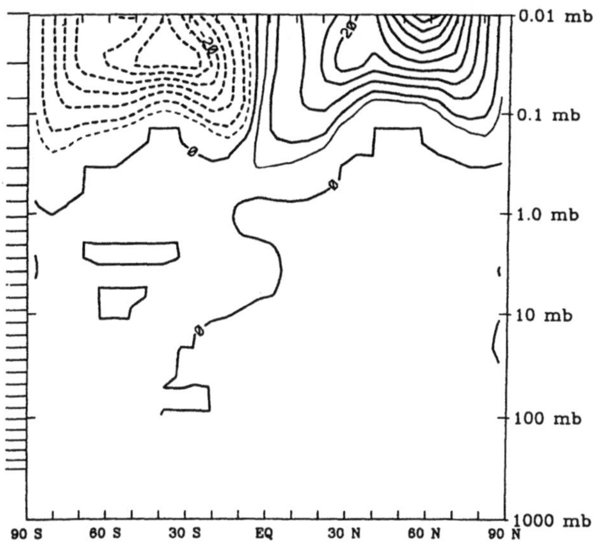


Fig. 10. The drag provided by the 10 wave Lindzen scheme described in the text with ε=0.001. Results obtained using the mean wind and temperature distributions from the Case 1 run averaged over the first half of July. Contour intervals as in Fig. 6.

A striking aspect of the diagnosed drag in all seasons is its considerable depth through the middle atmosphere. Limiting the drag even to the top 20 km of the model domain (*i.e.* Case 2) leads to a significant deterioration in the simulation. The simplest parameterization of the GW drag in terms of noninteracting quasi-linear waves is unable to reproduce the deep region of diagnosed drag. A critical issue for future research is whether a more realistic treatment of the subgrid scale gravity waves can produce significantly lower breaking heights for the waves and hence a sufficiently deep region of parameterized drag.

**References**

Fels, S.B., J.D. Mahlman, M.D. Schwarzkopf and R.W. Sinclair (1980) Stratospheric sensitivity to perturbations in ozone and carbon dioxide: Radiative and dynamical responses. J. Atmos. Sci., 37, 2265-2297.

Fleming, E.L., S. Chandra, M.R. Schoeberl and J.J. Barnett (1988) Monthly mean global climatology of temperature, wind, geopotential height, and pressure for 0-120 km. NASA Technical Memorandum 100697, 85 pp.

Garcia, R.R. and B.A. Boville (1994) "Downward control" of the mean meridional circulation and temperature distribution of the polar winter stratosphere. J. Atmos. Sci., 51, 2238-2245.

Hamilton, K. (1995a) Interannual variability in the Northern Hemisphere winter middle atmosphere in control and perturbed experiments with the SKYHI general circulation model. J. Atmos. Sci., 52: 44-66.

Hamilton, K. (1995b): Comprehensive modelling of the middle atmospheric climate: some recent results. Climate Dynamics, 11: 223-241.

Hamilton, K., R.J. Wilson, J.D. Mahlman and L.J. Umscheid (1995) Climatology of the GFDL SKYHI troposphere-stratosphere-mesosphere general circulation model. J. Atmos. Sci., 52: 5-43.

Holton, J.R. (1982) The role of gravity wave induced drag and diffusion in the momentum budget of the mesosphere. J. Atmos. Sci., 39, 791-799.

Lindzen, R.S. (1981) Turbulence and stress due to gravity wave breakdown. J. Geophys. Res., 86, 9707-9714.

Mahlman, J.D., J.P. Pinto and L.J. Umscheid, (1994) Transport, radiative, and dynamical effects of the Antarctic ozone hole: A GFDL "SKYHI" model experiment. J. Atmos. Sci., 51: 489-508.

Manabe, S., and J.D. Mahlman (1976) Simulation of seasonal and interhemispheric variations in the stratospheric circulation. J. Atmos. Sci, 33: 2185-2217

Naujokat, B., K. Labitzke, R. Lenschow, K. Petzoldt and R.-C. Wohlfart (1988) The stratospheric winter 1987/88: An unusually early major midwinter warming, Beilage zur Berliner Wetterakarte, no. 6, 20 pp..

# Seasonal Simulations with the Canadian Middle Atmosphere Model: Sensitivity to a combination of orographic and Doppler Spread Parameterizations of Gravity Wave Drag

N. McFarlane
Canadian Centre for Climate Modelling and Analysis
Victoria, British Columbia, Canada

C. McLandress
Institute for Space and Terrestrial Science
North York, Ontario, Canada

S. Beagley
York University
North York, Ontario, Canada

## Abstract

The Doppler spread parameterization is used to represent the effects of a background spectrum of non-orographic gravity waves in the Canadian Middle Atmosphere Model. Two versions of the Doppler spread parameterization are used, differing only in the range of the vertical wave number part of the input spectrum. Results from seasonal simulations made with combinations of Doppler spread and orographic gravity wave drag parameterizations are discussed and compared with those made with the basic version of the model which contains an orographic gravity wave drag parameterization but no explicit parameterization for non-orographic waves. Results obtained indicate that the Doppler spread parameterization brings about a number of improvements in simulations of the zonal mean structure of the middle atmosphere including closing of the mesospheric jets and reversal of the mean zonal winds above the mesopause in middle and high latitudes.

## 1. Introduction

The importance for the general circulation of the middle and lower atmosphere of the transport and deposition of angular momentum by vertically propagating gravity waves is now well known. Since the relevant portions of the gravity wave spectra can not be resolved by most existing atmospheric general circulation models (AGCMs) their effects on the resolved circulation must be parameterized. Gravity wave fields in nature have complex structure and this adds difficulty to the problem of parameterizing their effects in AGCMs.

Most of the practical parameterization schemes in current use employ a simplified form of linear wave theory in combination with a wave saturation hypothesis, such as that of Lindzen (1981) as a basis for representing the effects of dissipation and the associated exchange of momentum between the waves and the background flow (assumed resolved in AGCMs). This approach has been used in a fully three-dimensional context for parameterizing the effects of orographic gravity waves in AGCMs (Palmer et al, 1986; McFarlane, 1987) and in some cases possibly for other sources as well following the approach used earlier in mechanistic models of the middle atmosphere (Holton, 1983; Garcia and Soloman, 1985). However extensive observational work in recent years (e.g. Allen and Vincent, 1995; Fritts and VanZandt, 1993; Dewan and Good, 1986) suggest that gravity wave fields in the middle atmosphere have a typical spectral structure which is consistent with the existence of a continuous spectrum of waves such that the high vertical wave number portion has a slope that is nearly invariant.

NATO ASI Series, Vol. I 50
Gravity Wave Processes
Their Parameterization in Global Climate Models
Edited by Kevin Hamilton

In recent years theoretical models have been proposed to account for the observed spectral structure of gravity wave fields in the middle atmosphere and these have in turn led to new parameterization proposals which are just beginning to be applied practically as the basis of parameterization schemes in AGCMs. Among these are the schemes of Fritts and Lu (1993), a parameterization based on the Doppler Spread theory (Hines, 1996a,b and also in this volume), and one based on the recent work of Medvedev and Klaassen (1995) (also in this volume).

In this paper we show selected results from an application of one of these schemes, the Doppler Spread Parameterization (here designated DSP), in the Canadian Middle Atmosphere model. As this work is still preliminary and exploratory in nature we have chosen to use the DSP only to represent a broad background spectrum of gravity waves which fall into the category of extro-waves in the terminology used by Hines (1996a,b). In addition to the DSP we also include a separate parameterization of the effects of orographic gravity waves using the basic approach of McFarlane, 1987. In the work discussed here these two parameterizations were not directly coupled to each other, although an eventual goal is to do so, perhaps as suggested by Hines, 1996b.

## 2. Description of the Model, Parameterizations, and Numerical Experiments

### 2.1 The Canadian Middle Atmosphere Model.

The numerical model used in this study is and upward extension of the third generation spectral AGCM of the Canadian Centre for Climate Modelling and Analysis. While still developmental, the version of this model used as a basis for the upward extension to encompass the middle atmosphere is similar in its basic structure to the current operational second generation model documented by McFarlane et al (1992). However the third generation model has improved treatment of various physical parameterizations and an increased vertical resolution in the troposphere.

Here we provide only a brief description of the version of this model that has been constructed for use in middle atmosphere studies. It is described in greater detail by Beagley et al (1996). It is sufficient to note here that the vertical domain of the model is such as to encompass the region between the surface and the mesopause. A total of 50 levels is employed, arranged in such a way as to provide relatively high resolution in region of the planetary boundary layer with resolution decreasing with height to the lower stratosphere. From there to the top of the model it is roughly uniform with an effective resolution of roughly 3 km with the uppermost model level being located at an elevation of roughly 96km above sea level. The upper boundary of the model is at a finite pressure (.06 Pa) and a boundary condition of vanishing vertical mass flux is imposed there. In the spectral domain a triangular truncation at 32 waves (T32) is employed.

### 2.2 Parameterization of gravity wave drag.

In normal use the Canadian middle atmosphere model (CMAM) includes the orographic gravity wave drag parameterization of McFarlane (1987). However it does not include an explicit parameterization of the effects of gravity waves with non-zero phase speeds that are excited from other tropospheric sources. It does however employ a Rayleigh damping sponge layer in uppermost two density scale heights of the model. The damping coefficient increases quadratically with height from zero at a height of 80km above the surface to a maximum of $3.5 \times 10^{-5}$ sec$^{-1}$.

While the main purpose of the sponge layer is to absorb resolved upward propagating waves that may otherwise undergo spurious reflections, the drag force it produces mimics, imperfectly, those that are produced by more physically based gravity wave drag parameterizations. However as is well known, such Rayleigh damping schemes, being relaxational in nature, are unable to simulate those aspects of the gravity wave drag process that bring about certain well known features of large scale circulation in the middle atmosphere, such as the flow reversals in mean zonal wind field that are found at middle and high latitudes near the mesopause. The DSP, being physically based, can in principle improve the simulation of these features of the circulation.

The DSP will not be described in detail here as that is done in other papers in this volume (see those by Hines and McLandress). The scheme has been implemented in the CMAM following the basic approach suggested by Hines (1996a,b) in which the initial input spectrum is assumed azimuthally isotropic and is discretized using 8 azimuths. The total energy content in the initial spectrum is specified in terms of the total variance of the horizontal wind and the vertical wave number ($m$) part is taken to follow a simple power law as $m_j^s$ in the small $m$ part of the spectrum for each azimuth $j$, beginning at a specified lower value $m_{MIN}$ and terminating at a high wave number cut off defined below the turbopause as

$$m_{ji} = N_i / (\Phi_1 \sigma_j + \Phi_2 \sigma_h + V_j - V_{ji})$$

where the subscript $i$ denotes an initial value, defined at a specified reference level, $\sigma_j$ is the rms wind fluctuation in the $jth$ azimuth, $\sigma_h$ is the total rms wind fluctuation, and $V_j$ is the component of the background horizontal velocity in the $jth$ direction. As noted by Hines (1996a) this quantity must be constrained to decrease or remain constant with height and be further bounded in the region of the turbopause by a vertical wave number imposed by molecular viscosity. Evaluation of this quantity requires specification of an effective horizontal wave number which is taken to be a constant, independent of azimuth and given the value $k^* = 7. \times 10^{-6} (m^{-1})$. This value is also used in specification of the integral of the horizontal wave number spectrum, required to evaluate the vertical momentum flux of the gravity wave ensemble. (Hines, 1996a, see also Mengel et al, 1995; and McLandress, this volume).

In all of the test simulations that have been performed using the DSP the constant factors $\Phi_1$ and $\Phi_2$ have the values 1.5 and.3 respectively and $s$=1. Also the initial gravity wave source is located at the surface with an rms velocity ($\sigma_{hi}$) of 1 m/s. It is also assumed that the background wind vanishes at that level and this has the effect of eliminating intro-waves (Hines, 1996a).

### 2.3 Simulation experiments

A 3-year reference simulation (REF) with the basic version of the CMAM has been completed recently. Results of that simulation are discussed in some detail by Beagley et al (1996). Here selected results from this simulation are presented in order to illustrate some of the effects of introducing the DSP. As mentioned above the dominant drag mechanisms in the mesosphere in this simulation are due to the orographic gravity wave drag parameterization and Rayleigh friction.

The first seasonal simulations using the DSP were carried out with $m_{MIN}$ set to zero. It was hoped that with use of the DSP it would be possible to remove the Rayleigh damping. However, it was found that implementation of the DSP produced rather large effects near the top of the model. Some of these were possibly related to the amplification of the tides (see the paper by McLandress in this volume). In order to maintain computational stability at a reasonable value of the time-step the Rayleigh damping was retained for these experiments. A single year simulation (DSPZS) has been completed and selected results from it for the simulated time and zonal mean winds and temperatures for January and July are discussed below.

Near the top of the model the wave driving effects of the DSP are associated with obliteration of waves with small vertical wave numbers. However an argument can be made for limiting the amount of energy in this part of the spectrum, either by specifying a finite positive value for the minimum wave number and/or by using a larger value for the power, $s$. (Hines, 1996b). The third shorter (3 month) simulation that we will discuss below (denoted DSPNZS) retains the value of unity for $s$ but imposes a finite value of 1/(3km) for $m_{MIN}$.

The recent work of Shepherd et al (1996) has shown that the use of a Rayleigh damping sponge for the zonal mean flow may have the undesirable effect of significantly reducing the impact of any physically based drag scheme that is imposed in the upper part of the model. Motivated by the results discussed in that paper we have chosen also, in the DSPNZS simulation, to allow the Rayleigh damping to act only on the deviations of the horizontal velocity field from its zonally averaged value at a given height and latitude.

## 3. Results

The gravity-wave drag parameterizations in the CMAM are applied locally in space and time. Clearly this allows for the spatial variability of the orographic gravity wave excitation to be taken into account and affect the structure of the planetary waves as well as that of the mean flow. Previous studies (e.g. McLandress and McFarlane, 1993) have shown that the planetary wave structure in the middle atmosphere can be affected in important ways by this process. Although the source spectrum for the DSP is not spatially variable, the effects of the DSP on the background flow have considerable spatial and temporal variability associated with the fact the gravity waves represented by it propagate vertically through the spatially and temporally varying flow field simulated by the CMAM.

It is beyond the scope of this study to examine in a comprehensive way the effects of parameterizing the effects of gravity wave driving in the CMAM. However some of the interesting features of these effects can be illustrated by examining the simulated zonal mean wind and temperature fields. Here we confine attention to the simulations for January and July.

The mean zonal wind and temperature fields in the extra-tropical middle atmosphere vary considerably on inter-annual time scales in the observed climate. This is also true for that simulated by the basic version of the CMAM as is illustrated in Figure 1. which depicts the zonally and temporally averaged zonal wind fields for each of the three Januarys of the REF simulation. The wind structure is depicted for the entire vertical domain of the model, including the sponge layer. Note that, contrary to observations (not shown but see Fleming et al, 1990), the zonal winds do not reverse above the mesopause at high latitudes.

The zonal wind structure has distinct features in each of the Januarys of the simulation.The first two reflect the effects of warmings, one of which occurred in the lower stratosphere during the first January (upper panel), and a second one in the second January in the upper stratosphere (middle panel). Perhaps for that reason, the wind structure on the region below the 80 km level for the second January is qualitatively most similar to the observed mean zonal winds in the northern hemisphere (Figure 3). The other two January simulations have mesospheric jets in that region which are significantly different, being substantially stronger and closer to the pole.

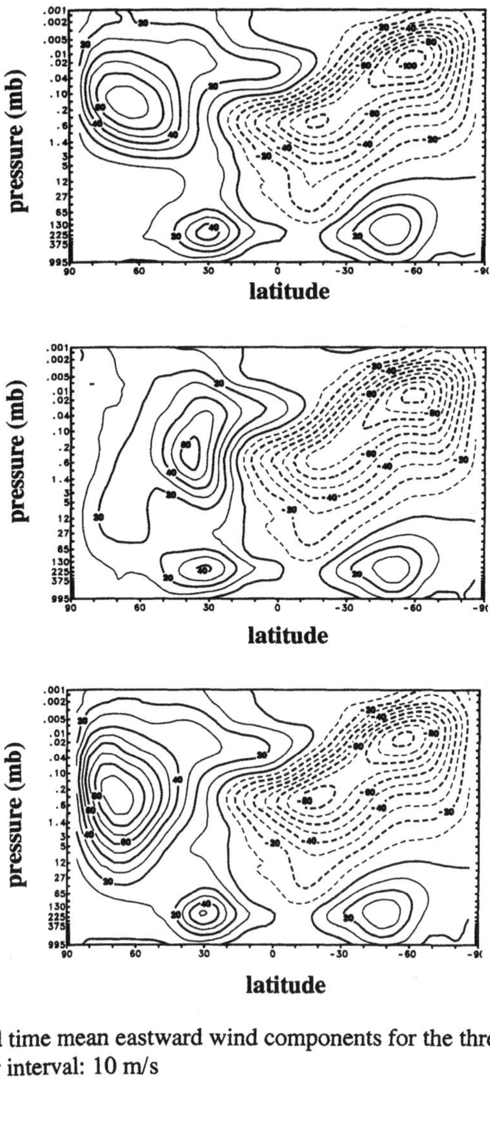


**Figure 1.** Zonal and time mean eastward wind components for the three Januarys of the REF simulation. Contour interval: 10 m/s

The mean zonal wind field for July do not show strong variation from year to year in the REF simulation. The mean of this field for the three year ensemble (Figure 2, upper) illustrates clearly that simulations made with this version of the CMAM have systematic biases in this season. In particular, mesospheric jet maximum in the southern hemisphere is too strong by a factor of two and located too close to the pole. The overall structure of the mean zonal winds in the summer hemisphere is more realistic though magnitudes are also somewhat too high. Also notable is the lack of a reversal in the mean zonal wind at high latitudes just above the mesopause. There are corresponding biases in the temperature field, shown in figure 2 for the region below the sponge layer Note that the stratopause in the southern hemisphere polar region is higher and colder than observed and the lower stratosphere in that region is also colder than observed (Figure 4).

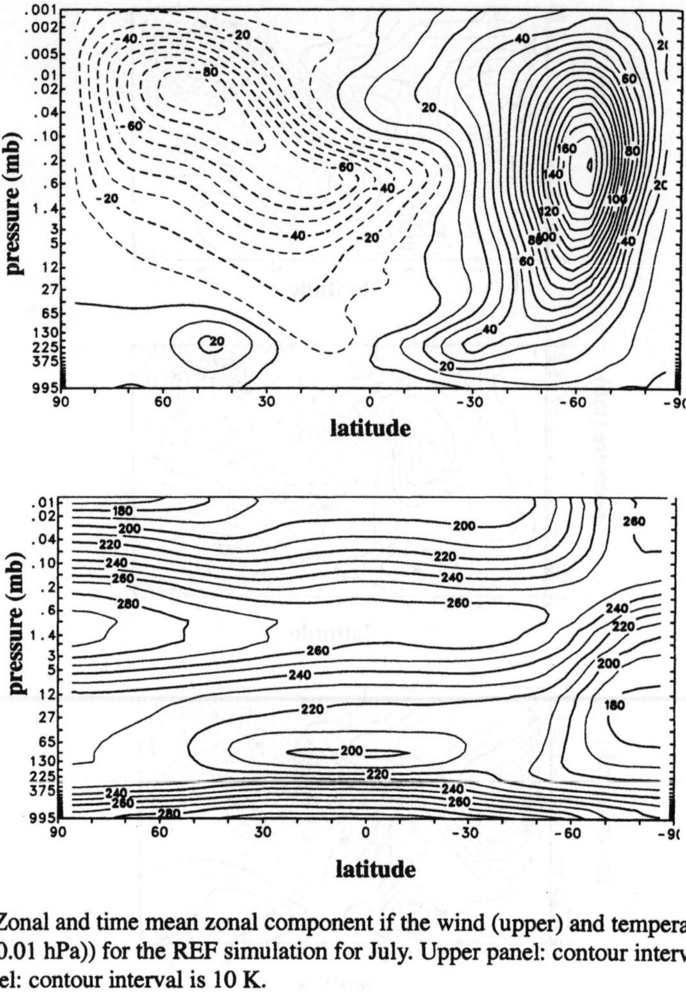


**Figure 2.** Zonal and time mean zonal component if the wind (upper) and temperature (lower, surface to 0.01 hPa)) for the REF simulation for July. Upper panel: contour interval is 10 m/s. Lower panel: contour interval is 10 K.

Also the axis of the westerly jet remains predominantly vertical with little evidence of the equatorward tilt with height as seen in the observations.

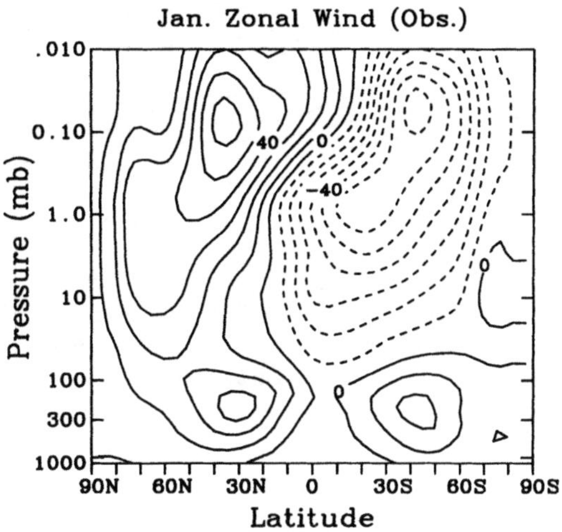


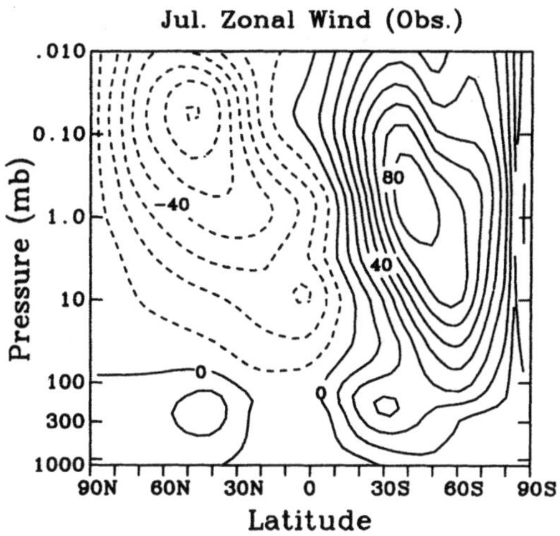


**Figure 3.** Observed mean zonal winds for January and July in the region between the surface and the mesopause from the CIRA86 data set (Fleming et al, 1990). Contour interval: 10m/s.

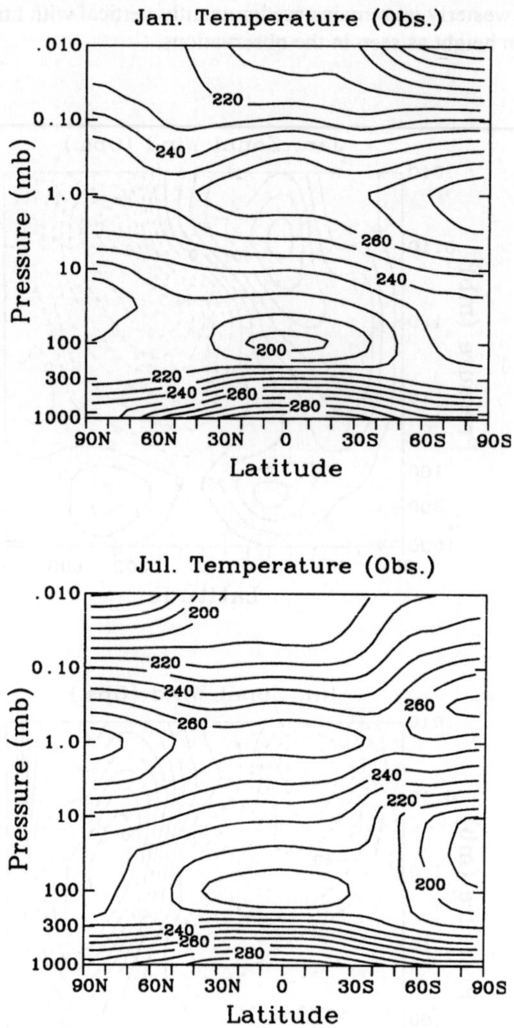


**Figure 4**: As in Figure 3 except for Temperature. Contour interval: 20 K

Simply adding in the DSP alleviates but does not remove some of the systematic bias in the zonal wind field. This is seen in Figure 5 which depicts the mean zonal flow field in the full vertical domain in the model for January and July of the single year of the DSPZS simulation.

A notable difference from the REF simulations is that introduction of the DSP does bring about a reversal of the winds just above the mesopause. In the northern hemisphere in January, in agreement with observations, this effect does not extend all of the way to the pole. Apart from this the zonal wind structure for January at high latitudes in the northern hemisphere is qualitatively similar to that of the first January of the REF simulation, although the strength of the mesospheric jet in the northern hemisphere is somewhat weaker. In view of the variability in this season seen in the REF simulation this difference may not be significant. The secondary wind maximum in the vicinity of the tropical mesopause in both January and July is associated with an enhancement of the mesospheric semi-annual oscillation in the zonal wind field that occurs when the DSP is used. This effect has also been noted in other studies which used the DSP (Manzini et al, this volume).

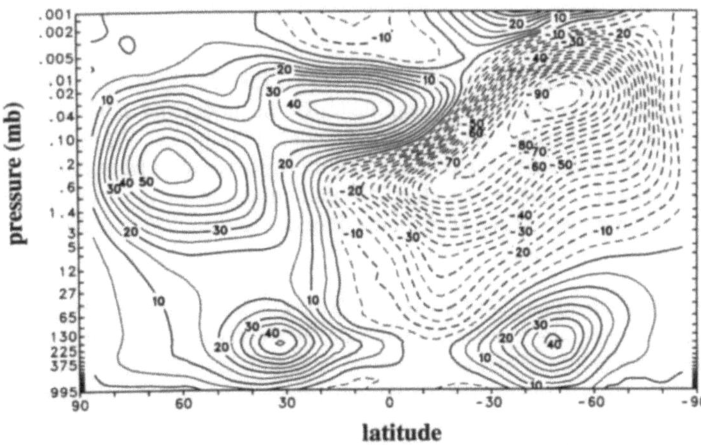


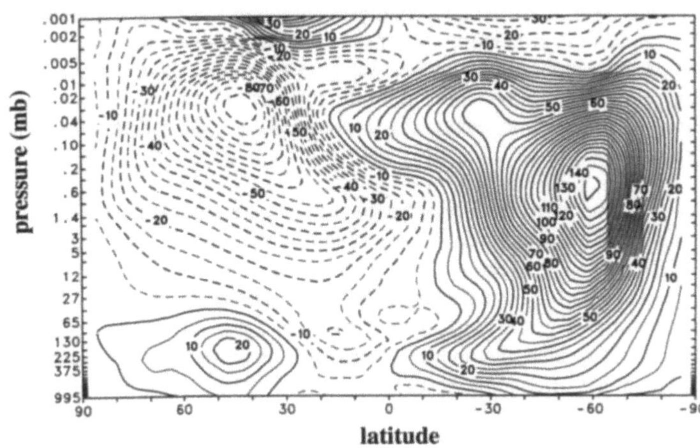


**Figure 5**. Mean zonal winds for January (top) and July (bottom) of the DSPZD simulation. Contour interval: 5m/s

The remaining most notable feature associated with introduction of the DSP is a modest (40 m/s) reduction in the strength of the zonal jet in the southern hemisphere in July. However this is only about half what is needed to bring the simulated flow into better agreement with observations.

The zonal mean drag forces associated with parameterized processes are shown in separated component form in Figures 6 and 7. In addition to the relatively large contributions from the DSP (more than 100 m/s/day above the mesopause) and the orographic gravity wave drag, there is also a relatively small contribution due to vertical diffusion and a substantial contribution due to Rayleigh damping in the sponge layer (lower left panels in Figures 6 and 7). The model also has a scale selective horizontal diffusion process. The contribution due to this process (not shown) is relatively small.

As indicated above, Rayleigh damping was retained in the DSPZS simulation in part because it tends to counteract the effects of the DSP near the top of the model and helps to maintain computational stability for the chosen time step for the model (15 minutes). While this counterveiling drag effect is apparent near the top of the model it is also apparent from Figures 6 and 7 that Rayleigh damping also complements the DSP, particularly in the region just above the mesopause in the summer hemispheres where the drag due to the DSP increases rapidly with height.

The recent study of Shepherd et al. (1996) examines the effect of coupling between a Rayleigh damping sponge layer and other (presumably physically based) drag forces that may have a maximum contribution below or within the sponge layer in middle latitudes. They show that, in the absence of Rayleigh damping, the meridional circulation which develops as part of the steady state response to and imposed drag force in the middle atmosphere is confined to the region below the imposed drag force. The associated temperature response is also confined to the region and has a dipole structure. Both of these effects are the expected features of the "downward control" response first discussed in detail by Haynes et al (1991). However, introduction of the sponge layer gives rise to a counterveiling drag force that maintains an upward component of the meridional circulation response in the region above the imposed forcing and an associated temperature response which is of the opposite sign and comparable magnitude to that in the region below it. More relevant still to the present study is the finding that when the imposed forcing is within the sponge layer the response to it is distorted and diminished by the Rayleigh damping.

The results referred to above from study of Shepherd et al. are obtained using a simplified model in which the imposed drag force is a specified function of height and latitude. In reality of course there is a coupling between the drag force and the response to it that comes about as a result of the dependence of the parameterized forces on the background flow. A further complicating factor is that the physical drag parameterizations act in a fully three dimensional sense in the CMAM and so also may induce a non-zonal response which is not discussed here but may be manifest in a change in the resolved planetary wave drag on the mean zonal flow (see for example McLandress and McFarlane, 1993). However, Shepherd et al. do also show results of numerical experiments with the CMAM, using an additional localised type of drag formulation which acts only in regions of strong winds, that are consistent with the results of the simplified model.

In view of the likelihood that the Rayleigh damping in the sponge layer distorts the response to physically based gravity-wave drag parameterizations that have been introduced into the model the most desirable way of dealing with this problem would of course be to remove it altogether. However, in view of the computational difficulties mentioned above this proved not to be practical. We have instead compromised by removing its effect on the zonal mean part of the flow as discussed above, and, as also discussed above we have imposed a finite lower cut-off wave number in specifying the initial vertical wave number spectrum while maintaining its

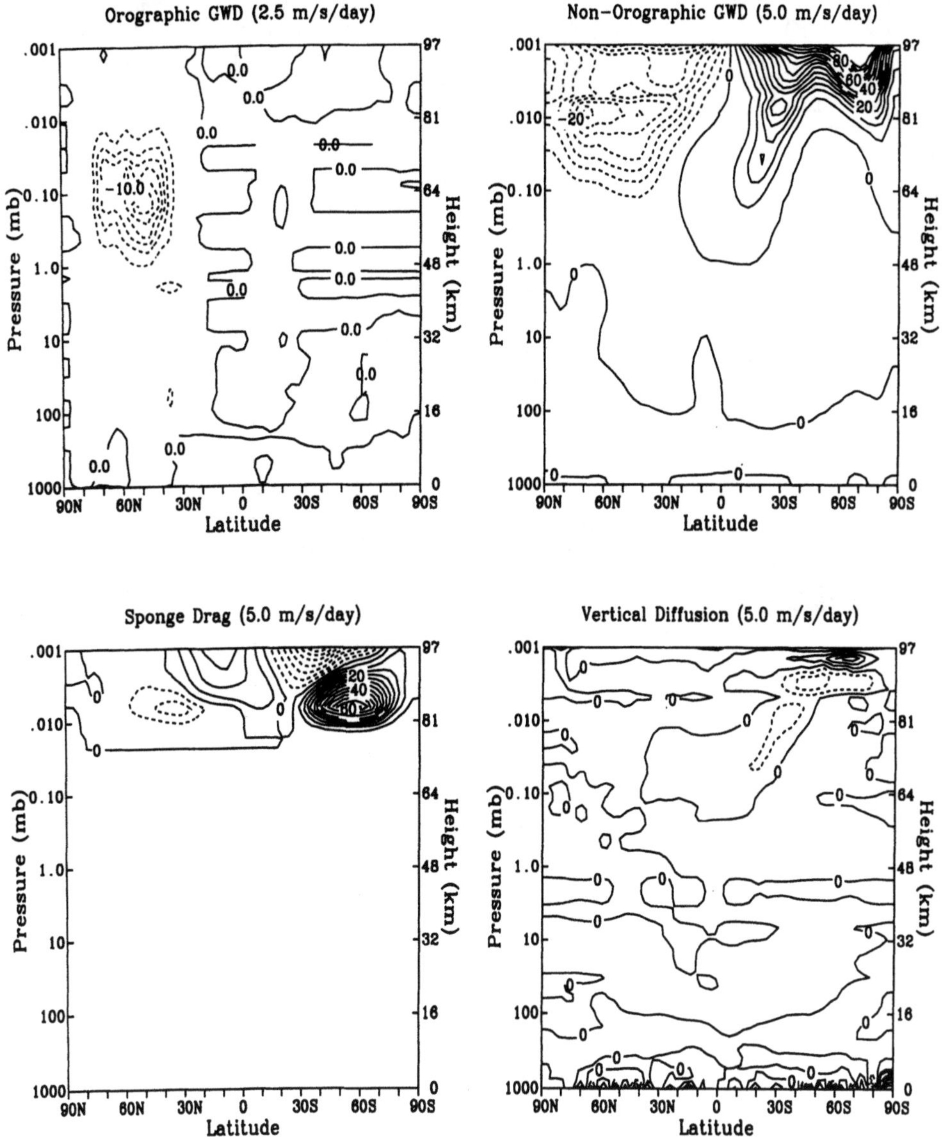


**Figure 6.** Mean zonal components of parameterized drag forces (per unit mass) for January of the DSPZS simulation.

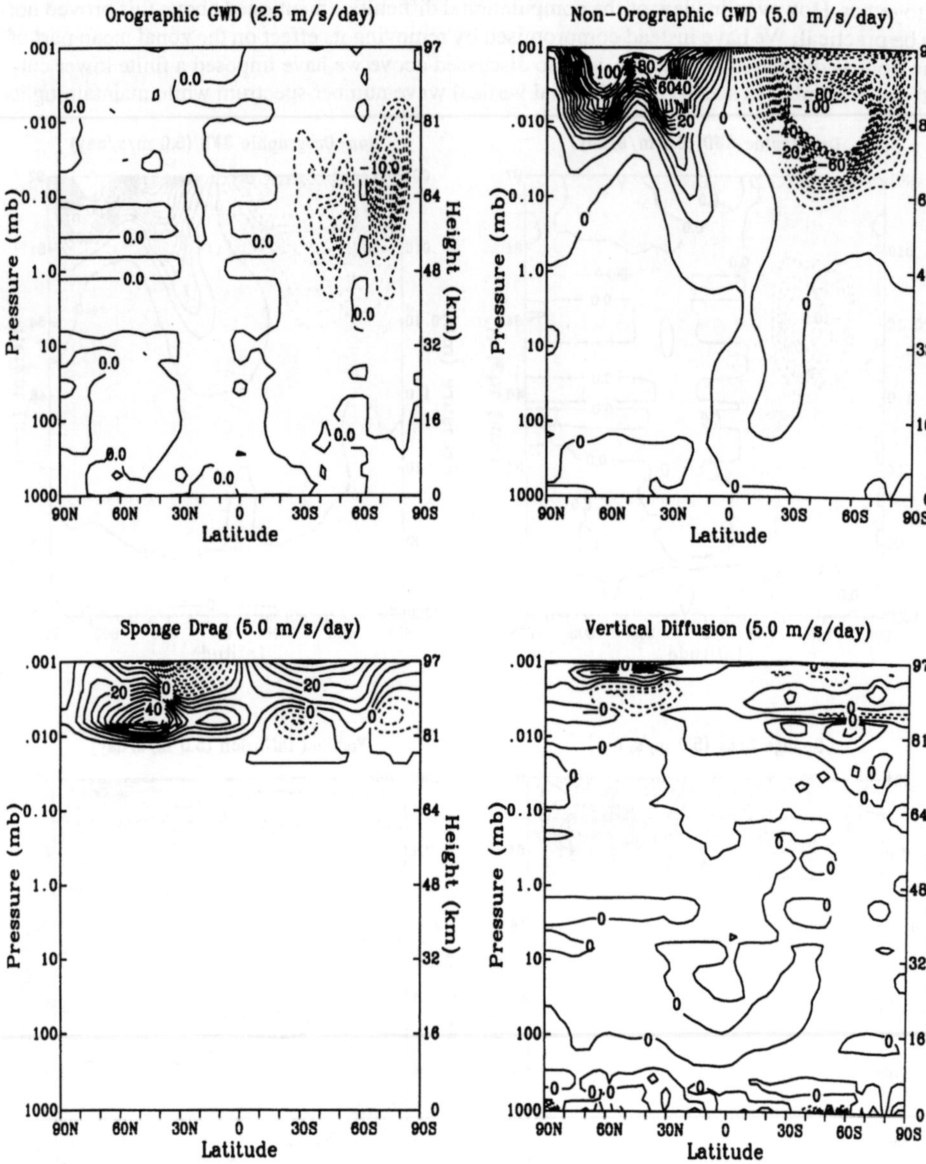


**Figure 7.** As in Figure 6 but for July.

total energy content. This has the added advantage of reducing the magnitude of the contribution from the DSP near the top of the model while enhancing it lower down.

The mean zonal wind for July for this case (DSPNZS), shown in the upper panel of Figure 8, is qualitatively similar to that for the DSPZS simulation. Although there is a slight reduction (relative to the DSPZS case) in the strength of the mesospheric winds at high latitudes in both hemispheres this may not be significant. The enhancement of the zonal wind maximum at the equator may be significant however.

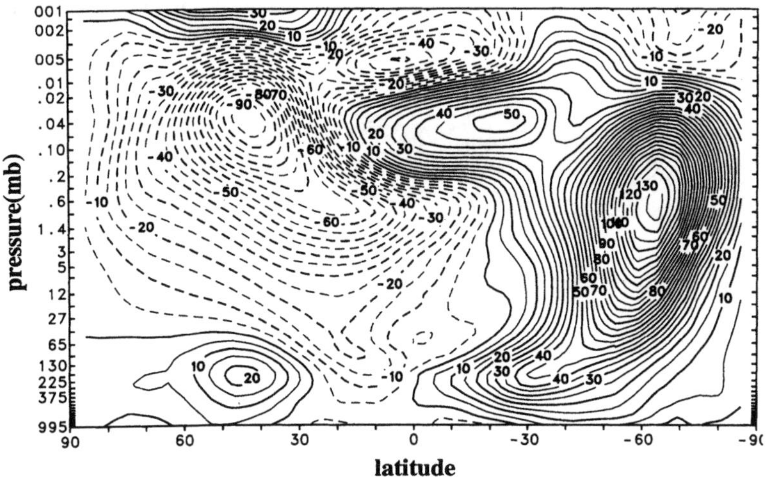


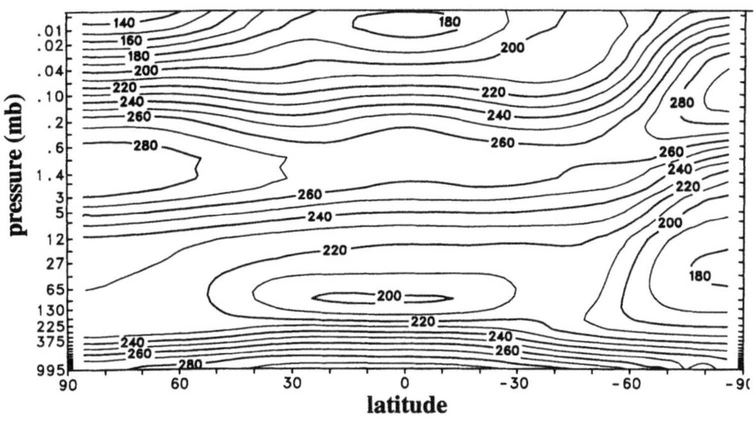


**Figure 8.** Mean zonal winds(top) and temperatures (bottom, as in Fig. 2) for July for the DSPNZS simulation. Contour intervals: winds, 5m/s; temperatures, 10K.

The temperature structure for this simulation (Figure 8, bottom) is more realistic in most respects, relative to the CIRA86 observations (Figure 4) than is that of the REF simulation. For example the stratopause at the south pole is both lower and warmer. In fact temperatures are slightly higher throughout the stratosphere in the south polar region. In contrast, the stratopause is slightly colder in the northern hemisphere polar region. The mesopause is substantially colder in the north polar region and in fact the slight warm bias there, relative to the

CIRA86 observations, in the REF simulation is replaced by a cold bias in the DSPNZS simulation.

The zonal mean of total drag for this simulation (Figure 9) is comprised almost entirely of contributions from the DSP and the orographic gravity-wave drag parameterization. Near and above the mesopause the contribution from the DSP dominates. As expected, imposing a finite lower limit on the range of the vertical wave number spectrum has the desired effect of reducing the magnitude of the DSP contribution near the top of the model while enhancing it slightly lower down.

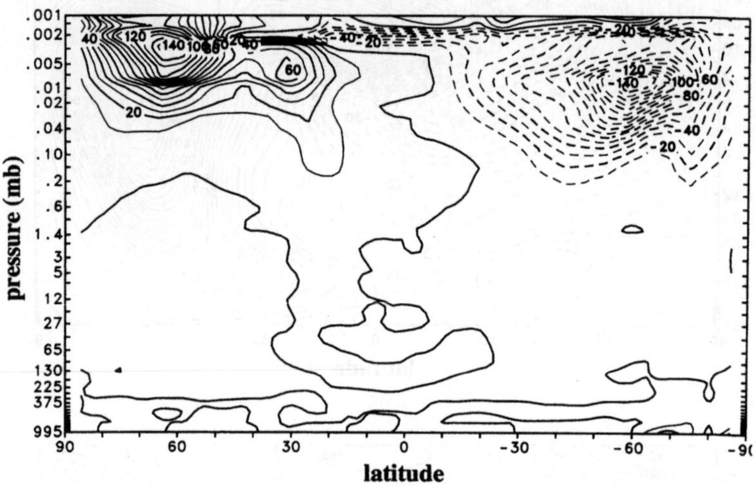


**Figure 9.** Total mean zonal drag force per unit mass for the DSPNZD simulation. Contour interval: 10 m/s/day

### 5. Summary

In this paper we have compared selected results of seasonal simulations made with the basic version of the Canadian Middle Atmosphere model with those obtained using an orographic gravity wave drag parameterization combined with two different versions of the Doppler spread parameterization proposed by Hines (1996a,b). In all cases the input spectrum for this parameterization is assumed to be continuous in the vertical wave number domain and azimuthally isotropic with a total kinetic energy content that is the same everywhere.

The basic version of the model uses only a combination of the orographic gravity wave parameterization and a Rayleigh friction sponge layer adjacent to its top. The simulated mean zonal winds and temperature fields obtained using this model have systematic errors in the middle atmosphere that are believed to be, in part at least, the result of neglecting part of the momentum flux deposition associated with dissipating gravity waves. These include an excessively high and cold stratopause and a cold lower stratosphere in polar regions in the winter season, most evident in the southern hemisphere during July. Another notable error is that the mean zonal wind field does not undergo a reversal in high latitudes just above the mesopause.

The versions of the Doppler spread parameterization used here are preliminary tests versions of what we expect will ultimately become a more comprehensive gravity wave drag parameterization based on the same general principles. The versions used here differ only in respect of the range of the initial vertical wave number spectrum. The first one includes all wave numbers between zero and an upper bounding cut-off wave number, while the second has a specified finite lower bound as well. These variations represent only two of a variety of possibilities that may be used for specification of the input spectrum. In principle this flexibility can be used to take account of the spatial variations in gravity wave sources and generation processes (for example as in Manzini et al, this volume).

Some improvements in the simulated mean zonal wind and temperature structure are obtained by simply implementing the first of the two DSP parameterizations without otherwise modifying in any way the other parameterizations, including the way in which the Rayleigh damping is applied in the sponge layer. These improvements include reversal of the winds above the mesopause, qualitatively in accord with observations, and warming of the stratosphere in the region of the south pole, leading to a more realistic location of the stratopause and a more realistic temperature structure in the entire stratosphere in that region.

However it is also found that most of the forcing due to the DSP occurs in the upper mesosphere, the strongest part of it being located within the sponge layer. This results in a complicated interplay between the DSP and the Rayleigh damping in the sponge layer that is of questionable desirability. The second version of the DSP was implemented in combination with a modified Rayleigh damping procedure which damps only departures of the local velocity field from the zonal mean. To date only relatively short simulations, for the period of April through July have been carried out using this version of the parameterization. The results which have been obtained indicate that additional improvements are produced when this version of the parameterization is used. In all about half of the high zonal mean wind bias apparent in July in the southern hemisphere is removed by implementing the second of the DSP parameterizations. However, what remains is still quite large. Some of it may be due to processes other than gravity wave drag that are missing or poorly represented in the model.

**Acknowledgements**

We thank Colin Hines for providing us with preliminary versions of his papers on the Doppler spread parameterization and for many helpful discussions during the course of this work. The computing resources needed to perform the simulations discussed in this paper were provided by the Atmospheric Environment Service of Canada on the NEC SX/3 super-computer of the Canadian Meteorological Centre.

## References

Allen SJ, Vincent RA (1995) Gravity wave activity in the lower atmosphere: Seasonal and latitudinal variations. J. Geophy. Res. 100:1327-1350

Beagley SJ, deGrandpre J, Koshyk JN, McFarlane NA, Shepherd TG (1996) The Canadian Middle Atmosphere Model: model description and climatology. Atmosphere - Ocean: submitted

Dewan EM, Good RE (1986) Saturation and the "universal" spectrum for vertical profiles of horizontal scalar winds in the atmosphere. J. Geophys. Res. 91:2742-2748

Fleming EL, Chandra S, Barnet JJ, Corney M (1990) Zonal Mean Temperature, Pressure,

Zonal Wind and Geopotential Height as Functions of Latitude. Adv. Space. Res. 10: (12)11-(12)59

Fritts DC, Lu W (1993) Spectral estimates of gravity wave energy and momentum fluxes, Part II: Parameterization of wave forcing and variability. J Atmos Sci 50:3695-3713

Hines CO (1996a) Doppler-spread parameterization of gravity-wave momentum deposition in the middle atmosphere, Part I: Basic formulation. J. Atmos Terr Phys (in press)

Hines CO (1996b) Doppler-spread parameterization of gravity-wave momentum deposition in the middle atmosphere, Part II: Broad and quasi monochromatic spectra and implementation. J Atmos Terr Phys (in press)

Holton JR (1983) The influence of gravity wave breaking on the general circulation of the middle atmosphere. J Atmos Sci 40:2497-2507

Garcia RR and Soloman S (1985) The effect of breaking waves on the dynamics and chemical composition of the mesosphere and lower thermosphere. J Geophys Res 90:3850-3868

Lindzen RS (1981) Turbulence and stress owing to gravity wave and tidal breakdown. J Geophys Res 86:9707-9714

McLandress C, McFarlane NA (1993) The effect of orographic gravity wave drag on quasi-stationary planetary wave dynamics in the northern hemisphere middle atmosphere. J Atmos Sci 50:1966-1990

McFarlane NA (1987) The effect of orographically excited gravity wave drag on the general circulation of the lower stratosphere and troposphere. J Atmos Sci 44:1775-1800

Medvedev AS, Klaassen GP (1995) Vertical evolution of gravity wave spectra and the parameterization of associated wave drag. J Geophys Res 100:25841-25853

Mengel JG, Mayr HG, Chan KL, Hines CO, Reddy CA, Arnold NF, Porter HS (1995) Equatorial oscillations in the middle atmosphere generated by small scale gravity waves. Geophys Res Lett 22:3027-3030

Palmer T, Shutts GJ, Swinbank R (1986) Alleviation of a systematic westerly bias in general circulation and numerical weather prediction models through an orographic gravity wave drag parameterization. Q J R Meteorol. Soc. 112:1001-1039

Shepherd TG, Semeniuk K, Koshyk JN (1996) Sponge-layer feedbacks in middle atmosphere models. J Geophys Res (in press)

# Middle Atmosphere Simulations with the ECHAM4 Model: Sensitivity to the Doppler Spread Gravity Wave Parameterization

E. Manzini
Max Planck Institute for Meteorology
Hamburg, Germany

N.A. McFarlane
Canadian Centre for Climate Modelling and Analysis
Victoria, B.C., Canada

C. McLandress
Institute for Space and Terrestrial Science
North York, ONT, Canada

## Abstract

The response of a general circulation model to the Hines Doppler spread parameterization of gravity wave momentum deposition is described. Two simulations were carried out using different specifications of the gravity wave rms wind speed at the tropopause. In the first simulation the rms wind speed is a function of the simulated precipitation. In the second a constant rms wind speed is used. For both simulations a realistic seasonal cycle of the zonal mean circulation is obtained. In particular, the simulated October circulation is considerably improved over previous results obtained using Rayleigh friction in the mesosphere. The major differences between the two simulations occur in the tropical middle atmosphere and in the austral autumn.

## 1. Introduction

Current general circulation models which simulate the middle atmosphere do not explicitly resolve small scale gravity waves. The absence of this small scale dynamical forcing is now generally recognized to be responsible for the extremely strong zonal winds that appear in the middle atmosphere of general circulation models, although to a different degree in different seasons and in different models (Hamilton et al.1995, Boville, 1995). Since long-term integrations at much finer horizontal resolutions are at present well beyond our computational capability, it is of primary interest to develop good parameterizations for the momentum deposition from the unresolved gravity waves.

NATO ASI Series, Vol. I 50
Gravity Wave Processes
Their Parameterization in Global Climate Models
Edited by Kevin Hamilton

The purpose of this paper is to report on the response of a general circulation model to the recently developed Doppler spread parameterization (referred to as DSP) of gravity wave momentum deposition (Hines, 1996a,b). In contrast to previous applications (Mengel et al. 1995, McFarlane et al. 1996, this volume) of the DSP which have assumed a uniform gravity wave source distribution, an attempt is made here to identify, and include in a simple manner, the seasonal and spacial variability of the sources associated with convective activity and frontal systems. The importance of latitudinal and seasonal variations of gravity wave sources is shown for instance by Allen and Vincent (1995). In the lower stratosphere over Australia they found that the gravity wave activity is larger at low latitudes from December to February and at high latitudes during the austral winter. They also suggested that this seasonal distribution of the gravity wave activity might be associated with convective activity at low latitudes and with the passage of cold frontal systems at high latitudes. Evidence of spatial patterns connected to tropical convection is also suggested by Wu and Water (1996, this volume) from MLS UARS temperature variances, which they attribute to gravity waves.

## 2. The model and the experiments

The model used here is the middle atmosphere version of the ECHAM4 spectral transform general circulation model with variable horizontal truncation and state-of-the-art physical parameterizations (Roeckner et al. 1992, 1996; Manzini and Bengtsson, 1996). In the standard version of the middle atmosphere ECHAM4 model (39 levels, top at 80 km), Rayleigh friction is used to dissipate upward propagating waves and also to act as a crude parameterization of the effects of gravity wave breaking on the zonal mean flow in the mesosphere. The Rayleigh friction coefficient increases upward from $(32\ \mathrm{d})^{-1}$ at 0.4 hPa to $(1\ \mathrm{d})^{-1}$ at the model top. This configuration of the middle atmosphere ECHAM4 model has been integrated for 10 years at T30 horizontal truncation.

In the current version of the model, the Rayleigh friction is replaced by the Doppler spread parameterization. In addition, an orographic gravity wave drag parameterization based on McFarlane (1987) is included. These parameterizations are applied independently. All the simulations performed employ monthly mean climatological sea surface temperature (Gates, 1992), a T30 horizontal truncation, and the diurnal cycle in the radiative transfer scheme.

Note that additional experiments using the DSP in an extended version of the middle atmosphere ECHAM4 model (top at 112 km) and also in the Canadian

Middle Atmosphere Model (top at 97 km, see McFarlane et al. 1996, this volume) have shown that numerical difficulties can appear when a sponge is not included at the model top. This occurs because of a positive feedback between the DSP and the diurnal solar tide (see also McLandress, 1996, this volume) and may be avoided by introducing a low wavenumber bound to the initial vertical wavenumber spectrum of the gravity waves. The fact that this problem does not appear in the current application of the DSP (without low wavenumber bound) is presumably explained by the fact that the upper boundary of the current ECHAM4 model is at 80 km, which is below the height where the diurnal tide maximizes (near 90 km).

### *a. The Doppler spread parameterization*

The formulation of the parameterization for a broad gravity wave spectrum follows Hines (1996a,b). The implementation of the DSP requires the specification of several parameters that describe the initial gravity wave spectrum, which may vary depending on the gravity wave sources considered. In the current application, the initial vertical wavenumber spectrum is assumed to follow a power law with a slope equal to one and the characteristic horizontal wavenumber is set to $(150 \text{ km})^{-1}$ (as in Mengel et al., 1995). The two simulations that will be discussed differ in the specification of the wave-induced rms wind speed. A variable distribution is prescribed in the first simulation, while a constant rms wind speed is used in the second. In both cases the initial spectrum is launched near the tropopause at about 15 km where it is assumed to be isotropic in the eight equally spaced azimuths considered (extrowaves and introwaves included). The specification of the fudge factors (parameters constrained by the theory) also follows from the recommendations of Hines (1996a). Note that the contribution to the vertical diffusion that arises from the gravity wave momentum deposition has not been included and that the momentum flux is set to zero at the upper boundary of the model.

The use of a variable gravity wave rms wind speed is a first crude attempt to take into account gravity wave sources which are associated with convective activity and frontal systems. For simplicity, the simulated precipitation is assumed to be indicative of the horizontal distribution and intensity of both these phenomena and therefore is used to construct the gravity wave rms wind field. This approach is consistent with the results of Allen and Vincent (1995). The spatial patterns of the gravity wave activity that they found over Australia clearly match those of tropical convection and mid-latitude storm tracks. In our simulation, the resulting rms wind speed consists of both a background term (constant in space and time) of $1.75 \text{ ms}^{-1}$ and a variable term that depends linearly on the total precipitation rate.

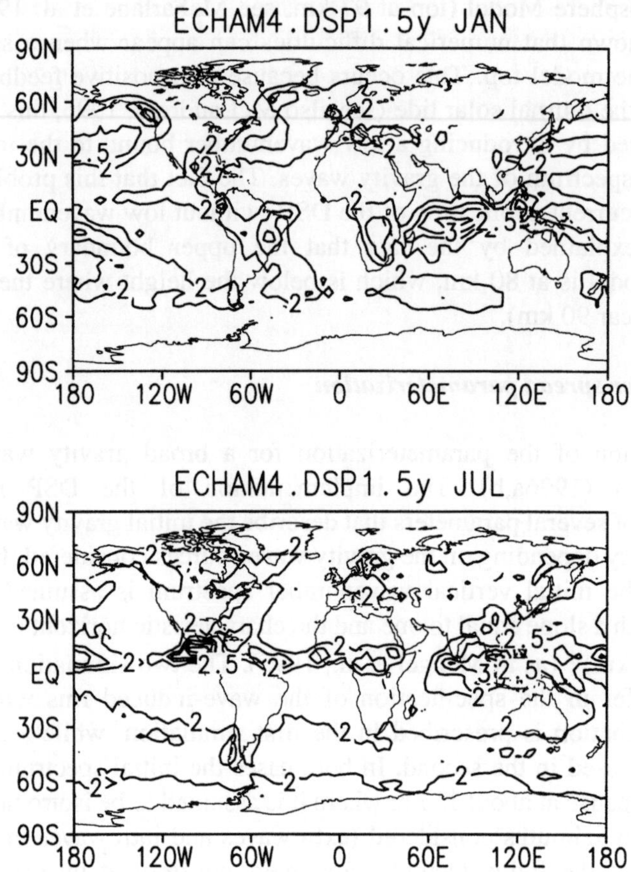


**Figure 1.** Gravity wave rms wind speed (DSP input parameter) at the launching height from the DSP1 simulation, 5 year average: January (upper panel) and July (lower panel). Contour: 0.5 $ms^{-1}$.

This precipitation dependent term is non-zero only for instantaneous total precipitation values larger than 5 $mmd^{-1}$. The total rms wind speed is also limited to a maximum of 6.5 $ms^{-1}$. These rms values are therefore comparable to the estimated rms wind speeds of 1.5 $ms^{-1}$ at high latitudes and 4 $ms^{-1}$ in the tropics from Allen and Vincent (1995), as well as to aircraft observations of gravity wave variances at the tropopause (Fritts and Nastrom, 1992).

A 5 year simulation (denoted DSP1) has been recently completed using the variable rms wind speed. In order to assess the sensitivity to the precipitation enhanced rms term, a 1-year simulation with the total rms wind speed fixed at the constant background value has been carried out (this experiment is denoted DSP2).

The 5-year monthly average rms wind speed of the gravity waves at the launching height obtained by coupling it with the precipitation (DSP1 simulation) is at most 3-4 $ms^{-1}$ over regions of high precipitation. Figure 1 shows that the regions of large rms wind speed, found in the intertropical convergence zone, are characterized by a seasonal migration from south (in January) to north (in July) of the Equator that closely parallel the precipitation field (not shown). Figure 1 also shows that other regions of enhanced rms wind speed are connected with the midlatitude storm tracks, for instance over the northwestern Atlantic and North Pacific in January, and in the $30^0$-$60^0$S latitude belt in July. Note that the tropical enhancement of the rms wind speed is associated with the convective part of the precipitation, while the mid-latitude enhancements correspond to the large scale precipitation. Although it of interest to couple parameterizations in a general circulation model in order to achieve a fully interactive system, some difficulties may be encountered in doing so. Systematic errors in the precipitation field may in fact misrepresent the distribution of the gravity wave sources. This is seen, for instance, over the Indian Ocean in January where the tropical precipitation (and hence the rms wind speed) is maximum in contrast to observations which show a maximum over Indonesia. However, the seasonal cycle of the precipitation is quite realistic in the ECHAM4 model and so justifies our approach of matching the rms gravity wave winds to it.

## 3. Zonal mean circulation

The dominant features of the January zonal mean zonal wind of the DSP1 experiment (Fig.2, upper left panel) are the winter polar jet in the lower stratosphere, the subtropical eastward jet in the mesosphere and the summer easterlies that increase poleward with height. The presence of strong westerlies in the subtropical mesosphere as well as the structure of the summer easterlies are realistic features (Fleming et al. 1990). Relative to Fleming et al., the strength of the winter polar jet is somewhat overestimated in the upper stratosphere. Note however that the Fleming et al. reference atmosphere is based only on few years of data in the stratosphere, therefore may not be an indicative estimate of the long term mean state, given the high variability of the Northern Hemisphere winter (Randel, 1992). The mesospheric summer easterlies are too weak, possibly because there is an excess of gravity wave drag at the model top.

Satisfactory aspects of the January zonal mean temperature (Fig.2 lower left)

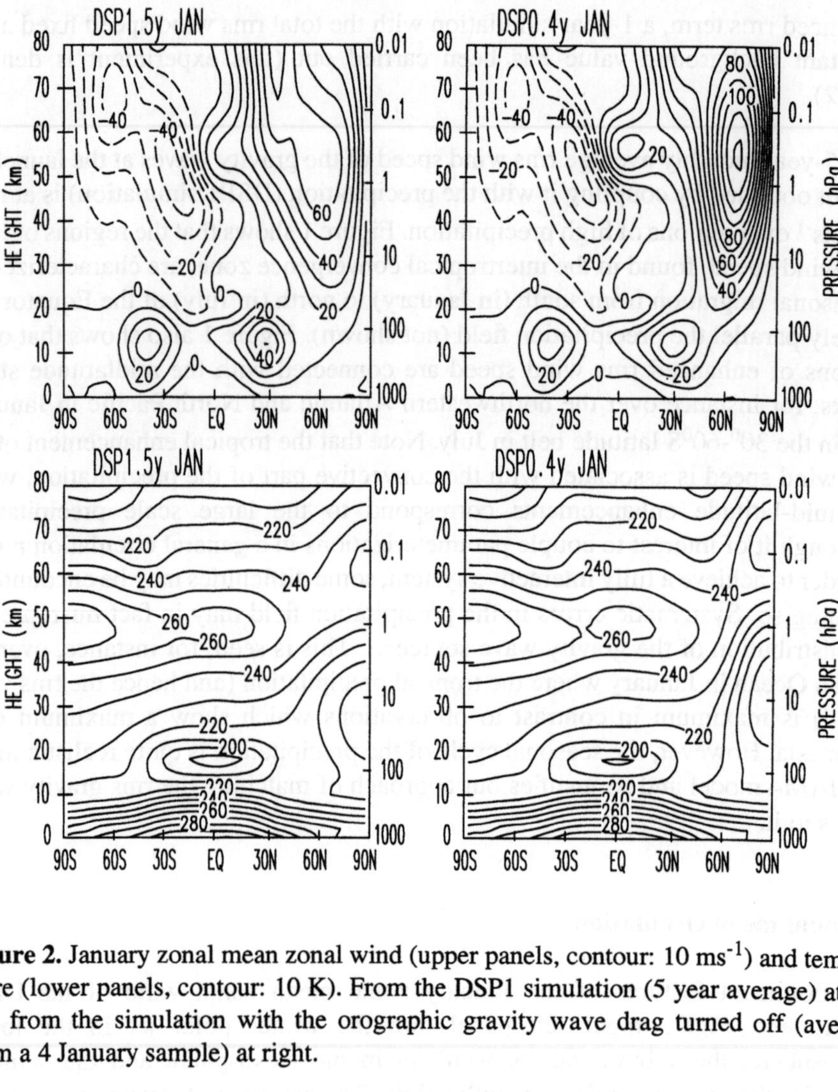


**Figure 2.** January zonal mean zonal wind (upper panels, contour: 10 $ms^{-1}$) and temperature (lower panels, contour: 10 K). From the DSP1 simulation (5 year average) at left and from the simulation with the orographic gravity wave drag turned off (average from a 4 January sample) at right.

include a warm stratopause located around 50 km in the summer hemisphere and at a slightly higher elevation in the Northern Hemisphere, a rapid temperature decrease with height above the summer stratopause and the consequent reversal of the temperature gradient at the model top (80 km), with colder temperatures at the summer pole in agreement with observations. The South Pole mesospheric temperature minimum and stratopause maximum are, however, underestimated in comparison to Fleming et al. (1990). A higher model top may therefore be required to improve the simulation of the mesospheric zonal mean temperature.

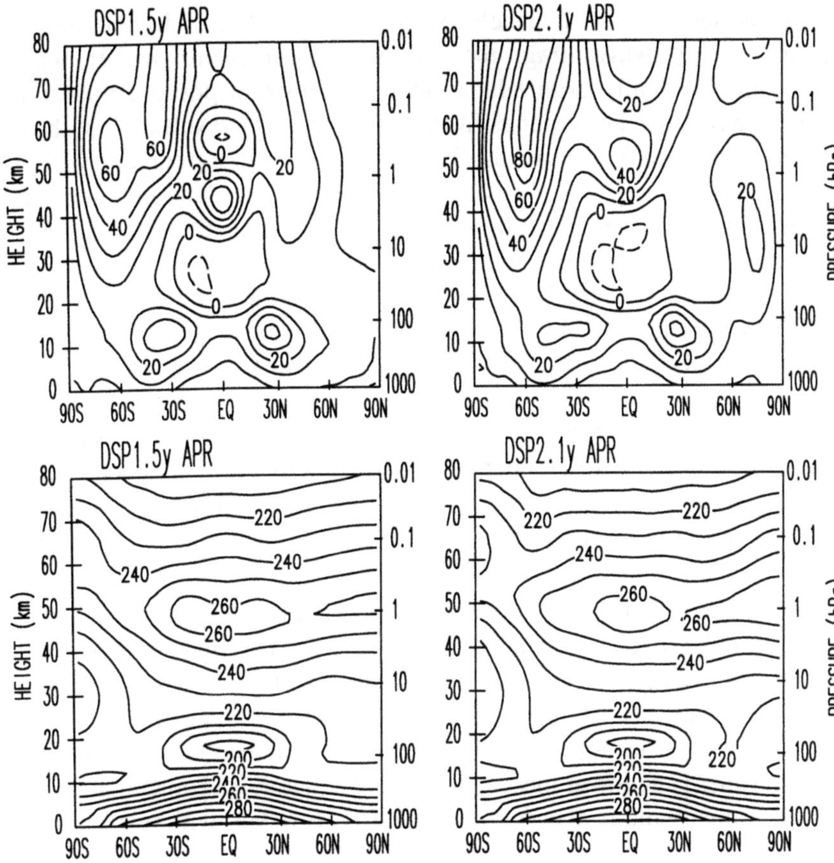


**Figure 3.** April zonal mean zonal wind (upper panels, contour: 10 $ms^{-1}$) and temperature (lower panels, contour: 10 K). From the DSP1 simulation (5 year average) at left and from the DSP2 simulation at right.

As found in other general circulation models, the excessively strong winter polar jet is associated with a cold temperature bias (of ~10 K) in the lower stratosphere.

The January zonal mean circulation of the 1-year DSP2 simulation (not shown) is quite similar to that for DSP1, with the exception of slightly stronger summer easterlies in the upper mesosphere. The agreement in the Northern Hemisphere may be fortuitous, however, because of the large interannual variability that occurs there.

The impact of the orographic gravity wave drag on the Northern Hemisphere winter circulation is seen in the 2 panels on the right of Fig.2. In the absence of orographic gravity wave drag, the Northern Hemisphere winter westerlies increase to unrealistically large values and the lower stratosphere temperature bias increases in magnitude and extends further upwards. As a result of this, the winter stratopause is also shifted upwards. These results are in agreement with earlier general circulation model sensitivity experiments that demonstrated that the orographic gravity wave drag plays a dominant role in the simulation of the Northern Hemisphere winter circulation of the middle atmosphere (Rind et al. 1988, and Boville, 1991, 1995). As expected, the summer (southern) hemisphere is not significantly affected by the removal of the orographic gravity wave drag due to the reversal of the zonal mean flow in the lower stratosphere.

In April (Fig.3, upper panels) the most interesting result is the sensitivity of the zonal wind in the Southern Hemisphere to the precipitation-enhanced rms term. Differences of up to 20 $ms^{-1}$ between the two integrations are found in the mesosphere, which are much larger than the typical interannual variability (few $ms^{-1}$ in DSP1) for this season. The weaker Southern Hemisphere westerlies in DSP1 are in much better agreement with observations (Fleming et al. 1990). Another difference also associated with the gravity wave rms wind distribution is the 10 km downward shift in DSP1 with respect to DSP2 of the location of an equatorial eastward jet. It is noted that during the 1-year DSP2 simulation, easterlies and westerlies are found to oscillate on a semiannual time scale in the upper stratosphere and lower mesosphere at the equator. A semiannual oscillation in the mean zonal wind is also found for DSP1. However, with respect to DSP2, the DSP1 oscillation is characterized by a more pronounced downward phase propagation of the westward phase and a 10 km downward shift of the first semiannual cycle. Contrary to the DSP2 results, the features found for DSP1 are not in agreement with the observed behavior of the semiannual oscillation at the stratopause, suggesting that the local forcing of the gravity wave rms winds may be excessive and possibly also that the azimuthal isotropy assumption is too simplistic.

The zonal mean temperature in April (Fig.3, lower panels) is similar in the two experiments, except for a slightly warmer stratopause in the DSP1 Southern Hemisphere, consistent with the weaker westerlies in that region.

As was found in January, the zonal mean circulation in July is not much affected by the precipitation enhancement of the gravity wave rms wind distribution and consequently only results from the DSP1 experiment are shown (Fig.4). In the Southern Hemisphere, the July zonal winds are somewhat too strong and the mesospheric jet is located too poleward, relative to the Fleming et al. (1990) reference atmosphere. The zonal mean temperature shows that the stratopause is

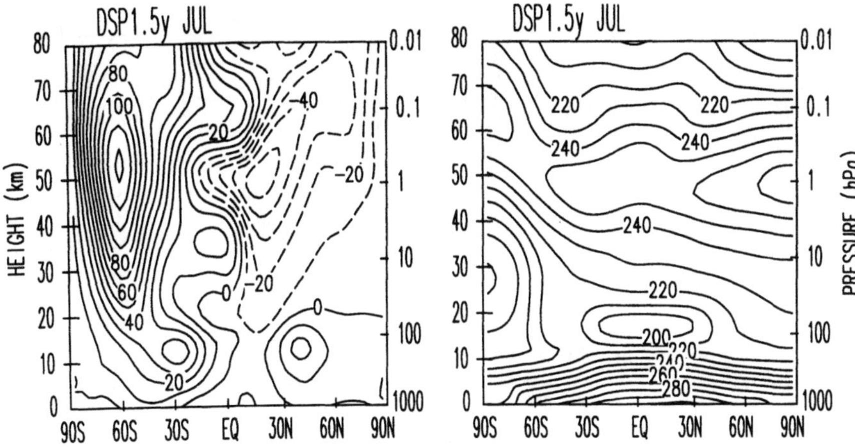


**Figure 4.** July zonal mean zonal wind (left panel, contour: 10 $ms^{-1}$) and temperature (right panel, contour: 10 K) from the DSP1 simulation (5 year average).

higher and colder than observed. The DSP1 results for the Northern Hemisphere summer are similar to those for the Southern Hemisphere discussed previously.

In October the Southern Hemisphere middle atmosphere circulation is characterized by a significant amount of interannual variability (Randel, 1988, 1992) that is associated with the upward propagation of planetary wave disturbances and the final breakdown of the polar vortex. Although the long term average shows a moderately weak eastward jet in the Southern Hemisphere lower stratosphere, individual years may therefore exhibit somewhat stronger or weaker zonal mean circulations. The October 5-year average zonal mean zonal wind for the DSP1 simulation (Fig.5, upper left panel) consists of a closed jet in the Southern Hemisphere middle atmosphere. While the averaged DSP1 October Southern Hemisphere westerlies are 10-20 $ms^{-1}$ too strong (in comparison to Fleming et al, 1990 and also Randel, 1992), on some occasions they may be quite weak and easterlies may appear in the upper mesosphere. A particularly weak circulation in the Southern Hemisphere was also found for the single October from the DSP2 simulation (not shown). Figure 5 (bottom left) also shows that in the Southern Hemisphere the stratopause in the DSP1 case descends from July to October and that a temperature minimum is present in the lower stratosphere, in qualitative agreement with observations.

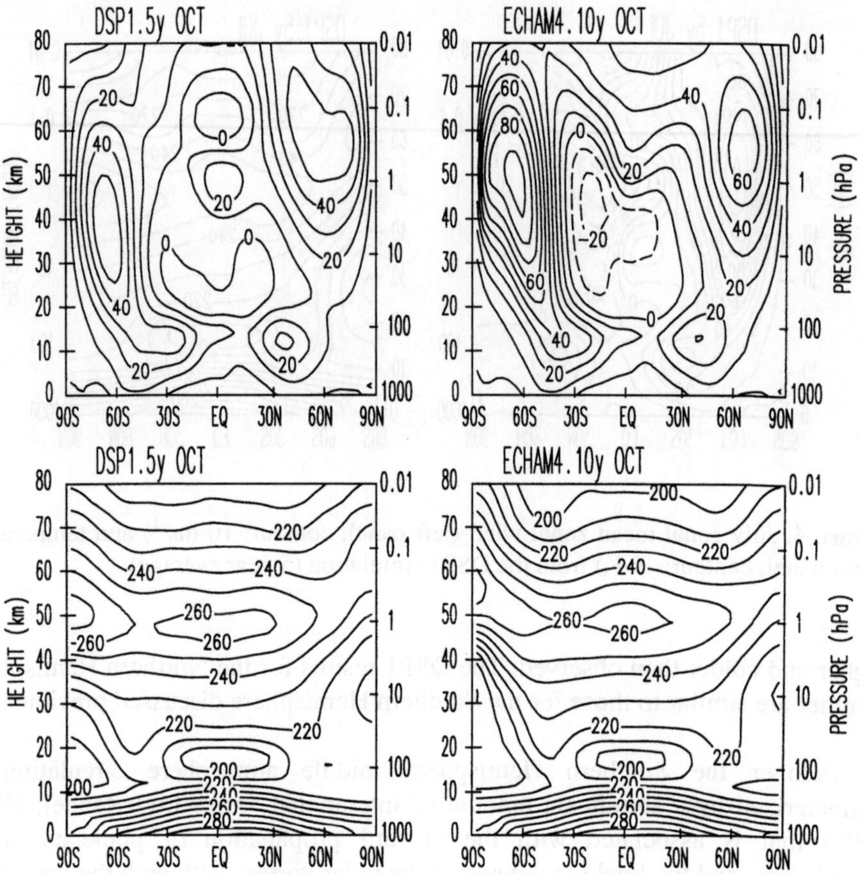


**Figure 5.** October zonal mean zonal wind (upper panels, contour: 10 $ms^{-1}$) and temperature (lower panels, contour: 10 K). From the DSP1 simulation (5 year average) at left and from the ECHAM4 simulation without DSP and orographic gravity wave drag but with Rayleigh friction in the mesosphere (10 year average).

It is interesting to note that both the DSP1 and DSP2 simulations of the Southern Hemisphere October westerlies show considerable improvement over previous middle atmosphere simulations, some of which used Rayleigh friction and/or lower horizontal resolution (e.g., Boville, 1995; Hamilton, 1995; Manzini and Bengtsson, 1996). As an example, the 10-year average October from the standard ECHAM4/T30L39 model integration with Rayleigh friction (with neither the DSP nor the orographic gravity wave drag parameterization) is shown on the right of Fig.5. In this case, westerlies are as strong as 90 $ms^{-1}$ at the stratopause in the

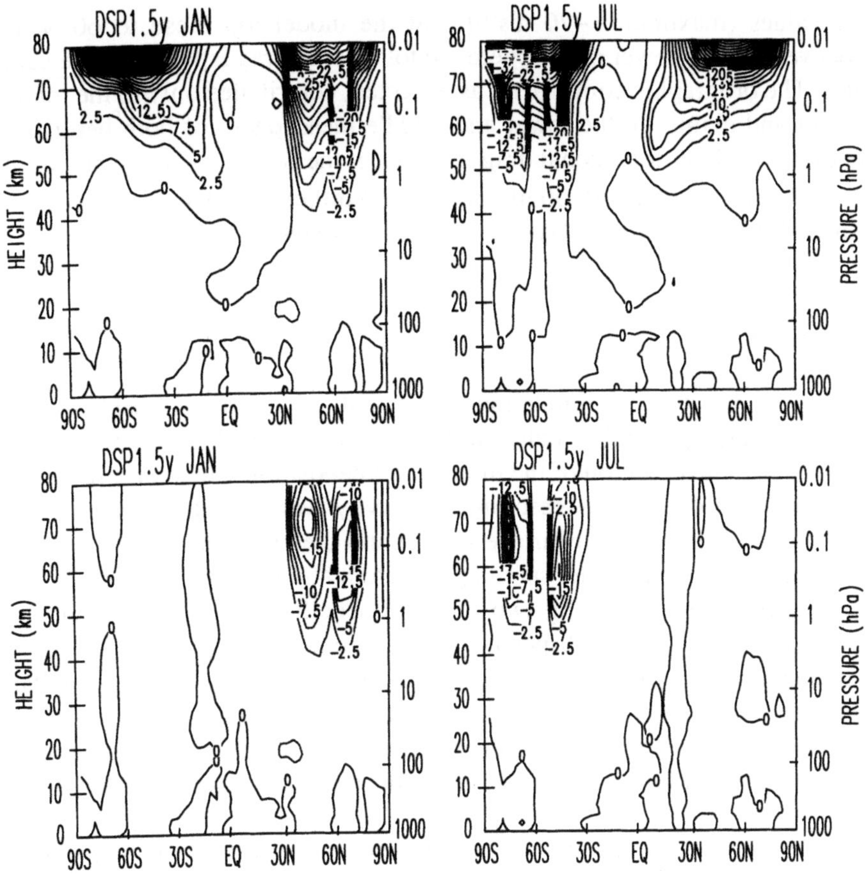


**Figure 6.** Total (upper panels) and orographic (lower panels) zonal mean zonal gravity wave drag. January (at left) and July (at right) 5 year average from the DSP1 simulation. Contour: 2.5 $ms^{-1}d^{-1}$.

extratropical Southern Hemisphere and temperatures below 200 K extend upward to almost 40 km at the South Pole. Although the orographic gravity wave drag is also absent in the ECHAM4 simulation, its role should be minor in the Southern Hemisphere, as indicated by the Boville (1995) results.

The 5 year average total zonal mean zonal gravity wave drag from the DSP1 simulation is shown in the upper panels of Fig.6. The corresponding orographic gravity wave drag components are shown below. The Northern and Southern Hemisphere distributions of the total zonal drag during summer are similar, with larger values (maximum: ~70 $ms^{-1}d^{-1}$) at the model top close to $60^0$N and S respectively. In the lower mesosphere the total zonal drag is larger in the summer tropics. The orographic component, as expected, contributes only to the total drag in winter and mainly in the lower and middle mesosphere of both hemispheres. Note that the winter total zonal drag is larger in the Southern Hemisphere (about -75 $ms^{-1}d^{-1}$ instead of about -50 $ms^{-1}d^{-1}$ in the Northern Hemisphere), presumably because of the smaller role played by planetary waves.

## 4. Conclusions

In this paper results from simulations made with the middle atmosphere version of the ECHAM4 general circulation model are presented. This model version employs a parameterization of the momentum deposition due to orographic gravity waves and a continuous spectrum of gravity waves. The orographic part is parameterized following McFarlane (1987) and the continuous spectrum part is represented using the Doppler spread parameterization of Hines (1996a,b). Simulations were carried out using two different specifications of the gravity wave induced rms horizontal wind speed, an input parameter of the DSP. In the DSP1 simulation the rms wind is assumed to be a function of the model precipitation, which is a simple way of taking into account gravity wave sources associated with convective activity and frontal systems. The DSP2 simulation employed instead a constant rms wind speed. The results are summarized as follows:

(1) In both simulations, the gravity wave drag parameterization alleviates, but does not remove, the cold temperature bias and associated zonal wind bias that are typically present in the winter season of both hemispheres. In January the orographic gravity wave drag plays a dominant role in the Northern Hemisphere.

(2) The structure of the westward summer jet was found to be realistic in both hemispheres and in both simulations, but its magnitude was somewhat too weak in the mesosphere, presumably because the drag from the DSP is too large. The summer zonal mean circulation and the upper mesosphere temperature may be improved by raising the top of the general circulation model to about 120 km, in order to allow for a better resolution of the gravity wave drag distribution at the mesopause.

(3) In October a major improvement of the simulation of the polar vortex breakdown was found for both simulations with respect to previous results that used a Rayleigh friction in the mesosphere. Although the simulations are still deficient in some respects, the DSP appears to be more effective in decelerating

the zonal mean winds. This behavior may be expected since the dissipation of the gravity wave spectrum in the DSP depends on vertical variations of the large scale winds. For a westward vertical wind shear in the upper mesosphere, westward propagating gravity waves are strongly dissipated, which accelerates the downward propagation of the summer easterlies. This mechanism is of course not possible using Rayleigh friction, which is linearly proportional to the zonal wind and therefore effective only in regions where the winds are strong.

(4) The autumn zonal mean circulation of the Southern Hemisphere (April) appeared to be significantly affected by the precipitation enhanced rms wind speed. The low sensitivity to the gravity wave source distribution that occurred in the other seasons may be caused by the relatively high background rms wind speed that was required to prevent the development of excessively strong zonal winds in the Southern Hemisphere at polar latitudes.

(5) A realistic quasi-biennial oscillation (QBO) in the lower stratosphere zonal mean wind was not evident during the 5-year DSP1 simulation. This result is in contrast with the findings of Mengel et al (1995). Their simulated QBO, however, was shown to be very sensitive to the magnitude of the vertical diffusion associated with the DSP. We have not included this effect in our experiments.

(6) The larger values of the gravity wave rms wind speed in the tropics over regions of convective activity (DSP1 experiment) may be a realistic feature, although it was detrimental to the simulation of the tropical middle atmosphere semiannual oscillation. It is suggested that this behavior at low latitudes may be improved by including space and time dependences of other input parameters of the DSP.


## Acknowledgments

The authors would like to thank Lennart Bengtsson, Colin O. Hines and Erich Roeckner for helpful discussions, Ulrich Schlese and colleges at DKRZ and MPI for technical assistance.


## References


Allen SJ, Vincent RA (1995) Gravity wave activity in the lower atmosphere: Seasonal and latitudinal variations J Geophys Res 100: 1327-1350

Boville BA (1991) Sensitivity of simulated climate to model resolution. J Clim 4: 469-485

Boville BA (1995) Middle atmosphere version of the CCM2 (MACCM2): Annual cycle and interannual variability J Geophys Res 100: 9017-9039.

Fleming EL, Chandra S, Barnett JJ, Corney M 1990 Zonal mean temperature, pressure, zonal wind and geopotential height as functions of latitude Adv Space Res 10: 1211-1259

Fritts DC, Nastrom GD (1992) Sources of mesoscale variability of gravity waves. Part II: Frontal, convective and jet stream excitation. J Atmos Sci 49: 111-127

Gates WL (1992) AMIP: The atmospheric model intercomparison project. Bull Am Met Soc 73: 1962-1970

Hamilton K (1995) Comprehensive simulation of the middle atmospheric climate: some recent results. Climate Dynamics 11: 223-241

Hamilton K, Wilson RJ, Mahlman JD, Umscheid LJ (1995) Climatology of the SKYHI troposphere-stratosphere-mesosphere general circulation model. J Atmos Sci 52: 5-43

Hines CO (1996a) Doppler Spread parameterization of gravity wave momentum deposition in the middle atmosphere. Part 1: Basic formulation. J Atmos Terr Phys (in press)

Hines CO (1996b) Doppler Spread parameterization of gravity wave momentum deposition in the middle atmosphere. Part 2: Broad and quasi monochromatic spectra and implementation. J Atmos Terr Phys (in press)

Manzini E, Bengtsson L (1996) Stratospheric climate and variability from a general circulation model and observations. Climate Dynamics 12: 615-639

McFarlane NA (1987) The effect of orografically exited gravity wave drag on the general circulation of the lower stratosphere and troposphere. J Atmos Sci 44: 1775-1800

McLandress C (1996) Sensitivity studies using the Hines and Fritts gravity wave drag parameterizations (this volume)

McFarlane NA, McLandress C, Beagly S (1996) Seasonal simulations with the Canadian Middle Atmosphere Model: Sensitivity to a combination of orographic and Doppler spreading gravity wave parameterizations (this volume)

Mengel JG, Mayr HG, Chan KL, Hines CO, Reddy CA, Arnold NF, Porter HS (1995) Equatorial oscillations in the middle atmosphere generated by small scales gravity waves. Geophys Res Let 22: 3027-3030

Randel WJ (1988) The seasonal evolution of planetary waves in the southern hemisphere and troposphere. Quart J Roy Meteor Soc 114: 1385-1409

Randel WJ (1992) Global atmospheric circulation statistics, 1000-1 mb. NCAR Technical Note TN-366+STR

Rind D, Suozzo R, Balachandran NK (1988) The GISS global climate middle atmosphere model. Part II: Model Variability due to interactions between planetary waves, the mean circulations, and gravity wave drag. J Atmos Sci 45: 371-386

Roeckner E, Arpe K, Bengtsson L, Brinkop S, Dümenil L, Esch M, Kirk E, Lunkeit F, Ponater M, Rockel B, Sausen R, Schlese U, Schubert S, Windelband M (1992) Simulation of the present-day climate with the ECHAM model: Impact of model physics and resolution. MPI Report: 93

Roeckner E, Oberhuber JM, Bacher A, Christoph M, Kirchner I (1996) ENSO variability and atmospheric response in a global coupled atmosphere-ocean general circulation model. Climate Dynamics (accepted)

Wu DL, Waters JW (1996) Observations of Gravity waves with the UARS Microwave Limb Sounder (this volume)

# The Mesosphere in the Extended UGAMP GCM

W. A. Norton
Atmospheric, Oceanic and Planetary Physics,
Department of Physics, Clarendon Laboratory, Parks Rd,
University of Oxford, Oxford OX1 3PU, U.K.
email: wan@atm.ox.ac.uk

J. Thuburn
Centre for Global Atmospheric Modelling,
Department of Meteorology, University of Reading,
2 Earley Gate, Whiteknights, Reading RG6 6AU, U.K.
email: swsthubn@met.reading.ac.uk

## Abstract

The simulation of the mesosphere is examined in a middle atmosphere GCM. Features examined include the zonal mean wind structure, the distribution of gravity wave drag, the meridional circulation, the evolution of the summer mesopause, the two-day wave, the four-day wave, and the diurnal tides.

## 1. Introduction

The dynamics of the mesosphere is significantly different from that of the stratosphere. With pressures some five orders of magnitude smaller in the mesosphere than in the troposphere, vertically propagating disturbances such as gravity waves and tides, which have small amplitude below 50 km, grow to very large amplitude in the mesosphere; and it has been recognized for many years (e.g., Houghton 1978) that momentum deposition by breaking gravity waves plays a key role in the mesospheric circulation.

However the role of breaking gravity waves is not the only difference from the stratosphere. Planetary waves in the stratosphere are caused mainly by upward propagation from the troposphere. However, in the mesosphere there is also the possibility that the local dynamics can generate planetary waves through instability of the mean flow (e.g., Plumb 1983). Another difference from the stratosphere is the reduced Brunt-Väissällä frequency. One consequence of this is that the flow fields induced by a potential vorticity (PV) anomaly of a given horizontal and vertical scale in the mesosphere will have a deeper vertical penetration than those induced by a PV anomaly of the same scale in the stratosphere. The dynamics of the mesosphere may be further complicated by the possibility of nonlinear interactions among different large-amplitude planetary scale waves, and between planetary scale waves and breaking gravity waves. One further difference from the stratosphere is that the radiative timescale in the mesosphere is much shorter, typically only a few days.

The mesosphere is of course of interest in its own right, but the mesosphere also influences the stratosphere. For instance, the propagation of planetary waves in the stratosphere is affected by the flow in the mesosphere. Furthermore the meridional circulation driven by breaking gravity waves in the mesosphere influences temperatures in the lower stratosphere (Garcia and Boville 1994) and transports trace gases from the mesosphere down into the middle stratosphere (Fisher *et al.* 1993).

NATO ASI Series, Vol. I 50
Gravity Wave Processes
Their Parameterization in Global Climate Models
Edited by Kevin Hamilton

There have been comprehensive observations of the upper mesosphere (above 70 km) over the last 30 years from rocketsonde and particularly radar. Many of the phenomena in this region of the mesosphere have been extensively studied from single station measurements. There have also been global measurements of the mesosphere from satellites, and in particular recently from the Upper Atmosphere Research Satellite (UARS).

The mesosphere has been very little studied in global circulation models (GCMs). Our attention was drawn to the mesosphere in the Extended UGAMP (U.K. Universities Global Atmospheric Modelling Programme) GCM when we discovered that the model contained a strong signal of the two-day wave (Norton and Thuburn 1996, hereafter NT96). This was studied in a low resolution (spectral truncation T21) version of the model. We have recently completed a 1200 day integration of the model at higher horizontal resolution (T42, approximately a gridspacing of 3°). In this paper we examine the representation of various phenomena in the mesosphere of this model integration.

The outline of the paper is as follows: in section 2 we describe the Extended UGAMP GCM, section 3 presents the January and July monthly mean circulation in the model, section 4 presents the evolution of the northern hemisphere summer mesopause, section 5 presents the structure and evolution of the two-day wave, section 6 presents the structure of the four-day wave, section 7 presents the structure of the tides, and section 8 offers some conclusions.

## 2. Model description

The Extended UGAMP Global Circulation Model (EUGCM) was developed from the cycle 27 forecast model of the European Centre for Medium Range Weather Forecasts (ECMWF). An early version of the EUGCM is described by Gray *et al.* (1993). A recent tropospheric version of the UGAMP GCM is described by Slingo *et al.* (1994). The EUGCM integrates the hydrostatic primitive equations over the globe from the ground to about 95–100 km (the exact height of the top model pressure level depends on the temperatures below). The model is 'spectral'; the main prognostic variables, vorticity, divergence, temperature, moisture and logarithm of surface pressure, are represented as truncated series of spherical harmonics at 47 different levels. The integration described here uses T42 horizontal resolution. The vertical hybrid sigma-pressure coordinate and vertical difference schemes are those described by Simmons and Burridge (1981), except for the vertical advection terms, which are calculated using a shape preserving scheme based on fourth order differences (Thuburn and McIntyre 1996). The vertical resolution in the middle atmosphere ranges from about 2 km in the lower stratosphere to about 3 km at the mesopause.

A number of processes that are not explicitly resolved are parameterized in the model. The schemes for large scale condensation, boundary layer processes, and surface processes are as described by Tiedtke *et al.* (1988), except that the boundary layer scheme is switched off above about 600 hPa. Deep and shallow convection are represented by the adjustment scheme of Betts and Miller (1993).

Two separate radiation schemes are used. In the troposphere and lower stratosphere up to about 74 hPa Morcrette's (1990) scheme is used. Above about 28 hPa the scheme of Shine (1987) and Shine and Rickaby (1989) is used. At intermediate levels the heating rate is given by a weighted mean of the heating rates from the separate schemes.

The representation of the effects of unresolved gravity waves has a crucial impact on the ability of a GCM to simulate the middle atmosphere. The EUGCM originally used the Palmer *et al.* (1986) gravity wave scheme. This scheme uses linear theory to estimate the gravity wave momentum flux at low levels due to flow over orography, and WKB theory and a simple saturation criterion to calculate the vertical profile of momentum flux due to wave propagation and breaking. Two particular shortcomings were noticeable in middle atmosphere simulations using this scheme. First, the orographically generated gravity waves represented by the scheme are stationary, so they are filtered by near zero mean winds in the tropical and summer hemisphere lower stratosphere and prevented from reaching higher altitudes. This led to an unrealistic summer and tropical mesospheric circulation. Second, very strong, localized occurrences of gravity wave drag in the winter stratospheric vortex led to unrealistic PV dipole features and eventually to disruption of the vortex.

To overcome the first shortcoming the gravity wave scheme was extended to include additional gravity waves with non-zero horizontal phase speeds. Observations of temperature and wind fluctuations in the middle atmosphere, using, e.g., rockets, radars and lidars, indicate the presence of gravity waves with a spectrum of horizontal phase speeds up to a few tens of metres per second. Some of these waves reach the summer mesosphere. These waves could be generated by a variety of mechanisms, including convection, envelope radiation from shear instabilities, adjustment of the large scale flow towards balance, and interactions among gravity waves. In the extended gravity wave scheme the additional waves propagate and break following essentially the same criteria as in the Palmer *et al.* scheme (but see below). However, in the absence of reliable quantitative theories for the generation of these waves, their phase speeds, amplitudes and source altitudes are simply specified in the GCM. Each wave in the scheme propagates as an individual monochromatic wave; there is no direct interaction between them. The EUGCM gives a more realistic simulation of the mesosphere even with just one eastward travelling gravity wave in each column in addition to the orographically generated wave (Jackson 1993). In the experiment described here 14 waves launched from 100 hPa are used in addition to the orographically generated wave. Their characteristics are given in the following table.

| number of waves | phase speed ($ms^{-1}$) | directions | amplitude ($Nm^{-2}$) |
|---|---|---|---|
| 2 | 0 | NS, EW | $10 \times 10^{-4}$ |
| 4 | 10 | N, E, S, W | $1.84 \times 10^{-4}$ |
| 4 | 20 | N, E, S, W | $0.68 \times 10^{-4}$ |
| 4 | 30 | N, E, S, W | $0.25 \times 10^{-4}$ |

To overcome the second shortcoming while still retaining most of the benefits of the orographic gravity wave drag scheme in the troposphere, the orographic gravity wave source strength is calculated as in the Palmer *et al.* scheme, but the drag due to this wave is exerted uniformly between the ground and the level $\sigma = 0.1$ (where $\sigma = p/p_*$ and $p_*$ is surface pressure) (e.g., Stephenson 1994).

Some initial experiments with the gravity wave scheme just described were encouraging, but showed a tendency for extremely strong shear zones to form, particularly on the upper equatorward flank of the summer easterly jet. The formation and maintenance of these shear zones involves a positive feedback: breaking gravity waves help to create the shear zones, while the change in Doppler-shifted phase speed of the gravity waves as they cross the shear zones enhances the breaking there. This kind of positive feedback probably occurs in the real atmosphere, and indeed it appears to be crucial in obtaining a realistic two-day wave in the GCM. However, the positive feedback is exaggerated in the GCM by the use of a discrete spectrum of non-interacting monochromatic gravity waves. These results led us to retune one of the parameters in the gravity wave scheme, reducing by a factor of 10 the critical value of the square of the wave amplitude above which waves break (i.e., the 'efficiency factor' is set to 0.1). This retuning causes the gravity waves to break less violently and a few kilometers lower in altitude. As justification for this otherwise arbitrary retuning, it could be argued that it makes some allowance for intermittency — the peak gravity wave momentum flux, which determines the breaking level, may be considerably greater than the average momentum flux over a model time step — and for constructive interference of the individual monochromatic waves. With this retuning of the gravity wave breaking criterion the strength of the tight shear zones in the GCM is more realistic.

## 3. Zonal mean circulation

Figures 1a,b show the monthly mean zonal winds for one January and one July from the EUGCM. These months are similar to other Januarys and Julys from the EUGCM except that the northern hemisphere stratosphere in January has large interannual variability because of stratospheric warmings. The winter westerly jets for January and July are both comparable in strength to those observed, and, like observations (e.g., CIRA 1986), slope equatorward with height with almost a separate westerly jet in the subtropical lower mesosphere. If we instead run the model with Rayleigh friction as the gravity wave drag scheme, we obtain winter jets that are vertically aligned. The sloping of the jets appears to be associated with the distribution of gravity wave drag having a somewhat localized maximum (see Figures 2a,b) that is not at the position of strongest winds.

The summer easterly jet has a realistic strength (maximum 80 $ms^{-1}$) in July, but is too strong (maximum over 100 $ms^{-1}$) in January. This is a consequence of the too strong westerly jet in the southern hemisphere lower stratosphere in January, which filters out many of the eastward traveling gravity waves that would otherwise break in the mesosphere and decelerate the easterly jet. This problem is worse in December when there is an even stronger easterly jet in the mesosphere and stronger lower stratospheric westerlies. This in turn may be related to the late breakdown of the southern hemisphere winter westerlies in the EUGCM, a problem shared by many other middle atmosphere GCMs.

The shaded regions in Figures 1a,b are where the meridional gradient of the zonal mean quasigeostrophic potential vorticity ($\bar{q}_y$) is strongly negative (less than $-1 \times 10^{-11}s^{-1}m^{-1}$). A region of negative $\bar{q}_y$ indicates that an instability may occur (e.g., Andrews *et al.* 1987). The strong horizontal and vertical curvature (and shear) of the summer and winter jets has produced regions of strong negative $\bar{q}_y$ in the upper part of the easterly jet, and above and poleward of the core in the westerly jets, particularly in the southern hemisphere. In section 4 we examine instability in the summer hemisphere associated with the two-day wave, and in section 5 we examine the possibility of instability in the winter hemisphere associated with the four-day wave.

Figures 1c,d show the corresponding monthly mean temperature fields. A number of extrema stand out for both January and July: a cold tropical tropopause and cold winter polar lower stratosphere; a warm summer stratopause and warm separated winter stratopause; and a very cold summer mesopause. We examine in more detail the summer mesopause in the next section.

Figures 2a,b show the corresponding monthly mean zonal gravity wave drag (note that now only the mesosphere is shown). In both summer hemispheres the maximum eastward force per unit mass is near 200 $ms^{-1}day^{-1}$. The gravity wave drag is stronger in the southern winter than in the northern winter, and the peak values occur at a lower altitude in the winter than in the summer. Note how the pattern of the drag follows the shape of the shear zones above the jets.

Figures 2c,d show for the same months, $1/(a\rho_0 \cos\phi)$ times the Eliassen-Palm (EP) flux divergence, i.e., the effective force per unit mass on the zonal mean flow due to resolved eddies. In the winter lower mesosphere of Figure 1c there is a negative region associated with dissipation of planetary waves forced from below. However the largest negative values and regions with strong positive values are in the summer hemispheres. This is an indication that there are large amplitude disturbances in these regions. The maximum values of EP flux divergence (around 30 $ms^{-1}day^{-1}$), although smaller than the values of gravity wave drag in the mesosphere, are not negligible, particularly in the tropics.

Figures 2e,f show the corresponding residual mean meridional wind ($\bar{v}^*$, see definition in An-

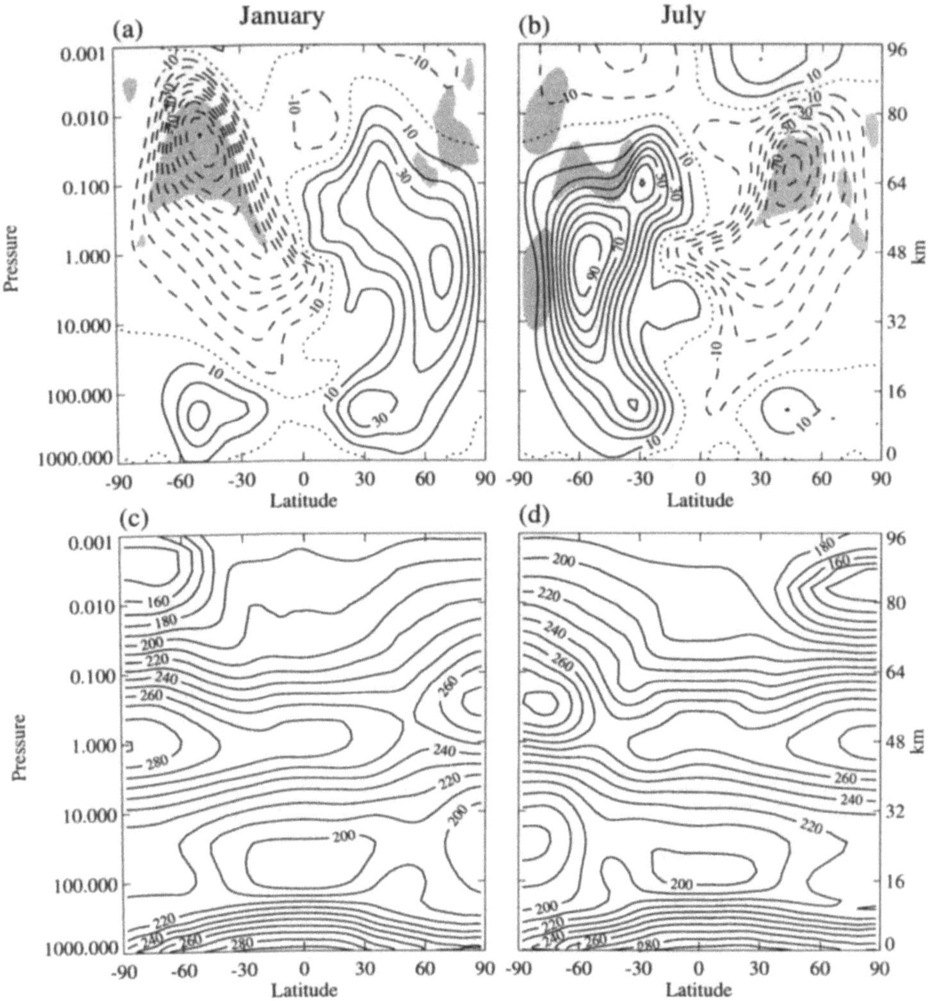


Figure 1: Zonal-mean zonal winds for (a) January (b) July, solid contours westerlies, dotted contour zero wind line, dashed contours easterlies, contour interval 10 $\mathrm{ms}^{-1}$. Shaded regions where $\bar{q}_y$ is less than $-1 \times 10^{-11}\mathrm{s}^{-1}\mathrm{m}^{-1}$. Zonal-mean temperatures for (c) January (d) July, contour interval 10 K.

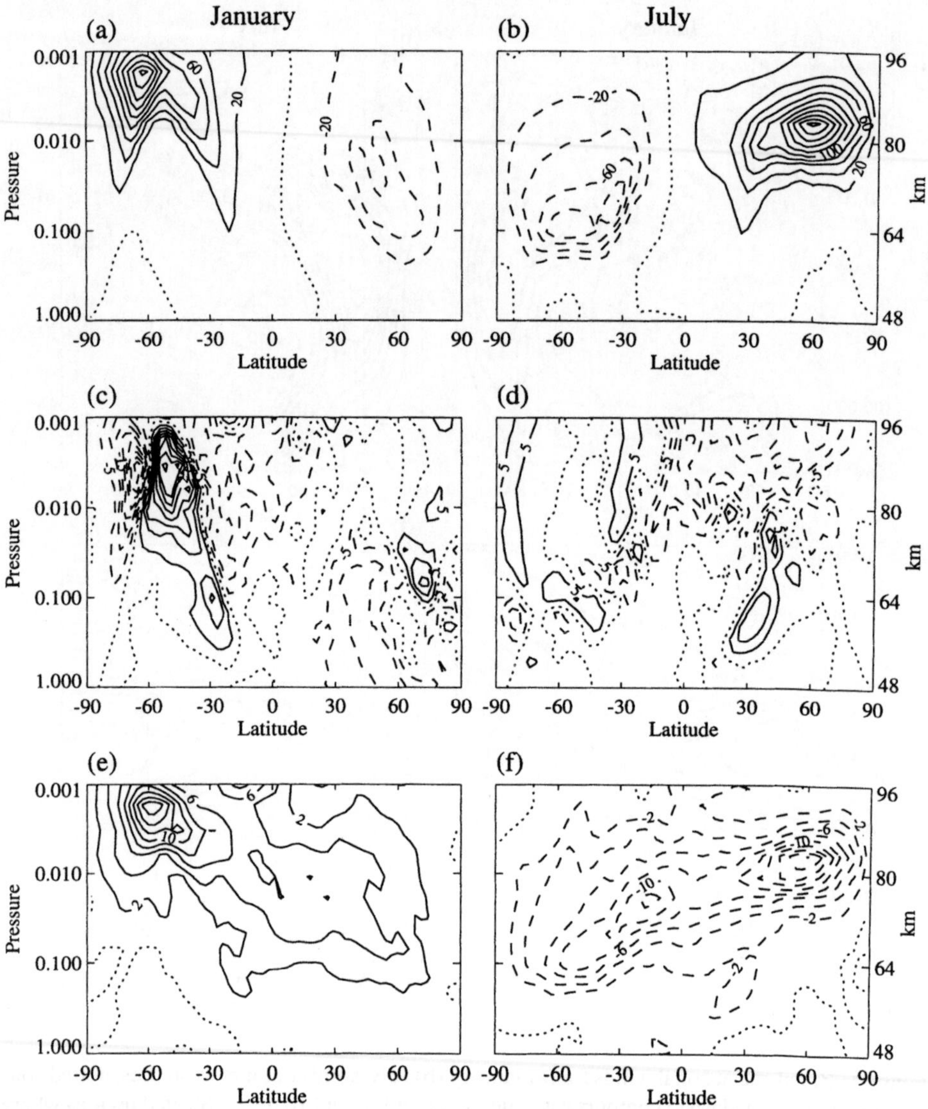


Figure 2: Zonal-mean zonal gravity wave drag for (a) January (b) July, contour interval 20 $ms^{-1}day^{-1}$. Eliassen-Palm flux divergence times $1/(a\rho_0 \cos\phi)$ for (c) January (d) July, contour interval 5 $ms^{-1}day^{-1}$. Residual mean meridional wind ($\bar{v}^*$) for (e) January (f) July, contour interval 2 $ms^{-1}$ . Solid contours positive values, dotted contour zero, dashed contours negative values.

drews *et al.* 1987). In the Transformed Eulerian Mean (TEM) formulation, this residual mean meridional wind can be considered to be driven by the gravity wave drag (Figures 2a,b) and the EP flux divergence (Figures 2c,d). The residual mean meridional wind shows strong flow from the summer to winter hemisphere with peak values of around 16 $ms^{-1}$ in a sharp maximum in the summer hemisphere. In the extratropics, the position and structure of residual mean meridional wind matches closely the gravity wave drag, indicating that gravity wave drag is predominantly driving this flow. The residual mean meridional wind is very similar to the zonal-mean meridional wind ($\bar{v}$), the maximum difference being 1-2 $ms^{-1}$ (unlike in the winter stratosphere, for example, where $\bar{v}$ and $\bar{v}^*$ can have opposite signs). Figures 2e,f are comparable to climatologies of $\bar{v}$ from radar (e.g., Vincent 1984), in particular the strength of the sharp maximum in meridional wind agrees well.

From mass continuity, to balance the meridional flow, there is a residual mean vertical velocity (not shown). This has rising motion at the summer pole, with peak values of around 6 cm $s^{-1}$ near 0.01 hPa, and descending motion at the winter pole, with peak values of around 3 cm $s^{-1}$ near 0.1 hPa. The vertical motion produces strong adiabatic cooling/warming at the summer/winter poles. This leads to a very cold summer mesopause and a warm winter mesosphere.

## 4. Evolution of the summer mesopause

We now examine more closely the summer mesopause and its relation to the gravity wave drag. Figure 3a shows the evolution over two annual cycles of the mesopause temperature (temperature minimum) at 69°N (solid line) and the maximum gravity wave drag at 60°N (dashed line, this is the latitude of maximum gravity wave drag). In the winter months the temperature minimum is at the lid of the model and so should not be considered to represent a realistic mesopause temperature. Figure 3a shows that the formation of a cold mesopause very closely matches the evolution of the gravity wave drag.

This seasonal variation in the gravity wave drag is due partly to changes in filtering of the gravity wave spectrum by winds at lower levels. Comparison of Figure 3a with Figure 3b shows that the eastward gravity wave drag increases rapidly as the maximum mean zonal wind above 100 hPa falls to less than approximately 20 $ms^{-1}$ so that fewer eastward travelling gravity waves are filtered out.

Figure 3c shows the evolution of the height of the mesopause at 69°N (solid line) and the height of the maximum gravity wave drag at 60°N (dashed line, shifted down 0.5 km for clarity). The mesopause height has a very sharp transition from the top of the model down to 83 km in April, and an equally sharp transition back up in late August. The position of maximum gravity wave drag shows very similar transitions. Other diagnostics show that the sharp descent of the mesopause and the gravity wave drag maximum is accompanied by a sharp descent of the shear zone above the easterly jet maximum; the positive feedback discussed in section 2 is crucial for this evolution.

The evolution of mesopause temperature and height in the model can be compared with observations from Andenes, Norway (69°N) presented in Lübken and von Zahn (1991). The summer mesopause in the model (135 K, 83 km) is slightly warmer and lower than observations (130 K, 88 km). The observations also show the very sharp transitions in mesopause height as in the model; thus indicating the important role in the real atmosphere of gravity wave filtering and gravity wave interaction with the mean flow.

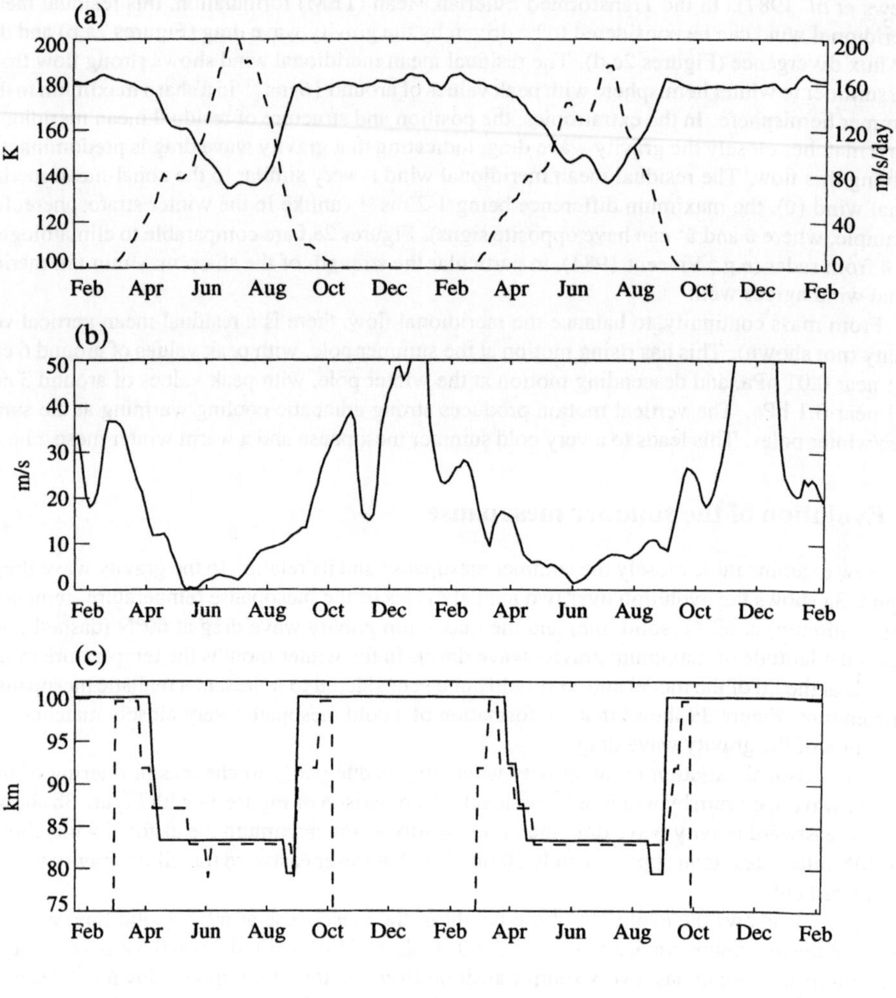


Figure 3: (a) Time series of mesopause temperature at 69°N (solid line, left hand scale) and gravity wave drag at 60°N (dashed line, right hand scale); (b) maximum mean zonal wind between 100 hPa and 0.1 hPa at 60°N; (c) time series of mesopause height at 69°N(solid line) and height of maximum gravity wave drag at 60°N (dashed line).

## 5. The two-day wave

Norton and Thuburn (1996) reported a strong signal of the two-day wave in a lower resolution (T21) version of the EUGCM. They found that the wave in the model closely resembled the observed two-day wave. They concluded that the wave has characteristics of both a normal mode and an unstable wave. Furthermore they found that gravity wave drag is essential for maintaining the unstable zonal-mean flow that enables the wave to exist. The negative $\bar{q}_y$ regions in the summer easterly jets in Figures 1a,b and the coincident patterns of strong positive EP-flux divergence in Figures 2c,d indicate that instability is again important in the higher resolution (T42) version of the model. In this section we present results of the two-day wave from this version of the model.

Figure 4 shows the PV distribution in the northern hemisphere on the 3000 K isentrope (near 0.1 hPa, 64 km) and the 7000 K isentrope (near 0.003 hPa, 88 km) on the 23 June. At both levels a striking wave 3 pattern is present, which propagates westward with a period close to two days. However there is a marked difference between the two levels. At 3000 K there is easterly mean flow with high PV regions near 30°N. At 7000 K there is westerly mean flow with long tongues of high PV extending from the high latitudes deep into the tropics. The regions of high PV are out of phase between the two levels showing the baroclinic structure of the wave.

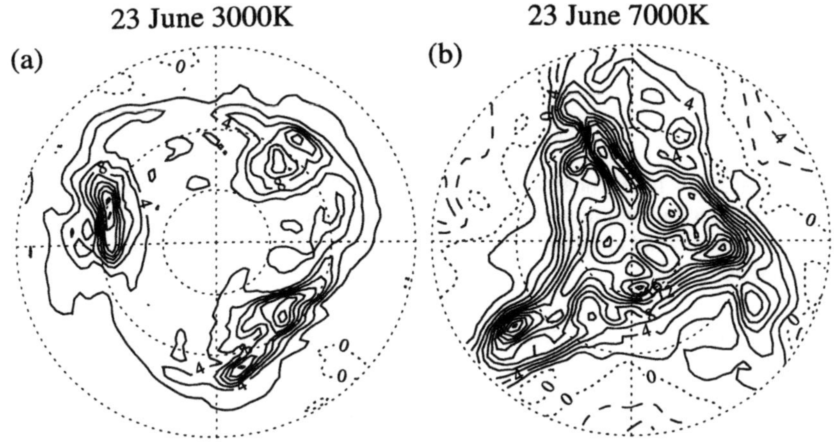


Figure 4: Northern hemisphere PV for 23 June (a) 3000 K, contour interval $2 \times 10^{-2}$ m$^2$ s$^{-1}$ kg$^{-1}$ K; (b) 7000 K, contour interval 2 m$^2$ s$^{-1}$ kg$^{-1}$ K. Latitude circles every 30°.

Figure 5 shows the temperature amplitude and phase (Figures 5a,b) and meridional wind amplitude and phase (Figures 5c,d) of the wave 3 westward propagating spectral component with period 2 days for 23–24 June. The peak in temperature amplitude in the summer hemisphere, minimum at the equator, and further peak in the winter hemisphere are very similar to the results in NT96 and are comparable to MLS observations reported in Wu *et al.* (1996). The temperature phase shows the baroclinic structure, with a rapid phase transistion near 0.1 hPa, and the winter hemisphere being out of phase with the summer. The amplitude of the meridional wind shows a strong peak just on the summer side of the equator in the upper mesosphere which is in agreement with observations from HRDI by Wu *et al.* (1993), although the peak values are somewhat stronger than have been observed from radar near the equator (Harris and Vincent 1993). The temperature and meridional wind amplitudes resemble those calculated for the gravest antisym-

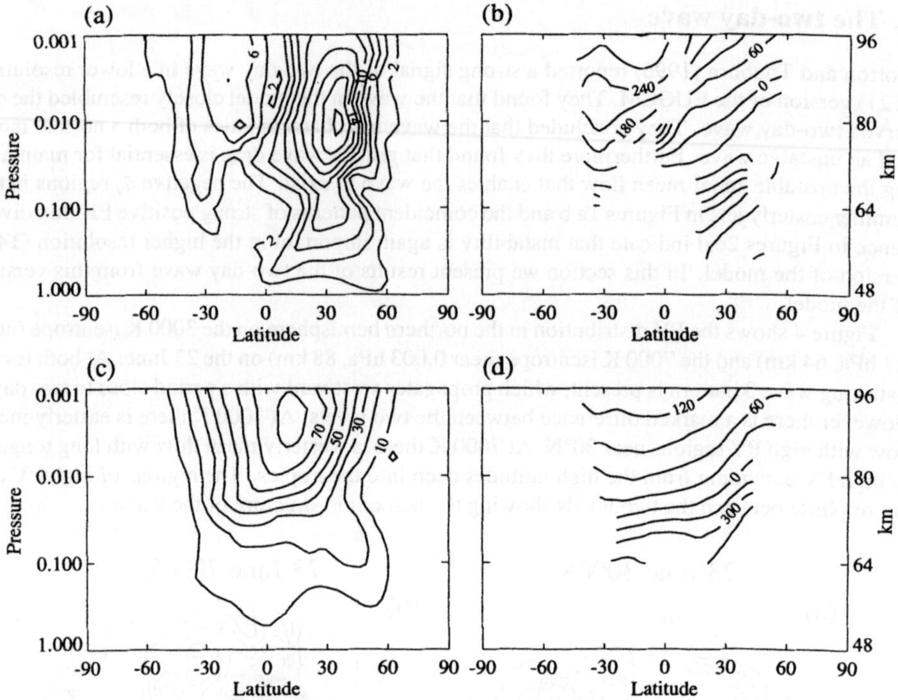


Figure 5: For 23-24 June, zonal wavenumber 3, two-day westward propagating spectral component: (a) temperature amplitude, contour interval 2 K; (b) temperature phase, contour interval 30°; (c) meridional wind amplitude, contour interval 10 $ms^{-1}$; (d) meridional wind phase, contour interval 30°.

metric wave 3 Rossby-gravity planetary normal mode for realistic solstice winds (Hagan *et al.* 1993).

Figure 6 shows the PV distribution in the northern hemisphere on the 3000 K isentrope and the 7000 K isentrope on the 26 July. At both levels there has been a transition to wave 4. This wavenumber transition was also seen in the low resolution version of the model in NT96. In contrast, the two-day wave in southern hemisphere summer tends not to strongly select inidividual wavenumbers. Figure 7 shows a typical situation in the southern hemisphere for 17 January. There is clearly some wave 3 present, but also other wavenumbers. The strong vortex near the bottom of Figure 7a can be distinguished for around 6 days at this time.

The more chaotic nature of the two-day wave in the southern hemisphere summer compared with northern hemisphere summer can also be seen in Figure 8 which shows the time series of meridional wind amplitude of two-day westward propagating spectral component in wave 3 (solid line) and wave 4 (dotted line) at the equator, 0.003 hPa (near 88 km). The dashed line is the amplitude of the diurnal tide at 15°N which will be discussed in section 7. Figure 8 clearly shows the semiannual variation in amplitude of the two-day wave. It also shows how there are much greater amplitude variations of wave 3 in December to February compared with June to August. Note also how the amplitude of wave 4 is often anticorrelated with the amplitude of wave 3.

One difference between the northern hemisphere and the southern hemisphere that has been

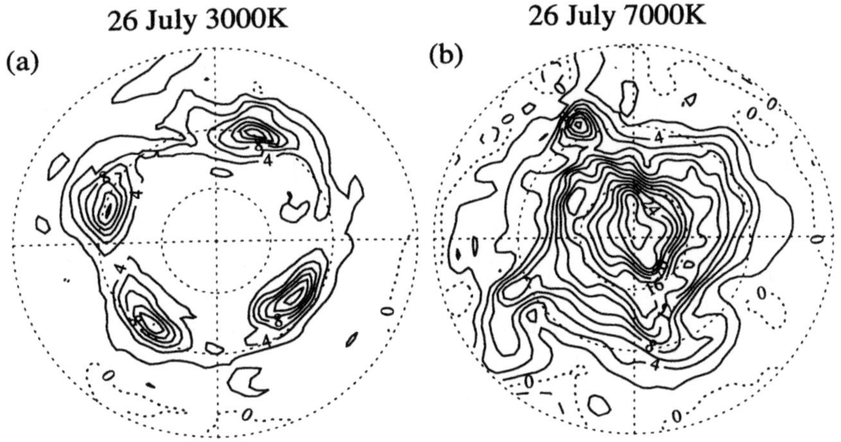


Figure 6: Northern hemisphere PV for 26 July (a) 3000 K; (b) 7000 K. Contour intervals same as in Figure 4

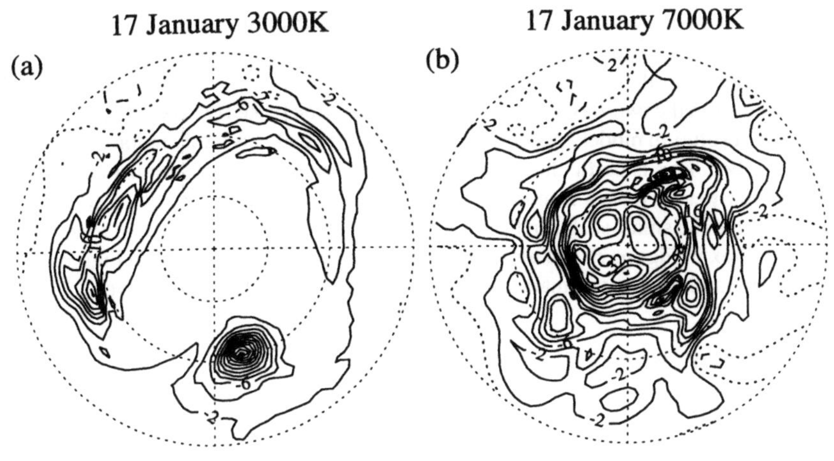


Figure 7: Southern hemisphere PV for 17 January (a) 3000 K; (b) 7000 K. Contour intervals same as in Figure 4

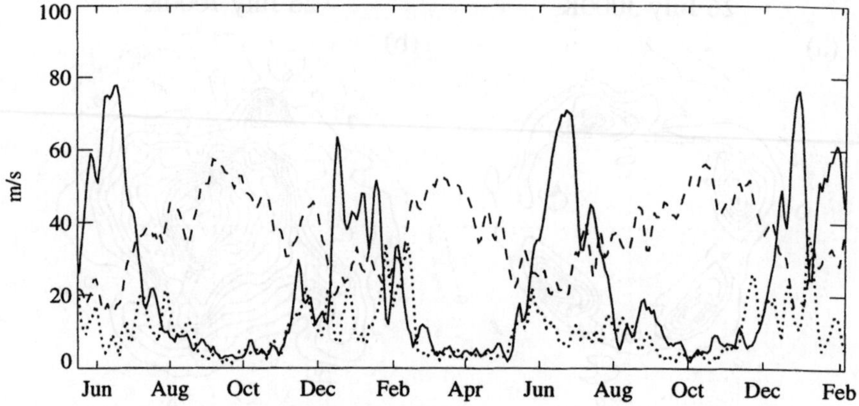


Figure 8: Meridional wind amplitude of two-day westward propagating spectral component at the equator, 0.003 hPa: wave 3 (solid line), wave 4 (dotted line). Dashed line is the amplitude of meridional wind diurnal tide at 15°N, 0.003 hPa.

noted from observations is the period of the two-day wave; in the northern hemisphere it is 51–52 hours while in the southern hemisphere it is 48–49 hours. In the EUGCM, the period of the wave in the northern hemisphere is around 45–47 hours, while in the southern hemisphere the peak spectral amplitude has a somewhat shorter period of around 42 hours, but with a greater spectral spread. The differences in period between the EUGCM and observations, and between the two hemispheres, probably reflect the different strengths of the mean easterly winds. In the EUGCM the summer easterly winds in the southern hemisphere are too strong (Figure 1), which probably accounts for the shorter period.

Compared with the two-day wave in the lower resolution model described by NT96, the wave in the higher resolution model appears to be more chaotic and less likely to lock onto an individual wavenumber. This is particularly true in the southern hemisphere summer and it is possible that the shorter period makes it more difficult for the wave to interact with a normal mode. It also appears that the small-scales in the higher resolution model play an active role, for example the strong vortex in Figure 7a and also the wavebreaking in the subtropics in Figures 4b and 6b. It is possible that these active small-scale features disrupt the large-scale wave.

## 6. The four-day wave

The four-day wave is a commonly observed feature of the winter polar upper stratosphere and mesosphere. It has been seen from satellites (e.g., Venne and Stanford 1979, and recently by MLS on UARS, Allen *et al.* 1996) and radar (e.g., Fraser *et al.* 1993). It has been identified as a warm pool of air which rotates eastward around the winter pole near the stratopause with an approximate four-day period. It is most commonly seen in the southern hemisphere but also has been identified in the northern hemisphere (Venne and Standford 1982). There have been several theoretical studies on the four-day wave (e.g., Manney and Randel 1993) which have connected it with instability of the upper part of the winter jet. To our knowledge, it has not been reported in a GCM.

Figure 9 shows a longitude versus time plot of temperature at 71°S 1 hPa from the EUGCM.

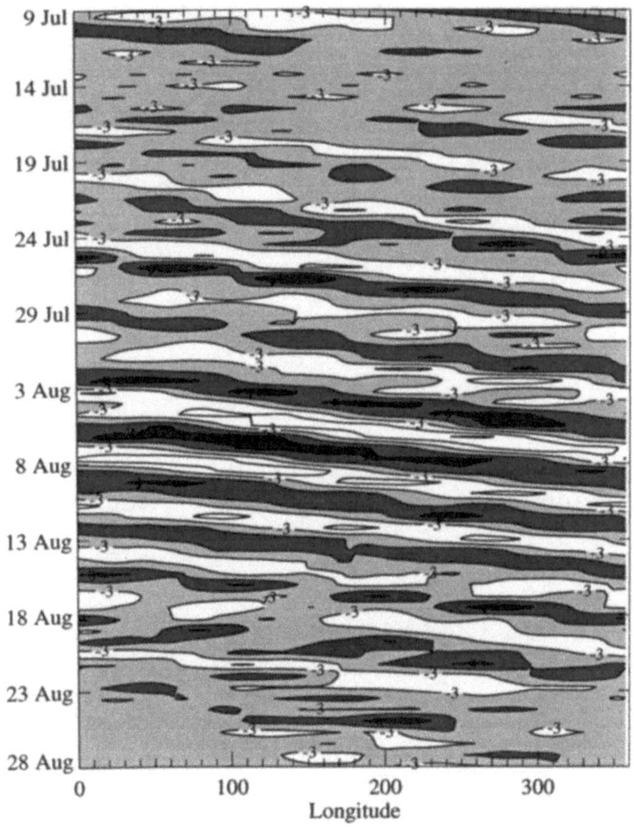


Figure 9: Time versus longitude plot of temperature at 71°S, 1 hPa showing periods less than 6 days. Contour interval 6 K, dark regions positive, light regions negative.

The data have been filtered to show waves with periods of less than 6 days. Figure 9 shows that in early August there is a large-scale eastward-propagating feature with period of around 3.3 days. This is a somewhat shorter period than observed for the four-day wave because the mean winds in the EUGCM are slightly stronger than observed for August. Figure 10 shows synoptic maps of PV on the 2050 K isentrope (this is located just above 1 hPa) at this time. The grey region inside the polar vortex shows that there is a high (magnitude) PV feature rotating around the pole. Figure 11 shows the temperature amplitude and phase (Figures 11a,b), meridional wind amplitude (Figure 11c), and PV amplitude (Figure 11d) of the wave 1 eastward propagating spectral component with period 3.3 days for 2-12 August. The PV values have been rescaled by the mean hemispheric value on isentropic levels. The temperature amplitude shows peaks near 1 hPa and 0.1 hPa with a minimum in between. The temperature phase is nearly constant near the two peaks with a very sharp transition in between. The meridional wind amplitude has a peak near the pole at 0.3 hPa but extends into the upper mesosphere. The PV amplitude shows a very localized peak just above 1 hPa in polar regions, and a smaller peak in midlatitudes.

The temperature amplitude of the wave in the EUGCM is remarkably similar to that obtained from MLS temperatures by Allen *et al.* (1996). The temperature structure of the wave is consistent with the localized PV structure with temperature peaks above and below the PV maxima.

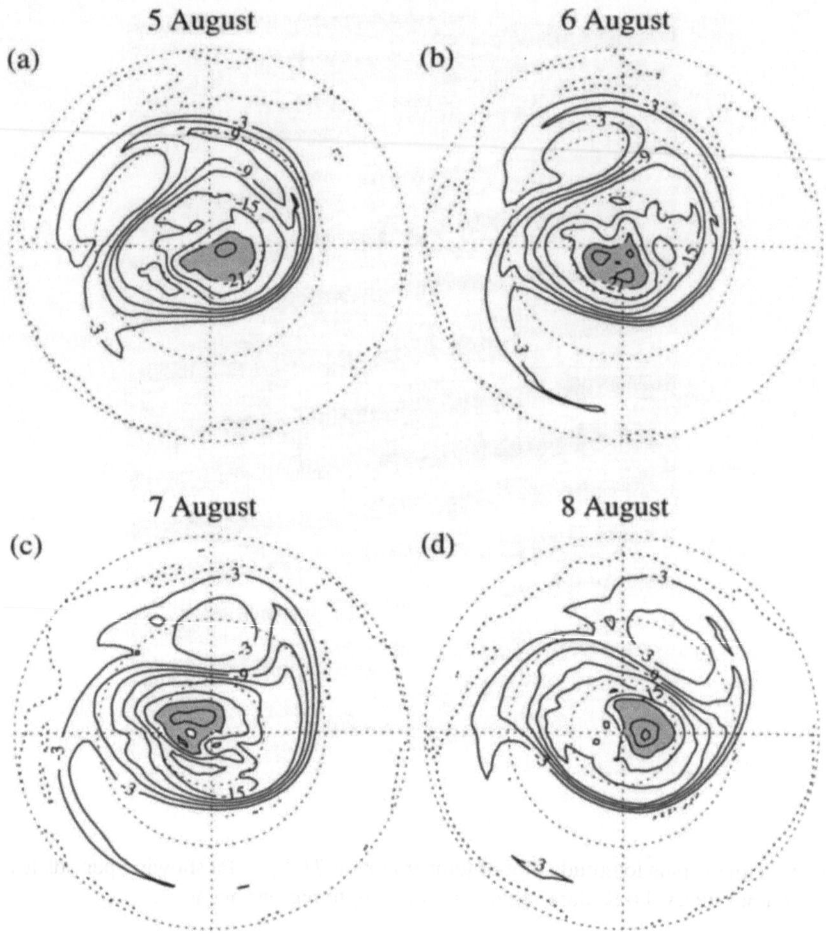


Figure 10: Southern hemisphere PV 5–8 August 2050 K, contour interval $3 \times 10^{-3}$ m$^2$ s$^{-1}$ kg$^{-1}$ K, shaded region less than -21 contour units.

As discussed in the introduction, with the low Brunt-Väissällä frequency in the mesosphere, the induced fields from the PV feature extend a long way into the mesosphere, as indicated by the amplitude of the wave in the meridional wind. The PV structure of the wave is very similar to that diagnosed by Allen *et al.* (1996) from MLS temperatures.

Diagnosing possible causal mechanisms in the EUGCM has turned out to be less straightforward for the four-day wave than for the two-day wave discussed in section 5. Theoretical studies (e.g., Manney and Randel 1993) have suggested that the four-day wave is generated through barotropic or baroclinic instability of the mean flow near the stratopause and have demonstrated that unstable modes exist in climatological mean wind fields. Such an unstable mode would show a characteristic pattern of positive EP flux divergence coincident with a region of negative $\bar{q}_y$. However, the existence of a region of negative $\bar{q}_y$ is not a sufficient condition for instability, especially when other processes like radiation, gravity wave drag, and other large amplitude waves may be important. Also, any downgradient PV flux or PV mixing in a region of negative $\bar{q}_y$ will give positive EP flux divergence, even without the presence of an unstable mode.

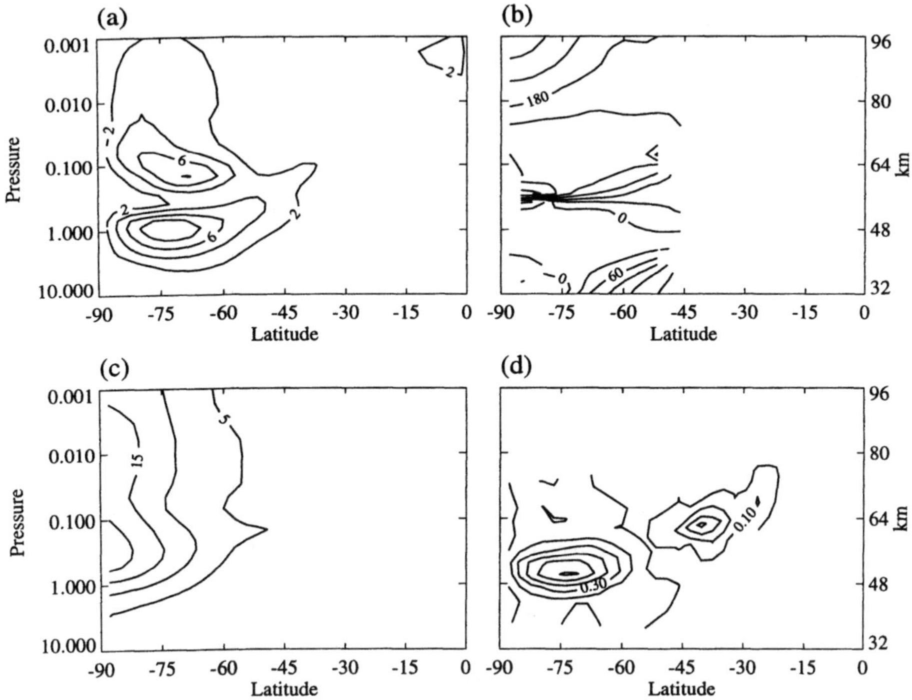


Figure 11: For 2-12 August, zonal wavenumber 1, eastward propagating spectral component with period 3.3 days: (a) temperature amplitude, contour interval 2 K; (b) temperature phase, contour interval 30°; (c) meridional wind amplitude, contour interval 5 $ms^{-1}$; (d) PV amplitude, scaled by the mean isentropic hemispheric value, contour interval 0.1

The EUGCM does indeed have regions of negative $\bar{q}_y$ on the poleward side and above the peak of the westerly jet in July (Figure 1b) and August. It also has positive EP flux divergence near 50°S 0.3 hPa coincident with the negative $\bar{q}_y$ (Figure 12). However, on this day (4 August) the EP flux vectors emanating from the region of positive EP flux divergence are very weak and tend to point downwards and equatorwards. On the other hand, the EP flux convergence into the region where the four-day wave is large is associated with much stronger upward EP flux from lower levels. This strong upward EP flux is part of a burst that lasts about 4–5 days and occurs just as the four-day wave amplifies (compare Figure 9). A large part of the burst of upward EP flux tilts equatorwards and converges around 35°S (where it produces wavebreaking on the edge of the polar vortex, see Figure 10) but a significant part propagates upwards in the polar region. These diagnostics must be interpreted with some caution as there are significant day-to-day variations in some of the details of $\bar{q}_y$, EP flux, and EP flux divergence. Nevertheless, these two pieces of evidence, namely the weakness and direction of the EP flux emanating from the region of negative $\bar{q}_y$, and the simultaneity of the strong burst of upward EP flux and the amplification of the four-day wave, together suggest that this occurrence of the four-day wave in the EUGCM may be caused by upward propagation of wave activity from below rather than in situ instability.

We suggest that the burst of upward EP flux, which is associated mainly with a quasi-stationary disturbance, produces a strong displacement of the core of the polar vortex, which subsequently

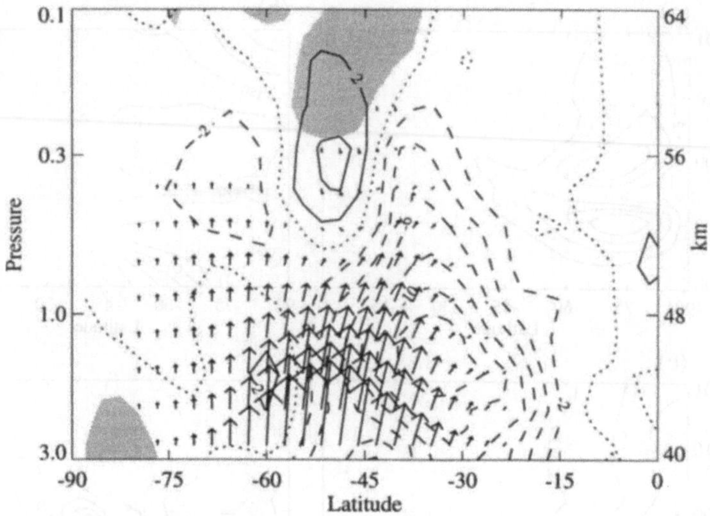


Figure 12: For August 4, EP flux vectors with magnitude indicated by length of arrow, and EP flux divergence, solid contours positive values, dotted contour zero, dashed contours negative values, contour interval $2\times10^{-4}$ Nm$^{-3}$. Shaded regions where $\bar{q}_y$ is less than $-1\times10^{-11}$s$^{-1}$m$^{-1}$.

precesses around the pole at a speed set by the strength of the mean wind. There is little irreversible distortion of PV contours within the vortex allowing the displaced vortex core to retain its identity for ten days or more.

We do not yet know how typical the above case is. We should not rule out the possibility that instability plays some role in other occurrences of the four-day wave either in the EUGCM or in the real world.

## 7. The diurnal tides

Figure 13 shows the monthly-mean amplitude in the meridional wind component of the migrating diurnal tides for January, April, July and October. The amplitude in the zonal winds (not shown) shows a similar structure but smaller by a factor of around 1.5. A prominent feature in Figure 13 is the reduction, by nearly a factor of two, in tidal amplitude at the solstice compared with the equinox. This semiannual variation in tidal amplitude is comparable to observations from UARS made by HRDI (Hays *et al.* 1994) and WINDII (McLandress *et al.* 1996).

Figure 13 also shows that the tidal amplitudes are nearly symmetric about the equator at the equinox, but are significantly distorted about the equator at the solstice. The observations of McLandress *et al.* (1996) also show this structural change, however most of these observations are at higher altitude.

One unrealistic feature of the tides in the EUGCM is the minimum in amplitude near 0.0015 hPa (93 km). This is probably the result of the tides reflecting off the model top.

Such a large semiannual variation in the tidal amplitude, and the interhemispheric asymmetry at the solstice, are not captured by linear tidal models, e.g., Hagan *et al.* (1995). Possible mechanisms for these variations in the real world and in the EUGCM, which may be represented inadequately or not at all in linear tidal models, include: nonlinear tidal breaking; interaction of the tides with breaking gravity waves; nonlinear interaction of the tides with other waves such as the two-day wave.

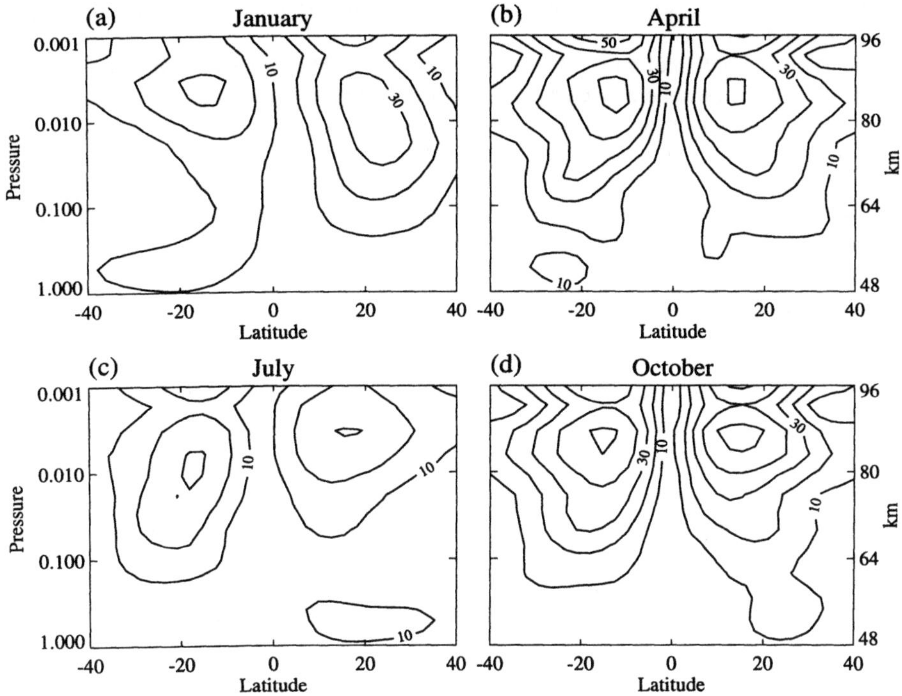


Figure 13: Meridional wind amplitude of diurnal tide. Contour interval 10 ms$^{-1}$ .

The possibility of a nonlinear interaction between the tides and the two-day wave is worthy of further investigation. The two-day wave certainly has large amplitude in the EUGCM, with the meridional wind near the equator getting locally in access of 100 ms$^{-1}$. The reduction in tidal amplitudes coincides closely with the amplification of the two-day wave: Figure 8 shows the meridional wind tidal amplitude at 15°N, 0.003 hPa (dashed line) along with the amplitude of the two-day wave (wave three solid line, wave four dotted line). Also, the reduced tidal amplitude below 0.01 hPa in the summer hemisphere (Figure 13) occurs exactly where there is strong nonlinear mixing associated with the two-day wave.

## 8. Conclusions

We have shown that the EUGCM reproduces many features of the observed mesosphere. The zonal mean winds and temperatures are quite realistic, particularly the seasonal evolution of the height and temperature of the high latitude mesopause. The EUGCM contains a strong signal of the two-day wave; it is forced through baroclinic instability of the summer easterly jet but also has characteristics of a planetary normal mode. The EUGCM also produces a four-day wave event. At that time there is a signature of EP flux divergence associated with a region of negative $\bar{q}_y$, but this appears to be too weak to explain the model's four-day wave as an unstable mode. A more likely explanation appears to be the upward propagation of planetary wave activity from below. Finally, the EUGCM includes a diurnal tide, not only with realistic amplitude and structure but also with a realistic semiannual variation in amplitude and a realistic interhemispheric asymmetry at the solstices.

The parameterization of gravity wave drag in the EUGCM is crucial for obtaining a realistic simulation of the mesosphere. We have clear evidence that it is important for the cold summer mesopause and for the unstable profile of the summer easterly jet that gives rise to the two-day wave. It is probably also important for the four-day wave though its effect on the structure of the westerly jet, and for the semiannual variation in tidal amplitude either directly through interactions between the tide and gravity waves or indirectly through interactions between the tide and the two-day wave.

## Acknowledgements

This work was supported by the U.K. Universities Global Atmospheric Modelling Programme which is funded by U.K. Natural Environmental Research Council.

## References

Allen, D. R., Stanford, J. L., Elson, L. S., Fishbein, E. F., Froidevaux, L., Waters, J. W., 1996: The 4-day wave as observed from the Upper Atmosphere Research Satellite Microwave Limb Sounder. *J. Atmos. Sci.*, submitted.

Andrews, D. G., Holton, J. R., Leovy, C. B., 1987: Middle Atmosphere Dynamics. Academic Press, 489pp.

Betts, A. K., Miller, M. J., 1993: The Betts-Miller scheme. In: The Representation of Cumulus Convection in Numerical Models of the Atmosphere. Eds K. A. Emanuel and D. J. Raymond, American Meteorological Society.

Fraser, G. J., Hernandez, G., Smith, R. W., 1993: Eastward-moving 2-4 day waves in the winter Antarctic mesosphere. *Geophys. Res. Lett.*, **20**, 1547–1550.

Fisher, M., O'Neill, A., Sutton, R., 1993: Rapid descent of mesospheric air into the stratospheric polar vortex. *Geophys. Res. Lett.*, **20**, 1267–1270.

Garcia, R. R., Boville, B. A., 1994: "Downward control" of the mean meridional circulation and temperature distribution of the polar winter stratosphere. *J. Atmos. Sci.*, **51**, 2238–2245.

Gray, L. J., Blackburn, M., Chipperfield, M. P., Haigh, J. D., Jackson, D. R., Shine, K. P., Thuburn, J., Zhong, W., 1993: First results from a 3-dimensional model of the middle atmosphere. *Adv. Space. Res.*, **13**, 363–372.

Hagan, M. E., Forbes, J. M., Vial, F., 1993: Numerical investigation of the propagation of the quasi-two-day wave into the lower thermosphere. *J. Geophys. Res.*, **98**, 23193–23205.

Hagan, M. E., Forbes, J. M., Vial, F., 1995: On modelling migrating solar tides. *Geophys. Res. Lett.*, **22**, 893-896.

Harris, T. J., Vincent, R. A., 1993: The quasi-two-day wave observed in the equatorial middle atmosphere. *J. Geophys. Res.*, **98**, 10481–10490.

Hays, P. B., Wu, D. L., Burrage, M. D., Gell, D. A., Grassl, H. J., Lieberman, R. S., Marshall, A. R., Morton, Y. T., Ortland, D. A., Skinner, W. R., 1994: Observations of the diurnal tide from space. *J. Atmos. Sci.*, **51**, 3077–3093.

Houghton, J. T., 1978: The stratosphere and the mesosphere. *Q. J. Roy. Meteorol. Soc.*, **104**, 1–29.

Jackson, D. R., 1993: Sensitivity of the extended UGAMP global circulation model to the specification of gravity wave phase speeds. *Q. J. Roy. Meteorol. Soc.*, **119**, 457–468.

Lübken, F.-J., von Zahn, U., 1991: Thermal structure of the mesopause region at polar latitudes. *J. Geophys. Res.*, **96**, 20841–20857.

Manney, G. L., Randel, W. J., 1993: Instability at the winter stratopause: a mechanism for the 4-day wave. *J. Atmos. Sci.*, **50**, 3928–3938.

McLandress, C., Shepherd, G. G., Solheim, B. H., 1996: Satellite observations of thermospheric tides: results from the Wind Imaging Interferometer on UARS. *J. Geophys. Res.*, **101**, 4093–4114.

Morcrette, J.-J., 1990: Impact of changes to the radiation transfer parameterizations plus cloud optical properties in the ECMWF model. *Mon. Wea. Rev.*, **118**, 847–873.

Norton, W. A., Thuburn, J., 1996: The two-day wave in a middle atmosphere GCM. *Geophys. Res. Lett.*, in press.

Palmer, T. N., Shutts, G. J., Swinbank, R., 1986: Alleviation of a systematic westerly bias in general circulation and numerical weather prediction models through an orographic gravity wave drag parametrization. *Q. J. Roy. Meteorol. Soc.*, **112**, 1001–1039.

Plumb, R. A., 1983: Baroclinic instability at the summer mesosphere: a mechanism for the quasi-two-day wave? *J. Atmos. Sci.*, **40**, 262–270.

Shine, K. P., 1987: The middle atmosphere in the absence of dynamical heat fluxes. *Q. J. Roy. Meteorol. Soc.*, **113**, 603–633.

Shine, K. P., Rickaby, J. A., 1989: Solar radiative heating due to absorption by ozone. In: Ozone in the Atmosphere. Eds R. D. Bojkov and P. Fabian, Deepak Publishing Co., Hampton Virginia.

Simmons, A. J., Burridge, D. M., 1981: An energy and angular-momentum conserving vertical finite-difference scheme and hybrid vertical coordinates. *Mon. Wea. Rev.*, **109**, 758–766.

Slingo, J., Blackburn, M., Betts, A., Brugge, R., Hodges, K., Hoskins, B., Miller, M., Steenman-Clark, L., Thuburn, J., 1994: Mean climate and transience in the tropics of the UGAMP GCM: Sensitivity to convective parametrization. *Q. J. Roy. Meteorol. Soc.*, **120**, 881–922.

Stephenson, D. B., 1994: The northern hemisphere tropospheric response to changes in the gravity wave drag scheme in a perpetual January GCM. *Q. J. Roy. Meteorol. Soc.*, **120**, 699–712.

Tiedke, M., Heckley, W. A., Slingo, J. M., 1988: Tropical forecasting at ECMWF: The influence of physical parameterization on the mean structure of forecasts and analyses. *Q. J. Roy. Meteorol. Soc.*, **114**, 639–664.

Thuburn, J., McIntyre, M. E., 1996: Numerical advection schemes, cross-isentropic random walks, and correlations between chemical species. Submitted to *J. Geophys. Res.*

Venne, D. E., Stanford, J. L., 1979: Observations of a 4-day temperature wave in the polar winter stratosphere. *J. Atmos. Sci.*, **36**, 2016–2019.

Venne, D. E., Stanford, J. L., 1982: An observational study of high-latitude stratospheric planetary waves in winter. *J. Atmos. Sci.*, **39**, 1026–1034.

Vincent, R. A., 1984: MF/HF radar measurements of the dynamics of the mesopause region - A review. *J. Atmos. Terr. Phys.*, **11**, 961–974.

Wu, D. L., Hays, P. B., Skinner, W. R., Marshall, A. R., Burrage, M. D., Lieberman, R. S., Ortland, D. A., 1993: Observations of the quasi 2-day wave from the High Resolution Doppler Imager on UARS. *Geophys. Res. Lett.*, **20**, 2853–2856.

Wu, D. L., Fishbein, E. F., Read, W. G., Waters, J. W., 1996: Excitation and evolution of the quasi-two-day wave observed in UARS/MLS temperature measurements. *J. Atmos. Sci.*, **53**, 728–738.

## Subject Index

## The ASI Series Books Published as a Result of Activities of the Special Programme on Global Environmental Change

This book contains the proceedings of a NATO Advanced Research Workshop held within the activities of the NATO Special Programme on Global Environmental Change, which started in 1991 under the auspices of the NATO Science Committee.

The volumes published as a result of the activities of the Special Programme are:

Vol. 1: **Global Environmental Change.**
Edited by R. W. Corell and P. A. Anderson. 1991.
Vol. 2: **The Last Deglaciation: Absolute and Radiocarbon Chronologies.**
Edited by E. Bard and W. S. Broecker. 1992.
Vol. 3: **Start of a Glacial.** Edited by G. J. Kukla and E. Went. 1992.
Vol. 4: **Interactions of C, N, P and S Biogeochemical Cycles and Global Change.**
Edited by R. Wollast, F. T. Mackenzie and L. Chou. 1993.
Vol. 5: **Energy and Water Cycles in the Climate System.**
Edited by E. Raschke and D. Jacob. 1993.
Vol. 6: **Prediction of Interannual Climate Variations.**
Edited by J. Shukla. 1993.
Vol. 7: **The Tropospheric Chemistry of Ozone in the Polar Regions.**
Edited by H. Niki and K. H. Becker. 1993.
Vol. 8: **The Role of the Stratosphere in Global Change.**
Edited by M.-L. Chanin. 1993.
Vol. 9: **High Spectral Resolution Infrared Remote Sensing for Earth's Weather and Climate Studies.**
Edited by A. Chedin, M.T. Chahine and N.A. Scott. 1993.
Vol. 10: **Towards a Model of Ocean Biogeochemical Processes.**
Edited by G. T. Evans and M. J. R. Fasham. 1993.
Vol. 11: **Modelling Oceanic Climate Interactions.**
Edited by J. Willebrand and D. L.T. Anderson. 1993.
Vol. 12: **Ice in the Climate System.** Edited by W. Richard Peltier. 1993.
Vol. 13: **Atmospheric Methane: Sources, Sinks, and Role in Global Change.**
Edited by M. A. K. Khalil. 1993.
Vol. 14: **The Role of Regional Organizations in the Context of Climate Change.**
Edited by M. H. Glantz. 1993.
Vol. 15: **The Global Carbon Cycle.**
Edited by M. Heimann. 1993.
Vol. 16: **Interacting Stresses on Plants in a Changing Climate.**
Edited by M. B. Jackson and C. R. Black. 1993.
Vol. 17: **Carbon Cycling in the Glacial Ocean: Constraints on the Ocean's Role in Global Change.**
Edited by R. Zahn, T. F. Pedersen, M. A. Kaminski and L. Labeyrie. 1994.
Vol. 18: **Stratospheric Ozone Depletion/UV-B Radiation in the Biosphere.**
Edited by R. H. Biggs and M. E. B. Joyner. 1994.
Vol. 19: **Data Assimilation: Tools for Modelling the Ocean in a Global Change Perspective.**
Edited by P. O. Brasseur and J. Nihoul. 1994.

Vol. 20: **Biodiversity, Temperate Ecosystems, and Global Change.** Edited by T. J. B. Boyle and C. E. B. Boyle. 1994.
Vol. 21: **Low-Temperature Chemistry of the Atmosphere.** Edited by G. K. Moortgat, A. J. Barnes, G. Le Bras and J. R. Sodeau. 1994.
Vol. 22: **Long-Term Climatic Variations – Data and Modelling.** Edited by J.-C. Duplessy and M.-T. Spyridakis. 1994.
Vol. 23: **Soil Responses to Climate Change.** Edited by M. D. A. Rounsevell and P. J. Loveland. 1994.
Vol. 24: **Remote Sensing and Global Climate Change.** Edited by R. A. Vaughan and A. P. Cracknell. 1994.
Vol. 25: **The Solar Engine and Its Influence on Terrestrial Atmosphere and Climate.** Edited by E. Nesme-Ribes. 1994.
Vol. 26: **Global Precipitations and Climate Change.** Edited by M. Desbois and F. Désalmand. 1994.
Vol. 27: **Cenozoic Plants and Climates of the Arctic.** Edited by M. C. Boulter and H. C. Fisher. 1994.
Vol. 28: **Evaluating and Monitoring the Health of Large-Scale Ecosystems.** Edited by D. J. Rapport, C. L. Gaudet and P. Calow. 1995.
Vol. 29: **Global Environmental Change Science: Education and Training.** Edited by D. J. Waddington. 1995.
Vol. 30: **Ice Core Studies of Global Biogeochemical Cycles.** Edited by R. J. Delmas. 1995.
Vol. 31: **The Role of Water and the Hydrological Cycle in Global Change.** Edited by H. R. Oliver and S. A. Oliver. 1995.
Vol. 32: **Atmospheric Ozone as a Climate Gas.** Edited by W.-C. Wang and I. S. A. Isaksen. 1995.
Vol. 33: **Carbon Sequestration in the Biosphere.** Edited by M. A. Beran. 1995.
Vol. 34: **Climate Sensitivity to Radiative Perturbations: Physical Mechanisms and Their Validation.** Edited by H. Le Treut. 1996
Vol. 35: **Clouds, Chemistry and Climate.** Edited by P. J. Crutzen and V. Ramanathan. 1996.
Vol. 36: **Diachronic Climatic Impacts on Water Resources.** Edited by A. N. Angelakis and A. S. Issar. 1996.
Vol. 37: **Climate Change and World Food Security.** Edited by T. E. Downing. 1996.
Vol. 38: **Evaluation of Soil Organic Matter Models.** Edited by D. S. Powlson, P. Smith and J. U. Smith. 1996.
Vol. 39: **Microbiology of Atmospheric Trace Gases.** Edited by J. C. Murrell and D. P. Kelly. 1996.
Vol. 40: **Forest Ecosystems, Forest Management and the Global Carbon Cycle.** Edited by M. J. Apps and D. T. Price. 1996.
Vol. 41: **Climate Variations and Forcing Mechanisms of the Last 2000 Years.** Edited by P. D. Jones, R. S. Bradley and J. Jouzel. 1996.
Vol. 42: **The Mount Pinatubo Eruption: Effects on the Atmosphere and Climate.** Edited by G. Fiocco, D. Fuà and G. Visconti. 1996.
Vol. 43: **Chemical Exchange Between the Atmosphere and Polar Snow.** Edited by E. W. Wolff and R. C. Bales. 1996.

Vol. 44: **Decadal Climate Variability: Dynamics and Predictability.** Edited by D. L. T. Anderson and J. Willebrand. 1996.

Vol. 45: **Radiation and Water in the Climate System: Remote Measurements.** Edited by E. Raschke. 1996.

Vol. 46: **Land Surface Processes in Hydrology: Trials and Tribulations of Modeling and Measuring.** Edited by S. Sorooshian, H. V. Gupta, and J. C. Rodda. 1997.

Vol. 47: **Past and Future Rapid Environmental Changes: The Spatial and Evolutionary Responses of Terrestrial Biota.** Edited by B. Huntley, W. Cramer, A. V. Morgan, H. C. Prentice, and J. R. M. Allen. 1997.

Vol. 48: **The Mathematics of Models for Climatology and Environment.** Edited by J. I. Díaz. 1997.

Vol. 49: **Third Millennium BC Climate Change and Old World Collapse.** Edited by H. N. Dalfes, G. Kukla, and R. Weiss. 1997.

Vol. 50: **Gravity Wave Processes: Their Parameterization in Global Climate Models.** Edited by K. Hamilton. 1997.

Vol. 51: **Sediment Records of Biomass Burning and Global Change.** Edited by J. S. Clark, H. Cachier, J. G. Goldammer, and B. Stocks. 1997.

MIX
Papier aus verantwortungsvollen Quellen
Paper from responsible sources
FSC® C105338

If you have any concerns about our products,
you can contact us on
**ProductSafety@springernature.com**

In case Publisher is established outside the EU,
the EU authorized representative is:
**Springer Nature Customer Service Center GmbH**
**Europaplatz 3, 69115 Heidelberg, Germany**

Printed by Libri Plureos GmbH
in Hamburg, Germany